Lecture Notes in Computer Science 13876

Founding Editors

Gerhard Goos
Juris Hartmanis

Editorial Board Members

The series Lecture Notes in Computer Science (LNCS), including its subseries Lecture Notes in Artificial Intelligence (LNAI) and Lecture Notes in Bioinformatics (LNBI), has established itself as a medium for the publication of new developments in computer science and information technology research, teaching, and education.

LNCS enjoys close cooperation with the computer science R & D community, the series counts many renowned academics among its volume editors and paper authors, and collaborates with prestigious societies. Its mission is to serve this international community by providing an invaluable service, mainly focused on the publication of conference and workshop proceedings and postproceedings. LNCS commenced publication in 1973.

Bruno Crémilleux · Sibylle Hess ·
Siegfried Nijssen
Editors

Advances in
Intelligent Data Analysis XXI

21st International Symposium on Intelligent Data Analysis, IDA 2023
Louvain-la-Neuve, Belgium, April 12–14, 2023
Proceedings

Editors
Bruno Crémilleux ⓘ
Université de Caen Normandie
Caen, France

Sibylle Hess ⓘ
Eindhoven University of Technology
Eindhoven, The Netherlands

Siegfried Nijssen ⓘ
UCLouvain
Louvain-la-Neuve, Belgium

ISSN 0302-9743   ISSN 1611-3349 (electronic)
Lecture Notes in Computer Science
ISBN 978-3-031-30046-2   ISBN 978-3-031-30047-9 (eBook)
https://doi.org/10.1007/978-3-031-30047-9

This Springer imprint is published by the registered company Springer Nature Switzerland AG
The registered company address is: Gewerbestrasse 11, 6330 Cham, Switzerland

Preface

We are pleased to introduce the proceedings of the 21th International Symposium on Intelligent Data Analysis (IDA 2023), taking place April 12–14, 2023, in Louvain-la-Neuve.

IDA is an international symposium presenting advances in the intelligent analysis of data. Distinguishing characteristics of IDA are its focus on novel, inspiring ideas, its focus on research, and its relatively small scale: all papers are presented in a single track, allowing for discussions that involve all participants. Next to the presentations based on the papers published in these proceedings, the conference features invited talks, a Frontier Prize, and a PhD forum.

The 2023 symposium received 91 submissions, of which 38 (42%) are now published in these proceedings. These publications were selected after a round of blind reviewing, in which each paper was allocated to 3 program committee members and 1 senior program committee member. Papers were evaluated on the basis of common scientific criteria such as technical quality and scholarship; however, the most focus was put on the level of novelty: we preferred small evaluations of novel ideas over thorough evaluations of more incremental ideas. This keeps in mind IDA's mission to promote potential breakthroughs and game-changing ideas over elaborating to the last detail.

One of the pleasures of chairing a conference is the opportunity to invite colleagues whose work we esteem highly. We are grateful to Francesco Bonchi (CENTAI Institute, Italy), Sarah Cohen-Boulakia (Université Paris-Saclay, France), Barbara Hammer (Bielefeld University, Germany) and Gaël Varoquaux (INRIA, France) for accepting our invitation to present recent work at the conference.

The symposium would not have been possible without the help of many people. We would like to acknowledge Ioanna Miliou for managing the PhD forum. We are grateful to Hendrik Blockeel for serving as Frontier Prize chair, and to Matthijs Van Leeuwen and Élisa Fromont for their help provided as Advisory Chairs. We thank Benoît Frenay for his help in obtaining sponsorship. The local organization would have been impossible without the extensive help provided by Vanessa Maons. We also thank Harold Kiossou for his help in preparing the proceedings, as well as the FNRS, NormaSTIC, and the ICTEAM institute of UCLouvain for providing financial support.

The quality of IDA 2023 was only possible due to the tremendous efforts of the Program Committee – our sincere thanks for all the great work and patience to make these proceedings possible. Our final thanks go to our authors, who submitted their inspiring work to the symposium. These proceedings will ensure that their work will be preserved.

February 2023

Siegfried Nijssen
Bruno Crémilleux
Sibylle Hess

Organization

General Chair

Siegfried Nijssen   UCLouvain, Belgium

Program Chairs

Bruno Crémilleux   Université de Caen Normandie, France
Sibylle Hess   Eindhoven University of Technology,
                       The Netherlands

Program Committee

Ehsan Aminian	Iran University of Science and Technology, Iran
Thiago Andrade	INESC TEC, Portugal
Ali Ayadi	Université de Strasbourg, France
Paulo Azevedo	University of Minho, Portugal
Mitra Baratchi	Universiteit Leiden, The Netherlands
Adrien Bibal	University of Colorado, USA
Hendrik Blockeel	KU Leuven, Belgium
Jose Borges	University of Porto, Portugal
Henrik Boström	KTH Royal Institute of Technology, Sweden
Paula Brito	University of Porto, Portugal
Dariusz Brzezinski	Poznan University of Technology, Poland
Mirko Bunse	TU Dortmund, Germany
Sebastian Buschjäger	TU Dortmund, Germany
Humberto Bustince	Universidad Pública de Navarra, Spain
Rui Camacho	University of Porto, Portugal
Peggy Cellier	INSA Rennes, France
Clément Chatelain	INSA Rouen Normandie, France
Dennis Collaris	Eindhoven University of Technology, The Netherlands
Paulo Cortez	University of Minho, Portugal
Bertrand Cuissart	Université de Caen Normandie, France
Tim d'Hondt	Eindhoven University of Technology, The Netherlands

Thi-Bich-Hanh Dao	University of Orleans, France
Remy Decoupes	INRAE, France
Xin Du	University of Edinburgh, UK
Wouter Duivesteijn	Eindhoven University of Technology, The Netherlands
Saso Dzeroski	Jozef Stefan Institute, Slovenia
Hussein El Amouri	Université de Strasbourg, France
Hadi Fanaee-T	Halmstad University, Sweden
Brígida Mónica Faria	Higher School of Health/Polytechnic of Porto (ESS/P.Porto), Portugal
Ad Feelders	Utrecht University, The Netherlands
Alberto Fernández	University of Granada, Spain
Sebastien Ferre	Université de Rennes 1, France
Carlos Ferreira	INESC TEC, Portugal
Francoise Fessant	Orange Innovation, France
Elisa Fromont	Université de Rennes 1, France
Benoît Frénay	Université de Namur, Belgium
Mikel Galar	Universidad Pública de Navarra, Spain
Esther Galbrun	University of Eastern Finland, Finland
Joao Gama	University of Porto, Portugal
Benoit Gaüzère	Normandie Université, INSA de Rouen, France
Franco Giustozzi	Université de Strasbourg, France
Rui Gomes	University of Coimbra, Portugal
Tias Guns	KU Leuven, Belgium
Zhijin Guo	University of Bristol, UK
Thomas Guyet	Inria Lyon, France
Barbara Hammer	Bielefeld University, Germany
Pedro Henriques Abreu	University of Coimbra, Portugal
Martin Holena	Czech Academy of Sciences, Czech Republic
Tomas Horvath	Eötvös Loránd University, Hungary
Frank Höppner	Ostfalia University of Applied Sciences, Germany
Dino Ienco	IRSTEA, France
Szymon Jaroszewicz	Polish Academy of Sciences, Poland
Baptiste Jeudy	Université Jean Monnet, France
Bo Kang	Ghent University, Belgium
Maiju Karjalainen	University of Eastern Finland, Finland
Frank Klawonn	Ostfalia University of Applied Sciences, Germany
Jiri Klema	Czech Technical University, Czech Republic
Arno Knobbe	Universiteit Leiden, The Netherlands
Maksim Koptelov	Normandy University, France
Georg Krempl	Utrecht University, The Netherlands
Charlotte Laclau	Télécom Paris, France

Christine Largeron	Université de Lyon, France
Anne Laurent	LIRMM, Université de Montpellier, France
Nada Lavrač	Jozef Stefan Institute, Slovenia
Vincent Lemaire	Orange Labs, France
João Mendes Moreira	INESC TEC, Portugal
Rosa Meo	University of Torino, Italy
Vera Migueis	University of Porto, Portugal
Ioanna Miliou	Stockholm University, Sweden
Alina Miron	Brunel University, UK
Mohamed Nadif	Université de Paris, France
Ana Nogueira	INESC TEC, Portugal
Slawomir Nowaczyk	Halmstad University, Sweden
Andreas Nürnberger	Otto-von-Guericke-Universität Magdeburg, Germany
Kaustubh Raosaheb Patil	Massachusetts Institute of Technology, USA
Ruggero G. Pensa	University of Torino, Italy
Pedro Pereira Rodrigues	University of Porto, Portugal
Lukas Pfahler	TU Dortmund, Germany
Nico Piatkowski	Fraunhofer IAIS, Germany
Marc Plantevit	EPITA, France
Yoeri Poels	Eindhoven University of Technology, The Netherlands
Lubos Popelinsky	Masaryk University, Czech Republic
Filipe Portela	University of Minho, Portugal
Ronaldo Prati	Federal University of ABC, Brazil
Mina Rafla	Orange Innovation, France
Luis Paulo Reis	University of Porto/LIACC, Portugal
Justine Reynaud	Université de Caen Normandie, France
Rita P. Ribeiro	University of Porto, Portugal
Celine Robardet	INSA Lyon, France
Duncan D. Ruiz	Pontifical Catholic University of RS - Porto Alegre, Brazil
Amal Saadallah	TU Dortmund, Germany
Akrati Saxena	Eindhoven University of Technology, The Netherlands
Jörg Schlötterer	University of Duisburg-Essen, Germany
Roberta Siciliano	University of Naples Federico II, Italy
Arno Siebes	Utrecht University, The Netherlands
Paula Silva	Porto University, Portugal
Prabhant Singh	Eindhoven University of Technology, The Netherlands
Malika Smail-Tabbone	University of Lorraine, France

Arnaud Soulet	Université de Tours, France
Myra Spiliopoulou	Otto-von-Guericke-University Magdeburg, Germany
Jerzy Stefanowski	Poznan University of Technology, Poland
Stephen Swift	Brunel University, UK
Shazia Tabassum	University of Porto, Portugal
Bouadi Tassadit	Université de Rennes 1, France
Maryam Tavakol	Eindhoven University of Technology, The Netherlands
Maguelonne Teisseire	IRSTEA, France
César Teixeira	University of Coimbra, Portugal
Sónia Teixeira	INESC TEC, Portugal
Alexandre Termier	Université de Rennes 1, France
Alicia Troncoso	Universidad Pablo de Olavide, Spain
Allan Tucker	Brunel University, UK
Peter van der Putten	Leiden University & Pegasystems, The Netherlands
Matthijs van Leeuwen	Leiden University, The Netherlands
Fabio Vandin	University of Padova, Italy
Cor Veenman	Netherlands Forensic Institute, The Netherlands
Julien Velcin	Université de Lyon, France
Bruno Veloso	University of Porto, INESC TEC, Portugal
Tom Viering	Delft University of Technology, The Netherlands
João Vinagre	Joint Research Centre - European Commission, Spain
Veronica Vinciotti	University of Trento, Italy
Sheng Wang	University of Bristol, UK
Hilde Weerts	Eindhoven University of Technology, The Netherlands
Pascal Welke	University of Bonn, Germany
David Weston	Birkbeck, University of London, UK
Zhaozhen Xu	University of Bristol, UK
Paul Youssef	University of Duisburg-Essen, Germany
Leishi Zhang	Canterbury Christ Church University, UK
Albrecht Zimmermann	Université de Caen Normandie, France
Indre Zliobaite	University of Helsinki, Finland

Additional Reviewers

Amorim, José P.
Davari, Narjes
Ferreira, Paulo
Gonçalves, André
Gourru, Antoine
Nicolaisen, Eliana

Salazar, Teresa
Santos, Joana Cristo
Sasse, Leonard
Silva, Catarina
Souag, Amina
Yang, Leshanshui

Contents

Contextual Word Embeddings Clustering Through Multiway Analysis: A Comparative Study

Mira Ait-Saada[1,2]([⊠]) and Mohamed Nadif[1]

[1] Centre Borelli UMR 9010, Université Paris Cité, 75006 Paris, France
{mira.ait-saada,mohamed.nadif}@u-paris.fr
[2] Groupe Caisse des Dépôts, 75013 Paris, France

Abstract. Transformer-based contextual word embedding models are widely used to improve several NLP tasks such as text classification and question answering. Knowledge about these multi-layered models is growing in the literature, with several studies trying to understand what is learned by each of the layers. However, little is known about how to combine the information provided by these different layers in order to make the most of the deep Transformer models. On the other hand, even less is known about how to best use these modes for unsupervised text mining tasks such as clustering. We address both questions in this paper, and propose to study several multiway-based methods for simultaneously leveraging the word representations provided by all the layers. We show that some of them are capable to perform word clustering in an effective and interpretable way. We evaluate their performances across a wide variety of Transformer models, datasets, multiblock techniques and tensor-decomposition methods commonly used to tackle three-way data.

Keywords: Word Embeddings · Word Clustering · Multiway Analysis

1 Introduction

Transformer-based contextual word embedding models have revolutionized the NLP state of the art. When fed with a word sequence, a Transformer model produces several different embeddings (one at each layer of the network) for each token in the sequence. Thus, in a word clustering context, for a dataset with n words and a model with b layers and latent dimension d, we can form a 3-way array of shape $n \times b \times d$ (Fig. 1) where each word is represented by b vectors.

The information captured at the different layers greatly varies: typically, the early layers may encode local syntactic phenomena while more complex semantic aspects are captured at the higher layers [1]. As a consequence it is no surprise that using as input the embeddings produced at different layers may result in very different performance levels on a given downstream task.

B. Crémilleux et al. (Eds.): IDA 2023, LNCS 13876, pp. 1–14, 2023.
https://doi.org/10.1007/978-3-031-30047-9_1

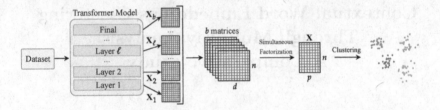

Fig. 1. Description of the proposed approach that leverages the whole Transformer model by performing multiway clustering using all of the layers' representations.

This problem can be overcome in a supervised context, where it is always possible to determine the layer that maximises a certain metric. In contrast, in the unsupervised setting it is difficult to determine the best layer in advance with no a *priori* information about the data. In the absence of ground truth, since there is no easy way of knowing which layer has given the best result, the solution is to design a technique guaranteeing a performance level better than that provided by the best layer, or at worst better than the mean of all layers's scores. More specifically, as a word described by several embedding vectors can be regarded as an input data observation described by several sets (blocks) of variables (dimensions), we propose to evaluate multiblock techniques [2], commonly used in chemiometrics and bioinformatics, in order to make the best use of information provided by the different layers in a word clustering context. Also, since the different layers can be seen as the slices of a 3D-tensor, we can see how do the tensor-decomposition (TD) techniques [3] behave.

We measure the results of these mutiblock- and tensor-based approaches on a word clustering task, and compare them with the performance obtained either with other ways of aggregating the information contained in all the layers such as *unfolding* (or *matricization*) or with layer-wise, non-multiway clustering methods where each layer is handled separately.

To the best of our knowledge, this is the first study that focuses on multiway analyses to make the best use of Transformer-based word representations. The main contributions of the paper are as follows:

- While the performance delivered by Transformer embeddings is mostly assessed in the context of supervised tasks, we study the behavior of these embeddings in the unsupervised setting, and provide a thorough investigation into how to make the best use of them for word clustering purposes, across a variety of Transformer models.
- We show that certain multiway methods, simultaneously considering the features provided by all the embedding layers, are capable of delivering a good performance.

Difference with Previous Contributions. In previous studies, we combined clustering and Transformer representations to find meaningful groups in text datasets. In [4], we first propose a study in order to gain insight about the black-box Transformer models using unsupervised techniques including clustering. To

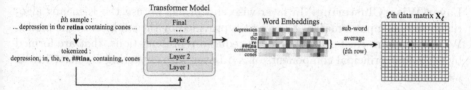

Fig. 2. Construction of the data matrix $\mathbf{X}_\ell$ from the contextual word embeddings provided by the ℓth layer. The context of the word "retina" is used to compute its embedding, which constitutes the ith sample vector representation in the $\mathbf{X}_\ell$ data matrix w.r.t. to the layer ℓ.

this end, we compare layers with each other and show several differences and similarities between layers of the same model. In the present study, we go further and show the complementarity of the different layers by combining them to perform the clustering task. In [5], we propose a framework to harness the whole set of b representations provided by a Transformer model using a clustering ensemble approach combined to a linear dimension reduction based on PCA and a whitening step. The ensemble approach consists in combining the b partitions obtained using each of the *document-level* representations. In contrast, in the present study, we carry out *word-level* clustering and more importantly, we combine the representations *before* computing the partitions, via multi-way techniques. Finally, in [6], we also perform *document* clustering and propose a graph-based three-way approach that simultaneously leverages several document representations such as document-term, static word representations and Transformer representations. In the current study, we rather leverage the representations provided by one given model and we use the representations in their raw form instead of representing then using an adjacency matrix [6].

2   Word Clustering Using Transformer Embeddings

For a given dataset of n words, and a model with b embedding layers, we obtain b different raw representations of size d, one from each layer. The dataset can then be represented by b matrices $\mathbf{X}_1, \dots, \mathbf{X}_b$ of size $n \times d$ as shown in Fig. 2.

One can then choose to use each of these matrices individually, the problem being that in an unsupervised context it is not possible to determine in advance which matrix is likely to help achieve the best performance. One alternative is to try to benefit from the information provided by all the layers, simultaneously.

Using each layer's output separately result in b different clustering partitions. As already said, considering the unsupervised setting of our task and the absence of true labels, it is impossible to determine in advance the best performing layer. In addition, it is worth noting that it is impossible to determine a unique layer for all datasets, since the best layer highly depends on the dataset used, and there is no satisfying universal choice. Therefore, we propose an alternative that benefits from all the network's layers, which we show to be very efficient while freeing us from the impossible task of choosing a unique layer to perform clustering.

Layer-Wise Clustering. In layer-wise clustering, we use the K-means algorithm with as input a matrix $\mathbf{X}_\ell$ of n rows provided by the ℓth layer of the model. We also perform clustering on the PCA-reduced representations, that are formed by the p first principal components of $\mathbf{X}_\ell$. Layer-wise clustering is referred to as LW.

Unfolding Layers' Representations. To try to unleash the potential of all the layers, a simple approach consists in concatenating the matrices $\mathbf{X}_\ell, \ell = 1, \ldots, b$ into a unique data matrix $\mathbf{X}$ of size $n \times (b \times d)$. For example, given a base model with 12 layers and 768 dimensions which provides 12 matrices $\mathbf{X}_b$ of size $n \times 768$, we obtain after the unfolding a matrix of size $n \times 9\,216$. We call UN the use of $\mathbf{X}$ directly by a clustering algorithm.

Multiblock Analysis. With a view to taking advantage of each layer of a given Transformer language model, we propose to harness multiblock methods: Consensus PCA (CPCA) [7], Generalized Canonical Correlation Analysis (GCCA) [8], Multiple Co-Inertia Analysis (MCIA) [9], Multiple Factor Analysis (MFA) [2,10], STATIS [11], Common Components and Specific Weights Analysis (CCSWA) [12]. These methods are designed to deal with simultaneous dimensionality reduction in b blocks (with different features describing the same observations) and are particularly popular in biological multiple omics data analysis. Thereby we argue that they can be very useful to tackle word clustering from three-way data.

In our case, we consider each layer ℓ as a block and each corresponding data matrix is represented by $\mathbf{X}_\ell$ of size $n \times d$ and $\mathbf{S}_\ell = \mathbf{X}_\ell \mathbf{X}_\ell^T$, where n is the number of samples in the dataset and d the number of features of the word representations (d is the same for all of the layers). The objective is to represent the b blocks by a unique matrix $\mathbf{W}$ of size $n \times p$ formed by p component vectors $\mathbf{q}_1, \mathbf{q}_2, \ldots, \mathbf{q}_p$ each one of size $n \times 1$ that optimally resumes the overall information present in the blocks. The MFA method can be seen as an extension of PCA adapted to multiblock data. It consists in applying a standard PCA on a data matrix $\mathbf{X}$ whose features are composed of all the variables (dimensions) weighted according to the block (layer) they belong to in order to balance the influence of each block of variables. The balance is achieved by normalizing each data block $\mathbf{X}_\ell$ using the first eigenvalue λ_1^ℓ. $\mathbf{X}$ is then obtained by concatenating the b resulting matrices. Another formulation of the problem is finding the vector $\mathbf{q}_1$ that is the most linked to all the weighted variables. More formally, MFA maximizes:

$$\sum_{\ell=1}^b \mathcal{L}_\ell(\mathbf{X}_\ell, \mathbf{q}_1) = \sum_{\ell=1}^b \frac{1}{\lambda_1^\ell} \sum_{j=1}^d cov^2(\mathbf{x}_{\ell.j}, \mathbf{q}_1) \tag{1}$$

where $\mathbf{x}_{l.j}$ corresponds to the jth variable of the data block $\mathbf{X}_\ell$ and $\mathbf{q}_1$ to the first component of MFA. The next component maximizes the same criterion while being orthogonal to $\mathbf{q}_1$.

The STATIS method is similar to MFA, and differs in the weighting step. With STATIS, instead of weighting the data blocks according to their corresponding eigenvalues, each data block is weighted using the first eigenvector $\mathbf{v}_1$ of the inner product matrix C of size $b \times b$ where each element $c_{\ell,\ell'}$ is computed as follows:

$$c_{\ell,\ell'} = trace(\mathbf{S}_\ell \times \mathbf{S}_{\ell'}) = \sum_{i=1}^{n}\sum_{k=1}^{n} s_{i,k,\ell}s_{i,k,\ell'}. \tag{2}$$

The weighting vector $\boldsymbol{\alpha}$ is computed by $\boldsymbol{\alpha} = \mathbf{v}_1 \times (\mathbf{v}_1^T \mathbb{1})^{-1}$ so that the elements of $\boldsymbol{\alpha}$ sum to one ($\mathbb{1}$ being the unit vector). In [13], the authors discuss the difference between CPCA, MCIA and GCCA and show their similarity, which lies in the optimized criterion that aims to reveal covariant patterns in and between the different blocks by finding scores vectors $\mathbf{q}_1^\ell$ for each block ℓ which are as much linked as possible to a global score vector $\mathbf{q}_1$: $\sum_{\ell=1}^{b} cov^2(\mathbf{q}_1^\ell, \mathbf{q}_1)$ where $\mathbf{q}_1^\ell$ and $\mathbf{q}_1$ are $n \times 1$ vectors. This criterion is maximized by the three multiblock methods, with different constraints. The same function is maximized to find the higher order components, using a deflated version of the current data blocks at each iteration. The deflating function used with CPCA is as follows:

$$\mathbf{X}_\ell^{(t+1)} = \mathbf{X}_\ell^{(t)} - \frac{\mathbf{q}_t \mathbf{q}_t'}{\mathbf{q}_t' \mathbf{q}_t} \mathbf{X}_\ell. \tag{3}$$

The CCSWA [12] method also computes the components iteratively. The first component $\mathbf{q}_1$ and their weights $\gamma_1^1, ..., \gamma_1^b$ are computed as to minimize the expression: $\sum_{\ell=1}^{b} \left\| \mathbf{S}_\ell - \gamma_1^\ell \mathbf{q}_1 \mathbf{q}_1^T \right\|^2$ where γ_1^ℓ represents the salience value of the ℓth block for the first component (common axis) represented by the vector $\mathbf{q}_1$. The CCSWA algorithm aims to find the parameters $\gamma_t^1, ..., \gamma_t^b$ and $\mathbf{q}_t$ for $t = 1, 2, ..., p$ that maximizes an overall loss function at each iteration t:

$$\sum_{\ell=1}^{b} \| \mathbf{S}_\ell - \sum_{k=1}^{t} \gamma_1^\ell \mathbf{q}_k \mathbf{q}_k^T \|^2. \tag{4}$$

This proportion belongs to $[0, 1]$ and the closer to 1 it is, the better is $\mathbf{X}_{\ell'}$ as a substitute for $\mathbf{X}_\ell$ (and *vice-versa*) to characterize the n samples of the dataset.

Whatever the method used, the p first common components $\mathbf{q}_1, ..., \mathbf{q}_p$ are used as input to a clustering algorithm as shown in Algorithm 1, thus leveraging all of the word representations provided by the multi-layered Transformer models. The common components can be obtained using the R package *FactoMineR* [14] for MFA and *mogsa* [15] for CPCA, GCCA, MCIA, and STATIS. For CCSWA, we used our Python implementation.

Tensor Decomposition. In this family of methods, data matrices $\mathbf{X}_\ell, \ell = 1, ..., b$ can be viewed as a three-way or three-modal tensor $\mathbf{X}$ of size $n \times d \times b$ where each matrix $\mathbf{X}_\ell$ (obtained with the ℓth layer of the model) is considered as a slice of the tensor. Two of the most popular TD techniques are CANDE-CAMP/PARAFAC (CP) [16] and Tucker decomposition [17]. For the sake of simplicity, we describe these two techniques for three-mode tensors, but they can be generalized for m-mode tensors ($m > 3$). A CP decomposition of rank p aims to

---

Algorithm 1: Multiway Clustering Procedure

Input: a dataset $\mathcal{D}$ of n words; a Transformer model $\mathcal{M}$ of b layers; a clustering algorithm $\mathcal{C}$; a multiway decomposition function $\mathcal{F}$, number of components p

Output: a clustering partition **p**

Step 1. Build data matrices, for each $\ell = 1 \ldots b$:

$\mathbf{X}_\ell \leftarrow \mathcal{M}(\mathcal{D}, \ell)$ as in Figure 2;

Step 2. Perform multiblock factorial decomposition:

$\mathbf{q}_1, \ldots, \mathbf{q}_p \leftarrow \mathcal{F}([\mathbf{X}_1, \ldots, \mathbf{X}_b], p)$;

$\mathbf{W} \leftarrow$ horizontal stacking of the p components $\mathbf{q}_1, \ldots, \mathbf{q}_p$;

Step 3. Perform clustering:

$\mathbf{p} \leftarrow \mathcal{C}(\mathbf{W})$;

return p

---

find three factor matrices **A**, **B** and **C** of size $n \times p$, $d \times p$ and $b \times p$ respectively, and $\boldsymbol{\lambda}$ that minimizes the approximation error of **X** by finding:

$$\min_{\mathbf{X}^*} \|\mathbf{X} - \mathbf{X}^*\| \text{ where } \mathbf{X}^* = \sum_{j=1}^{p} \lambda_j \mathbf{a}_j \circ \mathbf{b}_j \circ \mathbf{c}_j \tag{5}$$

where $\mathbf{a}_j$, $\mathbf{b}_j$ and $\mathbf{c}_j$ being the jth column of **A**, **B** and **C** respectively. The "$\circ$" stands for the outer product between two vectors. The outer product of the three vectors $\mathbf{a}_j$, $\mathbf{b}_j$ and $\mathbf{c}_j$ results in a three-way tensor with the sames dimensions as **X**. One of the most popular procedures to optimize this criterion is the Alternating Least Squares (ALS) [16,18]. The Tucker decomposition also computes three factor matrices **A**, **B** and **C**, this time of size $n \times p$, $d \times q$ and $b \times t$ respectively. Unlike in CP, the ranks of the three modes (p, q and t) are not necessarily equal. In addition to the factor matrices, a core tensor **G** of size $p \times q \times t$ is computed so as to optimize:

$$\min_{\mathbf{X}^*} \|\mathbf{X} - \mathbf{X}^*\| \text{ where } \mathbf{X}^* = \sum_{i=1}^{p} \sum_{j=1}^{q} \sum_{k=1}^{t} g_{ijk} \mathbf{a}_i \circ \mathbf{b}_j \circ \mathbf{c}_k \tag{6}$$

g_{ijk} being the intersection if the ith row (mode 1), the jth column (mode 2) and the kth tube (mode 3) of the tensor **G**. To solve this optimization problem, a popular approach named Higher-Order Orthogonal Iteration (HOOI) [19] and similar to ALS consists in fixing two factor matrices to compute the third one by using two-way singular vector decomposition (cf. [19] for further details).

In our experiments, we use for both CP and Tucker the matrix **A** as input to the clustering algorithm. The p columns of **A** can be seen as the principal components in the first mode of **X**. We use two Python implementations of ALS and HOOI to perform CP and Tucker decomposition respectively. The first implementation is provided by the *TensorD* [20] package, based on the *TensorFlow* framework, allowing GPU acceleration. The second is available in the *TensorLy* [21] package which flexibly leverages several CPU and GPU backends. The suffixes TensD and TensL in Table 2 stand for the packages *TensorLy* and *TensorD* respectively.

3 Experimental Study

Datasets and Models Used. The datasets used are described in Table 1 and contain words, their contexts and the corresponding topic. The first dataset, denoted as UFSAC3 is extracted from the UFSAC dataset [22] which consolidates multiple popular WSD datasets (such as SemEval and SensEval) into a uniform format. The examples are manually divided into three topics: Body, Botany and Geography. The second dataset, denoted as UFSAC4, is a slightly more difficult dataset; it includes a fourth class (words related to information technology) and is augmented with some polysemous examples (such as "lobes" which appears in Body and Botany with different contexts). The two last datasets yahoo4 and yahoo6, are extracted from the Yahoo! Answers dataset [23] by manually selecting sets of words for each category and some corresponding contexts.

Table 1. Datasets description: k denotes the number of clusters.

Datasets	n	k	Clusters' sizes
UFSAC3	583	3	body: 266, geo: 227, botany: 90
UFSAC4	691	4	body: 275, geo: 227, botany: 99, IT:90
yahoo4	528	4	finance: 150, music: 82, science-maths.: 152, computers-internet: 144
yahoo6	852	6	health: 201, finance: 150, computers-internet: 144, music: 82, science-maths.: 152, sports: 123

From each set of documents, we compute multiple contextual representations from 4 pre-trained models which are the base (12 layers) and large (24 layers) versions of BERT [24] and RoBERTa [25] with a vocabulary size of 28,996 for BERT and 50,265 for RoBERTa.

Clustering Evaluation. In all the experiments, we use K-means [26] with a known number of clusters (any other clustering algorithm can be used), which are run with 10 different initializations on different matrix representations of words occurrences. To validate the results produced by the clustering algorithm we relied on standard external measures devoted to assessing cluster quality, namely Normalized Mutual Information (NMI) [27], the Adjusted Rand Index (ARI) [28] and the clustering accuracy. When performing dimension reduction, a number of components p has to be fixed. Since we cannot tune the p parameter in our unsupervised setting, we set it to the same value ($p = 15$) for all experiments. This corresponds either to the number of principal components of PCA, the number of common components present in the $\mathbf{W}$ matrix computed by the multiblock techniques or the rank of the matrix $\mathbf{A}$ computed by the TD methods (cf. Section 2). Table 2 contains the NMI scores obtained by each method over the four Transformer models and the four datasets. For comparison purposes, we also include the results obtained when using fastText word embeddings or the representations obtained by the Transformers' layers individually (as described in the next section). FastText [29] was used as a representative of the static word embeddings family since it provides for the Out-of-Vocabulary issue which otherwise would distort comparisons.

Layer-Wise Clustering Experiments. We perform clustering on the PCA-reduced representations and present the results (NMI) in Fig. 3 only for two datasets, due to the lack of space. In Table 2, we call LW and LW-PCA the average of the scores obtained by all of the b clustering experiments run on each matrix $\mathbf{X}_\ell$ separately before and after applying PCA respectively.

Multiway Clustering Experiments. Since the obtained results may greatly vary with each layer, as clearly visible in Fig. 3, and since determining which one is the best is impossible in the absence of labels when dealing with real datasets, a sensible solution to overcome this incertitude is to use multiway techniques as described in Sect. 2 using the $\mathbf{X}_\ell$ data matrices $\ell = 1, \ldots, b$ of each model (results provided in Table 2).

4   Results and Discussion

In this section, we discuss the benefit of applying multiway methods in the unsupervised setting due to a diversity of the information provided by each layer. Thereby, from the representations of each layer used separately we first evaluate the performance in terms of word clustering. We aim at showing that the results might greatly vary according to the dataset.

To tackle this issue, for each dataset we propose to use all layers. We perform a dimensionality reduction in different ways and from a simplified representation, a clustering algorithm is applied and quality of word clustering is measured. Further, using user-friendly visualization tools, we show how the classes are arranged in a low-dimensional space, how to detect the most influential words and to measure the role of each layer.

Layer-Wise Clustering Results. Figure 3 shows that BERT presents better performance than RoBERTa, even if the latter is more recent and outperforms BERT in the supervised tasks. We can see in the same figures that BERT models achieve their best performance on the intermediate layers (which is in line with the findings issued in [30]) especially BERT-large which reaches a perfect NMI score on UFSAC3 from the 9th to the 14th layer, with a sudden drop of performance on the 5 last layers. Overall, we can see this decrease for all the models. Figure 3 also shows that reducing the number of dimensions to only 15 (which constitutes less than 2% of features for the base models and 1.5% for the large ones) leads to very competitive performance scores, achieving even better results than raw representations, which indicates a high redundancy in the pre-trained embeddings. Interestingly, in the case of RoBERTa the PCA reduction generally improves the performance of the last layers, which as said previously perform very poorly in their raw form.

Table 2. Compared performance (NMI scores) of multiway clustering on Transformers' over the four models and fastText. Values between parentheses correspond to the standard deviation.

Datasets	Method	BERT-base	BERT-large	RoBERTa-base	RoBERTa-large	fastText
UFSAC3	LW	0.89 (0.07)	0.81 (0.3)	0.71 (0.14)	0.64 (0.16)	0.92
	LW-PCA	0.91 (0.07)	0.8 (0.31)	0.78 (0.16)	0.66 (0.16)	0.92
	UN	0.96	1.0	0.94	0.66	–
	UN-PCA	**0.98**	1.0	**0.97**	0.65	–
	MFA	**0.98**	1.0	**0.97**	0.77	–
	STATIS	0.96	0.98	**0.97**	0.77	–
	CPCA	0.76	0.71	0.68	0.53	–
	GCCA	0.8	0.76	0.74	0.65	–
	MCIA	0.7	0.76	0.72	0.69	–
	CCSWA	0.58	0.58	0.77	0.68	–
	CP-TensD	0.66	0.28	0.01	0.64	–
	CP-TensL	0.92	0.66	0.57	0.56	–
	Tuck-TensD	0.68	0.75	0.78	0.83	–
	Tuck-TensL	0.73	0.81	0.89	0.69	–
UFSAC4	LW	0.86 (0.08)	0.72 (0.28)	0.67 (0.11)	0.6 (0.13)	0.77
	LW-PCA	0.88 (0.06)	0.75 (0.25)	0.74 (0.1)	0.63 (0.11)	0.84
	UN	0.96	0.97	0.67	0.65	–
	UN-PCA	**0.97**	**0.99**	0.76	0.65	–
	MFA	**0.97**	**0.99**	0.73	0.67	–
	STATIS	0.96	0.95	**0.90**	0.68	–
	CPCA	0.75	0.84	0.75	0.59	–
	GCCA	0.76	0.75	0.82	0.7	–
	MCIA	0.81	0.76	0.79	0.67	–
	CCSWA	0.73	0.82	0.76	**0.76**	–
	CP-TensD	0.43	0.59	0.06	0.46	–
	CP-TensL	0.61	0.55	0.48	0.61	–
	Tuck-TensD	0.75	0.85	0.84	0.62	–
	Tuck-TensL	0.81	0.86	0.79	0.65	–
yahoo4	LW	0.81 (0.07)	0.59 (0.24)	0.54 (0.17)	0.59 (0.14)	0.42
	LW-PCA	0.83 (0.07)	0.58 (0.25)	0.84 (0.04)	0.66 (0.15)	0.51
	UN	0.9	0.82	0.63	0.71	–
	UN-PCA	**0.93**	**0.91**	**0.91**	0.82	–
	MFA	0.92	**0.91**	**0.91**	**0.83**	–
	STATIS	0.91	0.89	0.9	0.82	–
	CPCA	**0.93**	0.74	0.87	0.82	–
	GCCA	0.88	0.71	0.85	0.79	–
	MCIA	0.85	0.76	0.86	0.8	–
	CCSWA	0.7	0.73	0.81	0.8	–
	CP-TensD	0.04	0.08	0.17	0.07	–
	CP-TensL	0.44	0.55	0.41	0.63	–
	Tuck-TensD	0.81	0.66	0.81	0.79	–
	Tuck-TensL	0.83	0.73	0.85	0.8	–
yahoo6	LW	0.73 (0.08)	0.58 (0.21)	0.61 (0.15)	0.62 (0.16)	0.42
	LW-PCA	0.73 (0.06)	0.57 (0.22)	0.79 (0.08)	0.69 (0.14)	0.42
	UN	0.92	0.8	0.72	0.79	–
	UN-PCA	0.94	0.81	0.81	0.81	–
	MFA	0.93	0.83	0.81	0.81	–
	STATIS	0.74	0.75	0.8	0.81	–
	CPCA	0.94	**0.9**	0.88	**0.84**	–
	GCCA	**0.96**	0.85	0.88	0.83	–
	MCIA	0.93	0.86	**0.9**	0.83	–
	CCSWA	0.82	0.7	0.65	0.76	–
	CP-TensD	0.53	0.18	0.09	0.11	–
	CP-TensL	0.6	0.74	0.57	0.46	–
	Tuck-TensD	0.87	0.71	0.78	0.81	–
	Tuck-TensL	0.85	0.66	0.79	0.81	–

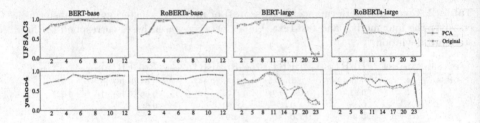

Fig. 3. Compared clustering performance (NMI) across layers of the original representations of base models (using all of the 768 dimensions of the $\mathbf{X}_\ell$ matrices) vs. reduced representations (using 15 principal components of $\mathbf{X}_\ell$), with $\ell = 1, \ldots, b$ ($b \in \{12, 24\}$).

Multiway Clustering Results. One first observation from Table 2 concerning the multiway clustering results is that multiblock methods seem more effective compared to TD techniques. Tucker seems more suitable for our data than CP. Besides, Table 2 shows that in the case of BERT (base and large), the multiblock techniques are very competitive. Moreover, MFA and UN-PCA on BERT-large have no difficulty in achieving the perfect score on UFSAC3. For RoBERTa-large, multiblock techniques are still better than the mean of all layers's scores. In general, the performance of MFA is much higher than the average performance of the layers used separately. This indicates that it is very effective for capturing the valuable information present in the $d \times b$ features of each model (9 216 for the base models and 24 576 for the large ones) in only 15 dimensions.

In Table 2 we observe that the raw unfolding (UN) is less powerful than when reduced with PCA (UN-PCA). This brings out the interest of factor decomposition and dimension reduction in improving performance in the clustering task while combining all layers. This difference in performance is in line with the observations made on Fig. 3 where the use of PCA noticeably improved the layer-wise clustering, which explains the enhancement observed with UN-PCA.

Visual Interpretation of Factorial Analysis Results. One additional advantage of Multiblock methods and specifically MFA is that they offer several visual interpretation possibilities that are presented in Figs. 4, 5 and 6.

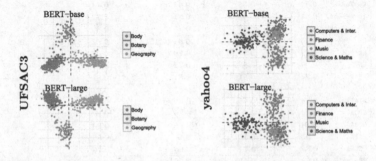

Fig. 4. 2D projections of UFSAC3 and yahoo4 words on the two first common components of MFA, colored according to the known topics.

Figure 4 illustrates the words' projections on the two first common factors of MFA from two datasets. We observe a good separation of the three clusters of UFSAC3, especially with BERT-large, which is in line with the scores provided in Table 2.

Figure 5 shows the contribution of the first 20 words to the first three components of MFA for BERT-base and RoBERTa-base. We see that for BERT-base, among the 20 most contributory words, the classes are perfectly distributed over the 3 first dimensions (Geography, Botany and Body contribute the most to dimensions 1, 2 and 3 respectively). For RoBERTa-base, the distribution is less homogeneous, the three classes contributing equally to the last dimension.

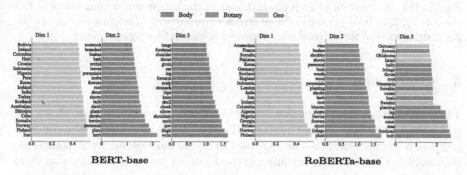

Fig. 5. The 20 most contributory words (of UFSAC3) to the first three dimensions of MFA applied to all the layers of BERT-base and RoBERTa-base.

Indices to Help with Interpretation. Figure 6 shows the contribution and NMI score of each layer with the 4 models. The contribution of a layer to a component is the sum of the contributions of its variables. More formally, the contribution of the ℓth layer to the jth component is computed as $\sum_{k \in \mathcal{G}_\ell} \alpha_\ell \times \mathbf{u}_{k,j}$ where $\mathbf{u}_{k,j}$ is the loading of the kth variable for the jth component, $\alpha_\ell = \frac{1}{\lambda_1^\ell}$ is the weight of the ℓth layer and $\mathcal{G}_\ell$ is the set of variables of the ℓth layer. The contribution of each variable takes values between 0 and 1 and sums to 1 for a given component. The values of contribution displayed in Fig. 6 are percentages. We observe from Fig. 6 that MFA seems to sum up the relevant information present in the layers of the models, without taking into account the most outlying (and less performing) layers, especially for BERT-base and BERT-large. For the other models, the drop of performance occurring at the end of some models coincides with a significant decrease of contribution. We can also see that for RoBERTa-base, the best performing layer (layer 5) is also the most contributory to the first component of MFA.

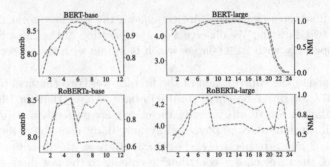

Fig. 6. The *contribution* of each block (layer) to the first common component of MFA (UFSAC3). The NMI (in red) of a layer ℓ is obtained with clustering separately the $\mathbf{X}_\ell$ matrix made of the original word representations. (Color figure online)

5   Conclusion

The present study has shown that a multiway-based approach to word clustering has a twofold benefit. It first removes the need to choose, among the different layers of a Transformer, the one supposed to be the most useful input, something impossible to determine in an unsupervised setting. Second, across a variety of Transformer models as well as datasets for which labels were available for evaluation, we have shown that multiway techniques can deliver a good performance without requiring the choice among layers, a topic widely discussed in the NLP community. Among these techniques, we retain MFA which turns out to be the most consensual, outscoring standard tensor-decomposition methods. This can be attributed to the fact that MFA aims at giving a balanced role to the layers. It is based on a factor analysis in which the variables are weighted and these weights are identical for the dimensions of the same layer (and vary from one layer to another) while offering a visualisation with many indices to help with interpretation. Thereby, we have shown for the first time the interest of using such a method in NLP.

References

1. Tenney, I., Das, D., Pavlick, E.: BERT rediscovers the classical NLP pipeline. In: Proceedings of the 57th Annual Meeting of the Association for Computational Linguistics, pp. 4593–4601. Florence, Italy (2019)
2. Escofier, B., et al.: Méthode pour l'analyse de plusieurs groupes de variables. application à la caractérisation de vins rouges du Val de Loire. Revue de statistique appliquée **31**(2), 43–59 (1983)
3. Kolda, T.G., Bader, B.W.: Tensor decompositions and applications. SIAM Rev. **51**(3), 455–500 (2009)
4. Ait Saada, M., Role, F., Nadif, M.: Unsupervised methods for the study of transformer embeddings. In: Abreu, P.H., Rodrigues, P.P., Fernández, A., Gama, J.

(eds.) IDA 2021. LNCS, vol. 12695, pp. 287–300. Springer, Cham (2021). https://doi.org/10.1007/978-3-030-74251-5_23

5. Ait-Saada, M., Role, F., Nadif, M.: How to leverage a multi-layered transformer language model for text clustering: an ensemble approach. In: Proceedings of the 30th ACM International Conference on Information & Knowledge Management (CIKM 2021), pp. 2837–2841 (2021)

6. Boutalbi, R., Ait-Saada, M., Iurshina, A., Staab, S., Nadif, M.: Tensor-based graph modularity for text data clustering. In: Proceedings of the 45th International ACM SIGIR Conference on Research and Development in Information Retrieval (SIGIR 2022), pp. 2227–2231 (2022)

7. Wold, S., Hellberg, S., Lundstedt, T., Sjostrom, M., Wold, H.: Proc. Symp. on PLS model building: theory and application. Frankfurt am Main (1987)

8. Carroll, J.D.: Generalization of canonical correlation analysis to three or more sets of variables. In: Proceedings of the 76th Annual Convention of the American Psychological Association, vol. 3, pp. 227–228. Washington, DC (1968)

9. Chessel, D., Hanafi, M.: Analyses de la co-inertie de K nuages de points. Revue de statistique appliquée **44**(2), 35–60 (1996)

10. Abdi, H., Williams, L.J., Valentin, D.: Multiple factor analysis: principal component analysis for multitable and multiblock data sets. Wiley Interdiscip. Rev. Comput. Stat. **5**(2), 149–179 (2013)

11. Escoufier, Y.: L'analyse conjointe de plusieurs matrices de données. Biométrie et temps **58**, 59–76 (1980)

12. Qannari, E.M., Wakeling, I., Courcoux, P., MacFie, H.J.: Defining the underlying sensory dimensions. Food Qual. Prefer. **11**(1–2), 151–154 (2000)

13. Hanafi, M., Kohler, A., Qannari, E.-M.: Connections between multiple co-inertia analysis and consensus principal component analysis. Chemom. Intell. Lab. Syst. **106**(1), 37–40 (2011)

14. Lê, S., Josse, J., Husson, F.: FactoMineR: a R package for multivariate analysis. J. Stat. Softw. **25**(1), 1–18 (2008)

15. Chen Meng. mogsa: Multiple omics data integrative clustering and gene set analysis (2019). R package version 1.20.0

16. Carroll, J.D., Chang, J.J.: Analysis of individual differences in multidimensional scaling via an n-way generalization of "Eckart-Young" decomposition. Psychometrika **35**(3), 283–319 (1970). https://doi.org/10.1007/BF02310791

17. Tucker, L.R.: Some mathematical notes on three-mode factor analysis. Psychometrika **31**(3), 279–311 (1966)

18. Harshman, R.A., et al.: Foundations of the PARAFAC procedure: models and conditions for an "explanatory" multimodal factor analysis (1970)

19. De Lathauwer, L., De Moor, B., Vandewalle, J.: On the best rank-1 and rank-$(R_1, R_2, \ldots, R_N)$ approximation of higher-order tensors. SIAM J. Matrix Anal. Appl. **21**(4), 1324–1342 (2000)

20. Hao, L., Liang, S., Ye, J., Zenglin, X.: TensorD: a tensor decomposition library in TensorFlow. Neurocomputing **318**, 196–200 (2018)

21. Kossaifi, J., Panagakis, Y., Anandkumar, A., Pantic, M.: TensorLy: tensor learning in Python. J. Mach. Learn. Res. **20**(26), 1–6 (2019)

22. Vial, L., Lecouteux, B., Schwab, D.: UFSAC: unification of sense annotated corpora and tools. In: Proceedings of the Eleventh International Conference on Language Resources and Evaluation (LREC 2018) (2018)

23. Zhang, X., Zhao, J., LeCun, Y.: Character-level convolutional networks for text classification. Adv. Neural. Inf. Process. Syst. **28**, 649–657 (2015)

24. Devlin, J., Chang, M.W., Lee, K., Toutanova, K.: BERT: pre-training of deep bidirectional transformers for language understanding. In: Proceedings of the 2019 Conference of the North American Chapter of the Association for Computational Linguistics: Human Language Technologies, vol. 1 (Long and Short Papers), pp. 4171–4186, Minneapolis, Minnesota (2019)
25. Liu, Y., et al.: RoBERTa: a robustly optimized BERT pretraining approach. arXiv preprint arXiv:1907.11692 (2019)
26. Lloyd, S.: Least squares quantization in PCM. IEEE Trans. Inf. Theory **28**(2), 129–137 (1982)
27. Strehl, A., Ghosh, J.: Cluster ensembles–a knowledge reuse framework for combining multiple partitions. J. Mach. Learn. Technol. **3**(Dec), 583–617 (2002)
28. Hubert, L., Arabie, P.: Comparing partitions. J. Classif. **2**(1), 193–218 (1985). https://doi.org/10.1007/BF01908075
29. Bojanowski, P., Grave, E., Joulin, A., Mikolov, T.: Enriching word vectors with subword information. Trans. Assoc. Comput. Linguist. **5**, 135–146 (2017)
30. Liu, N.F., Gardner, M., Belinkov, Y., Peters, M.E., Smith, N.A.: Linguistic knowledge and transferability of contextual representations. In: Proceedings of the 2019 Conference of the North American Chapter of the Association for Computational Linguistics: Human Language Technologies, vol. 1 (Long and Short Papers), pp. 1073–1094, Minneapolis, Minnesota (2019)

Transferable Deep Metric Learning
for Clustering

Mohamed Alami Chehboune[1,2(✉)], Rim Kaddah[2], and Jesse Read[1]

[1] Department of Computer Science, Ecole Polytechnique, Palaiseau, France
{mohamed.alami-chehboune,jesse.read}@polytechnique.edu
[2] IRT SystemX, Palaiseau, France
rim.kaddah@irt-systemx.fr

Abstract. Clustering in high dimension spaces is a difficult task; the usual distance metrics may no longer be appropriate under the curse of dimensionality. Indeed, the choice of the metric is crucial, and it is highly dependent on the dataset characteristics. However a single metric could be used to correctly perform clustering on multiple datasets of different domains. We propose to do so, providing a framework for learning a transferable metric. We show that we can learn a metric on a labelled dataset, then apply it to cluster a different dataset, using an embedding space that characterises a desired clustering in the generic sense. We learn and test such metrics on several datasets of variable complexity (synthetic, MNIST, SVHN, omniglot) and achieve results competitive with the state-of-the-art while using only a small number of labelled training datasets and shallow networks.

Keywords: Clustering · Transfer Learning · Metric Learning

1 Introduction

Clustering is the unsupervised task of assigning a categorical value $y_i \in \{1, \dots, k\}$ to each data point $x_i \in \mathbf{X}$, where no such example categories are given in the training data; i.e., we should map $\mathbf{X} = \{x_1, \dots, x_n\} \mapsto \mathbf{Y} = \{y_1, \dots, y_n\}$ with $\mathbf{X}$ the input matrix of n data points, each of dimension d; where $y_i = \kappa$ implies that data point x_i is assigned to the κ-th cluster.

Clustering methods complete this task by measuring similarity (the distance) between training pairs, using a similarity function $s(x_i, x_j) \in \mathbb{R}_+$. This similarity function should typically reflect subjective criteria fixed by the user. Basically, this means that the user decides what makes a good clustering. As mentioned in [6], "since classes are a high-level abstraction, discovering them automatically is challenging, and perhaps impossible since there are many criteria that could be used to cluster data (e.g., we may equally well cluster objects by colour, size, or shape). Knowledge about some classes is not only a realistic assumption, but also indispensable to narrow down the meaning of clustering". Taking the example of MNIST [11], one usually groups the same numbers together because

B. Crémilleux et al. (Eds.): IDA 2023, LNCS 13876, pp. 15–28, 2023.
https://doi.org/10.1007/978-3-031-30047-9_2

these numbers share the highest amount of features (e.g., mutual information based models do that). However one may want to group numbers given their roundness. In this case, we may obtain two clusters, namely straight shaped numbers (i.e., 1, 4,7) and round shaped numbers (i.e., all the others). Both clustering solutions are relevant, since each clustering addresses a different yet possible user subjective criteria (i.e., clustering semantics).

Finding an automated way to derive and incorporate user criteria in a clustering task based on intended semantics can be very hard. Nowadays, the wide availability of shared annotated datasets is a valuable asset and provides examples of possible user criteria. Hence, we argue that, given "similar" annotated data, classification logic can be used to derive a user criteria that one can apply to clustering similar non-annotated data. For example, we consider the situation where a human is placed in front of two datasets, each one consisting of letters of a certain alphabet she does not understand. The first dataset is annotated, grouping the same letters together. Only by seeing the first dataset, the person can understand the grouping logic used (grouping same geometrical shapes together) and replicate that logic to the second non annotated dataset and cluster correctly its letters.

In this paper, we are interested in tackling the problem of clustering data when the logic (i.e., user clustering criteria) is encoded into some available labelled datasets. This raises two main challenges, namely (1) find a solution that works well on the classification task but (2) ensure transferability in its decision mechanism so it is applicable to clustering data from a different domain.

We believe that addressing these challenges calls for the design of a scoring function that should be as general as possible to ensure transferability but is specific enough not to miss the user criteria. More specifically, the scoring function should be a comparing the logic used to produce a certain clustering to the one used to produce clusterings of the already seen training datasets. Using the concept of logic is useful as a logic is general enough to be used on any dataset and specific enough as is it is the main common property shared by all training dataset. Our goal is then to find a suitable metric that retrieves and encapsulate the seen concept for scoring a clustering outcome.

Moreover, modern applications require solutions that are effective when data is of high dimension (i.e., large d). While distance-based approaches are broadly used for clustering (e.g., Euclidean distance), we argue that they are not suitable for our problem since they would yield in data specific models in addition to their poor performance in high dimensional spaces due to the curse of dimensionality. To lower dimensionality, a solution is to perform instance-wise embeddings $x_i \mapsto z_i$, e.g., with an autoencoder. However this mechanism is still domain specific.

To achieve training on more general patterns, we think it is necessary to take the dataset in its entirety. Therefore, instead of learning a metric that compares pairs of data points in a dataset instance (like a similarity measure), a learned metric is applied to sets of data points so comparison is done between sets. The metric can be intuitively understood as a distance between the logic underlying a given clustering and the general logic that was used to produce clusterings in training datasets.

For this, we propose a solution where we use a graph autoencoder [9] to embed a set of data points into a vector of chosen dimension. Then, we use the critic part of a Wasserstein GAN (WGAN) [1] to produce a continuous score of the embedded clustering outcome. This critic represents the metric we seek. Thus, our main contributions are:

- We provide a framework for joint metric learning and clustering tasks.
- We show that our proposed solution yields a learned metric that is transferable to datasets of different sizes and dimensions, and across different domains (either vision or tabular) and tasks.
- We obtain results competitive to the state-of-the-art with only a small number of training datasets, relatively simple networks, and no prior knowledge (only an upper bound of the cluster number that can be set to a high value).
- Our method is scalable to large datasets both in terms of number of points or dimensions (e.g the SVHN dataset used in Sect. 4) as it does not have to compute pairwise distances and therefore does not heavily suffer when the number of points or dimensions increase.
- We test the metric on datasets of varying complexity and perform on par with the state-of-the-art while maintaining all the advantages cited above.

2 Related Work

Using auto-encoders before applying classic clustering algorithms resulted in a significant increase of clustering performance, while still being limited by these algorithms capacity. Deep Embedding Clustering (DEC) [19] gets rid of this limitation at the cost of more complex objective functions. It uses an auto-encoder along with a cluster assignment loss as a regularisation. The obtained clusters are refined by minimising the KL-divergence between the distribution of soft labels and an auxiliary target distribution. DEC became a baseline for deep clustering algorithms. Most deep clustering algorithms are based on classical center-based, divergence-based or hierarchical clustering formulations and hence bear limitations like the need for an *a priori* number of clusters.

MPCKMeans [2] is more related to metric learning as they use constraints for both metric learning and the clustering objective. However, their learned metrics remain dataset specific and are not transferable.

Constrained Clustering Network (CCN) [8], learns a metric that is transferable across domains and tasks. Categorical information is reduced to pairwise constraints using a similarity network. Along with the learned similarity function, the authors designed a loss function to regularise the clustering classification. But, using similarity networks only captures local properties instance-wise rather than global geometric properties of dataset clustering. Hence, the learned metric remains non fully transferable, and requires to adapt the loss to the domain to which the metric is transferred to.

In Deep Transfer Clustering (DTC) [6] and Autonovel [7], the authors tackle the problem of discovering novel classes in an image collection given labelled examples of other classes. They extended DEC to a transfer learning setting while estimating the number of classes in the unlabelled data. Autonovel uses self-supervised learning to train the representation from scratch on the union of labelled and unlabelled datasets then trains the data representation by optimizing a joint objective function on the labelled and unlabelled subsets of data. We consider these two approaches as our state of the art baselines.

3 Our Framework

To restate our objective, we seek an evaluation metric

$$r : \mathbb{R}^{n \times d} \times \mathbb{N}^n \to \mathbb{R}$$
$$(\mathbf{X}, \mathbf{y}) \mapsto r(\mathbf{X}, \mathbf{y}) \tag{1}$$

where $\mathbf{X} \in \mathbb{R}^{n \times d}$ is a dataset of n points in d dimensions and $\mathbf{y} \in \mathbb{N}^n$ a partition of $\mathbf{X}$ (i.e. a clustering of $\mathbf{X}$). Metric r should provide a score for *any* labelled dataset of any dimensionality; and in particular this score should be such that $r(\mathbf{X}, \mathbf{y})$ is high when the hamming distance between the ground truth labels $\mathbf{y}^*$ and $\mathbf{y}$ is small (taking cluster label permutations into account). This would mean that we could perform clustering on any given dataset, simply by solving an optimisation problem even if such a dataset had not been seen before.

Formally stated, our goal is: (1) to produce a metric r that grades the quality of a clustering such that $\mathbf{y}^* = \arg\max_{\mathbf{y}} r(\mathbf{X}, \mathbf{y})$; (2) Implement an optimisation algorithm that finds $\mathbf{y}^*$; (3) use (1) and (2) to perform a clustering on a new unrelated and unlabelled dataset. We use a collection $\mathcal{D} = \{\mathbf{X}_l, \mathbf{y}_l^*\}_{l=1}^{\ell}$ of labelled datasets as examples of correctly 'clustered' datasets, and learn r such that $\mathbb{E}[r(\mathbf{X}, \mathbf{y})]$ is high. In order to make r transferable between datasets, we embed each dataset with its corresponding clustering $(\mathbf{X}_l, \mathbf{y}_l)$ into a vector $\mathbf{z}_l \in \mathbb{R}^e$. More formally, the embedding function is of the form:

$$g : \mathbb{R}^{n \times d} \times \mathbf{Y} \to \mathbb{R}^e$$
$$(\mathbf{X}, \mathbf{y}) \mapsto \mathbf{z} \tag{2}$$

Therefore, the metric r is actually the composition of two functions g and c_θ (the scoring function from $\mathbb{R}^e$ to $\mathbb{R}$). Our training procedure is structured around 3 blocs A, B and C detailed in next sections and depicted in Fig. 1 and is summarised in the following main steps:

Bloc A. step 1 Select a labelled dataset $(\mathbf{X}, \mathbf{y}^*) \sim \mathcal{D}$

Bloc A. step 2 Given a metric function r (output from bloc B step 2, or initialised randomly), we perform a clustering of dataset $\mathbf{X}$: $\hat{\mathbf{y}} = \arg\max_{\mathbf{y}} r(\mathbf{X}, \mathbf{y})$

Bloc B. step 1 $\mathbf{y}^*$ and $\hat{\mathbf{y}}$ are represented as graphs where each clique represents a cluster.

Bloc B. step 2 Graph convolutional autoencoders perform feature extraction from $\hat{\mathbf{y}}$ and $\mathbf{y}^*$ and output embeddings $\hat{\mathbf{z}}$ and $\mathbf{z}^*$

Bloc C. step 1 The metric r is modelled by a WGAN critic that outputs evaluations of the clusterings: $r(\mathbf{X}, \mathbf{y}^*) = \mathbf{c}_\theta(\mathbf{z}^*)$ and $r(\mathbf{X}, \hat{\mathbf{y}}) = \mathbf{c}_\theta(\hat{\mathbf{z}})$

Bloc C. step 2 Train the model using the error between $r(\mathbf{X}, \mathbf{y}^*)$ and $r(\mathbf{X}, \hat{\mathbf{y}})$.

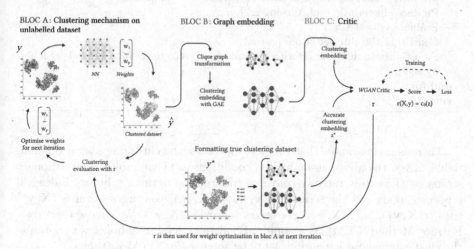

Fig. 1. Our framework's 3 components: the clustering mechanism (A), the GAE (B) and the WGAN (C). (A) takes an unlabelled dataset $\mathbf{X}$ as input and outputs a clustering $\hat{\mathbf{y}}$ that maximises a metric r. $\hat{\mathbf{y}}$ is then turned into a graph $\mathcal{G}(\mathbf{X}, \hat{\mathbf{y}})$ then into an embedding vector $\hat{\mathbf{z}}$ using (B). Same goes for the correctly labelled dataset, which is embedded as $\hat{\mathbf{z}}^*$. Then, (C), which is the metric itself, evaluates $\hat{\mathbf{z}}$ and $\mathbf{z}^*$ using c_θ and is trained to produce a new metric r which is then used for (A) in the next iteration.

3.1 Clustering Mechanism

We seek the most suitable optimisation algorithm for clustering given r. Considering a neural network that performs the clustering, we need to find its weights w such that the metric is maximised (see Eq. (3)). The type of algorithm to use depends on the nature of the metric r to optimise on.

$$\text{CEM}_r(\mathbf{X}) \xrightarrow{\text{finds}} w^* = \arg\max_{w} r(\mathbf{X}, \mathbf{y}^w) \tag{3}$$

where $\mathbf{y}^w$ is a clustering obtained with the weights w. The metric is assumed to hold certain properties, discussed in Sect. 3.3:

– **Unique Maximum:** A unique optimal clustering. r has a unique maximum.
– **Continuity**[1]: Any two clusterings $\mathbf{y}$ and $\mathbf{y}'$ should be similar if $r(\mathbf{y})$ and $r(\mathbf{y}')$ are close in $\mathbb{R}$ space. Hence, r has to satisfy a continuity constraint.

---

[1] As a reminder, Let T and U be two topological spaces. A function $f : T \mapsto U$ is continuous in the open set definition if for every $t \in T$ and every open set u containing $f(t)$, there exists a neighbourhood v of t such that $f(v) \subset u$.

Algorithm 1. CEM Algorithm

Input: Dataset $X \in \mathbb{R}^{n \times d}$; score function r; $\mu \in \mathbb{R}^d$ and $\sigma \in \mathbb{R}^d$; elite percentage to retain p; n samples of $w_i \sim \mathcal{N}(\mu, \text{diag}(\sigma))$; T number of iterations
for iteration $= 1$ **to** T **do**
    Produce n samples of neural network weights $w_i \sim \mathcal{N}(\mu, \text{diag}(\sigma))$
    Produce clusterings y_i of X using each w_i
    Evaluate $r_i = r(X, y_i)$
    Constitute the elite set of $p\%$ best w_i
    Fit a Gaussian distribution with diagonal covariance to the elite set and get a new μ_t and σ_t
end for
return: μ, w^*

There is no guarantee that the best metric for the clustering task is differentiable. Given the above assumptions, conditions are favourable for evolutionary strategies (ES) to iteratively converge towards the optimal solution. Indeed, if r is continuous and the series $((\mathbf{X}, \mathbf{y}_1), \ldots, (\mathbf{X}, \mathbf{y}_p))$ converges towards $(\mathbf{X}, \mathbf{y}^*)$ then $(r(\mathbf{X}, \mathbf{y}_1), \ldots, r(\mathbf{X}, \mathbf{y}_p))$ converges towards $r(\mathbf{X}, \mathbf{y}^*)$. We choose the Cross-Entropy Method (CEM) [3], a popular ES algorithm for its simplicity, to optimise the clustering neural network weights by solving Eq.(3) (Algorithm 1).

3.2 Graph Based Dataset Embedding

To capture global properties and be transferable across different datasets, we argue that it is necessary to input all the points of a dataset at once. Hence, instead of pairwise similarities between random pairs of points, we propose to get a representation of the relation between a bunch of neighbouring points. Thus, we represent each dataset by a graph structure $\mathcal{G}(\mathbf{X}, \mathbf{y})$ where each node corresponds to a point in $\mathbf{X}$ and where cliques represent clusters as shown in Fig. 1. This representation takes the form of a feature matrix X and an adjacency matrix A. Using X, and A, we embed the whole dataset into a vector $\mathbf{z} \in \mathbb{R}^e$. To do so, we use graph autoencoders (GAE). Our implementation is based on [9].

We obtain $z \in \mathcal{M}_{n,m}$ which is dependent of the shape of the dataset (where m is a user specified hyper-parameter). In order to make it independent from the number of points in $\mathcal{X}$, we turn the matrix z into a square symmetrical one $z \leftarrow z^T z \in \mathcal{M}_{m,m}$. The final embedding corresponds to a flattened version of the principal triangular bloc of $z^T z$, which shape is $\mathbf{e} = (\frac{m+1}{2}, 1)$. However, the scale of the output still depends on the number of points in the dataset. This could cause an issue when transferring to datasets with a vastly different number of data points. It should therefore require some regularisation; in order to simplify, we decided to use datasets with approximately the same number of points.

3.3 A Critic as a Metric

With embedded vectors of the same shape, we compare the clusterings proposed $\hat{z}$ and the ground truth ones z using the metric r. r is a function mapping an embedding vector $z \in \mathbb{R}^e$ to $\mathbb{R}$, we therefore parameterise it as:

$$r_\alpha(\mathbf{X}, \mathbf{y}) = r_\alpha(z) = \alpha_1 \phi_1(z) + \alpha_2 \phi_2(z) + ... + \alpha_h \phi_h(z) \tag{4}$$

where $\phi_j(z) \in \mathbb{R}$. As per [13], learning a viable metric is possible provided both the following constraints: (1) maximising the difference between the quality of the optimal decision and the quality of the second best; (2) minimising the amplitude of the metric function as using small values encourages the metric function to be simpler, similar to regularisation in supervised learning.

When maximising the metric difference between the two clusterings that have the highest scores, we get a similarity score as in traditional metric learning problems. The problem is formulated by Eq. (5) where $\mathcal{S}$ is a set of solutions (i.e., clustering proposals) found using r_α and $\mathbf{y}^*$ is the true clustering, $\mathbf{y}^{max}$ is the best solution found in $\mathcal{S}$: $\mathbf{y}^{max} = \arg\max_{\mathbf{y} \in \mathcal{S}} r_\alpha(\mathbf{X}, \mathbf{y})$.

$$\min_\alpha r_\alpha(\mathbf{X}, \mathbf{y}^*) - \max_\alpha \min_{\mathbf{y}' \in \mathcal{S} \backslash \mathbf{y}^{max}} r_\alpha(\mathbf{X}, \mathbf{y}^{max}) - r_\alpha(\mathbf{X}, \mathbf{y}')$$
$$\text{s.t} \quad \mathbf{y}^* = \arg\max_{\mathbf{y} \in \mathbf{Y}} r(\mathbf{y}) \tag{5}$$

To solve Eq. (5), we use a GAN approach where the clustering mechanism (i.e., CEM) plays the role of the generator while a critic (i.e., metric learning model) plays the role of the discriminator. In a classic GAN, the discriminator only has to discriminate between real and false samples, making it use a cross entropy loss. With this kind of loss, and in our case, the discriminator quickly becomes too strong. Indeed, the score output by the discriminator becomes quickly polarised around 0 and 1.

For this reason, we represent r as the critic of a WGAN [1]. This critic scores the realness or fakeness of a given sample while respecting a smoothing constraint. The critic measures the distance between data distribution of the training dataset and the distribution observed in the generated samples. Since WGAN assumes that the optimal clustering provided is unique, the metric solution found by the critic satisfies Eq. (5) constraints. r reaching a unique maximum while being continuous, the assumptions made in Sect. 3.1 are correctly addressed. To train the WGAN, we use the loss $\mathcal{L}$ in Eq. (6) where $\hat{z}$ is the embedding vector of a proposed clustering and z is the embedding vector of the desired clustering. Our framework is detailed in Algorithm 2.

$$\mathcal{L}(z^*, \hat{z}) = \max_\theta \mathbb{E}_{z^* \sim p}[f_\theta(z^*)] - \mathbb{E}_{\hat{z} \sim p(\hat{z})}[f_\theta(\hat{z})] \tag{6}$$

Algorithm 2. Critic2Metric (C2M)

Input: b: batch size, *epoch*: number of epochs; p: percentage of elite weights to keep; *iteration*: number of CEM iterations; *population*: number of weights to generate; $\mu \in \mathbb{R}^d$: CEM mean; $\sigma \in \mathbb{R}^d$: CEM standard deviation, θ: critic's weights

for $n = 1$ **to** *epoch* **do**

 for $k = 1$ **to** b **do**

  Sample $(\mathbf{X}_k, \mathbf{y}_k^*) \sim \mathcal{D}$ a correctly labelled dataset

  Generate ground truth embeddings $\mathbf{z}_{(\mathbf{X}_k, \mathbf{y}_k^*)} = GAE(\mathcal{G}(\mathbf{X}_k, \mathbf{y}_k^*))$

  Initialise clustering neural network weights $\{w_j\}_{j=1}^{population}$

  for $i = 1$ **to** *iteration* **do**

   for $j = 1$ **to** *population* **do**

    Generate clusterings $\hat{\mathbf{y}}_k^{w_j}$

    Convert $\hat{\mathbf{y}}_k^{w_j}$ into a graph

    $\mathbf{z}_{(\mathbf{X}_k, \hat{\mathbf{y}}_k^{w_j})} = GAE(\mathcal{G}(\mathbf{X}_k, \hat{\mathbf{y}}_k^{w_j}))$

    Evaluate: $r(\mathbf{X}_k, \hat{\mathbf{y}}_k^{w_j}) = c_\theta(\mathbf{z}_{(\mathbf{X}_k, \hat{\mathbf{y}}_k^{w_j})})$

   end

   Keep proportion p of best weights w_p

   $w^* \leftarrow \text{CEM}(w_p, \mu, \sigma)$

  end

  Generate clustering $\mathbf{y}_k^{w^*}$

  $\mathbf{z}_{(\mathbf{X}_k, \hat{\mathbf{y}}_k^{w^*})} = GAE(\mathcal{G}(\mathbf{X}_k, \hat{\mathbf{y}}_k^{w^*}))$

  Train critic as in [1] using $\mathbf{z}_{(\mathbf{X}_k, \hat{\mathbf{y}}_k^{w^*})}$ and $\mathbf{z}_{(\mathbf{X}_k, \mathbf{y}_k^*)}$

 end

end

---

4 Experiments

Table 1. Datasets description

Dataset family	Synthetic data				MNIST			Street view house numbers	Omniglot
Dataset	Blob	Moon	Circles	Aniso-tropic	MNIST-digits [11]	letters MNIST [4]	fashion MNIST [18]	SVHN [12]	Omniglot [10]
Snapshot									
Feature dimension	2	2	2	2	28 × 28	28 × 28	28 × 28	32 × 32	105 × 105
Maximum number of clusters	9 (custom)	9 (custom)	9 (custom)	9 (custom)	10	26	10	10	47
Size	200 (custom)	200 (custom)	200 (custom)	200 (custom)	60000	145600	60000	73257	32460

For empirical evaluation, we parameterise our framework as follows: The critic (block C in Fig. 1) is a 5 layer network of sizes 256, 256, 512, 512, and 1 (output) neurons. All activation functions are LeakyRelu ($\alpha = 0.2$) except last layer (no activation). RMSprop optimizer with 0.01 initial learning rate and a decay rate of 0.95. The CEM-trained neural network (bloc A in Fig. 1) has 1 hidden layer of size 16 with Relu activation, and a final layer of size $k = 50$ (the maximum number of clusters). The GAE (bloc B in Fig. 1) has 2 hidden layers; sized 32 and 16 for synthetic datasets, and 100 and 50 for real datasets.

We choose datasets based on 3 main criteria: having a similar compatible format; datasets should be large enough to allow diversity in subsampling configurations to guarantee against overfitting; datasets should be similar to the ones used in our identified baseline literature. All used datasets are found in Table 1.

For training, we construct n sample datasets and their ground truth clustering, each containing 200 points drawn randomly from a set of 1500 points belonging to the training dataset. Each one of these datasets, along with their clustering is an input to our model. To test the learned metric, we construct 50 new sample datasets from datasets that are different from the training one (e.g., if we train the model on MNIST numbers, we will use datasets from MNIST letters or fashion to test the metric). The test sample datasets contain 200 points each for synthetic datasets and 1000 points each otherwise. The accuracies are then averaged accross the 50 test sample datasets. To test the ability of the model to learn using only a few samples, we train it using 5 (few shots) and 20 datasets (standard), each containing a random number of clusters. For few shots trainings, we train the critic for 1 epoch and 10 epochs for standard trainings.

To evaluate the clustering, we use Normalised-Mutual Information (NMI) [16] and clustering accuracy (ACC) [20]. NMI provides a normalised measure that is invariant to label permutations while ACC measures the one-to-one matching of labels. For clustering, we only need that the samples belonging to the same cluster are attributed the same label, independently from the label itself. However, since we want to analyse the behaviour of the metric learned through our framework, we are interested in seeing whether it is permutation invariant or not. Hence, we need the two measures.

4.1   Results on 2D Synthetic Datasets

Analysis on synthetic datasets (see Table 1) proves that our model behaves as expected. We do not compare our results to any baseline since existing unsupervised methods are well studied on them. We train our model using exclusively samples from blobs datasets. We then test the learned metric on the 4 different types of synthetic datasets (blobs, anisotropic, moons and circles). Results are displayed in Table 2. We observe that the model obtains the best score on blobs since it is trained using this dataset. We can also notice that our model achieves high scores for the other types of datasets not included in training.

Table 2. Average ACC and NMI on synthetic test datasets.

Types of datasets	Standard training		Few shots training	
	ACC	NMI	ACC	NMI
Blobs	98.4%	0.980	97.3%	0.965
Anisotropic	97.9%	0.967	97.2%	0.945
Circles	91.7%	0.902	92.7%	0.900
Moons	92.1%	0.929	92.8%	0.938

Our model succeeds in clustering datasets presenting non linear boundaries like circles while blobs datasets used in training are all linearly separable. Hence, the model learns intrinsic properties of training dataset that are not portrayed in the initial dataset structure, and thus that the metric appears to be transferable.

Critic's Ablation Study. To test if the critic behaves as expected, i.e., grades the clustering proposals proportionally to their quality, we test it on wrongly labelled datasets to see if the score decreases with the number of mislabelled points. We consider 50 datasets from each type of synthetic datasets, create 50 different copies and mislabel a random number of points in each copy. A typical result is displayed in Fig. 2 and shows that the critic effectively outputs an ordering metric as the score increases when the number of mislabelled points decreases, reaching its maximum when there is no mislabelled point. This shows that the metric satisfies the constraints stated in Eq. 5.

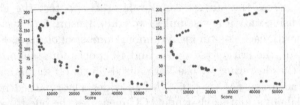

Fig. 2. Metric values (i.e., scores given by the critic) for several clusterings of a dataset. Plots are from an anisotropic dataset (left) and a moons dataset (right). In a 2 cluster case (right), the formula used to compute mislabelled points has been made sensitive to label permutation to verify if permuted labels can fool the critic. The critic assigns a high score either when all the labels match the given ground truth or when all the labels are permuted (which again does not affect the correctness of the clustering)

An interesting behaviour is shown in Fig. 2. Recall that since we are in the context of a clustering problem, we only need for the samples belonging to the same cluster to get the same label, independently from the cluster label itself. Thus, the formula used to compute mislabelled points has been made sensitive to label permutation to verify if permuted labels can fool the critic. For instance, in a 2 clusters case, one can switch the labels of all points in each cluster and still get the maximum score. Switching all labels makes all the points wrongly labelled compared to the given ground truth but nonetheless the clustering itself remains true. This explains the rounded shape in Fig. 2 where the used datasets in the right panel only consisted of 2 clusters. The critic assigns a high score either when all the labels match the given ground truth or when all the labels are permuted (which does not affect the correctness of the clustering).

4.2   Results on MNIST Datasets

MNIST datasets give similar results both in terms of ACC and NMI on all test datasets regardless of the used training dataset (see Table 3). Hence, the model effectively capture implicit features that are dataset independent. While standard training shows better results, the few shots training has close performance.

Table 3. Mean clustering performance on MNIST dataset.

Training Dataset	Testing Dataset					
	Numbers		Letters		Fashion	
	ACC	NMI	ACC	NMI	ACC	NMI
Numbers (standard)	72.3%	0.733	81.3%	0.861	65.2%	0.792
Numbers (few shots)	68.5%	0.801	79.0%	0.821	61.8%	0.672
Letters (standard)	75.9%	0.772	83.7%	0.854	67.5%	0.800
Letters (few shots)	69.8%	0.812	78.7%	0.806	60.9%	0.641
Fashion (standard)	70.6%	0.706	83.4%	0.858	72.5%	0.762
Fashion (few shots)	70.1%	0.690	82.1%	0.834	70.7%	0.697

Table 4. Critic based performance assessment: Best corresponds to the percentage of times the critic gives the best score to the desired solution. Top 3 is when this solution is among the 3 highest scores.

Training Dataset	Testing Dataset					
	Numbers		Letters		Fashion	
	Best	Top 3	Best	Top 3	Best	Top 3
Numbers (standard)	78.3%	92.5%	86.0%	97.5%	69.2%	87.2%
Numbers (few shots)	75.8%	82.1%	83.3%	92.0%	65.1%	83.9%
Letters (standard)	77.4%	89.2%	88.8%	96.4%	70.2%	86.7%
Letters (few shots)	73.1%	80.6%	85.1%	91.5%	61.0%	76.3%
Fashion (standard	70.1%	83.1%	85.0%	98.6%	76.9%	94.7%
Fashion (few shots)	67.9%	77.4%	83.5%	95.3%	70.2%	88.0%

Table 4 shows the percentage of times the critic attributes the best score to the desired solution. It shows that ES algorithm choice has a significant impact on the overall performance. Even with a metric that attributes the best score to the desired clustering, the CEM may be stuck in a local optimum and fails to reconstruct back the desired clustering. Hence, a better optimisation can enhance the performance shown in Table 3 closer to the one presented in Table 4.

4.3 Comparative Study

We compare our approach with baseline methods from the literature (Table 5). For some methods, we followed the procedure in [8] and used their backbone neural network as a pairwise similarity metric. Table 5a reports results when training on SVHN and testing on MNIST numbers. We obtain close ACC values to CCN and ATDA [14]. These methods uses Omniglot as an auxiliary dataset to learn a pairwise similarity function, which is not required for our model. Our model only uses a small fraction of SVHN, has shallow networks and does not require any adaptation to its loss function to achieve comparable results. Finally, other cited methods require the number of clusters as an a priori indication. We achieve comparable results without needing this information. When the loss adaptation through Omniglot is discarded (denoted source-only in Table 5a), or if the number of clusters is not given, their accuracy falls and our model surpasses them by a margin.

Table 5. Comparative clustering performance

Method	ACC	
	Loss Adaptation	Source Only
DANN [5]	73.9%	54.9%
LTR [15]	78.8%	54.9%
ATDA [14]	86.2%	70.1%
CCN [8]	89.1%	52%
Ours (standard)	–	84.3%
Ours (few shots)	–	81.4%

(a) Unsupervised cross-task transfer from SVHN to MNIST digits.

Method	ACC	NMI
k-means	18.9%	0.464
CSP [17]	65.4%	0.812
MPCK-means [2]	53.9%	0.816
CCN [8]	78.18%	0.874
DTC [6]	87.0%	0.945
Autonovel [7]	85.4%	–
Ours (standard)	83.4%	0.891

(b) Unsupervised cross-task transfer from Omniglot$_{train}$ to Omniglot$_{test}$ ($k = 100$ for all).

Table 5b reports results when training on Omniglot$_{train}$ and testing on Omniglot$_{test}$. Values are averaged across 20 alphabets which have 20 to 47 letters. We set the maximum number of clusters $k = 100$. When the number of clusters is unknown, we get an ACC score relatively close to DTC and Autonovel. Compared to these two approaches, our method bears several significant advantages:

- **Deep Networks:** DTC and Autonovel used Resnets as a backbone which are very deep networks while we only used shallow networks (2 layers maximum)
- **Pairwise similarity:** in Autonovel the authors used a pairwise similarity statistic between datasets instances which we aimed to avoid due to its significant computational bottleneck. Moreover, this metric is recalculated after each training epoch, which adds more complexity.
- **Vision tasks:** While DTC can only handle vision tasks, we present a more general framework which includes vision but also tabular datasets.
- **Number of classes:** DTC and Autonovel used the labelled dataset as a probe dataset, and estimates the number of classes iteratively, and when the labelled clusters are correctly recovered, they used the ACC metric to keep the best clustering. This approach is effective, but requires access to the labelled dataset at inference time to estimate the number of classes. This is a shortcoming (memory or privacy limitations). Our approach does not require the labelled dataset once the metric is learned. Our metric automatically estimates the number of clusters required to any new unlabelled dataset.

5 Conclusion

We presented a framework for cross domain/task clustering by learning a transferable metric. This framework consisted of ES methods, and GAE alongside a critic. Our model extracts dataset-independent features from labelled datasets that characterise a given clustering, performs the clustering and grades its quality. We showed successful results using only small datasets and relatively shallow architectures. Moreover, there is more room for improvement. Indeed, since our framework is composed of 3 different blocs (CEM, GAE, critic), overall efficiency can be enhanced by independently improving each bloc (i.e. replacing CEM).

In future work, we will study the criteria that determine why some auxiliary datasets are more resourceful than others given a target dataset. In our case, this means to study for instance why using the MNIST letters dataset as training allowed a better performance on Fashion MNIST than when using MNIST numbers. This would allow to deliver a minimum performance guarantee at inference time by creating a transferability measure between datasets.

Acknowledgements. We gratefully acknowledge Orianne Debeaupuis for making the figure. We also acknowledge computing support from NVIDIA. This work was supported by funds from the French Program "Investissements d'Avenir".

References

1. Arjovsky, M., et al.: Wasserstein generative adversarial networks. In: ICML, pp. 214–223 (2017)
2. Bilenko, M., Basu, S., Mooney, R.J.: Integrating constraints and metric learning in semi-supervised clustering. In: ICML, p. 11 (2004)
3. de Boer, P.T., Kroese, D.P., et al.: A tutorial on the cross-entropy method. Ann. Oper. Res. **134**(1), 19–67 (2005)
4. Cohen, G., et al.: EMNIST: Extending MNIST to handwritten letters. In: IJCNN (2017)
5. Ganin, Y., et al.: Domain-adversarial training of neural networks. JMLR **17**, 2096–2130 (2016)
6. Han, K., et al.: Learning to discover novel visual categories via deep transfer clustering (2019)
7. Han, K., Rebuffi, S.A., et al.: AutoNovel: automatically discovering and learning novel visual categories. PAMI 1 (2021)
8. Hsu, et al.: Learning to cluster in order to transfer across domains and tasks (2017)
9. Kipf, T.N., Welling, M.: Variational graph auto-encoders (2016)
10. Lake, B.M., Salakhutdinov, R., Tenenbaum, J.B.: Human-level concept learning through probabilistic program induction. Science **350**(6266), 1332–1338 (2015)
11. LeCun, Y., Cortes, C., Burges, C.: MNIST handwritten digit database (2010)
12. Netzer, Y., et al.: Reading digits in natural images with unsupervised feature learning. In: NIPS Workshop on Deep Learning and Unsupervised Feature Learning (2011)
13. Ng, A.Y., Russell, S.: Algorithms for inverse reinforcement learning. In: Proceedings of the 17th International Conference on Machine Learning, pp. 663–670 (2000)
14. Saito, K., Ushiku, Y., Harada, T.: Asymmetric tri-training for unsupervised domain adaptation. ICML, pp. 2988–2997 (2017)
15. Sener, O., Song, H.O., et al.: Learning transfer able representations for unsupervised domain adaptation. In: NIPS, pp. 2110–2118 (2016)

16. Strehl, A., Ghosh, J.: Cluster ensembles-a knowledge reuse framework for combining multiple partitions. JMLR **3**(Dec), 583–617 (2002)
17. Wang, X., Qian, B., Davidson, I.: On constrained spectral clustering and its applications. Data Min. Knowl. Discov. **28**, 1–30 (2014)
18. Xiao, H., Rasul, K., Vollgraf, R.: Fashion-MNIST: a novel image dataset for benchmarking machine learning algorithms (2017)
19. Xie, J., Girshick, R., Farhadi, A.: Unsupervised deep embedding for clustering analysis. In: ICML, pp. 478–487 (2016)
20. Yang, Y., Xu, D., et al.: Image clustering using local discriminant models and global integration. IEEE Trans. Image Process. **19**(10), 2761–2773 (2010)

Spatial Graph Convolution Neural Networks for Water Distribution Systems

Inaam Ashraf[(✉)] [ID], Luca Hermes[ID], André Artelt[ID], and Barbara Hammer[ID]

Center for Cognitive Interaction Technology, Bielefeld University, Bielefeld, Germany
{mashraf,lhermes,aartelt,bhammer}@techfak.uni-bielefeld.de

Abstract. We investigate the task of missing value estimation in graphs as given by water distribution systems (WDS) based on sparse signals as a representative machine learning challenge in the domain of critical infrastructure. The underlying graphs have a comparably low node degree and high diameter, while information in the graph is globally relevant, hence graph neural networks face the challenge of long term dependencies. We propose a specific architecture based on message passing which displays excellent results for a number of benchmark tasks in the WDS domain. Further, we investigate a multi-hop variation, which requires considerably less resources and opens an avenue towards big WDS graphs.

Keywords: Graphs · Graph Convolutional Neural Networks · Node Features Estimation · Water Distribution Systems · Pressure Estimation

1 Introduction

Transportation systems, energy grids, and water distribution systems (WDS) constitute parts of our critical infrastructure that are vital to our society and subject to special protective measures and regulations. As they are under increasing strain in the face of limited resources and as they are vulnerable to attacks, their efficient management and continuous monitoring is of great importance. As an example, the average amount of non-revenue water amounts to 25% in the EU [6], making the detection of leaks in WDS an important task. Advances in sensor technology and increasing digitalisation hold the potential for intelligent monitoring and adaptive control using AI technologies [5,13,25]. In addition to more classical AI approaches, deep learning technologies are increasingly being used to solve learning tasks in the context of critical infrastructures [4].

A common feature of WDS, energy networks and transport networks is that the data has a temporal and spatial character: Data is generated in real time according to an underlying graph, given by the power grid, the pipe network and the transport routes, respectively. Measurements are available for some nodes that correspond to local sensors, e.g. pressure sensors or smart meters. Based on this partial information, the task is to derive corresponding quantities at every node of the graph, to identify the system state or to derive optimal planning and control strategies. In this paper, we target the learning challenges of the first feature, inferring relevant quantities at each location of the graph based on few measurements. While classical deep learning models such as convolutional networks or recurrent models can reliably handle Euclidean data, graphs constitute non-Euclidean data that require techniques from geometric deep learning. Based on initial approaches dating back more than a decade [11,28], a variety of graph neural networks (GNNs) have recently been proposed that are able to directly process information such as present in critical infrastructure [2,3,7,14,29]. First applications demonstrate the suitability of GNNs for the latter [3,5,23].

Graphs from the domain of WDS or smart grids display specific characteristics (s. Fig. 3): as they are located in the plane, the node degree is small and the network diameter is large. These characteristics display a challenge for GNNs, as the problem of long-term dependencies and over-smoothing occurs [27,32]. In this contribution, we design a GNN architecture capable of dealing with these specific graph structures: We show that our spatial GNN is able to effectively integrate long-range node dependencies and we demonstrate the impact of a suitable transfer function and residual connections. As the required resources quickly become infeasible for big graphs, we also investigate the comparability of a sparse multi-hop alternative. All methods are evaluated for pressure prediction in WDS for a variety of benchmark networks, displaying promising results. The code has been made publicly available.[1]

2   Related Work

The task of pressure estimation at all nodes in a WDS from pressure values available at a few nodes has recently been dealt with [8]. The authors employed spectral graph convolutional neural networks (GCNs) and performed extensive experiments to demonstrate their approach. However, their methodology does not fully benefit from the available structural information of the graph; we provide further details on this in Sect. 4. We propose a spatial GCN based methodology that effectively utilizes the graph structure by using both node and edge features and thus produces significantly better results (s. Sect. 5).

A related task of state (pressure, flow) estimation in WDS based on demand patterns and sparse pressure information has been addressed [31]. The authors used hydraulics in the optimization objective since the task was to model the complex hydraulics used by the popular EPANET simulator [26] using GNNs. They present promising results only on relatively small WDS, the ability to scale to larger WDS is yet to be investigated. While their model solves the task of state estimation in WDS, their approach requires demand patterns from every consumer also during inference. In contrast, our proposed model relies on pressure values computed by the EPANET solver (based on demand patterns) only during the training process. During evaluation, our model estimates pressures solely based on sparse pressure values obtained from a few sensors. Further, it successfully estimates pressures even in case of noisy demands (s. Sect. 5).

---

[1] https://github.com/HammerLabML/GCNs_for_WDS.

GNNs were first introduced in the work [28] as an extension of recursive neural networks for tree structures [10]. Since then, a number of GCN algorithms have been developed, which can be classified in to spectral-based and spatial-based. The approach [2] introduced spectral GCNs based on spectral graph theory, which was followed by further work [3,12,14,18,20]. The counterpart are spatial GCNs which apply a local approximation of the spectral graph kernel [7,9,22,24,29,32]. These are also referred to as message passing neural networks.

Unlike convolutional neural networks (CNNs), spatial GCNs suffer from issues like vanishing gradient, over-smoothing and over-fitting, when used to build deeper models. Generalized aggregation functions, residual connections and normalization layers can address these issues and improve performance on diverse GCN tasks and large scale graph datasets [19].

To enable high-level embeddings in feed-forward neural networks, self normalizing neural networks (SNNs) were introduced [15] based on a special activation function called scaled exponential unit (SeLU). We combine residual connections [19] with SNNs since residual connections help solve the over-smoothing problem when we use multiple GCN layers, whereas self-normalizing property of SeLU enables the required information propagation in case of sparse features.

3 Methodology

The main contribution of our work is a spatial GCN capable of efficiently dealing with the specific graph characteristics as present in WDS. We address the estimation of missing node features based on sparse measurements. As we detail below, we employ multiple spatial GCN layers without suffering from typical problems of vanishing gradient, over-smoothing and over-fitting. For this purpose, we combine residual connections [19] with SeLU activation function [15]. To decrease model size, we leverage GCN layers with multiple hops realizing message passing between more distant neighbors comparable to [21]. Our model employs spatial GCNs using both node and edge features. Although WDS are characterized by spatio-temporal data, we did not find adding a temporal component to be beneficial and therefore keep our model purely spatial. The complete architecture is depicted in Fig. 1. Formally, a graph is represented as $G(V, X, E, F)$, where:

- $V = \{v_1, v_2, \ldots, v_N\}$ is the set of nodes
- $V_s \subset V$ is the set of sensor nodes
- $E = \{e_{vu} \mid \forall v \in V; u \in \mathcal{N}(v)\}$ is the set of edges
- $Y \in R^{N \times D}$ is the set of node features (ground truth), where $N = |V|$ and D is the number of node features
- $X \in R^{N \times D}$ is the set of sparse node features, where $\mathbf{x}_v = \begin{cases} \mathbf{y}_v, & \text{if } v \in V_s \\ 0, & \text{otherwise} \end{cases}$
- $F \in R^{M \times K}$ is the set of edge features, where $M = |E|$ and K is the number of edge features

Node features X are highly sparse as only nodes corresponding to sensors contain values while everything else is set to zero. Node and edge features are embedded by fully connected linear layers α and β:

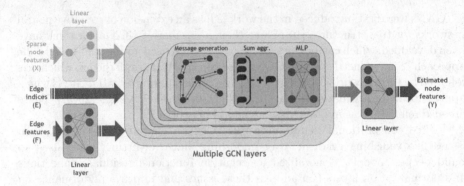

Fig. 1. Model architecture employing multiple GCN layers. Each GCN layer consists of message generation, sum aggregation and a final MLP.

$$\mathbf{g}_v^1 = \alpha(\mathbf{x}_v) \qquad v \in V, \quad \mathbf{x}_v \in X, \qquad \mathbf{g}_v \in G \qquad (1)$$
$$\mathbf{h}_{e_{vu}}^1 = \beta(\mathbf{f}_{e_{vu}}) \qquad e_{vu} \in E, \quad \mathbf{f}_{e_{vu}} \in F, \quad \mathbf{h}_{e_{vu}} \in H$$

We denote intermediate model activations as $G \in R^{N \times Z}$ for nodes and $H \in R^{M \times Z}$ for edges, where Z is the latent dimension. Multiple GCN layers convolve the information from the neighboring nodes for estimation of node features. Each GCN layer employs the three-step process of message generation, message aggregation and node feature update. In the lth layer, the edge features are updated by

$$\hat{\mathbf{h}}_{e_{vu}}^{(l)} = \mathbf{h}_{e_{vu}}^{(l)} + |\mathbf{h}_u^{(l)} - \mathbf{h}_v^{(l)}|. \qquad (2)$$

Adding the absolute difference between the current and neighbor nodes features empirically improves the learning. Then, messages are generated as follows:

$$\mathbf{m}_{e_{vu}}^{(l)} = \text{SeLU}\left(\mathbf{h}_u^{(l)} \parallel \hat{\mathbf{h}}_{e_{vu}}^{(l)}\right) \quad u \in \mathcal{N}(v), \qquad (3)$$

where $\cdot \parallel \cdot$ denotes vector concatenation. After concatenation, we employ the SeLU activation function [15] to all messages, which is given by:

$$\text{SeLU}(x) = \lambda \begin{cases} x & if\ x > 0 \\ \alpha e^x - \alpha & if\ x \leq 0 \end{cases} \qquad (4)$$

where λ and α are hyperparameters as in [15]. SeLU's self-normalizing nature greatly improves learning in the light of highly sparse values at the beginning of the training process. All messages from the neighbor nodes are sum-aggregated:

$$\mathbf{m}_v^{(l)} = \sum_{u \in \mathcal{N}(v)} \mathbf{m}_{e_{vu}}^{(l)} \qquad (5)$$

Similar to [19], we add residual connections to the aggregated messages and pass these through a Multi-Layer Perceptron (MLP):

$$\mathbf{h}_v^{(l+1)} = \text{MLP}\left(\mathbf{h}_v^{(l)} + \mathbf{m}_v^{(l)}\right) \qquad (6)$$

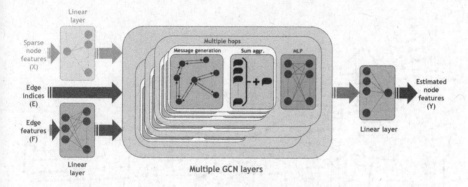

Fig. 2. Model architecture employing multiple multi-hop GCN layers.

The overall message construction, aggregation and update is [19]:

$$\mathbf{h}_v^{(l+1)} = \text{MLP} \left(\mathbf{h}_v^{(l)} + \sum_{u \in \mathcal{N}(v)} \text{SeLU} \left(\mathbf{h}_u^{(l)} \parallel \hat{\mathbf{h}}_{e_{vu}}^{(l)} \right) \right) \tag{7}$$

After employing multiple GCN layers, the resultant node embeddings are fed to a final fully-connected linear layer to estimate all node features.

$$\hat{\mathbf{y}}_v = \gamma(\mathbf{h}_v^L) \qquad v \in V, \quad \hat{\mathbf{y}}_v \in \hat{Y} \tag{8}$$

where $\hat{Y}$ is the estimated node features, L is the last GCN layer and γ is modeled by the linear layer. We use the L1 loss as objective function:

$$\mathcal{L}(\mathbf{y}, \hat{\mathbf{y}}) = \frac{1}{S \cdot N} \sum_{i=1}^{S \cdot N} |\mathbf{y}_i - \hat{\mathbf{y}}_i| \tag{9}$$

with N as the number of nodes and S as the number of samples in a mini-batch.

Multi-hop Variation. Given the sparsity and size of a graph, our methodology requires a comparably large number of GCN layers proportional to the size of graph. This reduces scalability to larger graphs. To reduce the number of parameters, we propose GCN layers with multiple hops as shown in Fig. 2. Specifically, message generation and aggregation are repeated in each GCN layer before passing it to the MLP:

$$\mathbf{h}_v^{(l)(p+1)} = \mathbf{h}_v^{(l)(p)} + \sum_{u \in \mathcal{N}(v)} \text{SeLU} \left(\mathbf{h}_u^{(l)(p)} \parallel \hat{\mathbf{h}}_{e_{vu}}^{(l)(p)} \right), \qquad p \in P \tag{10}$$

with P as number of hops. The embedding for the next layer is:

$$\mathbf{h}_v^{(l+1)} = \text{MLP} \left(\mathbf{h}_v^{(l)(P)} + \sum_{u \in \mathcal{N}(v)} \text{SeLU} \left(\mathbf{h}_u^{(l)(P)} \parallel \hat{\mathbf{h}}_{e_{vu}}^{(l)(P)} \right) \right) \tag{11}$$

This enables the model to gather information from neighbors that are multiple hops away, requiring fewer GCN layers.

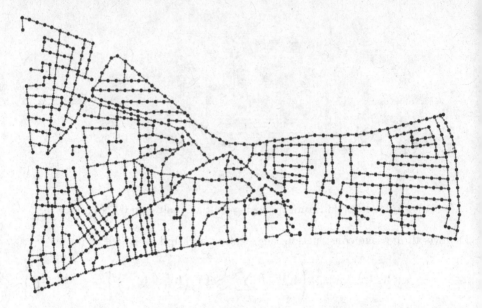

Fig. 3. L-Town Water Distribution System [30] – nodes in red have sensors. (Color figure online)

4   Experiments

The methodology can be applied to missing node feature estimation on any graph. Here, we investigate WDS, which are modelled as graphs by representing junctions as nodes and pipes between junctions as edges. WDS are especially challenging because pressure sensors are installed at only few nodes due to constraints (size of the system, cost, availability, practicality) [17], resulting in graphs with sparse feature information. Additionally, the node degree in WDS is usually low (s. Table 1). These properties can be observed in the popular L-Town WDS [30] shown in Fig. 3. Such characteristics require GNNs to model long-range dependencies between nodes to properly integrate the available information.

To the best of our knowledge, the task of node feature estimation in WDS using GNNs based on sparse features has only been dealt with by [8]. These researchers compared their model to a couple of non-GNN based baselines: The first baseline refers to the mean of known node features as value for unknown

Table 1. Major attributes of WDS.

WDS	Anytown	C-Town	L-Town	Richmond
Number of junctions	22	388	785	865
Number of pipes	41	429	909	79
Diameter	5	66	79	234
Degree (min, mean, max)	(1, 3.60, 7)	(1, 2.24, 4)	(1, 2.32, 5)	(1, 2.19, 4)

Table 2. Model Hyperparameters and Parameters.

Model		Anytown	C-Town	Richmond	L-Town	
ChebNet	No. of layers	4	4	4	4	
	Degrees (K_i)	[39, 43, 45, 1]	[200, 200, 20, 1]	[240, 120, 20, 1]	[240, 120, 20, 1]	
	No. of filters (F_i)	[14, 20, 27, 1]	[60, 60, 30, 1]	[120, 60, 30, 1]	[120, 60, 30, 1]	
	Parameters (million)	0.038	0.780	0.958	0.929	
m-GCN	No. of GCN layers	5	33	60	45	10
	No. of hops	1	2	3	1	5
	No. of MLP layers	2	2	2	2	2
	Latent dimension	32	32	48	96	96
	Parameters (million)	0.031	0.203	0.830	2.488	0.553

node features, the second baseline uses interpolated regularization [1]. The work [8] demonstrates that the GNN model significantly outperforms both baselines. Therefore, in our experiments, we compare our approach only to the GNN model, ChebNet, of [8]. We run two experiments on simulated data. First, we compare our approach to [8] on three WDS datasets Anytown, C-Town, and Richmond. Second, we conduct an in-depth evaluation on L-Town with extensive hyperparameter tuning.

4.1 Datasets

We use a total of four WDS datasets for our experiments: Anytown, C-Town, L-Town and Richmond[2,3] [30]. Major attributes of the WDS are listed in Table 1. We use the dataset generation methodology of [8] for three of the WDS (Anytown, C-Town, Richmond) and record 1000 consecutive time steps for each of the three networks. For each network, we use three different sparsity levels i.e. sensor ratios of 0.05, 0.1 and 0.2. We do not evaluate on sparsity levels of 0.4 and 0.8 as done in [8], which are more easy. We sample 5 different random sensor configurations for each sparsity level and each WDS instead of 20.

For the popular L-Town network, we use only a single configuration of sensors as designed by [30], which gives a sensor ratio of 0.0422. We use two different sets of simulation settings; one with smooth toy demands and the other close to actual noisy demand patterns. The simulations are carried out using EPANET [26] provided by Python package *wntr* [16]. The samples are generated every 15 min, resulting in 96 samples every day. We use one month of data for training (2880 samples) and evaluate on data of the next two months (5760 samples). The training data is divided in train-validation-test splits with 60-20-20 ratio.

4.2 Training Setup

The model parameters are summarized in Table 2. All models are implemented in Pytorch using Adam optimizer. For the ChebNet baseline [8], we set the learning rate of 3e-4 and weight decay of 6e-6. For our m-GCN models, we use learning

[2] https://engineering.exeter.ac.uk/research/cws/resources/benchmarks/#a8.
[3] https://www.batadal.net/data.html.

Table 3. Mean relative absolute errors across nodes and samples across 5 different sensor configurations for 3 different ratios of sensors.

WDS		Anytown			C-Town			Richmond		
Ratio	MRAE ×10^{-3}	ChebNet	m-GCN	Diff	ChebNet	m-GCN	Diff	ChebNet	m-GCN	Diff
0.05	All	54.19	53.15	−1.04	12.88	9.77	−3.11	4.34	2.17	−2.17
	Sensor	7.06	3.77	−3.28	7.50	4.61	−2.89	3.47	1.81	−1.66
	Non-sensor	56.44	55.50	−0.94	13.16	10.04	−3.12	4.38	2.19	−2.19
0.1	All	35.43	34.85	−0.57	8.16	5.47	−2.69	3.86	1.93	−1.93
	Sensor	6.66	7.19	0.53	7.10	4.83	−2.27	3.45	2.02	−1.43
	Non-sensor	38.3	37.62	−0.68	8.28	5.55	−2.73	3.90	1.92	−1.98
0.2	All	14.98	13.51	−1.47	7.05	5.58	−1.47	3.24	1.59	−1.65
	Sensor	5.40	3.06	−2.34	6.46	5.46	−1.00	3.03	1.62	−1.40
	Non-sensor	17.11	15.83	−1.28	7.20	5.61	−1.59	3.29	1.59	−1.71

rate of 1e-5 and no weight decay. We now describe the training setup of the ChebNet baseline and our m-GCN model for the two experiments, respectively.

For the first experiment the models are trained for 2000 epochs. We set an early stopping criteria such that it stops after 250 epochs if the change in loss is no larger than 1e-6. We configure ChebNet similar to [8]. Input is masked (i.e. set to zero) as per the sensor ratio and the binary mask is concatenated with the pressure values. Hence there are two node features. ChebNet can only use scalar edge fetaures, i.e. edge weights. Out of the three types of edge weights used by [8] (binary, weighted, logarithmically weighted), we use the binary weights since other types did not increase performance. For our model (m-GCN), we did not perform an extensive hyperparameter search since we achieved considerably better results than ChebNet model of [8] with a set of intuitive hyperparameter values. We use single hop configuration for Anytown and multi-hop architectures for C-Town and Richmond WDS. We only use masked pressure values as input i.e. one node feature. Further, we use two edge features namely pipe length and diameter.

For the second in-depth evaluation on L-Town, we dropped the second node feature for ChebNet since this significantly improved the results. We use the ChebNet model configuration used for Richmond WDS by the authors. We train our m-GCN model with two configurations; one with the default single hop and the second with multiple hops as listed in Table 2. For both m-GCN models, we add a third edge feature namely pressure reducing valves (PRVs) mask. PRVs are used at certain connections in a WDS to reduce pressure, hence these edges should be modeled differently. We use a binary mask to pass this information to the model that helps in improving the pressure estimation at neighboring nodes. We train all three models for 5000 epochs without early stopping.

5   Results

Comparison with Spectral GCN-Based Approach. First, we compare our model with the work of [8] using their datasets and training settings. The results of the experiments on Anytown, C-Town and Richmond WDS are shown in Table 3. Here, we evaluate on the basis of mean relative absolute error given by:

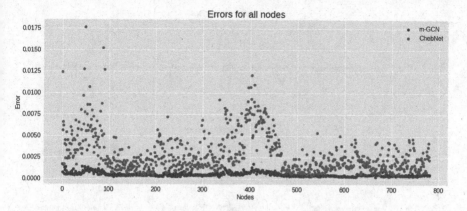

Fig. 4. Mean relative absolute error per node measured on the evaluation set of noisy data for L-Town WDS.

$$\text{MRAE}(\mathbf{y}, \hat{\mathbf{y}}) = \frac{1}{S \cdot N} \sum_{i=1}^{S \cdot N} \frac{|\mathbf{y}_i - \hat{\mathbf{y}}_i|}{\mathbf{y}_i} \qquad (12)$$

Since Anytown is a much smaller WDS, sensor ratios translate to very few sensors (0.05: 1 sensor, 0.1: 2 sensors, 0.2: 4 sensors). Hence, both models do not accurately estimate the pressures in these cases. The number of available sensors is comparatively bigger for both C-Town and Richmond WDS, even for the smallest ratio, thus naturally increasing performance. As can be seen, m-GCN outperforms ChebNet [8] by a considerable margin.

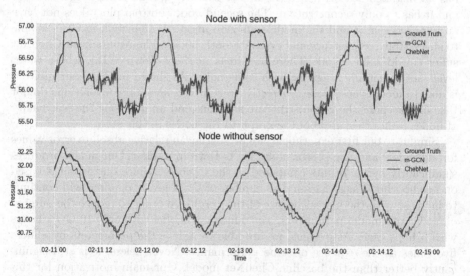

Fig. 5. Estimation results of m-GCN and ChebNet compared to ground truth on L-Town.

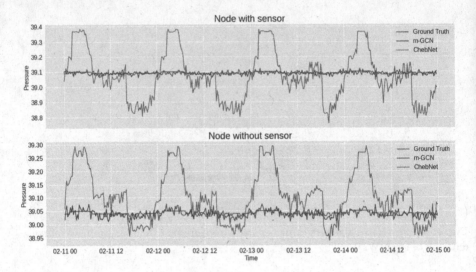

Fig. 6. Estimation results of m-GCN and ChebNet compared to ground truth on nodes from an area in L-Town with essentially stagnant pressure values.

Detailed analysis on L-Town We present more in-depth analysis for the evaluation results on L-Town. Mean relative absolute errors per node for ChebNet and single-hop m-GCN models are plotted in Fig. 4. Both models are trained on smooth data and evaluated on noisy realistic data. As can be seen, error values for m-GCN are much lower across all nodes compared to ChebNet. We plot time series of 4 days for a couple of nodes in Fig. 5. The first node (top plot) has an installed sensor, hence the model gets the ground truth value as input and it has to only reconstruct it. The second node (bottom plot) does not have an installed sensor and the model gets zero-input. As depicted, m-GCN is able to successfully reconstruct and estimate both nodes. The results from ChebNet suffer considerable errors. There are areas in the L-Town WDS, where water levels are essentially stagnant with some noise. As shown in Fig. 6 our m-GCN is able to model those nodes correctly. In contrast, spectral convolutions do not take into account the graph structure and thus end up imposing the seasonality of nodes from other areas of the graph to the nodes in this area.

Similar to our first experiment, we present mean relative absolute error values for all, sensor and non-sensor nodes for L-Town in Table 4. Our model produces significantly better results compared to the ChebNet. Since our model is based on neighborhood aggregation, the number of GCN layers required will continue to increase with the increasing size of the graphs. In order to reduce the number of layers and model parameters, we trained our model with only 10 GCN layers with 5 hops each. As evident, we are able to reduce the parameters by almost five times at the expense of some performance. Nevertheless, it is still significantly better than the baseline ChebNet model. Our main motivation for this is that the multi-hop approach makes the model more scalable to larger graphs.

Table 4. Mean relative absolute errors across nodes and samples on L-Town.

	Model	MRAE ($\times 10^{-3}$)		
		All	Sensor	Non-sensor
Smooth Data	ChebNet	2.55 ± 2.87	2.38 ± 3.55	2.55 ± 2.83
	m-GCN (45×1)	$\mathbf{0.39 \pm 0.37}$	$\mathbf{0.43 \pm 0.52}$	$\mathbf{0.39 \pm 0.36}$
	m-GCN (10×5)	0.83 ± 0.68	0.74 ± 0.59	0.83 ± 0.69
Noisy Data	ChebNet	2.92 ± 3.35	2.78 ± 4.02	2.93 ± 3.32
	m-GCN (45×1)	$\mathbf{0.54 \pm 0.75}$	$\mathbf{0.64 \pm 1.06}$	$\mathbf{0.53 \pm 0.73}$
	m-GCN (10×5)	0.90 ± 0.82	0.81 ± 0.74	0.90 ± 0.83

Further, it is a step towards developing a generalized version of the model that can work for different sensor configurations and/or different graph sizes without hyperparameter tuning and re-training.

6   Conclusion

We have proposed a spatial GCN which is particularly suited for graph tasks on graphs with small node degree and sparse node features, since it is able to model long-term dependencies. We have demonstrated its suitability for node pressure inference based on sparse measurement values as an important and representative task from the domain of WDS, displaying its behavior for a number of benchmarks. Notably, the model generalizes not only across time windows, but also from noise-less toy demand signals to realistic ones. In addition to a very good performance overall, we also proposed first steps to target the challenge of scalability to larger graphs by introducing multi-hop architectures with considerably fewer parameters as compared to fully connected deep ones. In future work, we will investigate the behavior for larger networks based on these first results. Moreover, unlike simulation tools in the domain, the GNN has the potential to generalize over different graphs structures including partially faulty ones. We will evaluate this capability in future work.

Acknowledgements. We gratefully acknowledge funding from the European Research Council (ERC) under the ERC Synergy Grant Water-Futures (Grant agreement No. 951424). This research was also supported by the research training group "Dataninja" (Trustworthy AI for Seamless Problem Solving: Next Generation Intelligence Joins Robust Data Analysis) funded by the German federal state of North Rhine-Westphalia, and by funding from the VW-Foundation for the project *IMPACT* funded in the frame of the funding line *AI and its Implications for Future Society*.

References

1. Belkin, M., Matveeva, I., Niyogi, P.: Regularization and semi-supervised learning on large graphs. In: Shawe-Taylor, J., Singer, Y. (eds.) COLT 2004. LNCS (LNAI), vol. 3120, pp. 624–638. Springer, Heidelberg (2004). https://doi.org/10.1007/978-3-540-27819-1_43
2. Bruna, J., Zaremba, W., Szlam, A.D., LeCun, Y.: Spectral networks and locally connected networks on graphs. CoRR (2014)
3. Defferrard, M., Bresson, X., Vandergheynst, P.: Convolutional neural networks on graphs with fast localized spectral filtering. In: NIPS, vol. 29, pp. 3844–3852 (2016)
4. Dick, K., Russell, L., Dosso, Y.S., Kwamena, F., Green, J.R.: Deep learning for critical infrastructure resilience. JIS 25(2), 05019003 (2019)
5. Eichenberger, C., et al.: Traffic4cast at NeurIPS 2021 - temporal and spatial few-shot transfer learning in gridded geo-spatial processes. In: Proceedings of the NeurIPS 2021 Competitions and Demonstrations Track, vol. 176, pp. 97–112. PMLR (2022)
6. EurEau: Europe's water in figures (2021)
7. Gao, H., Wang, Z., Ji, S.: Large-scale learnable graph convolutional networks. In: SIGKDD, pp. 1416–1424 (2018)
8. Hajgató, G., Gyires-Tóth, B., Paál, G.: Reconstructing nodal pressures in water distribution systems with graph neural networks (2021). https://doi.org/10.48550/ARXIV.2104.13619
9. Hamilton, W.L., Ying, R., Leskovec, J.: Inductive representation learning on large graphs. In: NIPS, pp. 1025–1035 (2017)
10. Hammer, B.: Learning with Recurrent Neural Networks. Leacture Notes in Control and Information Sciences, vol. 254. Springer, London (2000). https://doi.org/10.1007/BFb0110016
11. Hammer, B., Micheli, A., Sperduti, A.: Universal approximation capability of cascade correlation for structures. Neural Comput. 17(5), 1109–1159 (2005)
12. Henaff, M., Bruna, J., LeCun, Y.: Deep convolutional networks on graph-structured data. arXiv preprint arXiv:1506.05163 (2015)
13. Kammoun, M., Kammoun, A., Abid, M.: Leak detection methods in water distribution networks: a comparative survey on artificial intelligence applications. J. Pipeline Syst. Eng. Pract. 13(3), 04022024 (2022)
14. Kipf, T.N., Welling, M.: Semi-supervised classification with graph convolutional networks. In: International Conference on Learning Representations (ICLR) (2017)
15. Klambauer, G., Unterthiner, T., Mayr, A., Hochreiter, S.: Self-normalizing neural networks. In: NIPS 2017, pp. 972–981 (2017)
16. Klise, K.A., Murray, R., Haxton, T.: An overview of the water network tool for resilience (WNTR) (2018)
17. Klise, K.A., Phillips, C.A., Janke, R.J.: Two-tiered sensor placement for large water distribution network models. JIS 19(4), 465–473 (2013)
18. Levie, R., Monti, F., Bresson, X., Bronstein, M.M.: CayleyNets: graph convolutional neural networks with complex rational spectral filters. IEEE Trans. Signal Process. 67(1), 97–109 (2018)
19. Li, G., Xiong, C., Thabet, A., Ghanem, B.: DeeperGCN: All you need to train deeper GCNs (2020)
20. Li, R., Wang, S., Zhu, F., Huang, J.: Adaptive graph convolutional neural networks. In: Proceedings of the AAAI Conference on Artificial Intelligence, vol. 32 (2018)

21. Li, Y., Yu, R., Shahabi, C., Liu, Y.: Diffusion convolutional recurrent neural network: data-driven traffic forecasting (2017). https://arxiv.org/abs/1707.01926
22. Monti, F., Boscaini, D., Masci, J., Rodola, E., Svoboda, J., Bronstein, M.M.: Geometric deep learning on graphs and manifolds using mixture model CNNs. In: IEEE Conference on Computer Vision and Pattern Recognition, pp. 5115–5124 (2017)
23. Nandanoori, S.P., et al.: Graph neural network and Koopman models for learning networked dynamics: a comparative study on power grid transients prediction (2022)
24. Niepert, M., Ahmed, M., Kutzkov, K.: Learning convolutional neural networks for graphs. In: ICML, pp. 2014–2023. PMLR (2016)
25. Omitaomu, O.A., Niu, H.: Artificial intelligence techniques in smart grid: a survey. Smart Cities 4(2), 548–568 (2021)
26. Rossman, L., Woo, H., Tryby, M., Shang, F., Janke, R., Haxton, T.: EPANET 2.2 user's manual, water infrastructure division. CESER (2020)
27. Sato, R.: A survey on the expressive power of graph neural networks. CoRR (2020). https://arxiv.org/abs/2003.04078
28. Scarselli, F., Gori, M., Tsoi, A.C., Hagenbuchner, M., Monfardini, G.: The graph neural network model. IEEE Trans. Neural Netw. 20(1), 61–80 (2009)
29. Veličković, P., Cucurull, G., Casanova, A., Romero, A., Liò, P., Bengio, Y.: Graph Attention Networks. ICLR (2018)
30. Vrachimis, S.G., et al.: BattLeDIM: battle of the leakage detection and isolation methods. In: CCWI/WDSA Joint Conference (2020)
31. Xing, L., Sela, L.: Graph neural networks for state estimation in water distribution systems: application of supervised and semisupervised learning. J. Water Resour. Plann. Manage. 148(5), 04022018 (2022)
32. Xu, K., Hu, W., Leskovec, J., Jegelka, S.: How powerful are graph neural networks? arXiv preprint arXiv:1810.00826 (2018)

Data-Centric Perspective on Explainability Versus Performance Trade-Off

Amirhossein Berenji⬤, Sławomir Nowaczyk(✉)⬤, and Zahra Taghiyarrenani⬤

Center for Applied Intelligence Systems Research, Halmstad University,
Halmstad, Sweden
{amirhossein.berenji,slawomir.nowaczyk,zahra.taghiyarrenani}@hh.se

Abstract. The performance versus interpretability trade-off has been well-established in the literature for many years in the context of machine learning models. This paper demonstrates its twin, namely the data-centric performance versus interpretability trade-off. In a case study of bearing fault diagnosis, we found that substituting the original acceleration signal with a demodulated version offers a higher level of interpretability, but it comes at the cost of significantly lower classification performance. We demonstrate these results on two different datasets and across four different machine learning algorithms. Our results suggest that "there is no free lunch," i.e., the contradictory relationship between interpretability and performance should be considered earlier in the analysis process than it is typically done in the literature today; in other words, already in the preprocessing and feature extraction step.

Keywords: Explainable AI · SHAP · Intelligent Fault Diagnosis · Bearings · Hilbert Transform · Envelope Spectrum

1 Introduction

Rotary machines are one of the most crucial pieces of equipment in industrial production [9]; they consist of a huge number of components, including bearings. Even non-severe bearing faults disrupt the normal operation of rotating machines. Bearing fault is also among the frequent failure modes of rotary machines; 40% to 50% of all failures in rotating machinery are estimated to be due to bearing faults [20]. Therefore, bearing condition monitoring is of great importance.

The promising performance of pattern recognition techniques in machine condition monitoring use cases resulted in the creation of Intelligent Fault Diagnosis (IFD) – the application of artificial intelligence methods for machine fault diagnosis [13]. Although IFD-based solutions often achieve super-human performance in scientific settings, their application in the industrial sector is relatively limited due to a lack of transparency. Therefore, the employment of eXplainable Artificial Intelligence (XAI) methods to provide insight into their reasoning is of high priority.

B. Crémilleux et al. (Eds.): IDA 2023, LNCS 13876, pp. 42–54, 2023.
https://doi.org/10.1007/978-3-031-30047-9_4

Over the last decades, the interpretability versus performance trade-off from the *model perspective* – i.e., the fact that higher performance is often associated with higher complexity, and thus usually achieved by sacrificing the interpretability – has been well established [5]. While improved interpretability is not necessarily followed by reduced model performance, maintaining the latter while improving the former typically requires conscious effort, and often advanced techniques [21]. In this study, we pose a complementary question "does the application of preprocessing methods to make IFD pipelines more interpretable necessarily degrade their performance?"

The contribution of this work is to bring attention to an inherent decrease in classification performance caused by replacing the original data with an interpretable representation. A bearing fault diagnosis case study with and without counter-modulation transformation is an example of such a situation. To compare the original data versus an interpretable version of it, we evaluate two different preprocessing branches. One includes the Hilbert Transform as a demodulation technique, while the other excludes it. As pointed out by [2], bearing faults are easier to recognize – for a human expert – in the frequency spectrum of a demodulated signal. The classification accuracy achieved by the two branches, however, shows the opposite effect. Comparing the performance of the two representations clearly demonstrates that, for an artificial neural network, such human-interpretable features are subpar compared to raw data.

The rest of the paper is organized as follows: we first investigate relevant earlier work in Sect. 2. Afterward, in Sect. 3, a brief scientific background of the employed methods is provided. Next, in Sect. 4, the experimental setup is explained in detail, while the corresponding results are discussed in Sect. 5. Finally, in Sect. 6, we provide a discussion of the findings and conclude the paper.

2   Related Works

Explainability is on its way to becoming a must in IFD implementations. For example, in [3], authors introduced an unsupervised classification approach based on the attribution of explainability from an anomaly detection model. The effectiveness of this method is evaluated not only by the application of different models but also by an examination of different datasets. The authors took advantage of Shapely Additive Explanations (SHAP) to derive the feature importance scores. Similarly, in [19], authors evaluated the effectiveness of different XAI methods, including Gradient Class Activation Map (Grad-CAM), Layer-wise Relevance Propagation (LRP), and Local Interpretable Model-agnostic Explanations (LIME), to explain a shaft imbalance detection model. Another approach is to incorporate physics-inspired features, cf [6]. Authors applied a Frequency-RPM transformation to transform time domain signals to time-frequency representation; these representations are usually regarded as images, and therefore Convolutional Neural Networks (CNNs) are widely applied to manipulate these representations. Lastly, in [4], Grad-CAM is applied to derive explanations from a CNN model used to diagnose bearing faults. Short-Time Fourier Transform

(STFT) is used to extract the time-frequency representation of time-domain bearing acceleration signals. As the authors ignored the modulation phenomena in bearings, their derived explanations are not in good accordance with patterns expected physically; however, the authors then showed that patterns corresponding to different health states are repeatable and comparative.

Hilbert transform is frequently used to demodulate time domain signals. For example, in [11], authors used Hilbert transform for envelope extraction purposes, alongside cyclo-stationary analysis (to cope with non-stationary signals) to reveal fault frequency components in an air conditioning production assembly line. Moreover, Hilbert Transform is frequently used as the demodulation technique in bearing vibration analysis pipelines. As an example, in [16], authors used it alongside wavelet packet decomposition to extract the fault characteristics from the bearing acceleration signal. Similarly, authors of [22] showcased the effectiveness of the application of envelope analysis to reveal fault frequency components expected to observe in the Case Western Reverse University bearing dataset.

3 Background

3.1 Zoom FFT

Zoom FFT is a technique to improve frequency resolution within a specific frequency range [12]. Application of Zoom FFT not only reduces the length of the original signal to achieve the desired frequency resolution but also decreases the computational cost significantly [12]. Implementation of Zoom FFT consists of two main stages; the first one is the application of a group of operations to preprocess the original signal, while the second stage is the application of conventional FFT.

As illustrated in Fig. 1, the preprocessing stage starts with a multiplication of the original signal ($x[k]$ with a length of N) by the complex signal of $[\cos(2\pi f_c t) + i\sin(2\pi f_c t)]$, where f_c is the lower limit of the desired frequency range ($[f_c, f_c + B_p]$). It continues with low-pass filtering of the multiplication signal (x_{mu}), using the bandwidth of B_p. Afterward, the filtered signal is undersampled by M (known as decimation), resulting in a signal with the length of N/M. Next, zero padding is employed to fill in for the $N - (N/M)$ instances removed during the decimation process. Finally, the FFT is employed to derive a frequency domain signal, within the desired frequency range, out of the zero-padded signal.

3.2 Hilbert Transform to Extract Envelopes

Hilbert Transform (HT) of a signal is defined [7] as:

$$H[x(t)] = \tilde{x}(t) = \frac{1}{\pi} \int_{-\infty}^{\infty} \frac{x(t)}{t - \tau}\, d\tau \tag{1}$$

Therefore, we can define an analytic signal as a complex function in which the real part is the original signal, and its imaginary part is the HT [7]:

$$X(t) = x(t) + i\tilde{x}(t), \tag{2}$$

where $X(t)$ is the analytic signal, $x(t)$ is the original signal, and the $\tilde{x}(t)$ is the HT of the original signal. Similar to any other time variant complex function, the instantaneous amplitude of the analytic signal can be computed as:

$$A(t) = |X(t)| = \sqrt{x^2(t) + \tilde{x}^2(t)} \tag{3}$$

The instantaneous amplitude of the analytic signal varies slower than the original signal [7]. Therefore, the instantaneous amplitude function – also known as envelope – is a version of the original signal excluding high-frequency oscillations. Accordingly, the envelope extraction based on HT is considered a demodulation approach widely used in rotating machinery vibration analysis [8].

4 Experiments

4.1 Introduction to Datasets

Most of our experiments are done on the Case Western Reverse University (CWRU) bearing dataset; it includes four different bearing health states: normal, inner-race fault, outer-race fault, and ball problems. We focus our study on Drive-End (DE) bearings, as DE bearings are subjected to more mechanical stresses in real-world scenarios. Signals with 48000 12000 Hz sampling frequencies are available; however, we 12000 Hz sufficient. In this dataset, four levels of rotational speeds (1730 RPM, 1750 RPM, 1772 RPM, and 1797 RPM) are included, and we used them all to consider the challenge of variation in mechanical loading. The rotational speed is vitally important for bearing fault detection, as the occurrence of faults in the bearings is likely to exhibit dominant peaks at particular frequency components (fault characteristic components). These components are the multiplication of geometrically defined ratios by the rotational speed of the bearing. In Table 1, ratios of different faults[1] alongside the fault frequency component by the rotational speed are summarized.

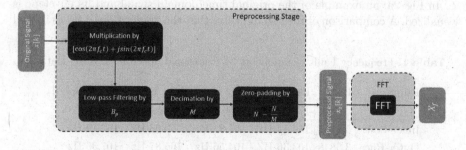

Fig. 1. Visual illustration of Zoom FFT

---

[1] Ratios from https://engineering.case.edu/bearingdatacenter/bearing-information.

Unfortunately, due to the modulation phenomena, the expected bearing fault components are not usually observable in frequency spectra; therefore, a demodulation step is essential to reveal the true fault frequency components.

To generalize our findings beyond a single dataset, we confirm our observations also using the Paderborn University (PU) bearing dataset [14]. We again focus on bearing fault classification, including normal and synthetically generated faults of the inner race and outer race. Moreover, we also considered mechanical loading variation by including both 900 and 1500 RPM shaft rotational speeds.

4.2   Data Preparation and Preprocessing

To study the effect of the application of HT on classification accuracy, we consider two preprocessing branches. Both preprocessing branches start with the initial step of splitting the original time domain signals to 2048 and 12800 points-long signals for CWRU and PU datasets, respectively. Following that, **on the first branch**, *raw*, we use a generally accepted pipeline for rotating machinery vibration analysis [15, 18, 23]. It starts with the application of a Hann window to avoid leakage error, and a Butterworth bandpass frequency filter (with a degree of 25 and cut-off frequencies of 2.5 Hz and 5500 Hz Hz for CWRU, and 2.5 and 31000 for PU) is employed to both remove the DC components and prevent aliasing. Afterward, we applied Fast Fourier Transform (FFT) algorithm to derive the frequency spectrum. The resulting frequency domain signals are 1024 points-long signals, covering 0 to 6000 Hz Hz and 0 32000 Hz for CWRU and PU datasets, respectively.

On the second branch, *envelope*, we take advantage of HT to extract the envelope from the raw time domain signal. Therefore, to derive a well-suited frequency resolution within the desired frequency range (0 to 1000 Hz Hz), Zoom FFT is employed. The choice of the frequency range is made to cover not only the frequency components corresponding to the faults but also their initial harmonics. Moreover, since 1024 points are used to apply the Zoom FFT technique, the resulting frequency domain signals are also 1024 points long. Similar to the raw branch, we also used the Butterworth bandpass frequency filter prior to the application of Zoom FFT; however, the second cut-off frequency 800 Hz.

In Fig. 2a, an example of the original time domain signal and its envelope is visualized. A comparison of the two indicates that the application of HT is indeed

Table 1. Frequency Fault Components by Rotational Speed for CWRU Dataset

Fault	Ratio	Fault Frequency Component by Rotational Speed			
		1730 RPM	1750 RPM	1772 RPM	1797 RPM
Inner-Race	5.4152	156.14 HZ	157.94 Hz	159.93 Hz	162.19 Hz
Outer-Race	3.5848	103.36 HZ	104.56 Hz	105.87 Hz	107.36 Hz
Ball	4.7135	135.91 HZ	137.48 Hz	139.21 Hz	141.17 Hz

capable of reducing the disturbance level in the time domain signal. Moreover, plots in Figs. 2b, 2c and 2d show that the envelope preprocessing branch is more powerful in revealing characteristic frequency components for bearing faults. It is worth noting that the red dashed lines in these plots highlight the expected fault frequency component, according to the values presented in Table 1. It is also to be noted that all the plots visualized in Fig. 2 come from the CWRU dataset; however, the insights from the PU dataset are analogous.

For the experiments, data is split so that 40% is the hold-out testing dataset, and 25% of the remaining data is used for validation purposes. Additionally, we employ min/max scaling to transform values of all the frequency components within the frequency spectra to the range from zero to one.

4.3   Training Classifiers

Our experiments start with the application of Multi-Layered Perceptrons (MLP) to classify signals from the CWRU dataset. Networks to classify signals from both preprocessing branches utilize the structure of 1024-512-256-128-64-4 as neurons per layer. For the training of the network on the data from the first preprocessing branch (the one with the application of FFT on raw time domain signals), a combination of 10^{-4} and 50 as the learning rate and the number of epochs, respectively, provides monotonic and smooth minimization of the categorical cross-entropy loss. Notably, the proposed architecture achieves repeatable 100% classification accuracy on the held-out test dataset. On the other hand, we experienced strong overfitting when training the same network on the second preprocessing branch (using envelope extraction and Zoom FFT). Our experiments showed that the highest classification accuracy is achieved using a learning rate of 10^{-5} and 150 epochs at the verge of overfitting. Nevertheless, perfect performance is not attainable anymore.

Additionally, to strengthen the claim of the ubiquity of the tradeoff and demonstrate that the difference in the performance of the two preprocessing branches is independent of the classification method and not specific to deep neural networks, we also trained a group of classic machine learning models – including Decision Tree (DT), Random Forest (RF) and Support Vector Machine (SVM) – utilizing data belonging to both preprocessing branches, on data from CWRU dataset. It is worth mentioning that all the hyper-parameters of these models were set to the default values of scikit-learn[2] library.

Finally, to generalize our findings beyond a single dataset, we decided to evaluate the classification performance of both preprocessing branches on the PU dataset. For the conventional preprocessing, we employed an MLP with the structure of 6400-2000-250-3 with the 10^{-5} and 200 as the learning rate and epochs, respectively. Similarly, for signals from the interpretable branch, the structure is 1024-256-64-3, and a learning rate of 10^{-5} with 250 epochs were utilized.

---

[2] https://scikit-learn.org/stable/.

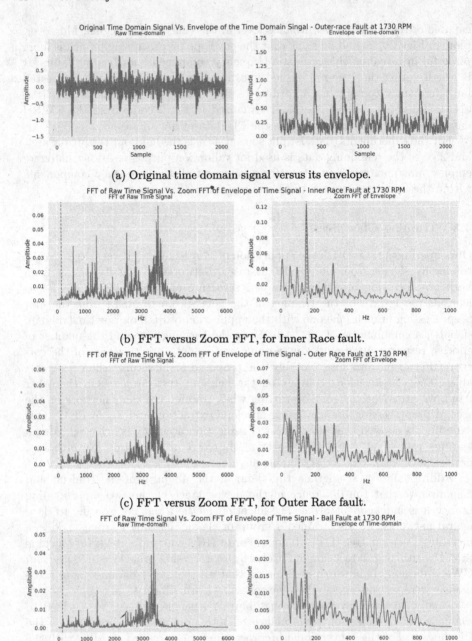

(a) Original time domain signal versus its envelope.

(b) FFT versus Zoom FFT, for Inner Race fault.

(c) FFT versus Zoom FFT, for Outer Race fault.

(d) FFT versus Zoom FFT, for Ball fault.

Fig. 2. Visual demonstration of the signals from each preprocessing branch

5 Results

Table 2 summarizes the classification performance of different methods for both datasets and preprocessing branches. We repeat each experiment 5 times to minimize the randomness effect of training. To be able to examine the misclassified observations one by one, we keep the train and test sets fixed across all the trials. Based on the results in this Table, the performance decrease caused by the substitution of *Raw FFT* data with the *Zoom FFT* is consistently seen for essentially all cases. The one exception is the DT's results on the PU dataset; however, since the performance of this method is overall very poor (barely any learning is done, and the result is essentially random), we do not consider this to be contradicting our claim.

To better understand the performance versus interpretability tradeoff showcased here, we analyze the observations consistently misclassified across all 5 trials. As presented in the rightmost columns of Table 2, for the MLP row on the CWRU dataset, 35 observations were misclassified every time. Compared to the minimum and the maximum number of misclassified observations over these trials (37 and 42, respectively), the number of consistently misclassified observations is quite significant. This brings up a hypothesis that the application of interpretability-enhancing preprocessing makes a portion of the data impossible to classify correctly. This phenomenon seems to originate in the fact that the envelope branch, specifically the HT, is making the signals more interpretable to humans by removing some (ostensibly) irrelevant features. Nevertheless, while the removed features are irrelevant to human practitioners, they can likely be helpful to machine learning models; by their removal, a noticeable decrease in the classification performance of the models is registered.

Next, we check whether this phenomenon is independent of the first cut-off frequency of the bandpass filter since this is the most important hyperparameter of the interpretable preprocessing branch. Frequency components before the first cut-off frequency are likely to get their magnitude reduced significantly; therefore, we find this value crucial for maintaining information. We study the effect of its variation on the classification accuracy and the number of misclassified observations. In Table 3, minimum and maximum classification accuracies of MLP and the number of misclassified observations over 5 trials of the experiments for different cut-off frequencies are provided. Across these results, no difference in overall performance is seen. Although we are likely to have around 35 observations constantly misclassified for any given value of the first cut-off frequency, it is to be noted that only 25 observations were *never* correctly classified across all the different frequency values.

Finally, in Table 4, the number of each combination of ground-truth and misclassified labels – of the 25 constantly misclassified observations, no matter what is the first cut-off frequency – is summarized. According to this Table, the ball problem is always either the ground truth or misclassified label, in all of these observations. This finding is a confirmation of the previously presumed hypothesis that the application of the interpretable preprocessing branch makes a portion of the data – in this case study, a relatively limited number of ball fault

Table 2. Classification performance of different methods on both datasets and two pre-processing branches, over 5 trials (the "C" column denotes the number of consistently misclassified observations).

Dataset	Method	Preprocessing	Classification Accuracy			# Misclassified		
			Min	Avg	Max	Min	C	Max
CWRU	DT	Raw FFT	0.9491	0.9525	0.9565	76	0	89
		ZoomFFT on Env	0.9376	0.9482	0.9605	69	5	109
	RF	Raw FFT	0.9977	0.9982	0.9989	2	2	4
		ZoomFFT on Env	0.9851	0.9859	0.9874	22	9	26
	SVM	Raw FFT	0.9994	0.9999	1.0000	0	0	1
		ZoomFFT on Env	0.9468	0.9469	0.9473	92	84	93
	MLP	Raw FFT	1.0000	1.0000	1.0000	0	0	0
		ZoomFFT on Env	0.9760	0.9769	0.9788	37	35	42
PU	DT	Raw FFT	0.7395	0.7629	0.7816	378	4	451
		ZoomFFT on Env	0.7556	0.7839	0.8018	343	26	423
	RF	Raw FFT	0.8914	0.8951	0.8983	176	54	188
		ZoomFFT on Env	0.8862	0.8889	0.8925	186	91	200
	SVM	Raw FFT	0.9041	0.9074	0.9110	154	135	166
		ZoomFFT on Env	0.8723	0.8776	0.8833	202	142	221
	MLP	Raw FFT	0.9365	0.9374	0.9393	105	70	110
		ZoomFFT on Env	0.8082	0.8109	0.8140	322	288	332

Table 3. Average classification accuracy, and number of repeatably misclassified observations, on CWRU using MLP, over 5 trials ("A", "M" and "C" stand for *accuracy, number of misclassified* and *constantly misclassified*, respectively)

Metric	First Cut-off Frequency (Hz)														
	2.5			10			20			30			40		
A	Min	Avg	Max	Min	Avg	Max	Min	Avg	Max	Min	Avg	Max	Min	Avg	Max
	97.60	97.69	97.88	97.37	97.40	97.42	97.31	97.40	97.48	96.97	97.17	97.31	96.97	97.22	97.42
M	Min	Max	C	Min	Max	C	Min	Max	C	Min	Max	C	Min	Max	C
	37	42	35	45	46	36	44	47	37	47	53	38	45	53	36

observations – impossible to classify correctly. By comparing Figs. 2b and 2c with Fig. 2d, we can see that – unlike inner race and outer race faults – the envelope preprocessing branch is not successful in revealing expected bearing fault characteristic frequency components. We believe that, alongside the missing dominant peak at the fault characteristic frequency components, the low-frequency peaks at the right subplot of Fig. 2b are the reasons why ball fault signals are often misclassified.

5.1 Application of SHAP to Explain Classifiers

SHAP is an explanation method originated from game theory literature [10], concerned with the calculation of an additive feature importance score [1]. The

Table 4. Types of misclassifications that occur in the envelope branch consistently, i.e., regardless of the cut-off frequency.

Ground-truth Label	Misclassified as	Count
Outer-Race Fault	Ball Problem	12
Ball Problem	Outer-Race Fault	12
Ball Problem	Normal	1

importance score of each feature is assessed by the comparison of the model performance when including and excluding the desired feature in different coalitions, computed as the weighted average of all possible differences [17].

$$\phi_i = \sum_{S \subseteq F \setminus \{i\}} \frac{|S|!(|F| - |S| - 1)!}{|F|!} [f_{S \cup \{i\}}(x_{S \cup \{i\}}) - f_S(x_S)], \tag{4}$$

where F is the set of all features, $f_{S \cup \{i\}}$ is the model trained with an arbitrary feature, and f_S is the model trained without that feature.

We employ SHAP (as implemented by [17]) to estimate the importance of every frequency component towards each prediction. In Fig. 3, three instances of frequency domain signals – each exemplifying a fault class – from both preprocessing branches (the conventional branch on the left and the interpretable one on the right) are visualized. The input data is shown in blue, and the corresponding SHAP explanations are in orange. The red dashed lines are the first, second, and third harmonics of the fault characteristic frequency components, according to Table 1.

The perfect alignment of peaks from both original signals and SHAP values at physically expected frequencies on the right-hand subplots of Figs. 3a and 3b shows that explanations from the envelope preprocessing branch match the expected physical patterns very well; the lack of the same on the left-hand subplots indicates that the opposite is true for the conventional, or raw, branch. Besides, the comparison of Fig. 3c with Figs. 3a and 3b shows that the agreement between explanations and the physically expected patterns varies with the type of fault. In other words, the interpretable processing branch is not capable of dealing with all the classes. While the explanations for inner and outer race faults are as expected, the ball faults are not. This can be seen in the right-hand subplot in Fig. 2d, where in contrast with inner race and outer race faults, no dominant peak can be observed for the ball fault. Moreover, low-frequency peaks are likely to make this bearing fault detection harder.

Moreover, while the model utilizing the conventional preprocessing branch is likely to perform perfectly, its explanations (left-hand plots visualized in Fig. 3) show no meaningful alignment with the physically expected patterns. This lack of agreement with the physics knowledge is the disadvantage of this model in comparison with its interpretable counterpart and will likely make it less trustworthy.

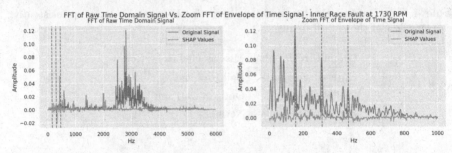

(a) FFT versus Zoom FFT and their SHAP values, for Inner Race Fault

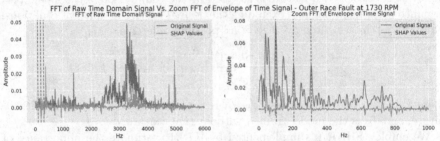

(b) FFT versus Zoom FFT and their SHAP Values, for Outer Race Fault

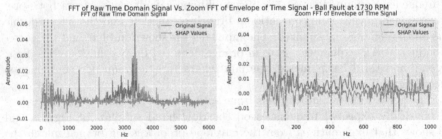

(c) FFT versus Zoom FFT and Their SHAP Values, for Ball Fault

Fig. 3. Examples of each fault from both branches, with SHAP values

6 Conclusions

In this study, we evaluated how the classification accuracy of bearing fault detection changes depending on including or excluding a counter-modulation technique. We ran experiments over two datasets and used four classification algorithms. Results show that while the demodulated pipeline offers higher interpretability, aligning better with the underlying physical phenomena, its classification performance is decreased noticeably. Therefore, we believe an inherent interpretability versus performance trade-off exists from the data-centric (alternatively to be called representation, feature extraction, or preprocessing) perspective. With complex enough problems, making the data representation interpretable involves simplifications that remove information – information that would otherwise be possible for machine learning algorithms to exploit. The effect is consistent for variations in the first cut-off frequency of the interpretable pre-processing branch, different datasets, and classification algorithms.

Our supplementary analysis shows that applying the envelope preprocessing branch affects a relatively minor portion of the data. We believe this is due to removing the features irrelevant to human analysts and simultaneously useful for AI models. The next step in pursuing this study is to understand the adversarial mechanism responsible for this decrease, hopefully leading to the discovery of transformations with a better balance between the two aspects.

Furthermore, since some of the misclassified samples differed between experiments with different first cut-off frequencies, this hyperparameter can be considered a factor in generating diverse datasets. It may be, therefore, possible to improve fault classification accuracy by using an ensemble of different datasets produced by varying the first cut-off frequencies.

Current results demonstrate the idea in a single domain. It is interesting to extend this research and explore this data-centric interpretability versus performance trade-off in other fields where well-understood interpretable transformations exist, such as computer vision or speech recognition.

Acknowledgements. This work was partially supported by Vinnova and by CHIST-ERA grant CHIST-ERA-19-XAI-012 from Swedish Research Council.

References

1. Arrieta, A.B., et al.: Explainable artificial intelligence (XAI): concepts, taxonomies, opportunities and challenges toward responsible AI. Inf. Fusion **58**, 82–115 (2020)
2. Bechhoefer, E.: A quick introduction to bearing envelope analysis. Green Power Monit. Syst. (2016)
3. Brito, L.C., Susto, G.A., Brito, J.N., Duarte, M.A.: An explainable artificial intelligence approach for unsupervised fault detection and diagnosis in rotating machinery. Mech. Syst. Signal Process. **163**, 108105 (2022)
4. Chen, H.Y., Lee, C.H.: Vibration signals analysis by explainable artificial intelligence (XAI) approach: application on bearing faults diagnosis. IEEE Access **8**, 134246–134256 (2020)
5. Došilović, F.K., Brčić, M., Hlupić, N.: Explainable artificial intelligence: a survey. In: 2018 41st International Convention on Information and Communication Technology, Electronics and Microelectronics (MIPRO), pp. 0210–0215. IEEE (2018)
6. Fan, Y., Hamid, S., Nowaczyk, S.: Incorporating physics-based models into data-driven approaches for air leak detection in city buses. In: ECML PKDD 2022 Workshops (2022)
7. Feldman, M.: Hilbert transforms. In: Braun, S. (ed.) Encyclopedia of Vibration, pp. 642–648. Elsevier, Oxford (2001)
8. Feldman, M.: Hilbert transform in vibration analysis. Mech. Syst. Signal Process. **25**(3), 735–802 (2011)
9. Han, D., Liang, K., Shi, P.: Intelligent fault diagnosis of rotating machinery based on deep learning with feature selection. J. Low Freq. Noise Vib. Active Control **39**(4), 939–953 (2020)

10. Holzinger, A., Saranti, A., Molnar, C., Biecek, P., Samek, W.: Explainable AI methods - a brief overview. In: Holzinger, A., Goebel, R., Fong, R., Moon, T., Müller, K.R., Samek, W. (eds.) xxAI 2020. LNCS, pp. 13–38. Springer, Cham (2022). https://doi.org/10.1007/978-3-031-04083-2_2

11. Lee, D.H., Hong, C., Jeong, W.B., Ahn, S.: Time-frequency envelope analysis for fault detection of rotating machinery signals with impulsive noise. Appl. Sci. **11**(12), 5373 (2021)

12. Lee, J.S., Yoon, T.M., Lee, K.B.: Bearing fault detection of IPMSMs using zoom FFT. J. Electr. Eng. Technol. **11**(5), 1235–1241 (2016)

13. Lei, Y., Yang, B., Jiang, X., Jia, F., Li, N., Nandi, A.K.: Applications of machine learning to machine fault diagnosis: a review and roadmap. Mech. Syst. Signal Process. **138**, 106587 (2020)

14. Lessmeier, C., Kimotho, J.K., Zimmer, D., Sextro, W.: Condition monitoring of bearing damage in electromechanical drive systems by using motor current signals of electric motors: a benchmark data set for data-driven classification. In: PHM Society European Conference, vol. 3 (2016)

15. Li, C., Zhang, W., Peng, G., Liu, S.: Bearing fault diagnosis using fully-connected winner-take-all autoencoder. IEEE Access **6**, 6103–6115 (2017)

16. Liu, Y.: Fault diagnosis based on SWPT and Hilbert transform. Procedia Eng. **15**, 3881–3885 (2011)

17. Lundberg, S.M., Lee, S.I.: A unified approach to interpreting model predictions. In: Advances in Neural Information Processing Systems, vol. 30, pp. 4765–4774. Curran Associates, Inc. (2017)

18. Meng, Z., Zhan, X., Li, J., Pan, Z.: An enhancement denoising autoencoder for rolling bearing fault diagnosis. Measurement **130**, 448–454 (2018)

19. Mey, O., Neufeld, D.: Explainable AI algorithms for vibration data-based fault detection: use case-adapted methods and critical evaluation. arXiv preprint arXiv:2207.10732 (2022)

20. Rajabi, S., Azari, M.S., Santini, S., Flammini, F.: Fault diagnosis in industrial rotating equipment based on permutation entropy, signal processing and multi-output neuro-fuzzy classifier. Expert Syst. Appl. **206** (2022)

21. Rudin, C., Radin, J.: Why are we using black box models in AI when we don't need to? A lesson from an explainable AI competition. Harvard Data Sci. Rev. **1**(2) (2019). https://hdsr.mitpress.mit.edu/pub/f9kuryi8

22. Wang, N., Liu, X.: Bearing fault diagnosis method based on Hilbert envelope demodulation analysis. In: IOP Conference Series: Materials Science and Engineering, vol. 436, p. 012009. IOP Publishing (2018)

23. Xia, M., Li, T., Liu, L., Xu, L., de Silva, C.W.: Intelligent fault diagnosis approach with unsupervised feature learning by stacked denoising autoencoder. IET Sci. Measur. Technol. **11**(6), 687–695 (2017)

Towards Data Science Design Patterns

Michael R. Berthold[1,4(✉)] (iD), Dashiell Brookhart[2], Schalk Gerber[3],
Satoru Hayasaka[2], and Maarit Widmann[3]

[1] KNIME, Zurich, Switzerland
`berthold@ieee.org`
[2] KNIME, Austin, USA
[3] KNIME, Konstanz, Germany
[4] Konstanz University, Konstanz, Germany
`http://www.knime.com`

Abstract. We propose data flow diagrams to model data science design
patterns and demonstrate, using a number of explanatory patterns, how
they can be used to explain and document data science best practices, aid
data science education, and enable validation of data science processes.

Keywords: Data Science · Design Patterns · Data Flow Diagrams ·
Blueprints · Best Practices

1 Introduction

Design Patterns have long been around. They first helped architects build well-
balanced houses by providing guidance on proper designs [2] and later became
more popular in software development, enabling programmers to build upon
well-established solutions patterns [6]. Design Patterns provide best practices
for commonly appearing problems. Providing a library of design patterns to
creators of complex systems allows to focus on the parts where new creative
solutions are needed and rely on proven solutions for the known parts.

For Data Science, various approaches to introduce variants of design patterns
have been published in recent years. Many have focused on the machine learning
aspect of data science and often tended to be code centric, similar to classic
software design patterns, or informal descriptions (see for example [8] and many
online articles using the term to show a handful of examples). Data Science is
more than just machine learning (or code), however. Data wrangling in itself is
an art and all of the challenges when it comes to deployment are another area
full of lesser-known recipes. Best practices exist for many of these problems but
are not available in one concise and consistent format.

In the following, we introduce a framework to represent design patterns for
the entire data science process using data flow diagrams. We believe that such
patterns can be useful in a variety of ways: in addition to establishing and
documenting best practices, just like in software engineering, we see three additional areas: teaching principles of data science, building data science process
recommendation and automation systems, and auditing real-world data science
processes for compliance and rule-violations.

B. Crémilleux et al. (Eds.): IDA 2023, LNCS 13876, pp. 55–64, 2023.
https://doi.org/10.1007/978-3-031-30047-9_5

1.1   Teaching Data Science

Assuming that the representation of data science design patterns is visually intuitive enough and covers data science best practices broadly, they can be used to teach principles of data science as well – independent of an actual implementation or tool, focusing on the data flow and not the underlying implementations or interfaces.

This is particularly beneficial, as using data science techniques requires both a profound understanding of what and how the used methods work as well as real-world experience with those methods. Often that clear understanding is confused with knowing every little detail of the actual implementation. But in order to apply those methods, understanding the <u>what</u> is much more important than the <u>how</u>. There is already plenty of knowledge and experience required to understand data aggregating techniques, modeling methods, and visualization concepts in order to apply them properly without requiring data scientists to understand implementational aspects as well.

1.2   Data Science Process Validation

Design Patterns for Data Science can also be used to automatically audit data science processes for violations of patterns, potentially pointing out flaws such as testing data bleeding into the training path or credentials shared outside the intended scope. Using an appropriate and visual representation makes the auditing process more intuitive as well.

1.3   Data Science Coaching and Auto-Creation

Design patterns can also be used to suggest continuations of data science processes under construction. This requires a sufficiently formal representation (e.g. using graph grammars) and, given that, one could imagine auto-generating hypotheses of complete data science process given a couple of side constraints.

2   Data Science Design Patterns

Data Science is fundamentally about creating a data flow—possibly with various parallel branches and maybe even with the occasional control flow aspect (consider a parameter optimization loop or iterative feature selection process). This is a fundamental difference from classic software development environments where the focus lies on the control flow. It is no surprise that even code-based development environments often offer a way to represent the code's underlying structure as a (data flow) diagram (e.g. Doxygen for UML diagrams or libraries that translate Python code into control or data flow diagrams).

We propose to formalize data science design patterns as *data flow* diagrams. This allows encapsulating and abstraction as well as focusing on relevant details. We will demonstrate this by showing a couple of examples illustrating their broad applicability. The idea to use data flow diagrams is obviously also not new—neither for programs nor for other types of logic (see e.g. [7]). However, to the best of our knowledge using data flow diagrams to describe data science design patterns is new.

2.1   General Data Science Flow

The most abstract data science design pattern is not particularly interesting, it just summarizes the five main steps, which can be broken down differently later on (Fig. 1).

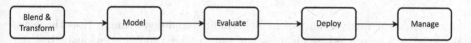

Fig. 1. The high level data science flow.

It is worth pointing out the difference to other types of Data Mining or Machine Learning process representations. The *Cross-industry standard process for data mining* (CRISP-DM, see [4] and Fig. 2) and others (such as SEMMA originally from SAS or the more pragmatic version in [3]) include also softer notions such as data and business understanding and constant need for refinement.

Other models of the data science life cycle, such as the one shown in Fig. 3 come closer but also here the focus includes meta steps around the actual data flow, for instance, model monitoring and retraining.

Those types of meta aspects of the process are not part of design patterns as introduced here. They are not concerned with the modeling of the actual data flow which, at the most abstract level, consists of data access, data blending and

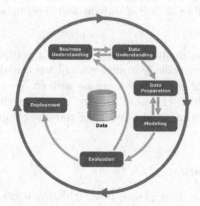

Fig. 2. The classic CRISP-DM data mining model (image from [4]).

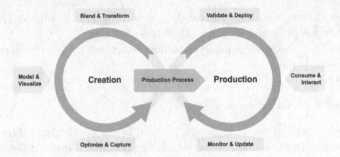

Fig. 3. The Data Science Life Cycle, including deployment and management.

transformation, analysis and/or model building, and deployment. How those are designed will be guided by business and data understanding, obviously, but the result is a data flow. The goal of data science design patterns is to help model and understand the resulting flow and not the process of arriving at it.

2.2   Model Training and Applying

Diving one step deeper into the data science flow, the most prominent design pattern represents the typical training a model and applying it to other data steps (see Fig. 4). Also here no surprising insights arise but it can already be used to validate and explain the need for separate training and testing data sets.

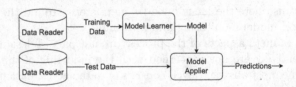

Fig. 4. The basic flow of training and applying a model.

As we will see later, in more complex training setups such as parameter optimization loops or ensemble learning, we will see this design pattern inside, potentially abstracting it as the *base learning pattern*. We can then easily model other requirements such as the need for three separate data sets for training, testing, and final validation of the resulting parameter-optimized or ensemble model.

2.3   Data Normalization

This one seems trivial at first glance but highlights a classic problem of early career data scientists. Figure 5 shows the standard pattern for normalization (ignore for a moment the little footnote of the normalizer node).

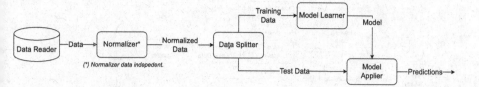

Fig. 5. Normalizing all data – training and testing.

This pattern can be modified so that the normalization routine moves into each of the two training and testing branches as shown in Fig. 6. Now the footnote becomes important. First-time users often run into the "little" mistake of applying normalizations in both branches but ignore that data properties are used for the normalization (e.g. min/max for numerical attributes) and the independent normalization steps, therefore, end up applying (often just slightly) different normalizations. In this visual representation, it becomes immediately obvious that making use of data properties in independent branches will cause problems.

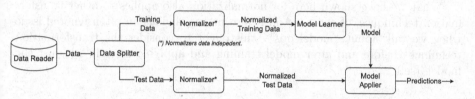

Fig. 6. Indepedent normalization of training and testing.

This naturally leads to the third pattern, shown in Fig. 7 which properly models how normalization properties need to be extracted in the training branch and then applied but not re-learned in the testing path. From this example, it also becomes clear why Fig. 5 has that little footnote. Although this flow would guarantee proper normalization also in the data-dependent case, it has some of the training (that is, the extraction of the normalization parameters) executed on the combined training and testing data set, effectively bleeding testing data properties into the training path.

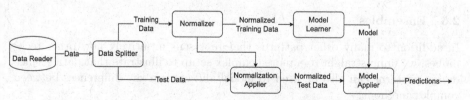

Fig. 7. Data dependent normalization of training and testing data.

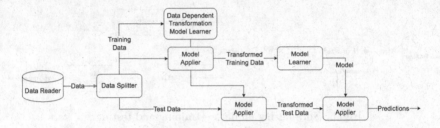

Fig. 8. Data dependent normalization of data using the model-apply pattern.

In analogy to the model training/applying pattern shown in the previous section, we really should model data-dependent normalizations as shown in Fig. 8. There is a "learning" component also in data-dependent normalization routines. It may be as simple as the extraction of the minimum and maximum value but it requires data interpretation similar to training a decision tree. If the underlying design pattern is modeled as shown here, all other validation and auditing techniques apply to data transformation steps just like they work for model learning and applying.

What we have shown here for normalization also applies to other transformation techniques, of course. And it also makes obvious an often ignored issue when we talk about "model ops"—this ought to consider the transformation techniques before and after model training and applying as well, not just the model itself.

2.4   Cross-Validation

Figure 9 shows the design patterns for classic cross-validation. It intuitively displays the key idea behind cross-validation: independent subsets of the entire data are used for testing, that is each row in the original data is used for testing exactly once. We use an additional element to indicate a flexible but pre-determined number of (in this case: parallel) branches.

This design pattern does not help determine the ideal split ratio, data scientists abstracting their flows to this level are still required to understand the implications of using different ratios. As discussed above, design patterns do not aim at simplifying the actual data science work but abstract away from the implementational details.

2.5   Ensembles

In addition to many other patterns that represent intuitively how other techniques, we jump straight to a more complex setup to illustrate that data science design patterns can also be used to model and explain key differences between complex methods.

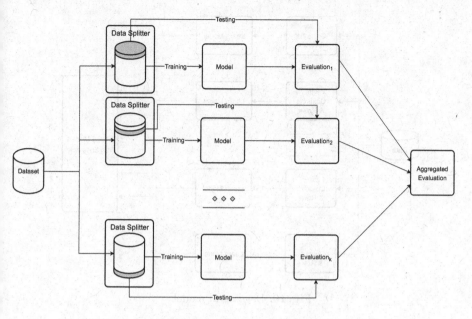

Fig. 9. Classic k-fold cross-validation.

Ensembles come in two key flavors: many (potentially even different) models trained in parallel or a sequence of models trained sequentially where each model focuses more on errors of the previous chain of models. Since boosting and bagging are generic wrapper methods it is worth modeling them explicitly as (reusable) design patterns.

Figure 10 shows the first variant. This design pattern clearly shows how different column and row sampling setups along each branch add diversity to the individual models that are trained on these subsets. Also here we model a flexible number of branches, similar to the cross-validation pattern above.

More interesting is boosting, as shown in Fig. 11. We directly see the cascading nature of the algorithm, where subsequent models focus on errors of their predecessors. Also note the reuse of the model-apply pattern shown earlier. Inside that "Learn Model Weight Data" box, we model an entire model-learning and data weighting step, which we then reuse in the diagram as one building block. Also here, a variable number of branches is used. Note how the model-apply pattern also shows up on the outside level: the boosting steps produce an (ensemble) model that is then applied just like any other model before. Abstracting from this pattern, makes this just fall back onto the base learner-apply pattern.

2.6 Factoring Out Transformations

A different example is shown in Fig. 12. It shows a rule that models how (data-independent) transformations can be factored out (or in) of a fusing step (which also needs to fulfill specific criteria, of course). Although this is at first glance

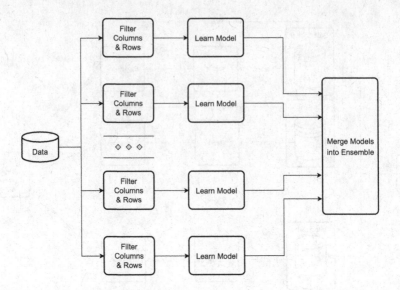

Fig. 10. Bagging of models.

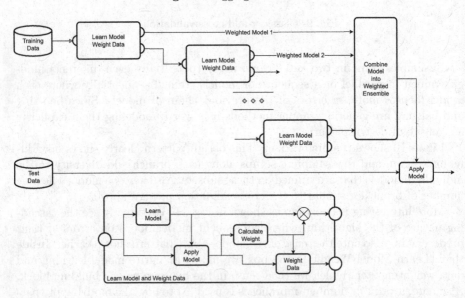

Fig. 11. Boosting of models.

rather obvious, modeling this formally allows for automatic systems to modify data flows while guaranteeing equivalent operation. This can be useful when the fusing step sits at the border of two technologies underneath the hood and a "workflow optimizer" automatically determines which technology to run the transformation step on. Such rules also allow comparing two data flows that are not structurally equivalent.

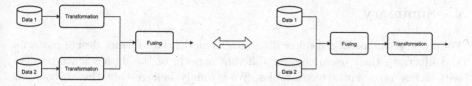

Fig. 12. Factoring out of transformation steps.

2.7 Key Elements

Any formulation requires a base alphabet. For the architectural examples cited early on the authors talk about a *pattern language* [1]. We are far from proposing a reasonably complete base alphabet, however for the examples shown here, the four consistent base elements are shown in Fig. 13.

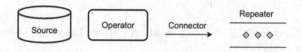

Fig. 13. Four base elements of a data science design pattern language.

We are using an element to represent a data source. Note that this can be any type of source (file, database, data lake) as well as type (tabular or unstructured data but also a model). Arrows model the data flow and boxes the actors/operations applied along that flow. For repeat operations, such as in the cross-validation or ensemble examples, the repeater element is used. Obviously, this is all but a starting point.

3   Outlook

The type of design pattern introduced here does not eliminate the need to under-stand the methods that are wrapped inside a node. If the data scientist does not understand what a decision tree does, design patterns will not make it easier for her to use that method. Design patterns help with modeling and understanding the data flow but do not make the actual application of the techniques any easier. However, they do help to focus on how those techniques are chained together.

Clearly, a complete list of patterns is outside the scope of this paper (and likely a moving target anyway) and what we presented here is a work in progress. Open questions remain. A well-defined base alphabet of symbols is needed so that formal procedures (e.g. graph transformation systems) can be applied for verification and automation purposes (similar to [5]). The ability to formalize anti-rules is also an open issue: how can we model, for example, two data sets that are overlap-free?

4   Summary

We propose the use of data flow diagrams to model data science design patterns and illustrate their usefulness for different aspects of the data science process with a few representative examples. We strongly believe that the abstraction layer of visual data flow diagrams enables data scientists to focus on the key aspect of their work: the flow of data without being dragged into implementation details as well as tool and API interfaces. The more transparent the entire process is and the more the data scientists can focus on the core of their work, the more intelligently they can make sense of and analyze their data.

Acknowledgments. Many other people were involved in earlier discussions around design patterns. We are particularly thankful to Dirk Streeb, Clara Biedermann, and Ahmad Varasteh who were part of the first brainstorming session in May 2022.

References

1. Alexander, C., Ishikawa, S., Silverstein, M.: A Pattern Language. Oxford University Press, Oxford (1977)
2. Alexander, C.: The pattern of streets. J. AIP **32**(3), 273–278 (1977)
3. Berthold, M.R., Borgelt, C., Hoeppner, F., Klawonn, F., Silipo, R.: Guide to Intelligent Data Science, 2nd edn. Springer, Heidelberg (2020). https://doi.org/10.1007/978-3-030-45574-3
4. CRISP-DM on Wikipedia. https://en.wikipedia.org/wiki/Cross-industry_standard_process_for_data_mining. Accessed 23 Oct 2022
5. Fischer, I., Koch, M., Berthold, M.R.: Proving properties of neural networks with graph transformations. In: IEEE International Joint Conference on Neural Networks, vol. 1, pp. 441–446 (1998)
6. Gamma, E., Helm, R., Johnson, R., Vlissides, J.: Design Patterns: Elements of Reusable Object-Oriented Software. Addison-Wesley (1994)
7. Hild, D.D.: Visual Languages and Computing Survey: Data Flow Visual Programming Languages. J. Vis. Lang. Comput. **3**, 69–101 (1992)
8. Lakshmanan, V., Robinson, S., Munn, M.: Machine Learning Design Patterns. O'Reilly Media, Inc. (2020)

Diverse Paraphrasing with Insertion Models for Few-Shot Intent Detection

Raphaël Chevasson[1(✉)], Charlotte Laclau[2], and Christophe Gravier[1]

[1] Université Jean Monnet Saint-Étienne, CNRS, Institut d Optique Graduate School
Laboratoire Hubert Curien UMR 5516, 42023 Saint-Étienne, France
{raphael.chevasson,christophe.gravier}@univ-st-etienne.fr
[2] Télécom Paris, Institut Polytechnique de Paris, Paris, France
charlotte.laclau@telecom-paris.fr

Abstract. In contrast to classic autoregressive generation, insertion-based models can predict in a order-free way multiple tokens at a time, which make their generation uniquely controllable: it can be constrained to strictly include an ordered list of tokens. We propose to exploit this feature in a new diverse paraphrasing framework: first, we extract important tokens or keywords in the source sentence; second, we augment them; third, we generate new samples around them by using insertion models. We show that the generated paraphrases are competitive with state of the art autoregressive paraphrasers, not only in diversity but also in quality. We further investigate their potential to create new pseudo-labelled samples for data augmentation, using a meta-learning classification framework, and find equally competitive result. In addition to proving non-autoregressive (NAR) viability for paraphrasing, we contribute our open-source framework as a starting point for further research into controllable NAR generation.

Keywords: Deep Learning · Natural language processing ·
Controllable text generation · Transformers · Non-autoregressive ·
Insertion models

1 Introduction

A *good* paraphraser should, for each source sentence, generate a batch of paraphrases which 1. are fluent, 2. have a similar meaning with the original source and 3. are sufficiently diverse between themselves and also between the source sentence. Since a classic language model only optimize for fluency, the two last requirements are harder to satisfy and require a special loss, architecture, or decoding scheme. For automatic text generation, predicting the next token autoregressively (left-to-right, one at a time) using transformers neural networks is the most popular approach. Other emerging methods, such as insertion-based models, can predict in a order-free way multiple tokens at a time, attracting a lot of attention recently due to the potential gain to inference time. However,

B. Crémilleux et al. (Eds.): IDA 2023, LNCS 13876, pp. 65–76, 2023.
https://doi.org/10.1007/978-3-031-30047-9_6

an understudied benefit of these models is their ability to constrain the generation to strictly include an ordered list of tokens, since they can build a sentence around them.

In this work, we investigate ways to leverage the generation constraints allowed by insertion-based text generation for diverse and slot-retaining generation for paraphrasing. We investigate the following scientific questions:

RQ1: Can insertion models be used as an efficient trade-off between fluency, semantic similarity and diversity in neural paraphrasing?
RQ2: What is the potential of such paraphraser as a data augmentation technique with respect to AR paraphrasers? Can this comparison inform us on the relative importance of fluency, similarity, and diversity for data augmentation using neural paraphrasing?

2 Related Work

As a text-to-text task, paraphrasing shares much similarity with translation and summarization. The most common approach is to use pretrained text-to-text models like BART [13] or T5 [21], fine-tuned on paraphrase corpus like MSRP [31], PAWS [28,30] or Quora [22][1]. A variation is to use off the shelf translation models to translate into many different languages (for diversity), then back-translating into the source language[2] [8,14,26], also known as round-trip translation (RTT). While RTT generates highly fluent sentences, it lacks diversity and does not guarantee that the meaning of the original sentence is preserved [4].

Other works opt to guaranty diversity, such as DivGAN [2] which forces the diverse sampling of a GAN latent via a diversity loss term. [20] also uses a GAN framework, but with several generators, and a compound loss with two discriminators ensuring paraphrases are distinguishable between themselves yet valid with respect to the source. ProtAugment [4] uses a variant of beam search with a diversity term [24] and randomly forbids unigrams from the source, forcing diversity at the expense of fluency. Rather than only using the paraphrases and their source labels as a fine-tuning corpus, it achieves semi-supervised learning with a compound loss that uses the paraphrases of labelled samples as positive examples, and paraphrases of unlabeled samples as negative examples to a prototypical learning objective. Blocking some particular unigrams to enhance diversity was also explored in [18] and coined as *dynamic blocking.*

Few works however focus on preserving meaning, which directly clashes with diversity. Diversity vs. fluency is the most important trade-off for textual data augmentation, and current generative models frequently loses key intent cues (such as important keywords or named entities), even when equipped with a copy mechanism. This is stressed in the Parrot framework [3], in which they add slots annotations to the training set. In [9], they extract those words at inference time, and they then average the logits from a reconstruction model trained on

---

[1] The Huggingface library [27] mostly uses this approach.
[2] The Fairseq library [19] also uses this approach.

sets of words with the logits from a RTT model. Our work explores a different path: from extracted slots keywords, we leverage the possibility of insertion-based models to enforce hard constraints in the generated texts. Noting that we can also opt to expand the hard constraints using synonyms of the slots keywords (under a stochastic process), we ultimately propose an insertion-based diverse paraphrasing framework (Sect. 3) which leads to more fluent yet more diverse paraphrases, and ultimately with impact on the meta-learning intent detection task (Sect. 4.3).

For an in-depth review of existing paraphrasers, we refer the interested reader to the survey of [32]. Used for data augmentation, such paraphraser have proven very efficient to improve classification where few labelled examples, even with very low fluency are available [12,25].

3 Diverse Paraphrases Generation

Insertion models generate a sentence by expanding an ordered list of words. Contrary to standard left-to-right autoregressive models, the input words are hard constraints – they are guaranteed to be included, ordered, and they are attended by all generated tokens (attention is bidirectional even at inference time). Relying on these properties, we propose a paraphrase generation scheme that promote diversity using an insertion model (see Fig. 1). We described each of the followed steps below.

3.1 Notations

Let $\mathcal{X}$ be a set of n source sentences, such that $\mathcal{X} = \{x^{(1)}, \cdots, x^{(n)}\}$. Let us consider $x^{(i)} \in \mathcal{X}$ one source sentence. For ease of reading, we use the simplified notation $x^{(i)} = x$ in the following. Let $x = (x_1, \cdots, x_\ell)$ be the sequence of tokens representing a source sentence, omitting the starting [CLS] and ending [SEP] tokens.

Our goal is to produce $Y = \{y_1, \cdots, y_m\}$, the set of m generated paraphrases for a given source sentence x. Note that each paraphrase y_j is potentially differing in length. We will note $\mathcal{Y}$ the complete paraphrase dataset, in contrast to Y, the batch of paraphrases from a particular source sentence x. Hence we have $Y^{(i)} \subseteq \mathcal{Y}, \forall x^{(i)} \in \mathcal{X}$.

3.2 Keywords Extraction

We first identify the k most important words from the source sentence in order to drive the downstream generation process. We call them keywords and note them $w = (w_1, \cdots, w_k)$. The notion of *most important* is defined as the greatest contribution to the sentence semantic. This is measured using the entropy of the word w_i in the sentence x, which can be approximated using a language model or a faster method like *tf-idf* and excluding stopwords – to name a few. Note that in our work, words are a list of tokens that should not be broken apart,

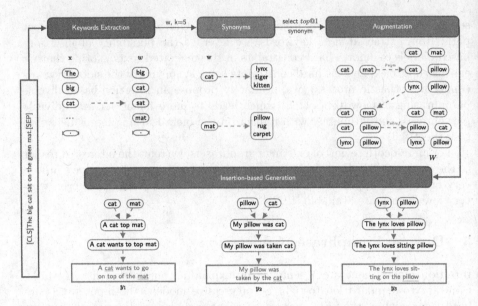

Fig. 1. Our method is two-fold. We first extract keywords and augment them via random shuffling and synonym replacement. Then, we generate paraphrases using them as an input. Those paraphrases are finally used as a data augmentation in a meta-learning framework.

like the tokens that form a word ("everyday") or an expression ("living room"). To stay close to the pre-training procedure of our non-autoregressive model, we follow [29] method, which consists in: (**1**) Splitting the sentence into words using an English, Regex-based word tokenizer[3]. (**2**) Applying the keyword extractor from [11], called YAKE, which was pre-trained to extract keywords, in an unsupervised way by leveraging text statistical features[4]. (**3**) Removing stopwords and duplicates, and keep k keywords, randomly chosen. We take care of treating them atomically throughout all our subsequent procedures. In addition, we empirically found that keeping the final punctuation sign, that would otherwise be removed as a stopword and preventing generation to its right[5], contributed to preserve the source sentence semantic. When our keywords extraction process fails to extract a sufficient number k of keywords, e.g. for very short sentences, we retain the k longest words. We found this strategy to be a surprisingly strong baseline, despite being very simple.

---

[3] segtok: github.

[4] YAKE: github.

[5] More precisely, keeping "![end]", "?[end]", and ".[end]" as keywords.

3.3 Keywords Augmentations

We augment the keyword list with two operations, namely shuffling and synonym replacement. For each keyword list w, we generate m augmentations, noted $W = (w_{i,j})_{1 \leq j \leq k}^{1 \leq i \leq m}$.

First, we generate a permutation for each line of W by swapping each keyword with another random word in the keyword sequence with a probability p_{shuf}. This step is meant to encourage the model to generate different alignments for better diversity. While this comes with the cost of a lower semantic alignment with the source sentence, it has been proven efficient for data augmentation [6].

Second, we replace each word with a random, but contextually relevant, synonym with a probability p_{syn}. We benchmarked several publicly available[6] synonymers using human evaluation and found EWISER [1], a transformer-based method that leverage the context around a word to ranks synonyms from its lemmatized WordNet synset, to be the most efficient one for our problem. A well-known downside of WordNet-based methods is the loss of semantic information resulting from the lemmatization; in practice, we found it to be acceptable for nouns or adverbs, but too significant for verbs, where the conjugation carry rich information. We reconjugate the verb synonyms to the most likely conjugation of the source verb, which qualitatively helps preserving the source semantic.

3.4 Constrained Paraphrases Generation

While we could have used a seq2seq NAR model like the Levenshtein Transformers [7], Insertion Transformers [23], or "Encode, Tag, Realize" [15], we test a different approach in this work, that is to drive our generation from a set of keywords to build around, rather than with cross-attention to the source sentence. This means we needed a language model rather than a seq2seq model. We use POINTER [29], which is to our knowledge, the only publicly available, large pre-trained NAR language model. POINTER is an insertion-based transformer model that build a sentence around an ordered list of given words. It is trained to predict the token to insert after each source token, doubling the sentence length in one iteration, and predicting [NoInsertion] everywhere when the sentence is fully built. The model heavily relies on a BERT backbone, and was fine-tuned for this progressive generation task from a pre-trained BERT checkpoint. We use this model for the generation of our diverse paraphrases Y, based on augmented keywords sequences W as an input.

We fine-tune the pre-trained POINTER model from [29] for each of the datasets used in our experiments (an unsupervised process). Unlike fine-tuning BERT on domain corpora, which not always provides improvements for the downstream tasks [17], this is a crucial step in our case. The paraphrase style depends on the domain at hand: for instance, the tone and turn of sentences in a dataset based on Wikipedia is different than the ones from a dataset made of user-generated

---

[6] We tested the lesker from nltk (→wordnet synset), bablify (→bablenet synset), getalp/disambiguate, and ewiser (→wordnet synset) wich vastly outperformed others.

queries to a chatbot. Otherwise, the model cannot use a long context to adapt the style as we do not provide the source sentence but only give it a few keywords as input.

4 Experiments

4.1 Datasets, Baselines and Metrics

There is no standard benchmark for NLP data augmentation, and comparing to different methods requires reproducing their results on a common dataset. Bearing this in mind, we choose to match our most strong and complex baseline settings. ProtAugment [4] was evaluated on the intent detection datasets of the DialoGLUE benchmark [16]. Summary statistics can be found in Table 1.

Other baselines are as follows. EDA [25] is a paraphrasing method that use random synonyms, additions, deletions and shuffles. We selected it for its widespread usage in NLP data augmentation and simplicity. AEDA [12] is a simpler variant of EDA that random inserts semicolons as its only aug-

Table 1. Summary statistics of each dataset.

Dataset	Classes	Samples	#tokens
Banking77	77	13,083	$11.7_{7.6}$
HWU64	64	11,036	$6.6_{2.9}$
Clinic150	150	22,500	$8.5_{3.3}$
Liu	54	25,478	$7.5_{3.4}$

mentation. It surprisingly often achieves higher results. RTT is a Round-Trip Translation scheme, which consists in translating from English to another language, then back to English. We construct 5 paraphrases using French, Spanish, Italian, German and Deutch intermediate languages, using public translation models from the Helsinki-NLP team for each language pair. Bart-uni [4] is finally our most challenging baseline. It is based on BART-base, a denoising transformer [13], and fine-tuned for paraphrasing on question-answering datasets in ProtAugment [4], which is the state-of-the-art intent detection framework, and which heavily relies on textual augmentation. We use their strongest configuration with unigram masking: At decoding times, we forbid source unigrams with a probability, which forces the beam search to diverge and create diverse generations.

We propose to automatically evaluate the quality of the generated paraphrases along two dimensions: the fluency and the diversity. We approximate the standalone fluency of the generated paraphrases with GPT2 language model perplexity (ppl). We use public checkpoints distylgpt2 from Hugging Face, and compute average and standard deviation in logarithmic space (meaning we use geometric mean and standard deviation). We estimate the diversity of a set of paraphrases that share the same source using the metric proposed by [10], coined as dist-2, which represents the number of distinct bigrams divided by the number of distinct tokens among all those paraphrases. As a more general and softer metric than encompass both fluency and semantic conservation, we log BLEURT and BERTScore between the paraphrase and it source reference.

4.2 Experimental Settings

Our code and paraphrases are publicly available[7].

Reproduction and Hyperparameters. We use publicly available paraphrases for RTT and Bart-uni baselines[8],[9] (5 per source), and run EDA and AEDA from their public repository[10],[11] using the default parameters. There was no default for AEDA number of augmentations, so we picked 9 to align with EDA. For our paraphrase generation, we chose to extract $k = 3$ keywords per sentence. In order to reduce the already long generation length, we picked the minimal number which still kept reasonable information. We generate $m = 5$ augmentations per sentence to match our best baseline. For p_{syn} and p_{shuf} we search the $\{0, 0.25, 0.5, 0.75, 1\}^2$ domain. To alleviate the computational requirements, we ran the complete search only on truncated datasets with the $1,000$ first sentences (which led to $5,000$ generated paraphrases), and reran the 4 more promising on the full dataset. The grid search maximizes the validation-set classification accuracy of the ProtAugment framework with our generated paraphrases over 5 cross-validation runs. We find $p_{syn} = 0.75$, and $p_{shuf} = 0.00$ (on BANKING77 and HWU64) or $p_{shuf} = 0.25$ (on Liu and Clinic150).

Insertion-Based Generation. Following Bart-uni, we fine-tuned our model with a single run on each of the 4 unlabeled training set. Our base model is POINTER wiki pretrained model[12], which is itself based on HuggingFace bert-large-uncased pretrained model for masked language modeling. We save and evaluate a checkpoint every 2^n and $100,000$ 8-batch training iteration, and find that training only on 500k samples (9 h on an Nvidia Titan RTX) is sufficient, our criterion being shorter sentence length, and no fluency/diversity degradation.

Application to Meta-learning Intent Detection. To evaluate the data augmentation potential of the paraphrases, we ran the ProtAugment meta-learning framework, using the paraphrases from each method as pseudo-labelled samples. We match [4,5] most challenging setup, with only 10 labelled samples per class and disjoint classes between the training, test and validations sets.

4.3 Analysis

[RQ1] **Paraphrase Quality.** Starting with a global and quantitative analysis (Fig. 2), our method reach a comparable fluency (ppl) while achieving a consistently higher inter-Y diversity (dist-2), which place us in the favorable

---

[7] https://github.com/RaphaelChevasson/DPIM.
[8] RTT: paraphrases.
[9] Bart-uni: paraphrases.
[10] EDA: official code.
[11] AEDA: official code.
[12] https://github.com/dreasysnail/POINTER.

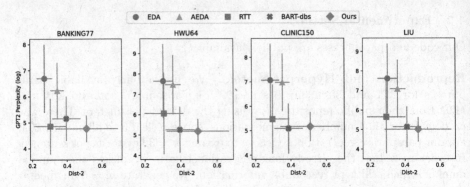

Fig. 2. Evaluation of the paraphrase quality. Lower-right corner is best as lower perplexity denotes better fluency and higher dist-2 value denotes more diversity.

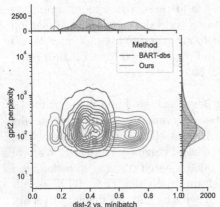

Fig. 3. Comparison of the distribution of the ppl vs. dist2 metric between `Bart-uni` and our approach for all paraphrases generated on Banking77.

Table 2. Average number of characters of the paraphrases $mean_{std}$.

	Banking77	HWU64	Liu	Clinic150
EDA	58_{40}	35_{15}	41_{16}	35_6
AEDA	63_{42}	36_{16}	44_{17}	38_{17}
RTT	55_{37}	34_{26}	40_{29}	34_{23}
Bart-uni	89_{31}	47_{12}	53_{14}	53_{16}
Ours	154_{67}	133_{46}	128_{44}	117_{41}

Table 3. Additional metrics, written with the $mean_{std}$ compact notation.

	Banking77		HWU64		Liu		Clinic150	
	BLEURT	BERTScore	BLEURT	BERTScore	BLEURT	BERTScore	BLEURT	BERTScore
RTT	$76.0_{13.1}$	$97.2_{2.4}$	$72.2_{15.2}$	$95.5_{3.5}$	$75.2_{15.7}$	$96.1_{3.4}$	$72.2_{13.5}$	$95.9_{2.9}$
Bart-uni	$36.7_{9.0}$	$85.3_{1.9}$	$33.4_{10.9}$	$84.5_{2.4}$	$32.3_{10.6}$	$84.2_{2.2}$	$34.7_{10.0}$	$85.2_{2.2}$
Ours	$43.3_{6.9}$	$86.7_{1.8}$	$41.1_{7.4}$	$85.2_{2.3}$	$38.2_{8.7}$	$85.2_{2.4}$	$42.8_{7.4}$	$85.8_{2.4}$

bottom-right corner of the tradeoff over every dataset. By focusing on the distribution over one dataset (Fig. 3), we see that for both methods, fluency and diversity are strikingly uncorrelated. Our method have a tighter range of perplexity, which we attribute to the model being able to bidirectional attend keywords

Table 4. A batch of diverse paraphrases from Banking77 where all 3 methods perform well.

Source	I'm interested in what your exchange rates are based on
RTT	1. I'm interested in what your exchange rates are 2. I'm interested in what exchange rates are based 3. I'm interested in what exchange rates are based on 4. I'm interested in your exchange rates 5. I'm interested in what your exchange rates are based on
Bart-uni	1. Interesting note: How much rate should we expect when comparing with other currency markets? 2. How much should we expect when considering Exchange rate variations between different currency zones? 3. How much should we expect when considering Exchange rate variations between different currency countries? 4. How much should we expect when considering Exchange rate options? 5. How much should we expect when considering Exchange rate variations between different currency regions/yearly ranges?
Ours	1. hi, i am extremely interested in getting an open interchange of exchange rates, but there have not been any established exchanges with exchange y 2. hi, i am extremely interested in purchasing, holding, exchanging and interchange. I noticed your website offer exchange rates, which are usually based on fair play 3. please, i am extremely interested in both in your exchange and exchange rates. I can make an interchange. Is no common ground for free play 4. hi, i am extremely interested in finding a way to interchange funds with your services. The charges on my account are usually a few per month from the foreign accounts unit, and i have not found what seems fair 5. hi, i am aware and extremely interested in some in changing my currency exchange policy. I have not established a balance but there are some strange charges. I am being charged a fee per sq pound at the correct unit exchange rate

from the beginning of the generation rather than being forced out of its comfort zone mid-generation at decode time like Bart-uni. We also have a wider range of diversity and distinguish about 5 emerging groups, which we attribute to the number of realizations of $p_{syn} + p_{shuf}$. BLEURT and BERTScore measurements shows that this advantage over Bart-uni extends to semantic conservation (Table 3), although still far from low-diversity methods like RTT.

We can further characterize this diversity by taking a qualitative look at a (x, Y) sample (Table 4). We note a strong bias in generation length w.r.t. the source sentence, which we quantify in Table 2. Our methods avoid a pattern often exhibited by Bart-uni: when the unigram blocking acts at the end of the beam search, it often leaves big chunks of Bart-uni sentences identical. From RTT to Bart-uni to Ours, the generation takes more liberty in interpreting and sometime adding information, sacrifice more semantic similarity, while opening more diversity. Without stronger constraints, our method is thus not suited for tasks where semantic conservation is key, but has a lot of potential for open-ended tasks like assisting creative writing or data augmentation.

Table 5. Classification accuracy $mean_{std}$. Best result(s) are in bold.

	Banking77	HWU64	Liu	Clinic150
EDA	$84.0_{1.3}$	$77.8_{2.3}$	$80.9_{1.9}$	$93.3_{0.7}$
AEDA	$82.4_{1.2}$	$78.0_{1.6}$	$80.3_{2.2}$	$93.1_{0.4}$
RTT	$83.4_{1.5}$	$78._{1.2}$	$80.5_{2.0}$	$93.1_{0.7}$
Bart-uni	$\mathbf{87.4_{0.6}}$	$\mathbf{83.2_{1.4}}$	$\mathbf{84.9_{1.8}}$	$\mathbf{95.8_{0.3}}$
Ours	$86.3_{0.8}$	$\mathbf{83.6_{1.6}}$	$\mathbf{84.4_{1.2}}$	$\mathbf{95.4_{0.5}}$

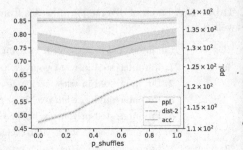

Fig. 4. Metrics ppl, dist-2 and accuracy as a function of $p_{shuffle}$ for Banking77.

[RQ2] Potential for Data Augmentation. For the latter, it seems hard to surmise if the fluency-diversity-conservation tradeoffs exhibited by our method will contribute sufficient but not excessive noise to the classifier and translate to a definitive improvement. By testing this empirically (Table 5), we observe a staggering improvement over most baseline, but not enough to surpass Bart-uni. Either our tradeoff advantage over this method does not carry to the classifier training signal, or the length bias negatively outweighs it.

By taking a closer look at the accuracy and tradeoff variations over our hyperparameters (Fig. 4), it seems that diversity really helps while fluency have a mixed impact; we cannot however make a definitive conclusion given the low variation compared to the variance over our 5 cross-validation runs.

Considering the statistical significance of accuracies in Table 5, we nevertheless attain state-of-the-art results in 3 out of the 4 datasets, which validate the potential of NAR data augmentation, and we largely exceed every other baseline. Considering our work is the first to tackle NAR data augmentation whereas the AR methods were explored and optimized in much more details by the community, this is very encouraging.

5 Limitations

Moving Parts. While our pipeline is simple to understand, the number of moving parts (python environments, implementation-wise) make it more difficult to setup than an end-to-end method.

Length of the Paraphrases. Our current generation model systematically provide sentences longer than the sources (see Table 2), only marginally reduced by our fine-tuning on unlabeled data. While it is possible to favors [no insertion] tokens to reduce sentence length, it just displaces the problem from being out-of-domain of the classifier training to being out-of-domain of the expander training. To solve the root problem without full cross-attention, we think providing the (augmented) source length as an input and using a model trained with oracle length, but still allowed to deviate from it, is the most promising direction.

6 Conclusion

In this work, we proposed an approach to replace AutoRegressive models by more flexible and controllable Non-AutoRegressive ones in the paraphrase generation task, with a deep dive in state-of-the-art (meta-learning) data augmentation for low-resource fine classification. Capitalizing on the open-endedness of these tasks, we proposed an extensible pipeline that achieve satisfactory results, while not even requiring cross-attention. We compared its behaviour against a number of diverse baselines, including two strong autoregressive ones, and found that the non-autoregressive model have a definite advantage for handling constrained diversity. We discussed the strengths and weaknesses of our method and open-source it to foster future research, as we ultimately think there is still untapped potential in the control offered by NAR generation methods.

Acknowledgments. This work was supported by the Futur & Ruptures program at Institut Mines-Télécom [2019–2022]. It was granted access to the HPC resources of IDRIS under the allocation 2022-AD011013590 made by GENCI.

References

1. Bevilacqua, M., Navigli, R.: Breaking through the 80% glass ceiling: raising the state of the art in word sense disambiguation by incorporating knowledge graph information. In: Proceedings of ACL, pp. 2854–2864. ACL (2020)
2. Cao, Y., Wan, X.: DivGAN: towards diverse paraphrase generation via diversified generative adversarial network. In: Findings of the Association for Computational Linguistics (EMNLP 2020), pp. 2411–2421. ACL, Online (2020)
3. Damodaran, P.: Parrot: paraphrase generation for NLU (2021)
4. Dopierre, T., Gravier, C., Logerais, W.: PROTAUGMENT: unsupervised diverse short-texts paraphrasing for intent detection meta-learning. In: Proceedings of ACL-IJCNLP, pp. 2454–2466 (2021)
5. Geng, R., Li, B., Li, Y., Zhu, X., Jian, P., Sun, J.: Induction networks for few-shot text classification. In: Proceedings of EMNLP-IJCNLP, pp. 3904–3913 (2019)
6. Goyal, T., Durrett, G.: Neural syntactic preordering for controlled paraphrase generation. In: Proceedings of ACL, pp. 238–252. ACL (2020)
7. Gu, J., Wang, C., Zhao, J.: Levenshtein transformer. In: NeurIPS, vol. 32 (2019)
8. Guo, Y., Liao, Y., Jiang, X., Zhang, Q., Zhang, Y., Liu, Q.: Zero-shot paraphrase generation with multilingual language models. arXiv preprint arXiv:1911.03597 (2019)
9. Guo, Z., et al.: Automatically paraphrasing via sentence reconstruction and round-trip translation. In: Zhou, Z.H. (ed.) Proceedings of IJCAI, pp. 3815–3821 (2021)
10. Ippolito, D., Kriz, R., Sedoc, J., Kustikova, M., Callison-Burch, C.: Comparison of diverse decoding methods from conditional language models. In: Proceedings of ACL, pp. 3752–3762 (2019)
11. Jorge, A., et al.: Text2story workshop - narrative extraction from texts. In: SIGIR Forum, pp. 150–152 (2018)
12. Karimi, A., Rossi, L., Prati, A.: AEDA: an easier data augmentation technique for text classification. In: Findings of ACL: EMNLP, pp. 2748–2754. ACL (2021)

13. Lewis, M., et al.: BART: denoising sequence-to-sequence pre-training for natural language generation, translation, and comprehension. In: Proceedings of ACL (2020)
14. Mallinson, J., Sennrich, R., Lapata, M.: Paraphrasing revisited with neural machine translation. In: Proceedings of EACL, pp. 881–893 (2017)
15. Malmi, E., Krause, S., Rothe, S., Mirylenka, D., Severyn, A.: Encode, tag, realize: high-precision text editing. In: Proceedings EMNLP-IJCNLP, pp. 5054–5065 (2019)
16. Mehri, S., Eric, M., Hakkani-Tür, D.: Dialoglue: a natural language understanding benchmark for task-oriented dialogue (2020)
17. Merchant, A., Rahimtoroghi, E., Pavlick, E., Tenney, I.: What happens to BERT embeddings during fine-tuning? In: Proceedings of the BlackboxNLP Workshop on Analyzing and Interpreting Neural Networks for NLP, pp. 33–44. ACL (2020)
18. Niu, T., Yavuz, S., Zhou, Y., Keskar, N.S., Wang, H., Xiong, C.: Unsupervised paraphrasing with pretrained language models. In: Proceedings of EMNLP, pp. 5136–5150. ACL (2021)
19. Ott, M., et al.: fairseq: a fast, extensible toolkit for sequence modeling. In: Proceedings of NAACL-HLT: Demonstrations (2019)
20. Qian, L., Qiu, L., Zhang, W., Jiang, X., Yu, Y.: Exploring diverse expressions for paraphrase generation. In: Proceedings of EMNLP-IJCNLP, pp. 3173–3182 (2019)
21. Raffel, C., et al.: Exploring the limits of transfer learning with a unified text-to-text transformer. J. Mach. Learn. Res. **21**(140), 1–67 (2020)
22. Sharma, L., Graesser, L., Nangia, N., Evci, U.: Natural language understanding with the Quora question pairs dataset (2019)
23. Stern, M., Chan, W., Kiros, J., Uszkoreit, J.: Insertion transformer: flexible sequence generation via insertion operations. In: Chaudhuri, K., Salakhutdinov, R. (eds.) Proceedings of ICML, vol. 97, pp. 5976–5985. PMLR (2019)
24. Vijayakumar, A., et al.: Diverse beam search for improved description of complex scenes. In: Proceedings of AAAI, vol. 32, issue 1 (2018)
25. Wei, J., Zou, K.: EDA: easy data augmentation techniques for boosting performance on text classification tasks. In: Proceedings of EMNLP-IJCNLP, pp. 6382–6388 (2019)
26. Wieting, J., Mallinson, J., Gimpel, K.: Learning paraphrastic sentence embeddings from back-translated bitext. In: Proceedings of EMNLP, pp. 274–285. ACL, Copenhagen, Denmark (2017)
27. Wolf, T., et al.: Transformers: state-of-the-art natural language processing. In: Proceedings of EMNLP: System Demonstrations, pp. 38–45. ACL (2020)
28. Yang, Y., Zhang, Y., Tar, C., Baldridge, J.: PAWS-X: a cross-lingual adversarial dataset for paraphrase identification. In: Proceedings of EMNLP-IJCNLP, pp. 3687–3692. ACL, Hong Kong, China (2019)
29. Zhang, Y., Wang, G., Li, C., Gan, Z., Brockett, C., Dolan, B.: POINTER: constrained progressive text generation via insertion-based generative pre-training. In: Proceedings of EMNLP, pp. 8649–8670. ACL (2020)
30. Zhang, Y., Baldridge, J., He, L.: PAWS: paraphrase adversaries from word scrambling. In: Proceedings of NAACL/HLT, pp. 1298–1308. ACL (2019)
31. Zhao, S., Wang, H.: Paraphrases and applications. In: Paraphrases and Applications-Tutorial Notes (Coling 2010), pp. 1–87 (2010)
32. Zhou, J., Bhat, S.: Paraphrase generation: a survey of the state of the art. In: Proceedings of EMNLP, pp. 5075–5086. ACL (2021)

LEMON: Alternative Sampling for More Faithful Explanation Through Local Surrogate Models

Dennis Collaris(✉)⬥, Pratik Gajane⬥, Joost Jorritsma⬥,
Jarke J. van Wijk⬥, and Mykola Pechenizkiy⬥

Eindhoven University of Technology, Eindhoven, The Netherlands
{d.a.c.collaris,p.gajane,j.jorritsma,j.j.v.wijk,m.pechenizkiy}@tue.nl

Abstract. Local surrogate learning is a popular and successful method for machine learning explanation. It uses synthetic transfer data to approximate a complex reference model. The sampling technique used for this transfer data has a significant impact on the provided explanation, but remains relatively unexplored in literature. In this work, we explore alternative sampling techniques in pursuit of more faithful and robust explanations, and present LEMON: a sampling technique that samples directly from the desired distribution instead of reweighting samples as done in other explanation techniques (e.g., LIME). Next, we evaluate our technique in a synthetic and UCI dataset-based experiment, and show that our sampling technique yields more faithful explanations compared to current state-of-the-art explainers.

Keywords: Machine learning · Explainable AI · XAI

1 Introduction

Explaining artificial intelligence (XAI) is important in high-impact domains such as credit scoring, employment and housing [5,9,12]. In these fields, incorrect model behavior may lead to additional direct costs, opportunity costs, as well as unfavorable bias and discrimination. XAI techniques can help identify and alleviate such problems [1]. Let us consider a real-world example: recent work has shown that, for commercial face classification services, accuracy of gender classification on dark-skinned females is significantly worse than on any other group [8]. This discrepancy was conjectured to be largely due to unrepresentative training datasets and imbalanced test benchmarks. However, using explanation techniques, it was shown that the classifiers made use of makeup as a proxy for gender in a way that did not generalize to the rest of the population [20].

A common approach to explain machine learning models is to create an explanatory, or *surrogate* model that mimics the reference model. As the surrogate is typically simpler, it can be used to understand the complex reference model. This enables us to understand any model (i.e., model-agnostic approach)

B. Crémilleux et al. (Eds.): IDA 2023, LNCS 13876, pp. 77–90, 2023.
https://doi.org/10.1007/978-3-031-30047-9_7

without having to alter that model (which could hurt performance). The extent to which this surrogate accurately approximates the reference model is called *faithfulness* (or fidelity).

There are two ways to obtain a surrogate model. The first is to *globally* mimic the reference model with an inherently simple surrogate model. However, due to this simplicity, the resulting surrogate can often not faithfully represent the reference model, which leads to inaccurate or incorrect explanations. Another approach is to consider only a small part of the complex reference model, and *locally* mimic that portion. Such surrogate models remain locally faithful to the reference model, while also being simple enough to understand. The current state-of-the-art techniques to explain individual predictions (e.g., LIME [21]) apply this approach by targeting only the part of the model that is relevant for that particular prediction. This process is illustrated in Fig. 1.

To generate such a surrogate, a simple model is trained on *transfer data*: a set of data points labeled by the reference model. This technique is well-known, but until recently was only applied to approximate models *globally*. For *local* explanations, samples from a constrained region are used to obtain a surrogate that is locally faithful, and simple enough to be considered interpretable [4, 21].

In this paper, we investigate transfer data sampling techniques for local surrogate models, and identify that the faithfulness of existing techniques may be impaired in high dimensionality. We explore alternative sampling techniques and introduce Local Explainable MOdel explanations using N-ball sampling (LEMON): an improved sampling technique that is more faithful and robust than the current state-of-the-art techniques by sampling directly from the desired distribution instead of reweighting samples (see Fig. 1).

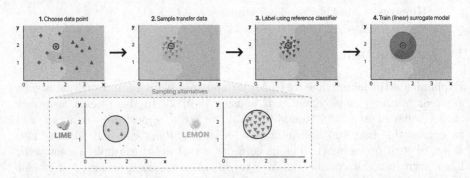

Fig. 1. The process of local surrogate learning: 1) choose a data point to be explained, 2) sample in the neighborhood of that point to obtain transfer data, 3) label the transfer data with the reference classifier, and 4) train a linear surrogate model on the data. Annotated are two examples of 20 samples generated with alternative sampling techniques (2): the one used in LIME [21], which reweights samples, and our proposed LEMON technique sampling directly from the desired distribution within a radius. Since more local samples are available with LEMON, the explanatory surrogate model is able to more faithfully represent the reference model.

2   Related Work

The idea of using transfer data to approximate a model globally was introduced by Craven et al. [10] and Domingos [11] and it has been used for model compression [3, 7, 18, 19, 22, 24], comprehensibility [4, 6, 21, 23] and generalization [18].

The types of surrogate models used vary widely. While for local explanation, linear regression is sufficient [21], global explanation requires more expressive surrogate models, e.g., shallow neural networks [22, 24], decision trees [6, 10], and rule sets [11, 23]. Furthermore, we identified two main categories of sampling techniques for surrogate learning used in previous work:

Synthetic sampling draws new samples from a distribution (e.g., uniform or normal), independently of the original data. For local techniques, this distribution is restricted to a predefined region of the feature space (i.e., the region of interest). The advantage of this approach is that we can sample as many transfer data points as desired. Most local explanation techniques use this approach.

Observation-based sampling uses the training data of the model as transfer data. When features in a dataset are correlated, certain values in feature space are less likely (or impossible) to occur compared to the correlated (or 'sensible') region of feature space. Observation-based sampling yields more samples in that sensible part of the feature space. However, the number of samples for transfer data is limited. Oversampling techniques like Naive Bayes Estimation (NBE) or MUNGE [7] can partially address this problem.

Which of these sampling techniques to use for surrogate learning is generally not considered thoroughly. For example, some authors make empirical claims such as "We have found that using the original training set works well" [18]. However, it is unclear what kind of benefit observation-based sampling yields compared to synthetic sampling, or how the chosen synthetic sample distribution affects the quality of explanations.

The vast majority of the reviewed papers focused on *global* approximations, in which the faithfulness (i.e., accuracy with respect to the reference model) of the surrogate model is compromised in order to simplify the surrogate and hence the resulting explanation, or reduce its memory footprint for model compression. The focus of this paper is on sampling for *local* surrogates instead. By only considering a small part of the reference model, and only *locally* mimicking that portion of the complex model, the surrogate remains faithful and simple. This approach is more recent and gained a lot of popularity with the introduction of the LIME explainability framework [21].

3   Issues with Sampling for Local Surrogates

To understand sampling for local surrogates, we consider LIME [21], as it is a widely popular local explanation technique, and for its clear and accessible usage of surrogate models. The transfer data in LIME are samples that are drawn from a fixed multivariate Gaussian distribution centered on the global mean of the training data. Here, fixed means that the distribution does not depend on the data point to be explained. Next, these samples are weighted based on their proximity to the data point to be explained. The locality of the technique is a result of this weighing. Then, a linear regression surrogate model is trained on these weighted samples, and the coefficients are presented as a "feature contribution" explanation that shows how important a feature is to a prediction: a small change in a feature with a high coefficient will lead to a large change in prediction, and hence can be considered important to the model.

The quality of a local surrogate is typically measured in *faithfulness*: the extent to which the local surrogate locally represents the reference model.

As a consequence of fixing the transfer data independently of the point to be explained, a notable drawback of systems such as LIME is that as the dimensionality of the data increases, the chances of obtaining samples close to the instance to be explained gets ever smaller. Hence, the robustness and faithfulness are significantly impaired for high-dimensional data. This is very similar to the known "curse-of-dimensionality" limitation of rejection sampling, in which most proposed points are not accepted as valid samples in high dimensions. In addition to faithfulness, Alvarez-Melis and Jaakkola [2] have demonstrated that using only few relevant samples (100 in their study) degrades the *robustness* of the explanation from LIME (i.e., very different explanations for similar inputs).

To experimentally verify this effect, we set up an experiment in which we can arbitrarily increase the dimensionality of the model without affecting other semantics of the machine learning setup. Consider the n-dimensional feature space $\mathcal{X} \subset \mathbb{R}^n$, and two classification models representing a hyperbox ($b(\mathbf{x})$) and hypersphere ($s(\mathbf{x})$) respectively:

$$b(\mathbf{x}) = \|\mathbf{x}\|_\infty \leq 1, \quad \text{and} \quad s(\mathbf{x}) = \|\mathbf{x}\|_2 \leq 1 \tag{1}$$

classifying $\mathbf{x} \in \mathcal{X}$ as either true or false. These models are simple enough to quickly change the dimensionality of the model, while being complex enough to resemble a realistic complex classification model that cannot perfectly be represented by the local surrogate model.

We chose the input data point $\mathbf{x} = [1, 0, 0, ...]$, a point on the surface of the decision boundary of the model. Next, a surrogate model is generated using LIME and four different kernel width parameters. We chose values from 0.1 to 0.4 to approximate the right side of the model only ($x_0 > 0$). By measuring the faithfulness of the surrogate models generated for different dimensional models, we assert whether the faithfulness is impaired in high-dimensional space.

For data point $\mathbf{x}$ and varying levels of dimensionality, we measure the faithfulness of the linear surrogate model, using the cosine similarity between the coefficient vector of the surrogate, and that of the best possible linear model in this setup: $f(\mathbf{x}) = \mathbf{x}_0$, coefficients shown in Fig. 2c. Contrary to more traditional faithfulness metrics (e.g., *RMSE* or R^2), this approach measures the agreement between the models without the need for additional sampling.

Figure 2 shows that for models with only a modest number of dimensions (i.e., 10–20 depending on the kernel width), the faithfulness of LIME is already significantly impaired, which can result in untrustworthy and misleading explanations. In addition, the explanations are not robust as indicated by the heavy

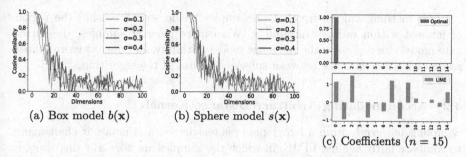

(a) Box model $b(\mathbf{x})$ (b) Sphere model $s(\mathbf{x})$

(c) Coefficients ($n = 15$)

Fig. 2. (a,b) LIME (5000 samples) is not faithful when explaining high dimensional models. Different lines represent different kernel widths σ. (c) the coefficients of the generated linear surrogate model (green) do not match the expected coefficients (blue). (Color figure online)

fluctuations. This happens because in high-dimensions, very few relevant samples are generated in the neighborhood of the point to be explained, and hence, the linear model is unable to approximate the behavior of the reference model. For a 15-dimensional box model, Fig. 2c shows the expected coefficients (blue) and coefficients of the linear surrogate from LIME (green). Even though only feature 0 has a substantial role for prediction, LIME incorrectly reports that many other features are relevant.

Note that LIME employs LASSO feature selection to reduce dimensionality ahead of explanation. However, this step is subject to the same limitations as outlined in this section. For simplicity, we disregard the feature selection option.

4   LEMON: Robust N-Ball Sampling

We introduce LEMON: Local Explainable MOdel explanations using N-ball sampling, which addresses the issues identified in Sect. 3. This technique samples directly from the desired distribution (defined by a distance-kernel function), instead of reweighting samples. This naturally yields data points where we need them: in the neighborhood (or *region of interest*) of the instance $\mathbf{x}$ to be explained.

4.1   Sampling from a Hypersphere

We first use sampling within a unit hypersphere followed by scaling the samples by radius r (region of interest), and translating the samples to be centered at $\mathbf{x}$.

Fishman [14] and Harman and Lacko [17] describe an efficient way to obtain points within an n dimensional hypersphere (i.e., n-sphere). If $Y \sim N(0,1)$, then $S_n = \frac{Y}{\|Y\|}$ is uniformly distributed on a unit n-sphere. Next, when we apply

$$S_n \cdot U^{\frac{1}{n}} \tag{2}$$

where U has the uniform distribution on $(0,1)$, we obtain the uniform distribution of a unit n-ball; the region enclosed by an n-sphere. Uniform samples from this distribution correspond to points uniformly distributed within the n-sphere.

This method will ensure that all samples reside strictly within the region of interest within radius r around $\mathbf{x}$. With more relevant samples, the surrogate model can represent the reference model faithfully, and output more robust results with less variance between subsequent runs of the algorithm.

4.2 Accommodating Arbitrary Distance Kernels

Sampling *uniformly* from a hypersphere is restrictive, and makes it challenging to compare fairly against LIME, in which the samples are normally distributed. In addition, different domains may require different distance metrics and kernels (e.g., cosine distance for text and L2 distance for images [21]). Hence, we expand our sampling technique to accommodate arbitrary distance kernels.

Let $K(r)$ denote a distance kernel on the domain $[0, r_{\max}]$, where the maximal distance $r_{\max} > 0$ may depend on the kernel. To sample points weighted by this kernel, note that the total weight of points at radius r is given by $c_n K(r) r^{n-1}$ for some dimension-dependent constant c_n. Thus, the cumulative distribution function (cdf) for the radius of a sample is

$$F(r) := \mathbb{P}\big(\|X\| \le r\big) = \frac{\int_0^r K(s) s^{n-1} \mathrm{d}s}{\int_0^{r_{\max}} K(s) s^{n-1} \mathrm{d}s}, \text{ for } r \le r_{\max}. \tag{3}$$

To sample using Eq. (3), we use inverse transform sampling [16]. However, an exact analytical integral of this density function may not always exist e.g., the Gaussian distance kernel used by LIME does not have a closed solution. Hence, we numerically approximate the inverse to sample from arbitrary distance kernels. Next, we show two examples of specific types of distance kernels that can be used with this technique.

Uniform Distance Kernel. The most basic distance kernel is the uniform kernel $K_{\mathrm{uniform}}(r) := 1$. We first show that substituting this distance kernel function into Eq. (3) yields the same cdf as the uniform sampling approach in Eq. (2). We get for $r \le r_{\max}$, $F(r) = \frac{\int_0^r s^{n-1} \mathrm{d}s}{\int_0^{r_{\max}} s^{n-1} \mathrm{d}s} = \frac{r^n/n}{r_{\max}^n/n} = (r/r_{\max})^n$. Ignoring the factor S_n that determines the angle from (2), we get for $r \le r_{\max}$,

$$\mathbb{P}\Big(U^{1/n} < \tfrac{r}{r_{\max}}\Big) = \mathbb{P}\Big(U < \big(\tfrac{r}{r_{\max}}\big)^n\Big) = \Big(\tfrac{r}{r_{\max}}\Big)^n = F(r).$$

In the equation above, the second equality follows from the fact that U has the uniform distribution on $(0, 1)$. Ergo, using this uniform distance kernel leads to uniformly distributed samples within a hypersphere of radius $r_{\max}$. An example of sampling using this distance kernel is shown in Fig. 3a.

Gaussian Distance Kernel. For a fair comparison with other methods, our sampling technique should also support the Gaussian distance kernel as used in LIME, defined as

$$K_{\mathrm{gaussian}}(r) := \exp\big(-r^2/(2\sigma^2)\big). \tag{4}$$

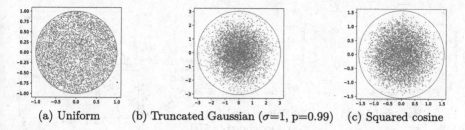

(a) Uniform   (b) Truncated Gaussian (σ=1, p=0.99)   (c) Squared cosine

Fig. 3. Samples of three distance kernels in 2D. Radius $r_{\max}$ is indicated in red. (Color figure online)

However, this distance kernel poses a problem: the Gaussian distribution is unbounded, while for our numeric approximations we require a kernel whose domain is bounded by some radius $r_{\max} < \infty$. For comparison to the Gaussian kernel used in LIME, we use a truncated distance kernel: we sample points from a Gaussian distribution with the same kernel standard deviation conditioned to be at most $r_{\max}$. Here, we choose $r_{\max}$ such that a fraction p of the sampled points resides within this radius. In Appendix A we show that

$$r_{\max} = \sqrt{2\sigma^2 \Gamma^{-1}\left(\frac{n}{2}, (1-p)\Gamma\left(\frac{n}{2}\right)\right)}, \tag{5}$$

Alternatively, we can start with a *predefined* radius $r_{\max}$ that defines the region that we would like to explain using a Gaussian distance kernel. This yields a σ^2 such that a fraction $p \in (0,1)$ of the sampled points resides within, i.e.,
$$\sigma^2 = \frac{r_{\max}^2}{2\Gamma^{-1}\left(\frac{n}{2}, (1-p)\Gamma\left(\frac{n}{2}\right)\right)}.$$
Using a truncated Gaussian distance kernel with these parameters enables us to generate samples that are distributed very closely to how samples in LIME are weighted, which enables us to fairly compare both techniques. An example of sampling using this distance kernel is shown in Fig. 3b.

5   Evaluation

In this section, we first revisit the first synthetic evaluation example introduced in Sect. 3. Next, to use a more realistic scenario, we compare LEMON and LIME on standardized UCI datasets and a variety of models. Source code for our experiments can be found here: https://github.com/iamDecode/lemon-evaluation.

5.1   Synthetic Scenario

In Sect. 3 we showed that the faithfulness of LIME is impaired for models trained on higher dimensional data (Fig. 2). We repeat this experiment with our LEMON sampling technique. We chose a truncated Gaussian kernel with the same σ as LIME, and an $r_{\max}$ computed using Eq. (5) with $p = 0.999$. This ensures we

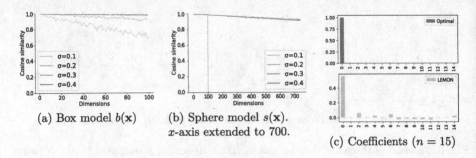

(a) Box model $b(\mathbf{x})$

(b) Sphere model $s(\mathbf{x})$.
x-axis extended to 700.

(c) Coefficients ($n = 15$)

Fig. 4. The same evaluation as performed on LIME shown in Fig. 2. (a,b) LEMON (5000 samples) remains faithful when explaining high dimensional models. Different lines represent different kernel widths σ. (c) the coefficients of the generated linear surrogate model (yellow) closely resemble the expected coefficients (blue). (Color figure online)

generate samples that are distributed very closely to how samples in LIME are weighted, such that we can fairly compare both techniques.

The results are shown in Fig. 4. In contrast to the results for LIME (Fig. 2), LEMON remains faithful to the reference model regardless of the dimensionality of the model. This is because more relevant samples are generated in the neighborhood of the point to be explained even in high dimensions. With more samples, the linear model is able to approximate the behavior of the reference model better than LIME. For a 15-dimensional box model, Fig. 4c shows the expected coefficients (blue) and coefficients of the linear surrogate from LEMON (green) are very close, as opposed to the coefficients of LIME shown in Fig. 2c. In addition, the results show smaller vertical fluctuations compared to LIME, indicating that the robustness of explanations from LEMON is affected less by variation in the transfer data.

5.2 Real-World Datasets

We used the Wine and Breast Cancer Wisconsin dataset from the UCI repository, and Diabetes dataset [13] which are ubiquitous in machine learning research. The datasets have a dimensionality of 13, 32, and 9 respectively, and contain only continuous features. For the reference models to be explained, we chose a Naive Bayes classifier, a Neural network with three layers of 100 neurons each, and a Random forest with 200 trees. As the kernel width may have a considerable impact on the explanation, we chose a wide range of kernel width parameter values $\sigma = 0.1, 0.3, 0.5, 1.0, 4.0$ and $\frac{3}{4}\sqrt{n}$. The latter is the default kernel width used in LIME, but is so large (> 1) that it can hardly be considered local. Next, we computed $r_{\max}$ for LEMON using Eq. (5) with $p = 0.999$.

To evaluate, it is not possible to directly compare the resulting surrogate model against a perfect surrogate model like we did for our synthetic scenario evaluation, because a perfect surrogate model for these classifiers is not known.

Instead, we chose to compute the Root Mean Square Error ($RMSE$) based on newly sampled evaluation data in the neighborhood of the point to be explained. For each data point, we generated $m = 50,000$ new samples in the area within radius $r_{\max}$ using Eq. (3) and an equivalent distance kernel to the ones used in LIME and LEMON. Next, we recorded the $RMSE$ between the predicted score of the reference $\hat{\mathbf{y}}^r$ and surrogate model $\hat{\mathbf{y}}^s$ for all m samples:

$$RMSE(\hat{\mathbf{y}}^r, \hat{\mathbf{y}}^s) = \sqrt{\frac{1}{m} \sum_{i=1}^{m} (\hat{\mathbf{y}}_i^r - \hat{\mathbf{y}}_i^s)^2}. \tag{6}$$

Note that due to the simple nature of the linear surrogate and complexity of the reference classifier, a perfect $RMSE = 0$ is implausible [15]. However, the metric does enable us to compare the relative faithfulness between LIME and LEMON. In Table 1 we show the mean $RMSE$ scores over all data points in the dataset.

Table 1. Average faithfulness scores ($RMSE$ on 50,000 samples, lower is better) of explanations generated for all instances in each of the 3 datasets, classified by 3 different ML models (Naive Bayes, Neural network and Random forest), using 6 different kernel width values. LEMON consistently achieves higher faithfulness compared to LIME.

	Kernel width (σ)	Naive Bayes		Neural network		Random forest	
		LIME	LEMON	LIME	LEMON	LIME	LEMON
Wine dataset	0.1	0.009	0.003	0.036	0.007	0.041	0.018
($n = 13$)	0.3	0.044	0.026	0.147	0.079	0.118	0.051
	0.5	0.103	0.071	0.283	0.143	0.186	0.082
	1.0	0.258	0.224	0.273	0.247	0.156	0.120
	$\frac{3}{4}\sqrt{n}$	0.652	0.303	0.543	0.271	0.376	0.124
	4.0	0.848	0.282	0.827	0.307	0.545	0.120
Diabetes dataset	0.1	0.018	0.016	0.017	0.015	0.072	0.036
($n = 9$)	0.3	0.057	0.031	0.051	0.026	0.141	0.053
	0.5	0.079	0.045	0.073	0.032	0.112	0.064
	1.0	0.120	0.110	0.068	0.063	0.104	0.088
	$\frac{3}{4}\sqrt{n}$	0.387	0.257	0.239	0.146	0.247	0.100
	4.0	0.686	0.349	0.452	0.192	0.419	0.096
Breast cancer dataset	0.1	0.011	0.006	0.222	0.102	0.038	0.015
($n = 32$)	0.3	0.052	0.030	0.401	0.208	0.103	0.038
	0.5	0.151	0.104	0.458	0.229	0.171	0.057
	1.0	0.490	0.263	0.585	0.312	0.265	0.072
	$\frac{3}{4}\sqrt{n}$	0.512	0.001	0.781	0.331	0.367	0.065
	4.0	0.504	0.002	1.162	0.305	0.358	0.065

These results show LEMON manages to consistently improve the faithfulness of the local surrogate model compared to LIME. This holds for each dataset, model and kernel width combination we have tested. On average, LEMON achieves 50.8% less *RMSE* compared to LIME. Next, we see that explanations generated with smaller kernel width tend to have a smaller *RMSE*. This is expected, because smaller regions naturally contain less intricate decision boundaries from the reference model, and smaller output gradients (e.g., the further we zoom in on a model, the better a linear model will fit its gradient).

There are a few exceptions, most notably the Naive Bayes classifier trained on the Breast cancer dataset. Here, the LEMON explanations get lower *RMSE* scores for very large kernel width values (> 1). While a smaller kernel width yields a faithful local surrogate, for larger kernel widths a linear surrogate may not be able to capture the complex behavior of the reference classifier. But if we increase the kernel width beyond the bounds of the original feature space (approximately $\sigma > 1$) the evaluation data points become out-of-distribution. In our example, the mean Euclidean distance of all Breast cancer training data to a point to be explained is 491.76, the mean Euclidean distance of all evaluation data with $\sigma = 1$ is 586.91 and $\sigma = \frac{3}{4}\sqrt{n}$ is 2409.37. The latter is almost five times larger than the training data. Hence, most predictions for evaluation data points are out-of-distribution model predictions, which yields unexpected results.

These (unrealistically) large kernel widths cause LIME to produce *RSME* scores exceeding 1 for certain dataset and model combinations (e.g., Neural network for the Breast cancer dataset). Smaller kernel width values should be chosen to ensure that LIME explanations remain faithful to the reference model.

The *RMSE* scores vary per dataset and per model, because both affect how much difference in predicted score (i.e., gradient) can be expected within the sampling region. For instance, in Naive Bayes models the predicted score changes smoothly for changes in the feature value. Hence, this model can be closely approximated with a linear model (especially for small kernel width values). The other two models are more complex, and hence cannot always be accurately approximated with a linear model (especially for larger kernel width values).

6 Discussion and Future Work

The LIME explanation framework includes an optional preceding feature selection step (using LASSO). One could argue that feature selection ahead of the explanation technique decreases the dimensionality, enabling LIME to be more suitable in higher dimensional space than we have shown in Sect. 3. However, the feature selection algorithm still needs to consider the full feature space in order to select features, which it cannot properly do without sufficient neighboring samples: it is subject to the same limitations. Hence, in our study we have disregarded the feature selection step in LIME as it makes evaluation and comparison more difficult. However, we expect similar results when including feature selection as part of both compared algorithms.

Supporting Observation-Based Sampling. Sampling with either a uniform or Gaussian distance kernel remains a *synthetic* approach: new samples are drawn regardless of the distribution of the original data. Thus, the surrogate model may be fitted using *out-of-distribution* data. To address this limitation, we cannot simply use a custom distance kernel in Eq. (3). The distance kernel is a kernel function applied to the distance r between a sample $\mathbf{x}_i$ and the instance to be explained $\mathbf{x}$, and r does not tell us enough about the location of that sample. Instead, we propose to find all original dataset samples $\{\mathbf{s} \in \mathcal{X} \mid \sqrt{(\mathbf{x} - \mathbf{s})^2} < r_{\max}\}$ within radius $r_{\max}$. Next, we approximate the density of these local samples with kernel density estimation (KDE) and sample points from the resulting estimated density function. This can be done by choosing a random point, and offsetting it by randomly drawn value from the KDE kernel function. This yields an alternative probability distribution on the ball of radius $r_{\max}$ around $\mathbf{x}$ to Eq. (3), but does not change the key idea behind LEMON.

Kernel Shape. Previous work and this paper assume that a spherical region around an instance is the best representation of a local neighborhood. However, some recent rule based techniques effectively use hyperboxes instead [23]. In addition, sampling towards the closest decision boundary may yield samples with a more salient gradient. It would be interesting to investigate what the relevance and effect is of the shape of the sampling region. Next, it would be interesting to see if we can extend our work to explain multiple instances at once by sampling from multiple distributions efficiently.

Measuring Faithfulness. We currently evaluated the explanations using faithfulness: the more closely the local surrogate model resembles the reference model, the better. However, there is no consensus on the best way to measure this. LIME itself calculates faithfulness based on the transfer data points the surrogate itself was trained on. This is problematic because, as we have shown in Sect. 3, LIME produces only few relevant samples in the neighborhood of the point to be explained. Hence, the surrogate model would be evaluated using few relevant samples, leading to misleading faithfulness scores. In our synthetic examples, we could circumvent this as the most optimal set of coefficients was known, and hence we use the cosine similarity between the most optimal coefficients and those from the local surrogate. However, in a realistic scenario, the most optimal coefficients are simply not known. For evaluating with real datasets (Sect. 5.2), we thus decided to use the $RMSE$ between the reference and surrogate model, computed on many (50,000) *newly* generated samples instead of the original transfer data.

As an alternative, we have considered the coefficient of determination (R^2) as a metric for faithfulness. This metric is used internally in LIME, and shows the proportion of the variance in the response variable of a regressor that can be explained by the predictor variables. However, we noted that for some (outlier) data points in our evaluation, almost all sampled data points get roughly the same predicted outcome from the reference classifier. In such case, the variance of the predicted outcomes is (very close to) 0. Computing the R^2 score with this data yields R^2 values of (close to) minus infinity, severely skewing the results.

Finally, faithfulness in itself does not guarantee the best possible explanation. There are many (often subjective) desiderata to consider when evaluating explanations, which are almost impossible to formalize due to their subjective nature. Hence, we do not claim to find an optimal explanation, just one closer to the behavior of the original model.

7   Conclusion

In this work, we explore alternative sampling techniques in pursuit of more faithful and robust explanations. To this end, we present LEMON: a sampling technique that outperforms current state-of-the-art techniques by sampling surrogate transfer data directly from the desired distribution instead of reweighting globally sampled transfer data. With both a synthetic evaluation, and evaluation with real-world datasets, we show that our sampling technique outperforms the state-of-the-art approaches in terms of faithfulness, measured in cosine similarity to the most optimal surrogate model, and $RMSE$ error between reference and surrogate model predictions respectively.

Acknowledgments. This work is part of the TEPAIV research project with project number 612.001.752, the NWO research project with project number 613.009.122, and the research programme Commit2Data, specifically the RATE Analytics project with project number 628.003.001, which are all financed by the Dutch Research Council (NWO).

A   Bounds on Gaussian Distance Kernel

Consider a point x and, for some $\sigma > 0$, equip every point at distance r from x with a weight given by the kernel

$$K(r) := \exp\big(-r^2/(2\sigma^2)\big). \tag{7}$$

We would like to find the radius of interest r_p such that the total weight of the points within distance r_p is at least a fraction p of the total weight. Since the surface of an n-dimensional ball is given by $c_n r^{d-1}$ for some dimension-dependent constant $c_n > 0$, we have to find the smallest r_p that satisfies the inequality

$$\frac{\int_0^{r_p} c_n r^{n-1} K(r) \mathrm{d}r}{\int_0^\infty c_n r^{n-1} K(r) \mathrm{d}r} \geq p \quad \Leftrightarrow \quad \int_0^{r_p} r^{n-1} K(r) \mathrm{d}r \geq p \int_0^\infty r^{n-1} K(r) \mathrm{d}r. \tag{8}$$

Rewriting the integrals,

$$(1-p) \int_0^\infty r^{n-1} K(r) \mathrm{d}r \,\&\geq \int_{r_p}^\infty r^{n-1} K(r) \mathrm{d}r. \tag{9}$$

Let $\Gamma(z,s) := \int_s^\infty t^{z-1}\exp(-t)\mathrm{d}t$ denote the incomplete gamma function and define the gamma function as $\Gamma(z) := \Gamma(z,0)$. Recall (7), so that the change of variables $t = r^2/(2\sigma^2)$ to both integrals in Eq. (9) yields

$$(1-p)\frac{1}{2}(2\sigma^2)^{\frac{n}{2}}\Gamma\left(\frac{n}{2}\right) \& \geq \frac{1}{2}(2\sigma^2)^{\frac{n}{2}}\Gamma\left(\frac{n}{2},\frac{r_p^2}{2\sigma^2}\right) \tag{10}$$

Writing $u \mapsto \Gamma^{-1}(\frac{n}{2},u)$, we have to find the smallest r_p such that

$$\Gamma^{(-1)}\left(\frac{n}{2},(1-p)\Gamma\left(\frac{n}{2}\right)\right) \leq \frac{r_p^2}{2\sigma^2}, \tag{11}$$

which is given by choosing

$$r_p = \sqrt{2\sigma^2\Gamma^{(-1)}\left(\frac{n}{2},(1-p)\Gamma\left(\frac{n}{2}\right)\right)} \quad \Leftrightarrow \quad \sigma^2 = \frac{r_p^2}{2\Gamma^{(-1)}\left(\frac{n}{2},(1-p)\Gamma\left(\frac{n}{2}\right)\right)}.$$

References

1. Abdollahi, B., Nasraoui, O.: Transparency in fair machine learning: the case of explainable recommender systems. In: Zhou, J., Chen, F. (eds.) Human and Machine Learning. HIS, pp. 21–35. Springer, Cham (2018). https://doi.org/10.1007/978-3-319-90403-0_2
2. Alvarez-Melis, D., Jaakkola, T.S.: On the robustness of interpretability methods. In: Workshop Human Interp. Mach. Learn., pp. 66–71 (2018)
3. Ba, J., Caruana, R.: Do deep nets really need to be deep? In: Adv. Neural Inf. Proc. Sys., pp. 2654–2662 (2014)
4. Baehrens, D., Schroeter, T., Harmeling, S., Kawanabe, M., Hansen, K., Müller, K.R.: How to explain individual classification decisions. J. Mach. Learn. Res. **11**, 1803–1831 (2010)
5. Barocas, S., Selbst, A.D.: Big data's disparate impact. California Law Rev. 671–732 (2016)
6. Bastani, O., Kim, C., Bastani, H.: Interpreting blackbox models via model extraction. arXiv preprint arXiv:1705.08504 (2017)
7. Buciluă, C., Caruana, R., Niculescu-Mizil, A.: Model compression. In: Int. Conf. Knowl. Discovery Data Mining, pp. 535–541. ACM SIGKDD (2006)
8. Buolamwini, J., Gebru, T.: Gender shades: intersectional accuracy disparities in commercial gender classification. In: Conf. Fairness, Accountability and Transparency, pp. 77–91. PMLR (2018)
9. Citron, D.K., Pasquale, F.D.: The scored society: due process for automated predictions. Wash. L. Rev. **89**, 1 (2014)
10. Craven, M., Shavlik, J.W.: Extracting tree-structured representations of trained networks. In: Adv. Neural Inf. Process. Sys., pp. 24–30 (1996)
11. Domingos, P.: Knowledge acquisition from examples via multiple models. In: Int. Conf. Machine Learn., pp. 98–106 (1997)
12. Edelman, B.G., Luca, M.: Digital discrimination: the case of airbnb.com (2014)
13. Efron, B., Hastie, T., Johnstone, I., Tibshirani, R.: Least angle regression (2004)

14. Fishman, G.: Monte Carlo: Concepts, Algorithms, and Applications. Springer, New York (2013). https://doi.org/10.1007/978-1-4757-2553-7
15. Garreau, D., Luxburg, U.: Explaining the explainer: a first theoretical analysis of lime. In: Int. Conf. AI Stat., pp. 1287–1296. PMLR (2020)
16. Gass, S.I., Fu, M.C. (eds.): Inverse transform method, p. 815. Springer (2013)
17. Harman, R., Lacko, V.: On decompositional algorithms for uniform sampling from n-spheres and n-balls. J. Multivar. Anal. **101**(10), 2297–2304 (2010)
18. Hinton, G., Vinyals, O., Dean, J.: Distilling the knowledge in a neural network. NIPS Deep Learning and Representation Learning Workshop. arXiv preprint arXiv:1503.02531 (2015)
19. Lopes, R.G., Fenu, S., Starner, T.: Data-free knowledge distillation for deep neural networks. NIPS Learn. Limited Labeled Data Workshop (LLD). arXiv preprint arXiv:1710.07535 (2017)
20. Muthukumar, V., Pedapati, T., Ratha, N., Sattigeri, P., Wu, C.W., Kingsbury, B.E.A.: Understanding unequal gender classification accuracy from face images. arXiv preprint arXiv:1812.00099 (2018)
21. Ribeiro, M.T., Singh, S., Guestrin, C.: Why should I trust you?: explaining the predictions of any classifier. In: Int. Conf. Knowl. Discovery Data Mining., pp. 1135–1144. ACM SIGKDD (2016)
22. Romero, A., Ballas, N., Kahou, S.E., Chassang, A., Gatta, C., Bengio, Y.: Fitnets: hints for thin deep nets. arXiv preprint arXiv:1412.6550 (2014)
23. Sanchez, I., Rocktaschel, T., Riedel, S., Singh, S.: Towards extracting faithful and descriptive representations of latent variable models. AAAI Spring Syposium Knowl. Represent. Reasoning **1**, 1–4 (2015)
24. Xu, Z., Hsu, Y.-C., Huang, J.: Training shallow and thin networks for acceleration via knowledge distillation with conditional adversarial networks. arXiv preprint arXiv:1709.00513 (2017)

GASTeN: Generative Adversarial Stress Test Networks

Luís Cunha[1,2]([✉]), Carlos Soares[1,2,3][iD], André Restivo[1,2][iD],
and Luís F. Teixeira[1,4][iD]

[1] Faculdade de Engenharia da Universidade do Porto, Porto, Portugal
{up201706736,csoares,arestivo,luisft}@fe.up.pt
[2] Laboratory for Artificial Intelligence and Computer Science (LIACC), Porto, Portugal
[3] Fraunhofer Portugal AICOS, Porto, Portugal
[4] INESC TEC, Porto, Portugal

Abstract. Concerns with the interpretability of ML models are growing as the technology is used in increasingly sensitive domains (*e.g.*, health and public administration). Synthetic data can be used to understand models better, for instance, if the examples are generated close to the frontier between classes. However, data augmentation techniques, such as Generative Adversarial Networks (GAN), have been mostly used to generate training data that leads to better models. We propose a variation of GANs that, given a model, generates realistic data that is classified with low confidence by a given classifier. The generated examples can be used in order to gain insights on the frontier between classes. We empirically evaluate our approach on two well-known image classification benchmark datasets, MNIST and Fashion MNIST. Results show that the approach is able to generate images that are closer to the frontier when compared to the original ones, but still realistic. Manual inspection confirms that some of those images are confusing even for humans.

Keywords: Global interpretability · Synthetic data · Generative adversarial networks · Responsible artificial intelligence

1 Introduction

As machine learning (ML) and artificial intelligence (AI) becomes widespread, techniques that enable their responsible usage are paramount, leading to a proliferation of the field known as responsible AI [2]. An example of these techniques are model cards [13]—a report documenting an ML model with information relevant to the user, such as the intended use case and evaluation metrics across different conditions.

As such, we believe information about the characteristics of data where the model is uncertain about its predictions should feature in model cards. For that purpose, we propose the generation of synthetic data that, while realistic, is

B. Crémilleux et al. (Eds.): IDA 2023, LNCS 13876, pp. 91–102, 2023.
https://doi.org/10.1007/978-3-031-30047-9_8

classified with low confidence by a given classifier. We address it with an approach based on generative adversarial networks (GANs) [7]. In GANs, a generator learns to sample realistic data by learning to fool a discriminator that is trained to distinguish real from fake samples. We employ the target classifier in the training process, using a measure of how far the generated data is to the frontier between the classes as an additional term for the generator's loss function. By identifying and analyzing examples that are in the frontier between classes, we hope to get a better understanding of the classifier. The proposed task has resemblances to generating adversarial attacks [9,18], *i.e.*, generating samples that are misclassified by a target classifier. However, our differs in two ways: firstly, we are looking for data that is classified with low confidence, instead of data that is misclassified; secondly, our aim is to generate data that can be useful to understand the model's behavior.

Since the most prominent success cases of GANs are in computer vision (CV), we empirically study the proposed approach on image classification using two popular benchmark datasets: MNIST [11] and Fashion MNIST [22]. Modern image classifiers are based on deep learning (DL), consisting of neural networks with convolutions as the fundamental building block. We apply our methodology to convolution neural network (CNN) based classifiers trained to have different predictive performances.

The contributions of this work are as follows:

1. A new approach for model interpretability that consists of generating data at the frontier between classes (*i.e.*, realistic data that is classified with low confidence by the model).
2. A variant of GANs that implements that approach, dubbed GASTeN.
3. An empirical validation of GASTeN.

2   Background and Related Work

2.1   Generative Adversarial Networks

Generative adversarial networks (GANs) [7] are a class of deep generative models that have been applied successfully to image generation tasks. The GAN framework is a two-player game between two neural networks, a generator (G) and a discriminator (D). The generator is a transformation that maps samples from a noise distribution $z \sim p_z$ to images that match the original data distribution p_d. The discriminator's role is to distinguish between real ($x \sim p_d$) and fake ($\hat{x} = G(z)$) samples, while the generator tries to fool it into classifying fake samples as real. Both networks are trained simultaneously. D is trained to maximize the probability of correctly distinguishing samples, and G to minimize that same probability. The original GAN formulation [7] is a *minimax* game described by the following equation:

$$\min_G \max_D \left[\mathbb{E}_{x \sim p_d} \left[log(D(x_{real})) \right] + \mathbb{E}_{z \sim p_z} \left[log(1 - D(G(z))) \right] \right]$$

Training GANs effectively presents some obstacles. An example of a challenge is that, as two neural networks are being updated simultaneously, it is not guaranteed that an equilibrium is reached and training does not converge. Another common scenario of GAN training failing is mode collapse [8], which occurs when G is not able to generate diverse images and maps different z values to the same output image.

Variants. To overcome challenges posed by GAN training, several variants have been proposed. The Deep Convolutional GAN (DCGAN) [15] introduces techniques and architectural guidelines that improve GAN training, such as using transposed convolutions in the generator and convolutions in the discriminator. Other variants, such as the Wasserstein GAN (WGAN) [1], introduce modifications to the loss functions used to train the networks. Variants that enable the conditional generation of samples have also been proposed [12,14].

Quantitative Evaluation. Another challenge of GANs is quantitatively evaluating the quality of the generated samples. The two most commonly used metrics in the literature are the Inception Score (IS) [16] and the Frechét Inception Distance (FID) [10]. Both metrics leverage an Inception image classifier pretrained on the ImageNet dataset [5]. The IS attributes the score based on two desired properties. Firstly, the quality of an image is reflected by how confidently the Inception network can classify it. Additionally, the set of generated images should be equally distributed across classes. The FID uses the Inception network to obtain embedded representations of the images. The Frechét Distance between the distribution of the representations of images in the original dataset and in the generated images is computed to obtain a measure of how realistic the generated images are. Compared to IS, FID is more sensitive to noise and able to detect scenarios where the same image is always generated for the same class [10]. However, it still has some drawbacks. For instance, as it relies on an Inception network pre-trained on the ImageNet dataset, it may fail to capture relevant features of images from datasets with different characteristics.

2.2 Adversarial Attacks

ML models have been found to be susceptible to adversarial attacks [9,18], *i.e.*, examples constructed to deceive the models. While our goal is to find examples that are classified with low confidence rather than misclassified, there are resemblances in both objectives, particularly in that both target a specific classifier.

The first approaches for generating adversarial attacks are perturbation-based, *i.e.*, consist of applying small perturbations to an image to produce an adversarial example that is misclassified by a target classifier [4,9,18]. The assumption that the adversarial example is a sample of the original class stems from the applied perturbation being small enough. Another class of attacks—unrestricted adversarial attacks [3]—lets go of the small perturbation constraint, and instead generates images from scratch. Approaches for generating unrestricted adversarial attacks using GANs have been proposed [6,17,19,20]. GANs have also been used to generate perturbation-based adversarial attacks [21,23].

3   GANs for Stress Testing

Our goal is to generate realistic data that is useful to understand classifiers and gain insights into their behavior. For that, we are interested in generating data that is close to the frontier between classes. To achieve that, we propose a GAN variation which we dub Generative Adversarial Stress Test Network (GASTeN). We use the GAN framework as the foundation to generate realistic images, and extend it with the target classifier (C). The classifier's output on the generated images is used as part of a new optimization objective for the generator. Our approach is similar to the technique for producing adversarial attacks proposed by Dunn *et al.* [6]. Both use the classifier's output to compute a new term of the function that the generator optimizes. However, the new term is combined with the generator loss of the original GAN differently. Figure 1 depicts the architecture of GASTeN.

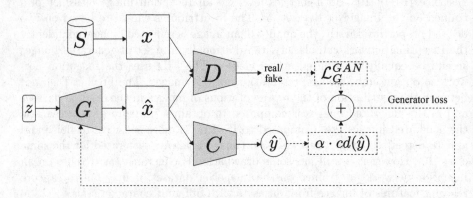

Fig. 1. Schematic overview of GASTeN.

The modifications introduced by GASTeN, not present in the classical GAN formulation, are the following:

Classifier (C): the classifier we wish to evaluate is introduced in the training loop. Its weights are not optimized and it is used only to classify the synthesized images $G(z)$. The outputs $C(G(z))$ are then used in the generator's loss function to guide its training. The usage of $C(G(z))$ to optimize G imposes the constraint that the classifier must be fully differentiable so that backpropagation can be applied. Thus, GASTeN is suitable for neural network-based classifiers, but not for some ML approaches that are non-differentiable.

Generator loss function: in addition to training G to deceive D, we introduce a new objective to the loss function of the generator. This term measures how far the generated images are from being classified as desired, and is a function of $C(G(z))$, which we dub *confusion distance* (*cd*). We factor the new term by a hyperparameter α and add it to the original generator loss function, resulting in Eq. 1. The hyperparameter α is a real-valued positive number.

$$\mathcal{L}_G^{GASTeN} = \mathcal{L}_G^{GAN} + \alpha \cdot cd(C(G(z))) \tag{1}$$

Since G's loss function is merely extended, GASTeN can, in theory, be used with any GAN variant as the base. As the role of D remains to distinguish between real and fake samples, no modification is introduced to its loss function.

Confusion distance: the *confusion distance* (*cd*) is a function of C's output. We focus on a binary classification setting in order to enable a better analysis of our algorithm's behavior. Thus, the output of C is a scalar between 0 and 1. A suitable confusion distance measures the distance from the classifier's prediction to the decision threshold. For a threshold of 0.5, the confusion distance results in $|C(G(z)) - 0.5|$. Despite focusing on non-conditional image generation and binary classification in this work, the algorithm is not tightly coupled to the proposed *cd* function. Different functions can be used to tackle other scenarios and objectives. For instance, the confusion distance could be inverted to look for images where the classifier has total confidence. Ultimately, GASTeN can be seen as a broad framework to generate images with properties that can be specified as a differentiable function.

In order to assess if it is more effective to adapt a generator that is already capable of generating realistic images than to train a generator from scratch using the modified loss function, we introduce a pre-train phase for GASTeN. The training process employed is now a two-step process. The first step consists of pre-training the chosen GAN architecture to learn a generator ($G_{original}$) that creates realistic images. In the second step, the classifier C is added to the process, and the generator $G_{original}$ and discriminator $D_{original}$ are further trained using the GASTeN loss function.

Since GASTeN uses a GAN architecture as its basis for image generation, using GASTeN requires the same careful selection of hyperparameters that is required when training GANs. Subtleties related to GAN training, such as difficulty to converge, also affect GASTeN. Additionally, GASTeN introduces two more hyperparameters, the weight α and the number of pre-train epochs.

4 Experimental Setup

4.1 Model Architectures

The used GAN follows the DCGAN [15] architecture, with the non-saturating GAN loss proposed by Goodfellow *et al.* [7]. Optimizer and training hyperparameters are set according to the original DCGAN work [15].

For classification, we use a CNN-based architecture. The architecture consists of two convolutional blocks but, despite its simplicity, can achieve low errors in the used datasets. Importantly, the classifiers are constructed using operations that are typical in DL-based image classifiers, such as convolutions and pooling. Each block contains a convolution and a max pooling operation. The first block uses nf number of filters in the convolutions, and the second uses $2 \cdot nf$ blocks. We vary nf in order to achieve classifiers with varying error rates, and thus study the behavior of our approach depending on the classifier quality. Classifiers with smaller nf values have less capacity.

4.2 Evaluation Metrics

To facilitate the analysis of the results obtained in the experiments, we use quantitative metrics for the two objectives that we aim to optimize, image quality and classifier confusion. To measure the quality of the generated samples, we resort to the FID. To evaluate the generator's ability to generate images that are classified with lower confidence by C, we measure the average value of the confusion distance. Henceforth, we refer to this metric as average confusion distance (ACD).

Additionally, we manually inspect generated samples by means of a visualization that includes information regarding the performance of C in the generated images. The used figure displays several examples in a manner that resembles a histogram. The visualization consists of ten groups of three columns, where the group in position i group contains images x such that $C(x) \in [(i-1)*0.1, i*0.1[$. The last group, in position 10, contains images such that $C(x) \in [0.9, 1]$.

4.3 Experiments

In order to test our approach, we perform experiments with two commonly used datasets in the CV literature: the MNIST [11] and the Fashion MNIST [22] datasets. As we are interested in binary classification problems, we derive binary subsets of the original datasets by selecting all images that belong to two classes.

Besides studying the behavior of our proposal against classifiers with different performances, we are interested in exploring the algorithm with different values for the hyperparameters. For α, we use values between 0 and 30 with an interval of 5. Using the value 0 allows direct comparison to the scenario where a regular GAN without a modified loss is trained. For the number of pre-train epochs we use values 0, 2, 5, and 1. All combinations of the mentioned hyperparameters are tested for each classifier. The source code for the experiments is available on GitHub[1].

---

[1] https://github.com/luispcunha/gasten.

5   Results and Discussion

5.1   Overview

We perform three runs of a fixed number of epochs for each combination of hyperparameters and classifier with different seed initializations. We analyze the algorithm by considering its performance after every epoch of the runs. Table 1 shows, for each classifier, the best FID such that, for the same G, ACD is less or equal to a given threshold. The thresholds used range from 0.5 to 0.1. Note that setting the threshold to 0.5 effectively removes the constraint since that is the maximum ACD value. So, the FID value in the column that refers to that threshold is the best FID achieved for the given dataset.

Table 1. Best FID obtained by a generator that has an ACD less or equal to a given threshold (values between 0.5 and 0.1 in the table header) for all datasets and classifiers. Results averaged over three runs. Bold entries highlight cases where the FID stays within a 100% increase for an ACD threshold smaller than the one obtained without modifications to the original GAN architecture.

	Dataset	$C.nf$	0.5	0.4	0.3	0.2	0.1
MNIST	7 v. 1	1	8.53 ± 0.12	$\mathbf{11.9 \pm 0.79}$	$78.5 \pm 17.$	$78.5 \pm 17.$	$106. \pm 20.$
		2	8.68 ± 0.11	40.0 ± 4.2	$101. \pm 14.$	$103. \pm 16.$	$127. \pm 27.$
		4	8.40 ± 0.084	88.6 ± 0.69	88.6 ± 0.69	$111. \pm 15.$	$123. \pm 3.4$
	8 v. 0	1	7.37 ± 0.47	7.37 ± 0.47	7.37 ± 0.47	$\mathbf{8.89 \pm 0.35}$	$74.6 \pm 28.$
		2	7.50 ± 0.53	$\mathbf{9.63 \pm 0.66}$	53.1 ± 3.7	53.1 ± 3.7	$136. \pm 17.$
		4	7.44 ± 0.48	$\mathbf{16.6 \pm 3.0}$	41.2 ± 3.0	41.2 ± 3.0	$97.2 \pm 51.$
	5 v. 3	1	6.68 ± 0.098	6.68 ± 0.098	6.68 ± 0.098	$\mathbf{8.10 \pm 0.36}$	26.1 ± 8.7
		2	6.63 ± 0.15	6.63 ± 0.15	$\mathbf{9.24 \pm 0.19}$	20.8 ± 0.90	$136. \pm 7.2$
		4	6.69 ± 0.14	$\mathbf{6.70 \pm 0.13}$	$\mathbf{18.0 \pm 2.1}$	25.8 ± 1.4	$136. \pm 5.7$
	9 v. 4	1	7.49 ± 0.036	7.49 ± 0.036	7.49 ± 0.036	$\mathbf{7.78 \pm 0.31}$	$\mathbf{20.0 \pm 0.65}$
		2	7.37 ± 0.38	7.37 ± 0.38	$\mathbf{8.12 \pm 0.36}$	26.7 ± 3.4	$123. \pm 14.$
		4	7.38 ± 0.12	$\mathbf{7.80 \pm 0.56}$	$\mathbf{22.9 \pm 1.1}$	30.9 ± 0.84	$163. \pm 10.$
FMNIST	Dress v. T-shirt/top	4	15.2 ± 0.23	$\mathbf{15.9 \pm 0.38}$	49.0 ± 7.3	60.8 ± 4.1	$128. \pm 4.7$
		8	15.3 ± 0.085	$\mathbf{16.0 \pm 0.57}$	46.3 ± 4.7	66.8 ± 5.2	$117. \pm 7.5$
		16	15.4 ± 0.075	$\mathbf{17.0 \pm 0.40}$	52.2 ± 1.7	$64.5 \pm 12.$	$133. \pm 2.3$
	Sneaker v. Sandal	4	16.2 ± 0.28	$\mathbf{17.4 \pm 0.95}$	56.9 ± 5.5	70.6 ± 5.6	$155. \pm 13.$
		8	16.2 ± 0.28	$\mathbf{19.6 \pm 1.3}$	65.4 ± 7.9	$113. \pm 9.2$	$153. \pm 13.$
		16	16.3 ± 0.22	$\mathbf{21.9 \pm 0.35}$	59.6 ± 2.3	$129. \pm 7.3$	$159. \pm 19.$

From Table 1, we note that confusing higher capacity classifiers requires images that are more distant from the original data distribution, thus, with higher FID values. There are, however, cases where we obtain higher FIDs when targetting worse classifiers. Examples of such exceptions are MNIST-8v0 with a threshold equal to or less than 0.3 and MNIST-7v1 with a threshold of 0.3 and 0.1.

The results also show that reducing ACD will inevitably lead to high FID values. Setting the threshold to 0.1, FID stays at values lower than 30 only for the MNIST-9v4 and MNIST-7v3 cases. That, however, happens only for the worst considered classifiers. Thus, it seems unlikely that it is always possible to obtain a G that almost always confuses the target classifier by generating images that, according to the FID measure, are realistic. Despite not achieving arbitrarily low ACD values while maintaining a satisfactory FID, there are cases where there is some decrease in ACD compared to the images generated without a modified G loss function. When compared to the same GAN architecture trained without the GASTeN modifications, there are cases where FID values stay within a 100% increase for an ACD threshold smaller than the ACD obtained without GASTeN modifications. Those cases are highlighted in bold in Table 1.

5.2   Visual Inspection

In addition to quantitatively evaluating our approach, we manually inspect generated images. Figure 2 shows samples obtained for digits 5 and 3. Some digits do not appear realistic, such as those highlighted by a red square with a dashed border. The examples highlighted in green (with a solid line square) are examples of digits that look plausible. The lower half of the digit looks like it could belong to either a 5 or a 3. However, the upper dash is not connected to the bottom part, making the image ambiguous (the digit could be either a three or five depending if the connecting stroke was drawn on the right or left, respectively). We argue that such examples could prove helpful in analyzing the classifier.

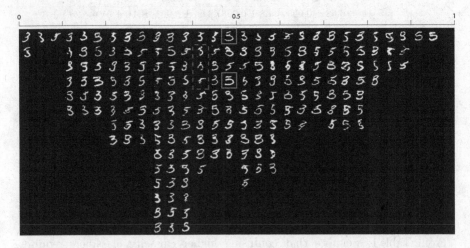

Fig. 2. Example of samples generated for the dataset consisting of digits 5 and 3.

Figure 3 depicts samples for digits 9 and 4. In this case, the goal of generating images that are closer to the border is achieved. Out of the 200 samples, none is classified between 0 and 0.1, and only one is classified between 0.9 and 1. Most generated images are classified between 0.5 and 0.6. Of the images that are confusing to the classifier, some are unrealistic and have noise-like marks disconnected from the actual digit (as the examples highlighted in a dashed red square). Others have characteristics that make them confusing, even for humans. The examples highlighted in green look like unfinished 9 s, *i.e.*, 9 s with the upper part disconnected, similar to hand-drawn 4 s.

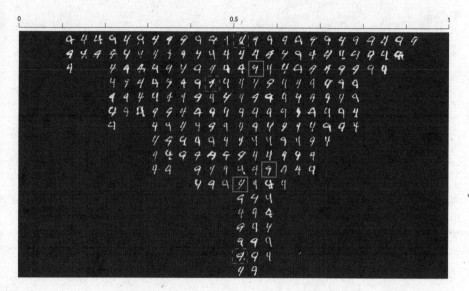

Fig. 3. Example of samples generated for the dataset consisting of digits 9 and 4.

5.3 Training Challenges

Despite being capable of achieving our goal, GASTeN exhibits erratic behavior during training against a classifier for digits 7 and 1. Figure 4 presents the evolution of the FID and ACD metrics during training. Unsurprisingly, the choice of α influences the algorithm's behavior. However, it does not allow for a fine-grained specification of the trade-off between image quality and classifier confusion. The influence for α values lower than 25 is subtle and is more noticeable for $\alpha = 30$. With $\alpha = 30$, the generator can confuse C by sacrificing the image's realness (the FID increases vastly), which is not aligned with our objectives. The algorithm behaves similarly among lower values of α: G ends up not being able to fool C, and the FID keeps improving, which means that the loss term responsible for image realness ends up dominating the other. For higher α values, such as 30, training frequently oscillates and does not converge. While during this oscillation interesting trade-offs can be achieved, as those reported previously, training does not converge to those trade-offs. This observation is in accordance with Dunn *et al.* [6], which point that the gradients for the realistic data objective and adversarial data objective may be conflicting. Similarly, the confusing data objective may conflict with the realistic data objective.

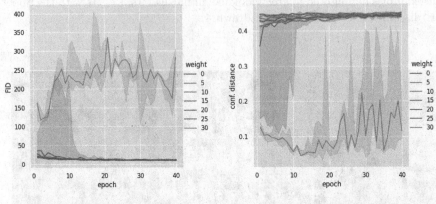

(a) FID evolution during training.   (b) ACD evolution during training.

Fig. 4. Metric evolution during GASTeN runs for different α weight parameters with 10 epochs of pre-train. Plotted values are the median over three runs, and the shadowed area ranges from the minimum and maximum values obtained.

5.4 Threats to Validity

We identify some limitations that may hinder the conclusions drawn from the empirical study of GASTeN. Due to time constraints, the number of hyperparameter combinations tested may not fully represent the behavior of GASTeN. Regarding the evaluation methodology, as mentioned in Sect. 2.1, the used FID metric has some limitations which can affect drawing conclusions from the quantitative evaluation. We tackle this limitation by inspecting the generated images. Despite that, visual inspection of images performed by us could be biased, and our evaluation could benefit from a human study to evaluate if the images generated by our approach are realistic and confusing.

6 Conclusions and Future Work

We propose the task of generating data that is realistic and lies in the frontier between classes for a given classifier, and a GAN-based approach for generating said data. Via an empirical study on the binary classification scenario using popular benchmark CV datasets, MNIST and Fashion MNIST, we demonstrate scenarios where GASTeN achieves our goal. Additionally, by visually inspecting generated samples, we show that GASTeN can synthesize images that look both plausible and ambiguous, even for humans. We also discuss the limitations and challenges of our approach. GASTeN is not always able to generate confusing data with satisfactory FIDs. Also, training frequently fails to converge or converges with the loss term related to data realism dominating the other.

For future work, we intend tackling the limitations of our proposal, for instance, by incorporating techniques from the approaches present in the adversarial attack literature. We also intend to scale the experiments to different GAN architectures, more complex datasets, different classifiers, the multi-class classification scenario, and different data types. Additionally, studying how to use the generated data for understanding the classifiers and contribute to model cards creation is required. Several directions can be pursued, such as using interpretability techniques.

Acknowledgements. This work was partially funded by projects AISym4Med (101095387) supported by Horizon Europe Cluster 1: Health, ConnectedHealth (n.º 46858), supported by Competitiveness and Internationalisation Operational Programme (POCI) and Lisbon Regional Operational Programme (LISBOA 2020), under the PORTUGAL 2020 Partnership Agreement, through the European Regional Development Fund (ERDF) and NextGenAI - Center for Responsible AI (2022-C05i0102-02), supported by IAPMEI, and also by FCT plurianual funding for 2020–2023 of LIACC (UIDB/00027/2020_UIDP/00027/2020).

References

1. Arjovsky, M., Chintala, S., Bottou, L.: Wasserstein generative adversarial networks, vol. 70, pp. 214–223. PMLR (2017). https://proceedings.mlr.press/v70/arjovsky17a.html

2. Barredo Arrieta, A., et al.: Explainable artificial intelligence (XAI): concepts, taxonomies, opportunities and challenges toward responsible AI. Inf. Fusion **58**, 82–115 (2020). https://doi.org/10.1016/j.inffus.2019.12.012

3. Brown, T.B., Carlini, N., Zhang, C., Olsson, C., Christiano, P., Goodfellow, I.: Unrestricted adversarial examples (2018). https://doi.org/10.48550/ARXIV.1809.08352

4. Carlini, N., Wagner, D.: Towards evaluating the robustness of neural networks. In: Proceedings of the IEEE Symposium on Security and Privacy, pp. 39–57 (2016). https://doi.org/10.48550/arxiv.1608.04644

5. Deng, J., Dong, W., Socher, R., Li, L.J., Li, K., Fei-Fei, L.: ImageNet: a large-scale hierarchical image database, pp. 248–255 (2010). https://doi.org/10.1109/CVPR.2009.5206848

6. Dunn, I., Pouget, H., Melham, T., Kroening, D.: Adaptive generation of unrestricted adversarial inputs (2019). https://doi.org/10.48550/arxiv.1905.02463

7. Goodfellow, I., et al.: Generative adversarial nets. In: Ghahramani, Z., Welling, M., Cortes, C., Lawrence, N., Weinberger, K. (eds.) Advances in Neural Information Processing Systems, vol. 27. Curran Associates, Inc. (2014). https://proceedings.neurips.cc/paper/2014/file/5ca3e9b122f61f8f06494c97b1afccf3-Paper.pdf

8. Goodfellow, I.J.: NIPS 2016 tutorial: generative adversarial networks. CoRR abs/1701.00160 (2017). https://arxiv.org/abs/1701.00160

9. Goodfellow, I.J., Shlens, J., Szegedy, C.: Explaining and harnessing adversarial examples. In: 3rd International Conference on Learning Representations, ICLR 2015 - Conference Track Proceedings (2014). https://doi.org/10.48550/arxiv.1412. 6572. https://arxiv.org/abs/1412.6572v3

10. Heusel, M., Ramsauer, H., Unterthiner, T., Nessler, B., Hochreiter, S.: GANs trained by a two time-scale update rule converge to a local Nash equilibrium, vol. 30. Curran Associates, Inc. (2017). https://proceedings.neurips.cc/paper/2017/ file/8a1d694707eb0fefe65871369074926d-Paper.pdf

11. Lecun, Y., Bottou, L., Bengio, Y., Haffner, P.: Gradient-based learning applied to document recognition. Proc. IEEE **86**(11), 2278–2324 (1998). https://doi.org/10. 1109/5.726791

12. Mirza, M., Osindero, S.: Conditional generative adversarial nets. CoRR abs/1411.1784 (2014). https://arxiv.org/abs/1411.1784

13. Mitchell, M., et al.: Model cards for model reporting. In: Proceedings of the Conference on Fairness, Accountability, and Transparency, FAT* 2019, pp. 220–229. Association for Computing Machinery, New York (2019). https://doi.org/10.1145/ 3287560.3287596

14. Odena, A., Olah, C., Shlens, J.: Conditional image synthesis with auxiliary classifier GANs, vol. 70, pp. 2642–2651. PMLR (2017). https://proceedings.mlr.press/v70/ odena17a.html

15. Radford, A., Metz, L., Chintala, S.: Unsupervised representation learning with deep convolutional generative adversarial networks. In: 4th International Conference on Learning Representations, ICLR 2016 - Conference Track Proceedings (2015). https://doi.org/10.48550/arxiv.1511.06434

16. Salimans, T., et al.: Improved techniques for training GANs, vol. 29. Curran Associates, Inc. (2016). https://proceedings.neurips.cc/paper/2016/file/ 8a3363abe792db2d8761d6403605aeb7-Paper.pdf

17. Song, Y., Shu, R., Kushman, N., Ermon, S.: Constructing unrestricted adversarial examples with generative models, vol. 31. Curran Associates, Inc. (2018). https:// proceedings.neurips.cc/paper/2018/file/8cea559c47e4fbdb73b23e0223d04e79- Paper.pdf

18. Szegedy, C., et al.: Intriguing properties of neural networks. In: 2nd International Conference on Learning Representations, ICLR 2014 - Conference Track Proceedings (2013). https://doi.org/10.48550/arxiv.1312.6199. https://arxiv.org/ abs/1312.6199v4

19. Tao, X., Hangcheng, L., Shangwei, G., Yan, G., Xiaofeng, L.: EGM: an efficient generative model for unrestricted adversarial examples. ACM Trans. Sens. Netw. (TOSN) (2021). https://doi.org/10.1145/3511893

20. Wang, X., He, K., Hopcroft, J.E.: AT-GAN: a generative attack model for adversarial transferring on generative adversarial nets. CoRR abs/1904.07793 (2019). https://arxiv.org/abs/1904.07793

21. Xiao, C., Li, B., Zhu, J.Y., He, W., Liu, M., Song, D.: Generating adversarial examples with adversarial networks. In: IJCAI International Joint Conference on Artificial Intelligence 2018-July, pp. 3905–3911 (2018). https://doi.org/10.48550/ arxiv.1801.02610

22. Xiao, H., Rasul, K., Vollgraf, R.: Fashion-MNIST: a novel image dataset for benchmarking machine learning algorithms. CoRR abs/1708.07747 (2017). https://arxiv. org/abs/1708.07747

23. Zhao, Z., Dua, D., Singh, S.: Generating natural adversarial examples. In: 6th International Conference on Learning Representations, ICLR 2018 - Conference Track Proceedings (2017). https://doi.org/10.48550/arxiv.1710.11342

Learning Permutation-Invariant Embeddings for Description Logic Concepts

Caglar Demir[✉] and Axel-Cyrille Ngonga Ngomo

Data Science Research Group, Paderborn University, Paderborn, Germany
{caglar.demir,axel.ngonga}@upb.de

Abstract. Concept learning deals with learning description logic concepts from a background knowledge and input examples. The goal is to learn a concept that covers all positive examples, while not covering any negative examples. This non-trivial task is often formulated as a search problem within an infinite quasi-ordered concept space. Although state-of-the-art models have been successfully applied to tackle this problem, their large-scale applications have been severely hindered due to their excessive exploration incurring impractical runtimes. Here, we propose a remedy for this limitation. We reformulate the learning problem as a multi-label classification problem and propose a neural embedding model (NERO) that learns permutation-invariant embeddings for sets of examples tailored towards predicting F_1 scores of pre-selected description logic concepts. By ranking such concepts in descending order of predicted scores, a possible goal concept can be detected within few retrieval operations, i.e., no excessive exploration. Importantly, top-ranked concepts can be used to start the search procedure of state-of-the-art symbolic models in multiple advantageous regions of a concept space, rather than starting it in the most general concept ⊤. Our experiments on 5 benchmark datasets with 770 learning problems firmly suggest that NERO significantly (p-value < 1%) outperforms the state-of-the-art models in terms of F_1 score, the number of explored concepts, and the total runtime. We provide an open-source implementation of our approach (https://github.com/dice-group/Nero).

Keywords: Description Logics · Concept Learning · Permutation Invariance

1 Introduction

Deep learning based models have been effectively applied to tackle various graph-related problems, including question answering, link prediction [19,30]. Yet, their predictions are not human-interpretable and confined within a fixed set vocabulary terms [9,11]. In contrast, Description Logics (DLs) provide means to derive human-interpretable inference in an infinite setting [1,16,27]. Deriving explanations for DLs concepts has been long understood [5]. For instance, explanations

© The Author(s), under exclusive license to Springer Nature Switzerland AG 2023
B. Crémilleux et al. (Eds.): IDA 2023, LNCS 13876, pp. 103–115, 2023.
https://doi.org/10.1007/978-3-031-30047-9_9

can be derived by using the subsumption hierarchy as a sequence of binary classifiers in a fashion akin to following a path in decision tree [4,35]. Utilizing DLs is considered as a possible backbone for explainable Artificial Intelligence (AI) [33]. Although DLs have become standard techniques to formalize Knowledge Base (KB) [19,20,29], the highly incomplete nature of KBs and impractical runtimes of symbolic models have been a challenge for fulfilling its potential. State-of-the-art Concept Learning (CL) models have been successfully applied to learn DL concepts from a KB and input examples [24,27]. Yet, their practical applications have been severely hindered by their impractical runtimes. This limitation stems from the reliance of myopic heuristic function that often incurs excessive exploration of concepts [16,22,35]. A DL concept is explored by retrieving its individuals and calculating its quality w.r.t. input KB and examples (see Sect. 2). As the size of an input KB grows, excessive exploration has been a computational bottleneck in practical applications. Here, we propose a remedy for this limitation. We reformulate the learning problem as a multi-label classification problem and propose NERO–a neural permutation-invariant embedding model. Given a set of positive examples E^+ and a set of negative examples E^-, NERO predicts F_1 scores of pre-selected DL concepts as shown in Fig. 1.

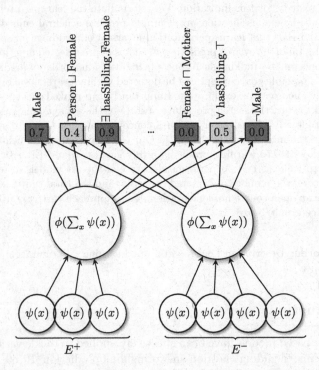

Fig. 1. Visualization of NERO. Boxes and values denote the pre-selected unique DL concepts and their predicted F_1 scores, respectively.

By ranking pre-selected DL concepts in descending order of predicted scores, a goal concept can be found by only exploring few top-ranked concepts. Importantly, top-ranked concepts can be used to initialize the standard search procedure of state-of-the-art models, if a goal concept is not found. By this, a state-of-the-art CL model is endowed with the capability of starting the search in more advantageous states, instead of starting it in the most general concept ⊤. Our experiments on 5 benchmark datasets with 770 learning problems indicate that NERO **significantly** (p-value < 1%) outperforms the state-of-the-art models in standard metrics such as F_1 score, the number of explored concepts, and the total runtime. Importantly, equipping NERO with a state-of-the-art model (CELOE) further improves F_1 scores on benchmark datasets with a low runtime cost. The results of Wilcoxon signed rank tests confirm that the superior performance of NERO is significant. We provide an open-source implementation of NERO, including pre-trained models, evaluation scripts as well as a web service.[1]

2 Background

Knowledge Base: A Knowledge Base (KB) is a pair $\mathcal{K} = (Tbox, Abox)$, where $Tbox$ is a set of terminological axioms describing relations between named concepts N_C [33]. A terminological axiom is in the form of $A \sqsubseteq B$ or $A \equiv B$ s.t. $A, B \in N_C$. $Abox$ is a set of assertions describing relationships among individuals $a, b \in N_I$ via roles $r \in N_R$ as well as concept membership relationships between N_I and N_C. Every assertion in $Abox$ must in the form of $A(x)$ and $r(x, y)$, where $A \in N_C$, $r \in N_R$, and $x, y \in N_I$. An example is visualized in Fig. 2.

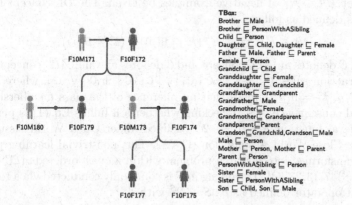

TBox:
Brother ⊑ Male
Brother ⊑ PersonWithASibling
Child ⊑ Person
Daughter ⊑ Child, Daughter ⊑ Female
Father ⊑ Male, Father ⊑ Parent
Female ⊑ Person
Grandchild ⊑ Child
Granddaughter ⊑ Female
Granddaughter ⊑ Grandchild
Grandfather ⊑ Grandparent
Grandfather ⊑ Male
Grandmother ⊑ Female
Grandmother ⊑ Grandparent
Grandparent ⊑ Parent
Grandson ⊑ Grandchild, Grandson ⊑ Male
Male ⊑ Person
Mother ⊑ Person, Mother ⊑ Parent
Parent ⊑ Person
PersonWithASibling ⊑ Person
Sister ⊑ Female
Sister ⊑ PersonWithASibling
Son ⊑ Child, Son ⊑ Male

Fig. 2. A visualization of Family KB with $Tbox$ and a subset of $Abox$. Colors denote concept assertions, while (·) and branching from (·) denote role assertions, respectively. (Color figure online)

Description Logics: Description Logics (DLs) are fragments of first-order predicate logic using only unary and binary predicates. The unary predicates, the binary predicates and constants are called concepts, roles and individuals, respectively [1]. DL have become standard techniques to formalize background knowledge for many application domains including Semantic Web [20,29]. Leveraging KBs defined over DLs has a potential of being a backbone for explainable AI [33]. Here, we consider KBs in the DL $\mathcal{ALC}$ (Attributive Language with Complements) [1] as in many other works (see Sect. 3). The model-theoretic semantics of $\mathcal{ALC}$ are given in Table 1.

Table 1. $\mathcal{ALC}$ syntax and semantics. $\mathcal{I}$ stands for an interpretation, $\Delta^{\mathcal{I}}$ for its domain.

Construct	Syntax	Semantics
Atomic concept	A	$A^{\mathcal{I}} \subseteq \Delta^{\mathcal{I}}$
Role	r	$r^{\mathcal{I}} \subseteq \Delta^{\mathcal{I}} \times \Delta^{\mathcal{I}}$
Top concept	$\top$	$\Delta^{\mathcal{I}}$
Bottom concept	$\bot$	$\emptyset$
Conjunction	$C \sqcap D$	$C^{\mathcal{I}} \cap D^{\mathcal{I}}$
Disjunction	$C \sqcup D$	$C^{\mathcal{I}} \cup D^{\mathcal{I}}$
Negation	$\neg C$	$\Delta^{\mathcal{I}} \setminus C^{\mathcal{I}}$
Existential restriction	$\exists\, r.C$	$\{x \mid \exists\, y.(x,y) \in r^{\mathcal{I}} \text{ and } y \in C^{\mathcal{I}}\}$
Universal restriction	$\forall\, r.C$	$\{x \mid \forall\, y.(x,y) \in r^{\mathcal{I}} \text{ implies } y \in C^{\mathcal{I}}\}$

Concept Learning: Let $\mathcal{K}$ over $\mathcal{ALC}$, the set $E^{+} \subset N_I$ of positive examples, and the set $E^{-} \subset N_I$ of negative examples be given. The DL concept learning problem is defined as follows

$$\forall p \in E^{+}, \forall n \in E^{-}\big(\mathcal{K} \models \mathtt{H}(p)\big) \wedge \big(\mathcal{K} \not\models \mathtt{H}(n)\big), \tag{1}$$

where $\mathtt{H} \in \mathcal{C}$ denotes an $\mathcal{ALC}$ concept and $\mathcal{C}$ denotes all valid $\mathcal{ALC}$ concepts under the construction rules: C:: $= \mathtt{A} \mid \neg \mathtt{C} \mid \mathtt{C} \sqcap \mathtt{C} \mid \mathtt{C} \sqcup \mathtt{C} \mid \exists r.\mathtt{C} \mid \forall r.\mathtt{C} \mid$, where $\mathtt{A} \in N_C$ and $r \in N_R$. $\mathcal{K} \models \mathtt{H}(p)$ implies that an inference of the class membership $\mathtt{H}(p)$ is a logical consequence of $\mathcal{K}$. Checking whether a $\mathtt{H}$ fulfills Eq. (1) is performed by a retrieval function $\mathcal{R} : \mathcal{C} \rightarrow 2^{N_I}$ defined under Open World Assumption (OWA) or Close World Assumption (CWA). This non-trivial learning problem is often transformed into a search problem within a quasi-ordered $\mathcal{ALC}$ concept space $(\mathcal{S}, \preceq)$ [7,13,27,34]. Traversing in $\mathcal{S}$ is commonly conducted via a top-down refinement operator defined as $\rho : \mathcal{S} \rightarrow 2^{\mathcal{S}}$ with

$$\forall \mathtt{A} \in \mathcal{S} : \rho(\mathtt{A}) \subseteq \{\mathtt{B} \in \mathcal{S} \mid \mathtt{B} \preceq \mathtt{A}\}. \tag{2}$$

State-of-the-art CL models begin their search towards a $\mathtt{H}$, after a search tree is initialized with the most general DL concept $(\top)$ as a root node. This search tree is iteratively built by selecting a node containing a quasi-ordered DL concept with the highest heuristic value and adding its qualifying refinements as its children into a search tree [27].

Heuristics: A heuristic function is the key to an efficient search in $\mathcal{S}$ towards a H [26]. The number of explored concepts and runtimes are used as proxy for the efficiency. Various heuristic functions have been investigated [26, 35]. Most heuristic functions of state-of-the-art models can be considered as myopic functions favoring syntactically short and accurate concepts. Hence, they are prone to stuck in a local optimum [35]. For instance, the heuristic function of CELOE is defined as

$$\phi_{\text{CELOE}}(A, B) = Q(B) + \lambda \cdot [Q(B) - Q(A)] - \beta \cdot |B|, \tag{3}$$

where $A \in \mathcal{S}$, $B \in \rho(A)$. $\beta > \lambda \geq 0$ and $Q(\cdot)$ denotes a quality function (e.g. F_1 score or accuracy). Through $Q(\cdot)$ and $| \cdot |$, the search is steered based on solely A and B towards more accurate and syntactically shorter concepts. $F_1(\cdot)$ is defined as

$$F_1(A) = \frac{| E^+ \cap \mathcal{R}(A) |}{| E^+ \cap \mathcal{R}(A) | + 0.5(| E^- \cap \mathcal{R}(A) | + | E^+ \setminus \mathcal{R}(A) |)}. \tag{4}$$

As the size of KB grows, runtimes of performing retrieval operations $\mathcal{R}(\cdot)$ increase [3, 4, 25]. Consequently, traversing in $\mathcal{S}$ becomes a computational bottleneck. Therefore, reducing the number of explored concepts plays an important role to tackle to tackle CL on KBs. Although state-of-the-art models (e.g. CELOE) apply redundancy elimination and expression simplification rules to reduce the number of explored concepts, impractical long runtimes of state-of-the-art models still prohibit large-scale applications [17]. Moreover, the selected assumption underlying $\mathcal{R}()$ also plays a role to tackle CL on large KBs. Due to the incomplete nature of KBs, OWA seems to be a more suitable assumption [31]. Yet, Using OWA often makes membership queries computationally more challenging [12, 26]. Consequently, CWA is often adopted in many recent works [16, 22, 34].

3 Related Work

A plethora of works have investigated learning DLs concepts from a KB and input examples. We refer to [1, 18, 23] for an introduction. Most symbolic systems differ in the usage of heuristic functions and the design of the refinement operators [2, 7, 12, 14, 21, 26, 27, 34]. DL-Learner [24] is regarded as the most mature and recent system for CL [32]. DL-Learner consists of several state-of-the-art models, including ELTL, OCEL, and CELOE. ELTL is based on a refinement operator for the DL $\mathcal{EL}$ and uses a heuristic function that favors syntactically short concepts. CELOE builds on OCEL and ELTL and it applies a more sophisticated heuristic function. CELOE is currently the best CL model available within DL-Learner and often outperforms many state-of-the-art models including OCEL and ELTL in terms of the quality of learned expression, number explored concepts, and runtimes [28, 35]. The aforementioned approaches apply redundancy elimination and expression simplification rules to reduce the number of explored concepts. Although applying redundancy elimination and expression simplification rules often reduce the number of explored concepts, these operations introduces more computation and long runtimes still prohibit large-scale applications [17]. Most recent works have focused on treating the impractical runtimes in CL. CLIP [22] is a neural approach that serves as an addition to refinement-based approaches and supports pruning the search space by predicting the length of a possible goal state. EvoLearner [16] represents a concept as an abstract syntax tree corresponding an individual of an evolutionary algorithm. The initial population of individuals is obtained via biased random walks originating from E^+. Westphal et al. [35] design a Simulated Annealing based meta-heuristic to balance the exploration-exploitation trade-off during the search process. In this work, we mainly evaluate NERo against CELOE provided in DL-Learner for two reasons: (1) DL-Learner is regarded as the most mature and recent system for CL [32] and (2) most recently developed models are often evaluated w.r.t. the quality of concepts as well as runtimes. Yet, not reporting the number of explored concepts does not permit us to quantify whether a possible improvement through NERo may stem from our novel idea or our efficient implementation. Consequently, in our experiments, we mainly compare NERo against CELOE in terms of number of explored concepts, quality of learned concepts as well as runtimes.

4   Methodology

Motivation: The goal in the CL problem is to find a DL concept $H \in \mathcal{C}$ maximizing Eq. (4). Here, we are interested in achieving this goal by learning permutation-invariant embeddings tailored towards predicting F_1 scores of pre-selected concepts. Through exploring top-ranked concepts at first, we aim to find a goal concept can only with few retrieval operations. If a goal state is not found within top-ranked concepts, the search tree of a state-of-the-art CL model can be initialized with top-ranked concepts and $\top$ concept along with corresponding heuristic values. By this, the standard search procedure can be started in more advantageous states, than the most general concept $\top$.

Approach: Equation (4) indicates that $F_1(\cdot)$ is invariant to the order of individuals in E^+, E^-, and $\mathcal{R}(\cdot)$. Previously, Zaheer et al. [36] have proven that all functions being invariant to the order in inputs can be decomposed into

$$f(\mathbf{x}) = \phi\Big(\sum_{x \in \mathbf{x}} \psi(x)\Big),$$ (5)

where $\mathbf{x} = \{x_1, \ldots, x_m\} \in 2^{\mathcal{X}}$ and $\phi(\cdot)$ and $\psi(\cdot)$ denote a set of input and two parameterized continuous functions, respectively. A permutation-invariant neural network defined via Eq. (5) still abides by the universal approximation theorem [36]. We conjecture that such neural network can learn permutation-invariant embeddings for sets of individuals (e.g. E^+ and E^-) tailored towards predicting F_1 scores of pre-selected concepts. Through accurately predicting F_1

scores of pre-selected DL concepts, possible goal concepts from pre-selected concepts can be detected without using $F_1(\cdot)$ and $\mathcal{R}(\cdot)$. With these considerations, we define NERO as follows

$$\text{NERO}(E^+, E^-) = \sigma\left(\phi\left(\sum_{x\in E^+} \psi(x)\right) - \phi\left(\sum_{x\in E^-} \psi(x)\right)\right), \qquad (6)$$

where $\psi(\cdot) : N_I \to \mathbb{R}^m$ and $\phi : \mathbb{R}^m \to [0,1]^{|\mathcal{T}|}$ denote an embedding look-up operation and an affine transformation, respectively. $\mathcal{T}$ represents the pre-selected DL concepts. The result of the translation operation denoted with $\mathbf{z} \in \mathbb{R}^m$ is normalized via the logistic sigmoid function $\sigma(\mathbf{z}) = \frac{1}{1+\exp(-\mathbf{z})}$. Hence, $\text{NERO} : 2^{N_I} \times 2^{N_I} \mapsto [0,1]^{|\mathcal{T}|}$ can be seen as a mapping from two sets of individuals to $|\mathcal{T}|$ unit intervals. NERO can be seen as a multi-task learning approach that leverages the similarity between multi-tasks, where a task in our case corresponds to accurately predicting the F_1 score of a pre-selected DL concept [8].

The importance of learning representations tailored towards *related* tasks has been well investigated [8,15]. Motivated by this, we elucidate the process of selecting DL concepts in Algorithm 1. We select such concepts that their canonical interpretations do not fully overlap (see the 4th line). As shown therein, NERO can be trained on knowledge base defined over any DLs provided that $\mathcal{R}(\cdot)$ and $\rho(\cdot)$ are given.

Algorithm 1. Constructing target DL concepts

Input: $\mathcal{R}(\cdot)$, $\rho(\cdot)$, d, maxlength **Output:** $\mathcal{T}$

1: $\mathcal{T} := \{\mathtt{C} \mid \mathtt{C} \in \rho(\top) \wedge |\mathtt{C}| \leq \text{maxlength} \wedge 0 < |\mathcal{R}(\mathtt{C})|\}$
2: **for** each $\mathtt{A} \in \mathcal{T}$ **do**
3:   **for** each $\mathtt{B} \in \mathcal{T}$ **do**
4:     **if** $\mathcal{R}(\mathtt{A}) \neq \mathcal{R}(\mathtt{B})$ **then**
5:       **for** each $\mathtt{X} \in \{\mathtt{A} \sqcap \mathtt{B}, \mathtt{A} \sqcup \mathtt{B}\}$ **do**
6:         **if** $|\mathcal{R}(\mathtt{X})| > 0 \wedge \mathcal{R}(\mathtt{X}) \notin \{\mathcal{R}(\mathtt{E}) \mid \mathtt{E} \in \mathcal{T}\}$ **then**
7:           Add $\mathtt{X}$ to $\mathcal{T}$.
8:         **end if**
9:         **if** $|\mathcal{T}| = d$ **then**
10:          **return** $\mathcal{T}$
11:         **end if**
12:       **end for**
13:     **end if**
14:   **end for**
15: **end for**
16: **if** $|\mathcal{T}| < d$ **then**
17:   Go to the step (2).
18: **end if**

Training Process: Let $\mathcal{D} = \{(E_i^+, E_i^-, \mathbf{y}_i)\}_{i=1}^N$ represent a training dataset, where a data point $(E^+, E^-, \mathbf{y})$ is obtained in four consecutive steps: (i) Sample C from $\mathcal{T}$ uniformly at random, (ii) Sample k individuals $E^+ \subset \mathcal{R}(C)$ uniformly at random, (iii) Sample k individuals $E^- \subset N_I \setminus E^+$ uniformly at random, and (iv) Compute F_1 scores $\mathbf{y}$ via Eq. (4) w.r.t. E^+, E^-, for $\mathcal{T}$. For a given $(E^+, E^-, \mathbf{y})$ and predictions $\hat{\mathbf{y}} := \text{NERO}(E^+, E^-)$, an incurred binary cross entropy loss. Important to note that after training process, permutation-invariant embeddings of any $\mathcal{ALC}$DL concepts can be readily obtained omitting the translation operation in NERO, e.g. embeddings of a DL concept (e.g. Male $\sqcap \exists$hasSibling.Female) can be obtained via $\phi\left(\sum_{x \in \mathcal{R}(\text{Male} \sqcap \exists \text{hasSibling.Female})} \psi(x)\right)$. In our project page, we provided a 2D visualization of learned embeddings for the Family KB.

5   Experiments

We based our experimental setup on [6,7,26] and used learning problems provided therein. An overview of the datasets is provided in Table 2. To perform extensive comparisons between models, additional learning problems are generated by randomly sampling E^+ and E^-. We ensured that none of the learning problems used in our evaluation has been used in the unsupervised training phase. In our experiments, we evaluated all models in $\mathcal{ALC}$ for Class Expression Learning (CEL) on the same hardware.

Table 2. An overview of class expression learning benchmark datasets.

Dataset	$|N_I|$	$|N_C|$	$|N_R|$
Family	202	18	4
Carcinogenesis	22372	142	21
Mutagenesis	14145	86	11
Biopax	323	28	49
Lymphography	148	49	1

We evaluated models via the F_1 score, the runtime and number of explored concepts. The F_1 score is used to measure the quality of the concepts found w.r.t. positive and negative examples, while the runtime and the number of explored concepts are to measure the efficiency. We measured the full computation time including the time spent prepossessing time of the input data and tackling the learning problem. Moreover, we used two standard stopping criteria for state-of-the-art models. (i) We set the maximum runtime to 10 s although models often reach good solutions within 1.5 s [27]. (ii) The models are configured to terminate as soon as they found a goal concept. In our experiments, we evaluate all models in $\mathcal{ALC}$ for CL on the same hardware. During training, we set $|\mathcal{T}| = 1000$, $N = 50$ and used Adam optimizer for NERO. We only considered top-100 ranked concepts to evaluate NERO.

6   Results

Results with Benchmark Learning Problems: Table 3 reports the concept learning results with benchmark learning problems. Table 3 suggests that equipping NeRo with the standard search procedure improves the state-of-the-art performance in terms of F_1 scores even further with a small cost of runtimes. CELOE and ELTL require at least **14.7**× more time than NeRo to find accurate concepts on Family. This stems from the fact that NeRo explores on average only 21 concepts, whereas CELOE explored 1429. On Mutagenesis and Carcinogenesis, NeRo finds more accurate concepts, while exploring less, hence, achieving better runtime performance. Runtime gains stem from the fact that NeRo explores at least **2.3**× fewer concepts.

Table 3. Results on benchmark learning problems. F_1, T, and Exp. denote F_1 score, total runtime in seconds, and the number of explored concepts, respectively. NeRo† denotes equipping NeRo with CELOE. ELTL does not report the Exp.

Dataset	NeRo†			NeRo			CELOE			ELTL	
	F_1	T	Exp.	F_1	T	Exp.	F_1	T	Exp.	F_1	T
Family	.987	.83	26	.984	.68	21	.980	4.65	1429	.964	4.12
Mutagenesis	.714	17.30	200	.704	13.18	100	.704	23.05	516	.704	21.04
Carcinogenesis	.725	32.23	200	.720	26.26	100	.714	37.18	230	.719	36.29

Important to note we did not use parallelism in NeRo and we reload parameters of NeRo for each single learning problem. To conduct more extensive evaluation, we generated total 750 random learning problems on five benchmark datasets. Since Lymphography and Biopax datasets do not contain any learning problems, they are not included in Table 3.

Results with Random Learning Problems: Table 4 reports the concept learning results with random learning problems. Table 4 suggests that CELOE explores at least **3.19**× more concepts than NeRo. Importantly, NeRo finds on-par or more accurate concepts, while exploring less. Here, we load the parameters of NeRo only once per dataset and are used to tackle learning problems sequentially. This resulted in reducing the total computation time of NeRo by **3 − 6**× on Family, Mutagenesis and Carcinogenesis benchmark datasets. Although NeRo can tackle learning problems in parallel (e.g. through multiprocessing), we did not use any parallelism, since CELOE and ELTL do not abide by parallelism [7]. Loading the learning problems in a standard mini-batch fashion and using multi-GPUs may further improve the runtimes of NeRo. These results suggest that NeRo can be more suitable than CELOE and ELTL on applications requiring low latency.

Table 4. Random learning problems with different sizes per benchmark dataset. Each row reports the mean and standard deviations attained in 50 learning problems. $|E|$ denotes $|E^+| + |E^-|$.

Dataset	$|E|$	NeRo			CELOE			ELTL	
		F_1	T	Exp.	F_1	T	Exp.	F_1	T
Family	10	.913 ± .06	.16 ± .51	74 ± 43	.903 ± .06	11.61 ± 3.58	5581 ± 2375	.718 ± .01	4.45 ± 2.84
	20	.807 ± .04	.16 ± .49	100 ± 00	.795 ± .05	13.28 ± 1.47	7586 ± 645	.678 ± .02	3.59 ± 1.27
	30	.775 ± .03	.15 ± .41	100 ± 00	.760 ± .03	13.24 ± 1.42	7671 ± 575	.672 ± .01	3.46 ± 1.59
Lymphography	10	.968 ± .07	.12 ± .43	75 ± 41	.968 ± .07	6.63 ± 4.29	5546 ± 5169	.733 ± .09	3.07 ± .30
	20	.828 ± .04	.13 ± .40	100 ± 00	.826 ± .05	13.01 ± 1.23	11910 ± 1813	.678 ± .02	3.08 ± .50
	30	.780 ± .04	.13 ± .01	100 ± 00	.780 ± .04	13.02 ± 1.69	13138 ± 2601	.672 ± .01	3.09 ± .72
Biopax	10	.859 ± .08	.19 ± .71	86 ± 34	.806 ± .07	13.26 ± 1.94	4752 ± 2153	.685 ± .06	3.71 ± .10
	20	.793 ± .05	.19 ± .52	100 ± 00	.746 ± .04	13.63 ± .10	4151 ± 748	.668 ± .06	3.72 ± .10
	30	.749 ± .03	.18 ± .52	100 ± 00	.718 ± .02	13.91 ± .44	3843 ± 963	.668 ± .06	3.90 ± .22
Mutagenesis	10	777 ± .05	3.47 ± 1.61	100 ± 00	.753 ± .06	20.27 ± 1.39	546 ± 613	.670 ± .02	10.29 ± .40
	20	.746 ± .05	3.09 ± 1.75	100 ± 00	.712 ± .02	20.38 ± 1.30	430 ± 28	.667 ± .00	10.73 ± 1.10
	30	.721 ± .03	2.89 ± 1.60	100 ± 00	.700 ± .02	20.39 ± 1.06	429 ± 38	.667 ± .00	11.74 ± .97
Carcinogenesis	10	.768 ± .06	5.39 ± 2.98	98 ± 14	.764 ± .06	29.90 ± 1.02	401 ± 125	.673 ± .05	19.99 ± .67
	20	.722 ± .03	5.40 ± 1.87	100 ± 00	.713 ± .02	30.30 ± .19	318 ± 152	.667 ± .00	20.00 ± 1.11
	30	.704 ± .05	4.70 ± 2.78	100 ± 00	.697 ± .02	29.99 ± .58	319 ± 43	.667 ± .00	20.38 ± .85

Table 5. Performance comparison with different number of explored concepts. Each row reports the mean and standard deviations attained in 50 learning problems.

Dataset	$|E|$	NeRo-1		NeRo-10		NeRo-1000	
		F_1	T	F_1	T	F_1	T
Family	10	.906 ± .07	.08 ± .06	.910 ± .05	.09 ± .06	**.916 ± .06**	.81 ± .50
	20	.793 ± .05	.08 ± .05	.806 ± .04	.09 ± .05	**.807 ± .04**	1.17 ± .50
	30	.742 ± .05	.08 ± .05	.773 ± .03	.09 ± .05	**.775 ± .03**	1.15 ± .50
Lymphography	10	.882 ± .07	.08 ± .06	.905 ± .05	.08 ± .06	**.916 ± .06**	.77 ± .50
	20	.793 ± .05	.07 ± .05	.827 ± .04	.08 ± .05	**.828 ± .04**	1.03 ± .50
	30	.738 ± .05	.07 ± .06	.777 ± .04	.08 ± .05	**.780 ± .03**	1.00 ± .60
Biopax	10	.853 ± .08	.09 ± .06	.856 ± .05	.97 ± .59	**.868 ± .08**	1.31 ± .80
	20	.779 ± .05	.09 ± .06	.791 ± .04	.10 ± .62	**.793 ± .04**	1.35 ± .60
	30	.708 ± .07	.09 ± .06	.742 ± .04	.10 ± .63	**.749 ± .03**	1.39 ± .60
Mutagenesis	10	.733 ± .07	**.32 ± 2.03**	.785 ± .06	.57 ± 1.81	**.803 ± .06**	36.98 ± 5.81
	20	.689 ± .08	**.31 ± 1.99**	.734 ± .05	.52 ± 1.82	**.751 ± .04**	34.29 ± 5.91
	30	.673 ± .08	**.31 ± 2.03**	.712 ± .04	.49 ± 1.77	**.728 ± .03**	32.88 ± 6.13
Carcinogenesis	10	.717 ± .09	**.41 ± 2.53**	.740 ± .09	.89 ± 2.89	**.783 ± .02**	56.941 ± 9.55
	20	.680 ± .06	**.40 ± 2.49**	.707 ± .05	.82 ± 2.45	**.731 ± .02**	57.205 ± 5.08
	30	.610 ± .11	**.41 ± 2.83**	.671 ± .06	.77 ± 2.66	**.716 ± .02**	52.872 ± 7.58

Results with Limited Exploration: Table 5 reports concept learning results with limited exploration on five benchmark datasets. Table 5 suggests that NeRo-10 often outperforms CELOE and ELTL (see Table 4) in all metrics even when exploring solely 10 top-ranked concepts.

Significance Testing: To validate the significance of our results, we performed Wilcoxon signed-rank tests (one and two-sided) on F_1 scores, runtimes and the number of explored concepts. Our null hypothesis was that the performances of NERO and CELOE come from the same distribution. We were able to reject the null hypothesis with a p-value $< 1\%$ across all the datasets, hence, the superior performance of NERO is statistically significant.

6.1 Discussion

Our results uphold our hypothesis: F_1 scores of DL concepts can be accurately predicted by means of learning permutation-invariant embeddings for sets of individuals. Through considering top-ranked DL concepts at first, the need of excessive number of retrieval operations to find a goal concept can be mitigated. Throughout our experiments, NERO consistently outperforms state-of-the-art models w.r.t. the F_1 score, the number of explored concepts and the total computational time. Importantly, starting the standard search procedure on these top-ranked concepts further improves the results. Hence, NERO can be applied within state-of-the-art models to decrease their runtimes. However, it is important to note that Lehmann et al. [26] have previously proved the completeness of CELOE in the CL problem, i.e., for a given learning problem, CELOE finds a goal expression if it exists provided that there are no upper-bounds on the time and memory requirements. Although these requirements are simply not practical, equipping NERO with the search procedure of CELOE is necessary to achieve the completeness in CL.

7 Conclusion

We introduced a permutation-invariant neural embedding model (NERO) to efficiently tackle the description logic concept learning problem. For given learning problem, NERO accurately predicts F_1 scores of pre-selected description logic concepts in a multi-label classification fashion. Through ranking concepts in descending order of predicted F_1 scores, a goal concept can be learned within few retrieval operations. Our experiments showed that NERO outperforms state-of the art models in 770 concept learning problems on 5 benchmark datasets w.r.t. the quality of predictions, number of explored concepts and the total computational time. Equipping NERO with the standard search procedure further improves the F_1 scores across learning problems and benchmark datasets.

We believe that incorporating neural models in concept learning problems is worth pursuing further. In future, we will work on using NERO on more expressive description logics and integrating embeddings for concepts in non-myopic heuristics [10].

Acknowledgments. This work has been supported by the European Union's Horizon Europe research and innovation programme (GA No 101070305), by the Ministry of Culture and Science of North Rhine-Westphalia within the project SAIL (GA No NW21-059D), and the Deutsche Forschungsgemeinschaft (GA No TRR 318/1 2021 - 438445824).

References

1. Baader, F., Calvanese, D., McGuinness, D., Patel-Schneider, P., Nardi, D., et al.: The Description Logic Handbook: Theory, Implementation and Applications. Cambridge University Press, Cambridge (2003)
2. Badea, L., Nienhuys-Cheng, S.H.: A refinement operator for description logics. In: ILP (2000)
3. Bin, S., Bühmann, L., Lehmann, J., Ngomo, A.C.N.: Towards SPARQL-based induction for large-scale RDF data sets. In: ECAI (2016)
4. Bin, S., Westphal, P., Lehmann, J., Ngonga, A.: Implementing scalable structured machine learning for big data in the sake project. In: 2017 IEEE International Conference on Big Data (Big Data), pp. 1400–1407. IEEE (2017)
5. Borgida, A., Franconi, E., Horrocks, I., McGuinness, D.L., Patel-Schneider, P.F.: Explaining ALC subsumption. In: ECAI, pp. 209–213 (2000)
6. Bühmann, L., Lehmann, J., Westphal, P.: DL-learner-a framework for inductive learning on the semantic web. J. Web Semant. **39**, 15–24 (2016)
7. Bühmann, L., Lehmann, J., Westphal, P., Bin, S.: DL-learner structured machine learning on semantic web data. In: Companion Proceedings of the Web Conference 2018, WWW 2018, Republic and Canton of Geneva, Switzerland, pp. 467–471. International World Wide Web Conferences Steering Committee (2018)
8. Caruana, R.: Multitask Learning. Springer, Heidelberg (1998)
9. Demir, C., Moussallem, D., Heindorf, S., Ngomo, A.C.N.: Convolutional hypercomplex embeddings for link prediction. In: Asian Conference on Machine Learning, pp. 656–671. PMLR (2021)
10. Demir, C., Ngomo, A.: Drill-deep reinforcement learning for refinement operators in ALC. CoRR abs/2106.15373 223, 224 (2021)
11. Dettmers, T., Minervini, P., Stenetorp, P., Riedel, S.: Convolutional 2D knowledge graph embeddings. In: AAAI (2018)
12. Fanizzi, N., d'Amato, C., Esposito, F.: DL-FOIL concept learning in description logics. In: Železný, F., Lavrač, N. (eds.) ILP 2008. LNCS (LNAI), vol. 5194, pp. 107–121. Springer, Heidelberg (2008). https://doi.org/10.1007/978-3-540-85928-4_12
13. Fanizzi, N., Rizzo, G., d'Amato, C.: Boosting DL concept learners. In: Hitzler, P., et al. (eds.) ESWC 2019. LNCS, vol. 11503, pp. 68–83. Springer, Cham (2019). https://doi.org/10.1007/978-3-030-21348-0_5
14. Fanizzi, N., Rizzo, G., d'Amato, C., Esposito, F.: DLFoil: class expression learning revisited. In: Faron Zucker, C., Ghidini, C., Napoli, A., Toussaint, Y. (eds.) EKAW 2018. LNCS (LNAI), vol. 11313, pp. 98–113. Springer, Cham (2018). https://doi.org/10.1007/978-3-030-03667-6_7
15. Goller, C., Kuchler, A.: Learning task-dependent distributed representations by backpropagation through structure. In: Proceedings of International Conference on Neural Networks (ICNN 1996), vol. 1, pp. 347–352. IEEE (1996)
16. Heindorf, S., et al.: EvoLearner: learning description logics with evolutionary algorithms. In: WWW. ACM (2022)
17. Hitzler, P., Bianchi, F., Ebrahimi, M., Sarker, M.K.: Neural-symbolic integration and the semantic web. Semant. Web **11**(1), 3–11 (2020)
18. Hitzler, P., Krotzsch, M., Rudolph, S.: Foundations of semantic web technologies (2009)
19. Hogan, A., et al.: Knowledge graphs. ACM Comput. Surv. (CSUR) **54**(4), 1–37 (2021)

20. Horrocks, I., Patel-Schneider, P.F., Van Harmelen, F.: From SHIQ and RDF to OWL: the making of a web ontology language. J. Web Semant. **1**(1), 7–26 (2003)
21. Iannone, L., Palmisano, I., Fanizzi, N.: An algorithm based on counterfactuals for concept learning in the semantic web. Appl. Intell. **26**(2), 139–159 (2007)
22. Kouagou, N.J., Heindorf, S., Demir, C., Ngomo, A.C.N.: Learning concept lengths accelerates concept learning in ALC. In: Groth, P., et al. (eds.) ESWC 2022. LNCS, vol. 13261, pp. 236–252. Springer, Cham (2022). https://doi.org/10.1007/978-3-031-06981-9_14
23. Krötzsch, M., Simancik, F., Horrocks, I.: A description logic primer. arXiv preprint arXiv:1201.4089 (2012)
24. Lehmann, J.: DL-learner: learning concepts in description logics. J. Mach. Learn. Res. **10**, 2639–2642 (2009)
25. Lehmann, J.: Learning OWL Class Expressions, vol. 22. IOS Press (2010)
26. Lehmann, J., Auer, S., Bühmann, L., Tramp, S.: Class expression learning for ontology engineering. J. Web Semant. **9**(1), 71–81 (2011)
27. Lehmann, J., Hitzler, P.: Concept learning in description logics using refinement operators. Mach. Learn. **78**(1–2), 203 (2010)
28. Lehmann, J., et al.: DL-learner manual (2016)
29. Michel, F., Turhan, A.-Y., Zarrieß, B.: Efficient TBox reasoning with value restrictions—introducing the $\mathcal{FL}_ower$ reasoner. In: Fodor, P., Montali, M., Calvanese, D., Roman, D. (eds.) RuleML+RR 2019. LNCS, vol. 11784, pp. 128–143. Springer, Cham (2019). https://doi.org/10.1007/978-3-030-31095-0_9
30. Nickel, M., Murphy, K., Tresp, V., Gabrilovich, E.: A review of relational machine learning for knowledge graphs. Proc. IEEE **104**(1), 11–33 (2015)
31. Rudolph, S.: Foundations of description logics. In: Polleres, A., et al. (eds.) Reasoning Web 2011. LNCS, vol. 6848, pp. 76–136. Springer, Heidelberg (2011). https://doi.org/10.1007/978-3-642-23032-5_2
32. Sarker, M.K., Hitzler, P.: Efficient concept induction for description logics. In: Proceedings of the AAAI Conference on Artificial Intelligence, vol. 33, pp. 3036–3043 (2019)
33. Schockaert, S., Ibanez-Garcia, Y., Gutierrez-Basulto, V.: A description logic for analogical reasoning. In: Proceedings of IJCAI 2021, pp. 2040–2046 (2021)
34. Tran, A.C., Dietrich, J., Guesgen, H.W., Marsland, S.: Parallel symmetric class expression learning. J. Mach. Learn. Res. **18**, 64:1–64:34 (2017)
35. Westphal, P., Vahdati, S., Lehmann, J.: A simulated annealing meta-heuristic for concept learning in description logics. In: Katzouris, N., Artikis, A. (eds.) ILP 2021. LNCS, vol. 13191, pp. 266–281. Springer, Cham (2022). https://doi.org/10.1007/978-3-030-97454-1_19
36. Zaheer, M., Kottur, S., Ravanbakhsh, S., Poczos, B., Salakhutdinov, R.R., Smola, A.J.: Deep sets. In: Advances in Neural Information Processing Systems, vol. 30. Curran Associates, Inc. (2017)

Diffusion Transport Alignment

Andrés F. Duque[1], Guy Wolf[2], and Kevin R. Moon[1(✉)]

[1] Utah State University, Logan, UT, USA
kevin.moon@usu.edu.com
[2] Université de Montréal, Mila - Quebec AI Institute, Montréal, Canada
guy.wolf@umontreal.ca

Abstract. The integration of multimodal data presents a challenge in cases where the study of a given phenomena by different instruments or conditions generates distinct but related domains. Many existing data integration methods assume a known one-to-one correspondence between domains of the entire dataset, which may be unrealistic. Furthermore, existing manifold alignment methods are not suited for cases where the data contains domain-specific regions, i.e., there is not a counterpart for a certain portion of the data in the other domain. We propose Diffusion Transport Alignment (DTA), a semi-supervised manifold alignment method that exploits prior knowledge of between only a few points to align the domains. After building a diffusion process, DTA finds a transportation plan between data measured from two heterogeneous domains with different feature spaces, which by assumption, share a similar geometrical structure coming from the same underlying data generating process. DTA can also compute a partial alignment in a data-driven fashion, resulting in accurate alignments when some data are measured in only one domain. We empirically demonstrate that DTA outperforms other methods in aligning multiview data in this semi-supervised setting. We also show that the alignment obtained by DTA can improve the performance of machine learning tasks, such as domain adaptation, inter-domain feature mapping, and exploratory data analysis, while outperforming competing methods.

Keywords: Manifold alignment · Semi-supervised learning · Manifold learning

1 Introduction

In many data science applications, data may be collected from different measurement instruments, conditions, or protocols of the same underlying system. Examples include single cell RNA sequence and ATAC sequence measurements of the same group of cells [30], text documents translated into different languages [24], brain images from multiple neuroimaging techniques [33], and images of a scene captured from different views [17]. In such settings, researchers are often interested in integrating data from the different domains to enhance our understanding of the system as well as the relationships between the different domains.

B. Crémilleux et al. (Eds.): IDA 2023, LNCS 13876, pp. 116–129, 2023.
https://doi.org/10.1007/978-3-031-30047-9_10

Integrating the data may also lead to improved downstream analysis, such as classification, if there is domain-specific information about the task.

Multi-view data integration is usually performed assuming knowledge of one-to-one correspondences, i.e., the data comes in a paired fashion between domains. One of the simplest methods for this setting is Canonical Correlation Analysis (CCA), a linear approach that finds a projection that maximizes the correlation between the two domains [31]. Kernel CCA extends this to nonlinear projections via the kernel trick [5,13]. Alternating diffusion [18] and integrated diffusion [19] are nonlinear alignment methods based on the robust manifold learning algorithm Diffusion Maps [8]. For an overview of other approaches see [14,21].

A popular way to integrate distinct domains is manifold alignment. First introduced in the seminal works [15] and [16], this family of methods seeks to find projections of the multiple domains into a common latent space where inter-domain relationships can be captured. Manifold alignment can be performed in various scenarios, depending on how much information is provided about the correspondence between different domains. The edge case, usually referred to as *unsupervised manifold alignment*, arises in the absence of any relationship known a priori between the domains as in [3,4,11,12,29,35]. Some of the data integration approaches described previously, such as CCA, may be viewed as belonging to the opposite edge case of *supervised manifold alignment*.

In contrast, other problems can be categorized as *semi-supervised manifold alignment*, where some degree of correspondence between domains is assumed to be known. In some cases, a one-to-one correspondence is known for only a few of the data points. This is the case in [16], which uses the Laplacian eigenmaps loss function in both domains while penalizing mismatches of known correspondences in the embedding. In [34], the authors first learn a latent representation for each domain using a variation of Laplacian eigenmaps [2]. Then, they use Procrustes analysis in the common embedding space to find a transformation that aligns the matching observations, which subsequently is applied to the rest of the data. Similarly, the approach proposed in [20] finds a low dimensional embedding generated by diffusion maps [8] and then performs an affine transformation to align the known correspondences. More recently, a generative adversarial network called manifold alignment GAN (MAGAN) was introduced in [1]. MAGAN is based on a similar architecture as cycleGAN [38], which learns functions that map from one domain to another. However, the authors of MAGAN showed that cycleGAN and similar approaches tend to superimpose rather than align the data manifolds, resulting in incorrect alignments between distinct groups. To mitigate this issue, MAGAN incorporates a correspondence loss between the known correspondences enforcing a consistent alignment.

Alternatively, the correspondence information may be available at the feature level. MAGAN can be applied to this case with a correspondence loss imposed on the shared features. Other approaches use class labels in both domains as the correspondence knowledge, as in [36] where the labels act as anchors points for the alignment. This was further expanded to a kernelized version in [32].

In this work we focus on the semi-supervised problem where we assume a known one-to-one correspondence between domains is available for a few of the data points. Our method, called Diffusion Transport Alignment (DTA), starts by building a diffusion process [8] that connects measurements in different domains via the known correspondences. In this fashion, DTA transforms both domains to a shared feature space, allowing us to extract inter-domain distances. Finally, DTA solves a partial optimal transport problem to determine a coupling between data samples from one domain and their counterparts in the other domain. The obtained coupling can be further used to improve the performance of downstream analysis. For instance, one may be interested in learning a mapping between both domains, but the known correspondences are insufficient to successfully train a regression model. Another use-case is to perform unsupervised multi-domain analysis with methods as in [22] or [18], which require one-to-one correspondences between all points in all domains. DTA is also useful for domain adaptation, where a model is trained on a source domain and then applied to a target domain.

In summary, our contributions are as follows: 1) We develop a manifold alignment method, DTA, that outperforms current methods in recovering inter-domain relationships. 2) DTA can perform a data-driven partial alignment when a subset of the data is domain-specific, preventing spurious couplings between domains. 3) We demonstrate how DTA can leverage limited correspondence knowledge to improve the performance in other tasks, such as regression and domain adaptation.

2   Diffusion Transport Alignment

Consider a multi-domain data collection of a data generating process where two different views in potentially different feature spaces $\Phi_1 \in R^{n \times q}$ and $\Phi_2 \in R^{m \times p}$ are measured, containing observations $\{x_i\}_{i=1}^n$, and $\{y_i\}_{i=1}^m$, respectively. We wish to learn a correspondence between both domains in a semi-supervised setting, where one-to-one correspondence is known for a set of observations denoted by $\mathcal{C}$. That is, for each $c \in \mathcal{C}$ we have access to its features in both domains.

As a motivating example, consider a classification problem where both domains contain labeled data points for some shared classes. The two domains may contain distinct information that is relevant for classification. An example of this is in single cell data with both RNA-sequencing and ATAC-sequencing measurements. In this case, training on the aligned data will lead to improved performance compared with training on the domains separately. As another example, researchers may be interested in the relationships between variables measured in separate domains. Aligning the domains enables a larger dataset to obtain more accurate estimates of relationship measures such as the correlation coefficient or mutual information.

The fundamental idea of DTA consists of learning a diffusion process in each particular domain, and then leverage the known correspondences as anchor points to find a common feature representation. Ultimately, this allows us to extract an inter-domain distance measure, providing a dissimilarity among the

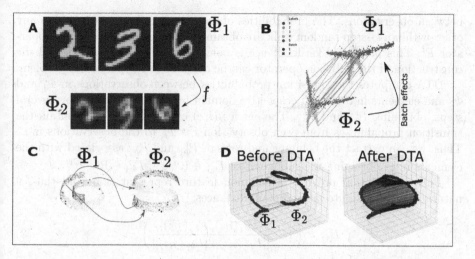

Fig. 1. Motivating examples for DTA. In all of these examples we have data measured in two distinct domains Φ_1 and Φ_2, and we possess a small subset of matching observations $\mathcal{C}$. This corresponds to the scenario where obtaining corresponding measurements may be costly, e.g. via expert annotation. The goal of DTA is to leverage the small subset of known correspondences to align the remaining observations. **A) Distorted MNIST digits.** Here Φ_1 consists of the original MNIST digits, while Φ_2 consists of distorted images after applying multiple transformations: rotation, downscaling, and Gaussian blurring. To learn a parametric function that maps from one domain to the other, the small set of correspondences is not enough. Thus, we need to find a greater set of matching data. **B) Splatter simulation with batch effects** [37]. A common problem when dealing with biological data is the distortion produced by the measurement protocols, introducing what is known as batch effects. Accurate alignment would overcome theses batch effects. **C) Swiss roll and S curve.** This case presents the ideal scenario where the two domains are a smooth mapping from a common latent space. Black points indicate correspondences with three of them (red arrows) highlighted. **D) Two helixes.** Here we use a dataset from [32] and display the effect of DTA after leveraging the known correspondences to align both manifolds.

observations in both domains. The diffusion operators over each domain, denoted as P_{Φ_1} and P_{Φ_2}, are built by a standard approach. First, we compute an affinity matrix with an α-decay kernel [27]:

$$K_{k,\alpha}(x_i, x_j) = \frac{1}{2}\exp\left(-\frac{||x_i - x_j||^\alpha}{\sigma_k^\alpha(x_i)}\right) + \frac{1}{2}\exp\left(-\frac{||x_i - x_j||^\alpha}{\sigma_k^\alpha(x_j)}\right), \quad (1)$$

where $\sigma_k(x_i)$ is the k-nearest neighbor distance of x_i and $\alpha > 0$. This kernel has two hyper-parameters α and k, which provide a trade-off between connectivity in the graph and local geometry preservation. Methods that employ this kernel are typically robust to the choice of these hyper-parameters [27]. The diffusion operator P is then computed by row-normalizing the kernel matrix. In this way P can be viewed as a probability transition matrix, representing a Markov chain between observations. The probabilities of transitioning from one point to any other within a $t-$step random walk are obtained by powering the diffusion operator P^t. This particular kernel choice is not required for our method, and the construction of the diffusion operator can be adapted to the particular problem.

DTA computes the transition probabilities between observations in Φ_1 and Φ_2 and elements in $\mathcal{C}$ in their respective domain by diffusing the process several steps, obtaining $P^t_{\Phi_1}$ and $P^t_{\Phi_2}$. The entries (i,c) of $P^t_{\Phi_k}$ with $c \in \mathcal{C}$ contain the transition probabilities from each observation $i \in \Phi_k$ to the observations in $\mathcal{C}$. Thus, we can extract the columns and rows of $P^t_{\Phi_1}$ and $P^t_{\Phi_2}$ associated with the elements in $\mathcal{C}$, obtaining the submatrices: $\Gamma_{\Phi_1} \in \mathrm{R}^{n \times |\mathcal{C}|}$, $\Gamma_{\Phi_2} \in \mathrm{R}^{m \times |\mathcal{C}|}$.

This construction provides a common feature representation, and thus, a natural way to compute inter-domain distances:

$$D_{ij} = \left(1 - \frac{\langle \Gamma_{\Phi_1}(i,:), \Gamma_{\Phi_2}(j,:) \rangle}{||\Gamma_{\Phi_1}(i,:)|| ||\Gamma_{\Phi_2}(j,:)||} \right). \tag{2}$$

We resort to cosine over euclidean distances since it resulted in superior performance.

The matrix D contains inter-domain distances, but does not provide a direct alignment of the domains. The final step in DTA is to solve a partial optimal transport problem with D as the cost matrix:

$$\min_{T} \quad \sum_{i=1}^{n} \sum_{j=1}^{m} D_{ij} T_{ij}$$

$$\text{s.t.} \quad \sum_{i=1}^{n} T_{ij} \leq q_j, \ \forall j \in \{1, \dots, m\}; \quad \sum_{j=1}^{m} T_{ij} \leq v_i, \ \forall i \in \{1, \dots, n\} \tag{3}$$

$$\sum_{i=1}^{n} \sum_{j=1}^{m} T_{ij} = M; \quad T_{ij} \geq 0, \ \forall i \in \{1, \dots, n\}, \forall j \in \{1, \dots, m\}.$$

Optimal transport has been extensively used in data science [28], and is a common tool for transfer learning and domain adaptation [6,9,10,25]. It provides a principled framework to compute a distance between probability distributions, also known as the Wasserstein distance, by finding the minimal effort required to "transport" the mass of one distribution to another. Our formulation deviates from the original optimal transport problem by constraining the total mass M to be transported. As we show in Sect. 3.1, M can be selected in a data-driven fashion, permitting alignments that respect domain-specific regions that are not present in the other domain.

The user-defined parameters q_j and v_i indicate the mass assigned to each observation. For instance, to find a hard assignment from each observation in Φ_1 to Φ_2, and if $n \leq m$, we can set $v_i = 1/n$, $q_j = 1/n$ and $M = 1$, which is the case for the experiments in Sect. 3. Soft assignments can be obtained by different choices of masses. Alternatively an entropy regularization $\epsilon \sum_{i,j} T_{ij} \log(T_{ij})$ can be added to the objective function. In this work we focus on hard assignments since we want to learn one-to-one correspondences. Nevertheless, we state the general formulation, which is useful when there is less confidence in the existence of one-to-one correspondences.

The coupling T contains the information required to combine both manifolds. After a *min-max* normalization denoted by $\tilde{T}$, we can find a projection of a given sample $x_i \in \Phi_1$ on Φ_2 by its barycentric projection $x_i \mapsto \sum_j \tilde{T}_{ij} y_j$. Alternatively, we can build a cross-modality similarity matrix $W_{\Phi_1 \Phi_2} = (W_{\Phi_1} \tilde{T} + \tilde{T} W_{\Phi_2})$, where W_{Φ_k} are the similarities in each domain (computed using Eq. (1) in this paper). Using a similar construction as in [16] we can build a joint manifold learning loss:

$$\mathcal{L} = \mu \sum_{ij} ||f_i - f_j|| W_{\Phi_1}^{ij} + \mu \sum_{ij} ||g_i - g_j|| W_{\Phi_2}^{ij} + (1 - \mu) \sum_{ij} ||f_i - g_j|| W_{\Phi_1 \Phi_2}^{ij}.$$

(4)

The parameter μ controls the preservation of the intra-domain geometry. The solution of (4) provides a shared embedding where f and g represent the embedding coordinates for both domains. They are the generalized eigenvectors of the graph Laplacian matrix associated with the joint similarity matrix:

$$W = \begin{bmatrix} \mu W_{\Phi_1} & (1 - \mu) W_{\Phi_1 \Phi_2} \\ (1 - \mu) W_{\Phi_1 \Phi_2}' & \mu W_{\Phi_2} \end{bmatrix}.$$

(5)

DTA differs from [16] in several ways. First, their method starts by solving (4), with a T matrix instead of $W_{\Phi_1 \Phi_2}$, which encodes only the *a priori* known correspondences, containing a 1 in entry (i, j) if $x_i \in \Phi_1$ corresponds to $y_j \in \Phi_2$ and 0 otherwise. Inter-domain correspondences for the rest of the data are obtained in the latent space produced by the solution. In contrast, DTA first finds a matrix T that couples all the data, and then builds the inter-domain similarities based on these correspondences. Second, using only T in (4) assigns a 0 similarity between x_i and the neighbors of y_j. We argue that a more natural way to construct the off-diagonal matrices of W is to include the neighbors of y_j as being similar to x_i as well, motivating our particular construction of $W_{\Phi_1 \Phi_2}$.

3 Experimental Results

To demonstrate DTA's effectiveness in finding a coupling between domains, we compare DTA with semi-supervised manifold alignment (SSMA) [16], manifold alignment with Procrustes analysis (MA-PA) [34], and MAGAN [1]. For consistency, we use the same α-decay Kernel in Eq. (1) for the graph-based methods DTA, SSMA, and MA-PA, with $\alpha = 10$ and $k = 10$. For MAGAN we use the same architecture provided by the author's code[1]. MAGAN's architecture is composed of two generators, one mapping from Φ_1 to Φ_2 and the other in the opposite direction, and two discriminators, one for each domain. The model is

---

[1] https://github.com/KrishnaswamyLab/MAGAN/tree/master/MAGAN.

trained via a *min-max* game between the generators and discriminators, with a cycle consistency loss [38], and a correspondence loss that tries to preserve the known correspondences. We found that MAGAN usually needs an extra penalization parameter ρ in the correspondence loss to improve its performance, which was not included in the original paper.

Given the nature of the problem, it is difficult to tune the hyper-parameters present in each method. Thus, we set the same values for each method across all the experiments. This leave us with one hyperparameter t for DTA, which we set equal to 10 for all the experiments. SSMA and MA-PA require a predefined number of dimensions for the latent space. We selected all eigenvectors associated with non-zero eigenvalues. We set $\rho = 1000$ for MAGAN.

We used four simulated datasets shown in Fig. 1. **MNIST-Double**: one domain contains the original MNIST digits, while the other is constructed by downscaling the images to 14×14 pixels, applying a rotation, and adding Gaussian blurring. **SWISSR-SCURVE**: starting from a common 2D latent space we apply two different transformations resulting in the well known swiss roll and s-curve manifolds embedded in a 3D space. **STL10**: a popular dataset for computer vision [7]. The first domain contains the original images, and we generated the second by applying brightness, gray scaling, and Gaussian blurring. We performed feature extraction using the 512 outputs after the last convolution layer in ResNet-18. **SPLATTER-BE**: we simulated single-cell RNA-sequencing data using Splatter [37]. The difference between Φ_1 and Φ_2 is due to batch effects, which often arise in biological experiments. For real data, we used the single-cell dataset from the *Multimodal Single-Cell Data Integration* challenge, NeurIPS competition track 2021. The data contains two sets with jointly measured observations for both domains, providing us ground truth information about the coupling between domains. The first set measures gene expression (RNA) and protein abundance (ADT), while the second measures RNA and chromatin accessibility (ATAC). The samples are taken from different donors and batches. We selected batches "s1d1" in both sets for our experiments. Both RNA and ATAC domains are preprocessed, reducing their dimensionality to 1000 features via truncated SVD.

Inter-domain Feature Mapping. Our first comparison metric is the regression performance when mapping between the two domains. When the prior known correspondences are insufficient to successfully train a model, we can improve the training data by expanding the correspondences using each of the considered manifold alignment methods. For DTA, we use hard assignments where for each observation in Φ_1 we assign an unique counterpart in Φ_2. The correspondences in SSMA and MA-PA are computed as suggested in [16], where the assigned counterpart for each observation in Φ_1 corresponds to its nearest sample from Φ_2 in the shared latent space. For MAGAN, once the model is trained, we map the data from the first domain into the second using one of the generators. The assigned correspondence is the closest sample. The newly found correspondences serve as the training data for the regression task.

To reduce the dependency on a given regression model, we trained both a fully-connected neural network and a Kernel Ridge Regression (KRR) model. Since the true one-to-one correspondences are accessible to us, the regression models are also trained with the complete data, as well as the *a priori* known correspondences. This provides a baseline to show the improvement due to the new information acquired after each of the manifold alignment models, and how well they perform compared to the full correspondence case.

The results are summarized in Table 1 with the test MSE values for each model as well as for the regression trained using all of the correct correspondences. DTA is the most consistent method as it almost always outperforms the other methods across different datasets and different levels of prior known correspondences.

Domain Adaptation. Now we compare the methods on a domain adaptation problem. Table 2 contains the test error for two k-nearest neighbor classifiers, with $k = 1$ and $k = 10$. The classification models are trained on Φ_2 and then tested on the barycentric projections of Φ_1 onto Φ_2. The matrix $\tilde{T}$ is computed for SSMA, MA-PA, and MAGAN from the assigned correspondences as described above. An alternative approach for SSMA and MA-PA is to train and test the classification on the shared latent representation. For MAGAN the testing can be computed in the generator mapping from Φ_1 to Φ_2. Overall, DTA outperforms the other methods as it typically has the best performance and is in second otherwise. In contrast, while other methods occasionally outperform DTA on some datasets (e.g. MAGAN on MNIST-Double), these methods perform worse on other datasets.

Fraction of Samples Closer than the True Match (FOSCTTM). Lastly, a common metric to measure the goodness of alignment was proposed in [23] and further employed by [4,12] among others. The idea is to measure the proportion of observations that are closer to the true match after alignment, and average over the entire dataset. Thus, the lower this number, the better are the samples aligned with their counterparts in the opposite domain. Since this metric can be measured in different spaces after alignment, we include three different cases in Table 3. After alignment, we can compute the distances after computing the barycentric projection in the ambient space. Alternatively, it is possible to find a low dimensional representation after computing the spectral embedding using the matrix W, and find the neighbors and distances in this new representation. In particular, we computed the FOSCTTM metric in both, the 2 and 10 dimensional embeddings.

Table 1. Regression MSE average over 10 runs. When both models (Neural network and KRR) are trained with all the ground truth correspondences a lower MSE is obtained, and if only the *a priori* known correspondences are used the worst results are obtained for the majority of scenarios.

Dataset	Model	Test MSE (Neural Network)				Test MSE (KRR)			
		1%	2%	5%	10%	1%	2%	5%	10%
MNIST-Double	AllData	0.000	0.000	0.000	0.000	0.000	0.000	0.000	0.000
	PriorInfo	0.012	0.008	0.003	0.001	0.011	0.006	0.002	0.000
	DTA	**0.006** (2)	**0.004** (2)	<u>**0.002**</u> (1)	<u>**0.002**</u> (1)	**0.005** (2)	**0.003** (2)	<u>**0.002**</u> (1)	<u>**0.001**</u> (1)
	MA-PA	0.012 (3)	0.009 (3)	0.006 (3)	0.004 (3)	0.012 (3)	0.009 (3)	0.005 (3)	0.003 (3)
	MAGAN	<u>**0.002**</u> (1)	<u>**0.002**</u> (1)	**0.003** (2)	**0.002** (2)	<u>**0.001**</u> (1)	<u>**0.002**</u> (1)	**0.002** (2)	**0.001** (2)
	SSMA	0.013 (4)	0.010 (4)	0.007 (4)	0.005 (4)	0.012 (4)	0.009 (4)	0.006 (4)	0.004 (4)
RNA-ADT	AllData	0.109	0.108	0.109	0.109	0.104	0.104	0.104	0.105
	PriorInfo	0.718	0.519	0.330	0.243	0.304	0.204	0.177	0.173
	DTA	<u>**0.130**</u> (1)	<u>**0.131**</u> (1)	<u>**0.125**</u> (1)	<u>**0.124**</u> (1)	<u>**0.115**</u> (1)	<u>**0.116**</u> (1)	<u>**0.112**</u> (1)	<u>**0.112**</u> (1)
	MA-PA	0.230 (4)	0.190 (4)	0.147 (4)	0.137 (4)	0.235 (4)	0.180 (4)	0.125 (4)	0.117 (3)
	MAGAN	0.175 (3)	**0.143** (2)	**0.133** (2)	0.133 (3)	0.162 (3)	**0.129** (2)	0.121 (3)	0.122 (4)
	SSMA	**0.170** (2)	0.163 (3)	0.136 (3)	0,130 (2)	**0.148** (2)	0.140 (3)	**0.118** (2)	**0.115** (2)
RNA-ATAC	AllData	0.369	0.369	0.369	0.370	0.346	0.346	0.346	0.346
	PriorInfo	0.522	0.472	0.431	0.399	0.406	0.376	0.361	0.355
	DTA	<u>**0.422**</u> (1)	**0.404** (2)	0.404 (3)	0.397 (3)	<u>**0.419**</u> (1)	<u>**0.401**</u> (1)	0.397 (3)	0.388 (3)
	MA-PA	**0.430** (2)	<u>**0.403**</u> (1)	<u>**0.386**</u> (1)	<u>**0.387**</u> (1)	0.460 (3)	**0.402** (2)	<u>**0.373**</u> (1)	<u>**0.368**</u> (1)
	MAGAN	0.661 (4)	0.664 (4)	0.648 (4)	0.544 (4)	0.661 (4)	0.662 (4)	0.643 (4)	0.537 (4)
	SSMA	0.443 (3)	0.410 (3)	**0.399** (2)	**0.396** (2)	**0.456** (2)	0.403 (3)	**0.383** (2)	**0.374** (2)
SPLATTER-BE	AllData	0.372	0.396	0.391	0.401	0.376	0.376	0.376	0.377
	PriorInfo	0.440	0.424	0.413	0.405	0.457	0.470	0.414	0.398
	DTA	**0.388** (1)	<u>**0.377**</u> (1)	<u>**0.397**</u> (1)	<u>**0.406**</u> (1)	<u>**0.377**</u> (1)	<u>**0.376**</u> (1)	<u>**0.377**</u> (1)	<u>**0.377**</u> (1)
	MA-PA	0.410 (3)	0.409 (3)	**0.408** (2)	0.409 (3)	0.401 (3)	0.403 (3)	0.393 (3)	0.390 (3)
	MAGAN	0.466 (4)	0.518 (4)	0.466 (4)	0.481 (4)	0.483 (4)	0.527 (4)	0.475 (4)	0.498 (4)
	SSMA	**0.407** (2)	**0.408** (2)	0.408 (3)	**0.409** (2)	**0.387** (2)	**0.386** (2)	**0.386** (2)	**0.386** (2)
STL10	AllData	0.373	0.374	0.374	0.378	0.321	0.322	0.323	0.325
	PriorInfo	0.564	0.530	0.497	0.467	0.557	0.534	0.476	0.433
	DTA	<u>**0.470**</u> (1)	<u>**0.461**</u> (1)	<u>**0.458**</u> (1)	<u>**0.454**</u> (1)	<u>**0.460**</u> (1)	<u>**0.444**</u> (1)	<u>**0.438**</u> (1)	<u>**0.433**</u> (1)
	MA-PA	0.532 (3)	0.503 (3)	0.479 (3)	**0.468** (2)	0.554 (3)	0.507 (3)	0.471 (3)	0.452 (3)
	MAGAN	0.552 (4)	0.537 (4)	0.562 (4)	0.498 (4)	0.564 (4)	0.532 (4)	0.546 (4)	0.469 (4)
	SSMA	**0.503** (2)	**0.484** (2)	**0.476** (2)	0.469 (3)	**0.489** (2)	**0.474** (2)	**0.464** (2)	**0.451** (2)
SWISSR-SCURVE	AllData	0.002	0.003	0.001	0.001	0.000	0.000	0.000	0.000
	PriorInfo	0.682	0.648	0.263	0.151	0.610	0.311	0.036	0.004
	DTA	**0.043** (2)	<u>**0.015**</u> (1)	<u>**0.003**</u> (1)	<u>**0.001**</u> (1)	**0.036** (2)	<u>**0.008**</u> (1)	<u>**0.001**</u> (1)	<u>**0.000**</u> (1)
	MA-PA	<u>**0.018**</u> (1)	0.064 (3)	0.044 (3)	0.021 (4)	<u>**0.014**</u> (1)	0.061 (3)	0.043 (3)	0.017 (4)
	MAGAN	0.620 (4)	0.546 (4)	0.088 (4)	**0.004** (2)	0.682 (4)	0.513 (4)	0.088 (4)	**0.002** (2)
	SSMA	0.267 (3)	**0.039** (2)	**0.012** (2)	0.006 (3)	0.204 (3)	**0.027** (2)	**0.010** (2)	0.003 (3)

Table 2. Domain adaptation classification accuracy results under different correspondence percentages. Overall DTA achieves the best results as it is consistently in the top two.

Dataset	Model	KNN-1				KNN-10			
		1%	2%	5%	10%	1%	2%	5%	10%
MNIST-Double	DTA	**0.79** (2)	**0.87** (2)	**0.92** (2)	**0.94** (2)	**0.79** (2)	**0.85** (2)	**0.88** (2)	**0.89** (2)
	MA-PA	0.65 (3)	0.75 (3)	0.80 (3)	0.84 (3)	0.64 (3)	0.75 (3)	0.78 (3)	0.81 (3)
	MAGAN	**0.96** (1)	**0.95** (1)	**0.95** (1)	**0.97** (1)	**0.89** (1)	**0.88** (1)	**0.88** (1)	**0.89** (1)
	SSMA	0.42 (4)	0.55 (4)	0.65 (4)	0.75 (4)	0.42 (4)	0.56 (4)	0.65 (4)	0.73 (4)
RNA-ADT	DTA	**0.67** (1)	**0.68** (1)	**0.73** (1)	**0.73** (1)	**0.67** (1)	**0.67** (1)	**0.72** (1)	**0.72** (1)
	MA-PA	0.61 (3)	0.64 (3)	**0.70** (2)	**0.71** (2)	0.52 (4)	0.58 (4)	0.61 (4)	0.63 (4)
	MAGAN	0.61 (4)	0.62 (4)	0.69 (4)	0.65 (4)	**0.60** (2)	**0.61** (2)	**0.66** (2)	**0.66** (2)
	SSMA	**0.64** (2)	**0.66** (2)	0.69 (3)	0.70 (3)	0.58 (3)	0.60 (3)	0.63 (3)	0.65 (3)
RNA-ATAC	DTA	**0.66** (1)	**0.72** (1)	**0.77** (1)	**0.78** (1)	**0.61** (1)	**0.67** (1)	**0.70** (1)	**0.71** (1)
	MA-PA	**0.61** (2)	**0.70** (2)	**0.76** (2)	**0.76** (2)	0.54 (3)	**0.62** (2)	**0.66** (2)	**0.66** (2)
	MAGAN	0.30 (4)	0.32 (4)	0.42 (4)	0.53 (4)	0.31 (4)	0.33 (4)	0.44 (4)	0.54 (4)
	SSMA	0.59 (3)	0.65 (3)	0.70 (3)	0.72 (3)	**0.56** (2)	0.61 (3)	0.63 (3)	0.65 (3)
SPLATTER-BE	DTA	**0.83** (1)	**0.84** (1)	**0.84** (1)	**0.83** (1)	**0.79** (1)	**0.80** (1)	**0.80** (1)	**0.80** (1)
	MA-PA	**0.65** (2)	**0.57** (2)	**0.61** (2)	0.61 (3)	**0.65** (2)	**0.57** (2)	**0.62** (2)	**0.61** (2)
	MAGAN	0.30 (4)	0.30 (4)	0.42 (4)	0.46 (4)	0.31 (4)	0.30 (4)	0.43 (4)	0.47 (4)
	SSMA	0.51 (3)	0.54 (3)	0.58 (3)	**0.61** (2)	0.51 (3)	0.54 (3)	0.57 (3)	0.61 (3)
STL10	DTA	**0.75** (1)	**0.80** (1)	**0.81** (1)	**0.82** (1)	0.71 (2)	**0.75** (1)	**0.76** (1)	**0.76** (1)
	MA-PA	**0.73** (2)	**0.73** (2)	**0.74** (2)	**0.72** (2)	**0.74** (1)	**0.73** (2)	**0.74** (2)	**0.73** (2)
	MAGAN	0.51 (4)	0.61 (3)	0.56 (4)	0.71 (3)	0.52 (4)	0.63 (3)	0.59 (4)	0.72 (3)
	SSMA	0.53 (3)	0.61 (4)	0.65 (3)	0.69 (4)	0.53 (3)	0.61 (4)	0.65 (3)	0.69 (4)

Table 3. FOSCTTM average over 10 runs. DTA consistently achieves the best or second best performance.

Dataset	Model	10-dim Emb.		2-dim Emb.		Barycentric proj.	
		1%	10%	1%	10%	1%	10%
MNIST-Double	DTA	**0.01** (2)	**0.00** (1)	**0.03** (2)	**0.01** (1)	**0.05** (2)	**0.01** (2)
	MA-PA	0.14 (3)	0.01 (3)	0.08 (3)	0.03 (3)	0.14 (3)	0.04 (3)
	MAGAN	**0.01** (1)	0.00 (2)	**0.02** (1)	0.01 (2)	**0.01** (1)	**0.01** (1)
	SSMA	0.26 (4)	0.18 (4)	0.28 (4)	0.22 (4)	0.22 (4)	0.06 (4)
RNA-ADT	DTA	**0.20** (1)	**0.14** (1)	**0.11** (1)	**0.10** (1)	**0.10** (1)	**0.09** (1)
	MA-PA	0.40 (3)	0.22 (3)	0.19 (3)	0.26 (3)	0.16 (4)	0.12 (4)
	MAGAN	**0.25** (2)	**0.22** (2)	**0.14** (2)	**0.12** (2)	**0.12** (2)	**0.10** (2)
	SSMA	0.40 (4)	0.36 (4)	0.43 (4)	0.41 (4)	0.13 (3)	0.10 (3)
RNA-ATAC	DTA	**0.29** (1)	0.20 (2)	**0.17** (1)	**0.13** (1)	**0.37** (1)	**0.33** (1)
	MA-PA	**0.36** (2)	**0.19** (1)	**0.25** (2)	**0.27** (2)	**0.38** (2)	**0.33** (2)
	MAGAN	0.49 (4)	0.41 (4)	0.44 (3)	0.32 (3)	0.46 (4)	0.41 (4)
	SSMA	0.44 (3)	0.34 (3)	0.45 (4)	0.42 (4)	0.38 (3)	0.35 (3)
SPLATTER-BE	DTA	**0.14** (1)	**0.13** (1)	**0.14** (1)	**0.14** (1)	**0.27** (1)	**0.26** (1)
	MA-PA	**0.30** (2)	**0.22** (2)	**0.22** (2)	**0.20** (2)	**0.32** (2)	0.34 (3)
	MAGAN	0.42 (4)	0.31 (3)	0.44 (4)	0.33 (3)	0.40 (4)	**0.32** (2)
	SSMA	0.42 (3)	0.39 (4)	0.42 (3)	0.44 (4)	0.37 (3)	0.34 (4)
STL10	DTA	**0.07** (1)	0.05 (2)	**0.10** (1)	**0.07** (1)	**0.17** (1)	0.13 (2)
	MA-PA	**0.24** (2)	0.10 (3)	**0.18** (2)	0.14 (3)	**0.21** (2)	0.16 (3)
	MAGAN	0.27 (3)	**0.05** (1)	0.24 (3)	0.08 (2)	0.23 (3)	**0.11** (1)
	SSMA	0.36 (4)	0.32 (4)	0.40 (4)	0.36 (4)	0.26 (4)	0.17 (4)
SWISSR-SCURVE	DTA	**0.01** (1)	**0.00** (1)	0.02 (2)	**0.00** (1)	0.03 (2)	**0.00** (1)
	MA-PA	**0.05** (2)	0.00 (3)	**0.02** (1)	0.01 (3)	**0.01** (1)	0.02 (4)
	MAGAN	0.15 (4)	**0.00** (2)	0.19 (4)	**0.01** (2)	0.17 (4)	**0.00** (2)
	SSMA	0.14 (3)	0.08 (4)	0.15 (3)	0.09 (4)	0.13 (3)	0.01 (3)

Overall, DTA achieves the best results in this metric for the various types of comparisons. MAGAN performs considerably well for MNIST-Double, but it tends to have the worst performance in the more complex single-cell datasets.

3.1   Partial Alignment

Here we show the ability of DTA to perform partial alignment. Figure 2 demonstrates this scenario where the data in one or both domains is not completely represented in the other. If, for instance, we use MAGAN to perform the alignment, the nature of its *min-max* game will map samples from one domain into high density regions of the other. This causes false positive correspondences, and an incorrect alignment for some portions of the data. In contrast, DTA can handle this scenario in a data-driven way. The idea is to select a value of M in (3), that corresponds to the mass from Φ_1 that has an actual counterpart in Φ_2. We select M using the normalized transportation cost: $NTC = \frac{\sum_{ij} D_{ij} T_{ij}}{M}$.

After selecting a grid of values for M ranging from 0 to 1, we solve (3) for each particular value and compute its corresponding NTC. The transportation cost for observations far away from the known correspondences (i.e. points that are present in only one of the domains) starts to increase rapidly after a certain threshold that likely corresponds to the case where all of the shared points have been aligned. Thus the selected mass M to be transported is computed by identifying a knee point in the NTC vs M plot (Fig. 2B).

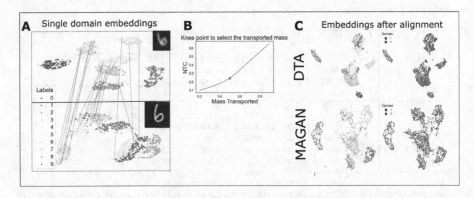

Fig. 2. Partial alignment. We subset both domains of the MNIST-Double dataset such that both domains contain specific regions with no counterpart in the other domain. **A)** Domain specific 2D UMAP [26] embeddings and dashed lines connecting the *a priori* known correspondences. **B)** Knee plot used to indentify the optimal mass M to be transported. **C)** Joint embedding of both domains after alignment, colored by labels and domain membership. DTA is able to retain domain-specific regions separate, while combining successfully the true counterparts. In contrast, MAGAN maps regions of Φ_1 to non-corresponding counterparts in Φ_2.

A quantitative evaluation of DTA and MAGAN in this scenario is presented in Table 4. After finding the *min-max* normalized coupling matrix T, we compute W via (5) and transform it to a distance matrix used in a kNN classifier. The test accuracy values are reported and, as expected, the results show how MAGAN maps observations close to incompatible regions on Φ_2, deteriorating the performance of the classifier.

Table 4. Test accuracy for the partial alignment experiments. DTA outperforms MAGAN.

Dataset	Model	KNN-1				KNN-10			
		1%	2%	5%	10%	1%	2%	5%	10%
MNIST-Double (P)	DTA	**0.821**	**0.861**	**0.882**	**0.887**	**0.900**	**0.917**	**0.924**	**0.926**
	MAGAN	0.583	0.663	0.720	0.743	0.753	0.801	0.827	0.836
RNA-ADT (P)	DTA	**0.820**	**0.831**	**0.844**	**0.849**	**0.910**	**0.910**	**0.912**	**0.919**
	MAGAN	0.627	0.655	0.675	0.679	0.692	0.719	0.726	0.726

4   Conclusion

We introduced Diffusion Transport Alignment (DTA), a manifold alignment method that exploits prior known correspondences between two related domains. We showed that DTA is superior to previous state-of-the-art manifold alignment methods by various metrics of comparison. DTA is able to recover meaningful connections that can be leveraged for downstream analysis tasks that may be otherwise difficult to perform. We also showed that partial manifold alignment can be handled by DTA, reducing the likelihood of falsely connecting points between domains, whereas previous methods are not naturally equipped to tackle this case.

Acknowledgments. This research was supported in part by Canada CIFAR AI chair [G.W.], in part by NSERC under Discovery Grant 03267 [G.W.], in part by the NIH under Grant R01GM135929 [G.W.], and in part by the NSF under Grant 2212325 [K.M.].

References

1. Amodio, M., Krishnaswamy, S.: MAGAN: aligning biological manifolds. In: International Conference on Machine Learning, pp. 215–223. PMLR (2018)
2. Belkin, M., Niyogi, P.: Laplacian eigenmaps for dimensionality reduction and data representation. Neural Comput. **15**(6), 1373–1396 (2003)
3. Cao, K., Bai, X., Hong, Y., Wan, L.: Unsupervised topological alignment for single-cell multi-omics integration. Bioinformatics **36**, 48–56 (2020)
4. Cao, K., Hong, Y., Wan, L.: Manifold alignment for heterogeneous single-cell multi-omics data integration using Pamona. Bioinformatics **38**(1), 211–219 (2022)
5. Chang, B., Kruger, U., Kustra, R., Zhang, J.: Canonical correlation analysis based on Hilbert-Schmidt independence criterion and centered kernel target alignment. In: International Conference on Machine Learning, pp. 316–324. PMLR (2013)

6. Chapel, L., Alaya, M.Z., Gasso, G.: Partial optimal transport with applications on positive-unlabeled learning. arXiv preprint arXiv:2002.08276 (2020)
7. Coates, A., Ng, A., Lee, H.: An analysis of single-layer networks in unsupervised feature learning. In: Proceedings of the Fourteenth International Conference on Artificial Intelligence and Statistics, pp. 215–223. JMLR Workshop and Conference Proceedings (2011)
8. Coifman, R.R., Lafon, S.: Diffusion maps. Appl. Comput. Harmon. Anal. **21**(1), 5–30 (2006)
9. Courty, N., Flamary, R., Habrard, A., Rakotomamonjy, A.: Joint distribution optimal transportation for domain adaptation. In: Advances in Neural Information Processing Systems, vol. 30 (2017)
10. Courty, N., Flamary, R., Tuia, D.: Domain adaptation with regularized optimal transport. In: Calders, T., Esposito, F., Hüllermeier, E., Meo, R. (eds.) ECML PKDD 2014. LNCS (LNAI), vol. 8724, pp. 274–289. Springer, Heidelberg (2014). https://doi.org/10.1007/978-3-662-44848-9_18
11. Cui, Z., Chang, H., Shan, S., Chen, X.: Generalized unsupervised manifold alignment. In: Advances in Neural Information Processing Systems, vol. 27 (2014)
12. Demetci, P., Santorella, R., Sandstede, B., Noble, W.S., Singh, R.: SCOT: single-cell multi-omics alignment with optimal transport. J. Comput. Biol. **29**(1), 3–18 (2022)
13. Gao, G., Ma, H.: Multi-modality movie scene detection using kernel canonical correlation analysis. In: Proceedings of the 21st International Conference on Pattern Recognition (ICPR2012), pp. 3074–3077. IEEE (2012)
14. Gravina, R., Alinia, P., Ghasemzadeh, H., Fortino, G.: Multi-sensor fusion in body sensor networks: state-of-the-art and research challenges. Inf. Fusion **35**, 68–80 (2017)
15. Ham, J.H., Lee, D.D., Saul, L.K.: Learning high dimensional correspondences from low dimensional manifolds (2003)
16. Ham, J., Lee, D., Saul, L.: Semisupervised alignment of manifolds. In: International Workshop on Artificial Intelligence and Statistics, pp. 120–127. PMLR (2005)
17. Hu, J., Hong, D., Zhu, X.X.: MIMA: mapper-induced manifold alignment for semi-supervised fusion of optical image and polarimetric SAR data. IEEE Trans. Geosci. Remote Sens. **57**(11), 9025–9040 (2019)
18. Katz, O., Talmon, R., Lo, Y.L., Wu, H.T.: Alternating diffusion maps for multi-modal data fusion. Inf. Fusion **45**, 346–360 (2019)
19. Kuchroo, M., Godavarthi, A., Tong, A., Wolf, G., Krishnaswamy, S.: Multimodal data visualization and denoising with integrated diffusion. In: 2021 IEEE 31st International Workshop on Machine Learning for Signal Processing (MLSP), pp. 1–6. IEEE (2021)
20. Lafon, S., Keller, Y., Coifman, R.R.: Data fusion and multicue data matching by diffusion maps. IEEE Trans. Pattern Anal. Mach. Intell. **28**(11), 1784–1797 (2006)
21. Lahat, D., Adali, T., Jutten, C.: Multimodal data fusion: an overview of methods, challenges, and prospects. Proc. IEEE **103**(9), 1449–1477 (2015)
22. Lindenbaum, O., Yeredor, A., Salhov, M., Averbuch, A.: Multi-view diffusion maps. Inf. Fusion **55**, 127–149 (2020)
23. Liu, J., Huang, Y., Singh, R., Vert, J.P., Noble, W.S.: Jointly embedding multiple single-cell omics measurements. In: Algorithms in bioinformatics:... International Workshop, WABI..., Proceedings. WABI (Workshop), vol. 143. NIH Public Access (2019)
24. Liu, Z., Wang, W., Jin, Q.: Manifold alignment using discrete surface RICCI flow. CAAI Trans. Intell. Technol. **1**(3), 285–292 (2016)

25. Lu, Y., Chen, L., Saidi, A.: Optimal transport for deep joint transfer learning. arXiv preprint arXiv:1709.02995 (2017)
26. McInnes, L., Healy, J., Melville, J.: UMAP: uniform manifold approximation and projection for dimension reduction. arXiv preprint arXiv:1802.03426 (2018)
27. Moon, K.R., et al.: Visualizing structure and transitions in high-dimensional biological data. Nat. Biotechnol. **37**(12), 1482–1492 (2019)
28. Peyré, G., Cuturi, M., et al.: Computational optimal transport: with applications to data science. Found. Trends® Mach. Learn. **11**(5–6), 355–607 (2019)
29. Stanley, J.S., III., Gigante, S., Wolf, G., Krishnaswamy, S.: Harmonic alignment. In: Proceedings of the 2020 SIAM International Conference on Data Mining, pp. 316–324. SIAM (2020)
30. Stuart, T., et al.: Comprehensive integration of single-cell data. Cell **177**(7), 1888–1902 (2019)
31. Thompson, B.: Canonical Correlation Analysis: Uses and Interpretation, vol. 47. Sage, Thousand Oaks (1984)
32. Tuia, D., Camps-Valls, G.: Kernel manifold alignment for domain adaptation. PLoS ONE **11**(2), e0148655 (2016)
33. Vieira, S., Pinaya, W.H.L., Garcia-Dias, R., Mechelli, A.: Multimodal integration. In: Machine Learning, pp. 283–305. Elsevier (2020)
34. Wang, C., Mahadevan, S.: Manifold alignment using procrustes analysis. In: Proceedings of the 25th International Conference on Machine Learning, pp. 1120–1127 (2008)
35. Wang, C., Mahadevan, S.: Manifold alignment without correspondence. In: Twenty-First International Joint Conference on Artificial Intelligence (2009)
36. Wang, C., Mahadevan, S.: Heterogeneous domain adaptation using manifold alignment. In: Twenty-Second International Joint Conference on Artificial Intelligence (2011)
37. Zappia, L., Phipson, B., Oshlack, A.: Splatter: simulation of single-cell RNA sequencing data. Genome Biol. **18**(1), 1–15 (2017)
38. Zhu, J.Y., Park, T., Isola, P., Efros, A.A.: Unpaired image-to-image translation using cycle-consistent adversarial networks. In: Proceedings of the IEEE International Conference on Computer Vision, pp. 2223–2232 (2017)

Mind the Gap: Measuring Generalization Performance Across Multiple Objectives

Matthias Feurer[1]([✉]), Katharina Eggensperger[1], Edward Bergman[1],
Florian Pfisterer[2,3], Bernd Bischl[2,3], and Frank Hutter[1,4]

[1] Albert-Ludwigs-Universität Freiburg, Freiburg im Breisgau, Germany
{feurerm,eggenspk,bergmane,fh}@cs.uni-freiburg.de
[2] Ludwig-Maximilians-Universität München, Munich, Germany
{florian.pfisterer,bernd.bischl}@stat.uni-muenchen.de
[3] Munich Center for Machine Learning, Munich, Germany
[4] Bosch Center for Artificial Intelligence, Renningen, Germany

Abstract. Modern machine learning models are often constructed taking into account multiple objectives, e.g., minimizing inference time while also maximizing accuracy. Multi-objective hyperparameter optimization (MHPO) algorithms return such candidate models, and the approximation of the Pareto front is used to assess their performance. In practice, we also want to measure generalization when moving from the validation to the test set. However, some of the models might no longer be Pareto-optimal which makes it unclear how to quantify the performance of the MHPO method when evaluated on the test set. To resolve this, we provide a novel evaluation protocol that allows measuring the generalization performance of MHPO methods and studying its capabilities for comparing two optimization experiments.

1 Introduction

Multi-objective hyperparameter optimization (MHPO; Feurer and Hutter, 2019; Karl et al., 2022; Morales-Hernández et al., 2021) and multi-objective neural architecture search (MNAS; Benmeziane et al., 2021; Elsken et al., 2019b) are becoming increasingly important and enable moving beyond the purely performance-driven selection of machine learning (ML) models. Important additional objectives are, for example, model size, inference time and the number of operations (Elsken et al., 2019a), interpretability (Molnar et al., 2020), feature sparseness (Binder et al., 2020), or fairness (Chakraborty et al., 2019; Cruz et al., 2021; Schmucker et al., 2021). To evaluate and compare multi-objective methods, papers often report the hypervolume indicator of the Pareto front approximation as a measure of optimization performance.

However, as we show in this paper, an ML model that is located on the approximated Pareto front on the validation set can become a dominated model on the test set and vice versa. This phenomenon makes it impossible to compute the hypervolume indicator using the canonical *train-validation-test* evaluation protocol (Raschka, 2018). To remedy this, we propose a novel evaluation protocol that takes such models into account in order to lay a solid foundation for

B. Crémilleux et al. (Eds.): IDA 2023, LNCS 13876, pp. 130–142, 2023.
https://doi.org/10.1007/978-3-031-30047-9_11

multi-objective hyperparameter optimization. In addition, we also conduct an initial study in which we use this evaluation protocol to compare the hyperparameter optimization of two machine learning algorithms.

This paper is structured as follows. First, in Sect. 2, we give background on multi-objective optimization. In Sect. 3, we then discuss the problem of evaluating generalization performance on a test set and the problems of a naive solution. We go on and describe our new protocol in Sect. 4 and exemplify it in Sect. 5. Then, we describe how multi-objective generalization was (not) measured in related work in Sect. 6 before concluding the paper in Sect. 7.

We provide Python code to reproduce our experiments at https://github.com/automl/IDA23-MindTheGap.

2   Background

In the remainder of this paper, we follow the notation from Karl et al., (2022) and aim to minimize the multi-objective function $c : \Lambda \to \mathbb{R}^M$ defined as $\min_{\lambda \in \Lambda} c(\lambda) = \min_{\lambda \in \Lambda}(c_1(\lambda), \ldots, c_M(\lambda))$, where each $c_i : \Lambda \to \mathbb{R}$ denotes the cost of hyperparameter configuration (HPC) $\lambda \in \Lambda$ according to one cost metric $i \in (1, \ldots, M)$. Since, typically, there is no total order on the space of objectives $\mathbb{R}^M$, and hence there usually is no single best objective value, we now consider *Pareto-dominance* and *Pareto-optimality* instead. Given a function $c : \Lambda \to \mathbb{R}^M$, we define a binary relation '$\prec$' on $\mathbb{R}^M \times \mathbb{R}^M$. Given two cost vectors $\zeta^{(1)}, \zeta^{(2)} \in \mathbb{R}^M$, defined as $\zeta^{(1)} = c(\lambda^{(1)})$ and $\zeta^{(2)} = c(\lambda^{(2)})$, we say $\zeta^{(1)}$ *dominates* $\zeta^{(2)}$, written as $\zeta^{(1)} \prec \zeta^{(2)}$, if and only if

$$\forall k \in \{1, \ldots, M\} : \zeta_k^{(1)} \le \zeta_k^{(2)} \ \wedge \ \exists l \in \{1, \ldots, M\} : \zeta_l^{(1)} < \zeta_l^{(2)}.$$

We similarly define a dominance relationship for configurations λ: A configuration $\lambda^{(1)}$ dominates another configuration $\lambda^{(2)}$, so $\lambda^{(1)} \prec \lambda^{(2)}$, if and only if $c(\lambda^{(1)}) \prec c(\lambda^{(2)})$. The *non-dominated* set of solutions, the Pareto front $\mathcal{P}$, is then given by $\mathcal{P} = \{\zeta \in c(\Lambda) \mid \nexists \zeta' \in c(\Lambda) \text{ s.t. } \zeta' \prec \zeta\}$ and conversely, the Pareto set as the pre-image of $\mathcal{P}$: $\mathcal{P}_\Lambda = c^{-1}(\mathcal{P}) = \{\lambda \in \Lambda \mid \nexists \lambda' \in \Lambda \text{ s.t. } \lambda' \prec \lambda\}$.

An MHPO algorithm then aims to return the best approximation of the Pareto front, trading off all given objectives. To obtain candidate HPCs $\lambda^{(i)}$, an MHPO algorithm iteratively generates and evaluates HPCs $\{\lambda^{(1)}, \lambda^{(2)}, \ldots, \lambda^{(T)}\}$. In the next step, the MHPO algorithm[1] compares the performance of all evaluated solutions to obtain the subset of HPCs $\tilde{\mathcal{P}}_\Lambda \subseteq \{\lambda^{(1)}, \lambda^{(2)}, \ldots, \lambda^{(T)}\}$ approximating the Pareto set and thereby also the Pareto front.[2]

The literature provides several quality metrics for Pareto-optimal sets focusing on different aspects (Emmerich and Deutz, 2018; Zitzler et al., 2000, 2003):

---

[1] In principle, this is agnostic to the capability of the HPO algorithm to consider multiple objectives. Any HPO algorithm (including random search) would suffice since one can compute the Pareto-optimal set post-hoc.

[2] The true Pareto front is only *approximated* because there is usually no guarantee that an MHPO algorithm finds the optimal solution. Furthermore, there is no guarantee that an algorithm can find all solutions on the true Pareto front.

(1) Approximation quality of the Pareto front, (2) a good (often uniform) distribution of solutions, and (3) diversity w.r.t. to the values for each metric. Here, we consider the commonly used hypervolume indicator (Karl et al., 2022), i.e., the volume of the objective space covered by the dominating solutions w.r.t. a reference point. The hypervolume indicator mostly considers (1), and it can be used to capture the performance of an MHPO experiment in a single value.

While we discuss our work in the context of MHPO, the background, problem, and proposed solution also apply to other multi-objective optimization problems which involve separate validation and test sets, such as neural architecture search (NAS; Elsken et al., 2019b), Automated Machine Learning (AutoML; Hutter et al., 2019), or ensemble learning.

3   Evaluating Generalization

Having discussed how to evaluate an MHPO method in general, we now turn to the problem that in ML, the predictive performance is usually measured on unseen test data. To highlight the challenges, we summarize the standard evaluation protocol, describe a previously unknown failure mode, hypothesize a naive solution and point out two potential issues of such a naive solution.

In MHPO, we typically tune the hyperparameters of an ML model on a supervised ML task, e.g., classification, with dataset $\mathcal{D} = [(\boldsymbol{x}, y)^{(1)}, \ldots, (\boldsymbol{x}, y)^{(d)}]$. We consider minimizing data-based costs c_i that estimate an empirical risk w.r.t. to the entire data distribution, e.g., the empirical risk, fairness metrics, or explainability scores (in contrast to model-based costs, such as inference time or model size). Because we only have access to a finite sample from the entire data distribution, we estimate this risk using the canonical *train-validation-test* protocol (Raschka, 2018), which trains models on the *train* portion of the data, and *validation* & *test* costs are estimated on the respective data splits (one could also use other protocols, such as cross-validation with a test set). Empirically estimating *validation* and *test* costs induces separate estimation errors. Validation set quantities are used for MHPO and to approximate the Pareto front. This approximation is then evaluated on the *test* set to obtain an unbiased estimate of the generalization error and also to measure the performance of MHPO.

Problems in the Multi-objective Setting. Due to these separate estimation errors, an HPC deemed Pareto-optimal on the validation set is not necessarily Pareto-optimal on the test set.[3] We visualize this in Fig. 1. On the left-hand side, the Pareto front approximation generalizes well to the test set. In the middle, all HPCs are still Pareto-optimal but switch order, which could lead to unexpected performance degradation when selecting an HPC to deploy in practice. However, on the right-hand side, two HPCs are no longer Pareto-optimal, i.e., the Pareto front approximation does not generalize to the test set and contains dominated

---

[3] This is due to a shift in distributions when going from the validation set to the test set due to random sampling. The HPC might then no longer be optimal due to overfitting.

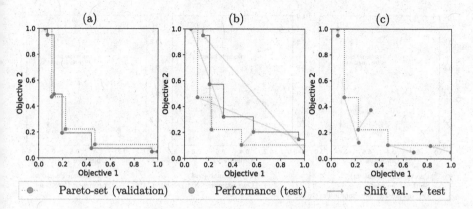

Fig. 1. We visualize validation (orange) and test performance (green) of the Pareto set, as found on the validation data set. Considering test performance, (a) all configurations are non-dominated, (b) the configurations are still Pareto-optimal, but switch order, and (c) the configurations are no longer Pareto-optimal. (Color figure online)

solutions. We would like to highlight that the two problems depicted in the two right-most plots were so far not discussed in the literature, yet, their existence thwarts the evaluation of MHPO algorithms.

A Naive Solution. Discarding dominated solutions based on the test set – which would *not* be possible in practice because we can only access test labels once in the end to compute final performance – would enable us to compute the hypervolume indicator. For this, we can either consider all evaluated HPCs (as common in assessing the performance of multi-objective methods) or expect the MHPO method to return a reasonable subset (which it believes to be Pareto-optimal). Then, we evaluate these HPCs on the test set, compute the Pareto front approximation based on the test scores, and finally calculate the hypervolume indicator. Unfortunately, this raises the following two issues.

Issue 1: Overestimation. We discard dominated points and thus overestimate the true hypervolume of the returned solutions, i.e., ignore solutions that are no longer part of the Pareto set, as displayed in the left-hand-side plot in Fig. 2. In practice, a user could pick one of the discarded solutions (based on its validation performance) and observe a worse performance than what we computed as the generalization performance of the optimization method.

Issue 2: Test data leakage. An adversarial MHPO method could exploit this procedure by returning as many models as possible and thus implicitly selecting its Pareto-optimal set based on the test set, as visualized in the middle of Fig. 1. While such a system would seemingly obtain a good score, its benefit in practice is limited.

These two issues emphasize the need for a new evaluation protocol that can detect these issues so that we can develop MHPO methods that return reliable Pareto front approximations.

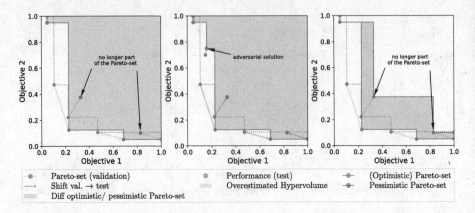

Fig. 2. We visualize validation (orange) and test performance (green/pink) of the Pareto set, as found on the validation data set. Left: We show that ignoring dominated points on the test set leads to an overestimation of the hypervolume indicator. Middle: We show how adversarial MHPO can return points that lead to an increased hypervolume on the test data. Right: We show our proposed optimistic Pareto-set (green), pessimistic Pareto-set (pink), and the approximation gap between the optimistic and pessimistic Pareto-set (pink area). (Color figure online)

4   A New Protocol to Measure Generalization

We propose a new evaluation protocol to assess the performance and robustness of an MHPO method reliably. We introduce the concept of *optimistic* and *pessimistic* approximations of the Pareto front. We visualize this on the right-hand side of Fig. 2. Given an approximation of a Pareto front $\tilde{\mathcal{P}}$ that was computed using the validation split of a data set, we formally define the *optimistic* Pareto front as

$$\mathcal{P}_{optimistic} = \{\zeta \in c(\tilde{\mathcal{P}}) \mid \nexists\ \zeta' \in c(\tilde{\mathcal{P}})\ \text{s.t.}\ \zeta' \prec_{test} \zeta\}, \tag{1}$$

where $\prec_{test}$ denotes a dominance relationship between costs ζ and ζ' evaluated on the test set instead of the validation set. Similarly, we define the *pessimistic* Pareto front as

$$\mathcal{P}_{pessimistic} = \{\zeta \in c(\tilde{\mathcal{P}}) \mid \nexists\ \zeta' \in c(\tilde{\mathcal{P}})\ \text{s.t.}\ \zeta \prec_{test} \zeta'\}. \tag{2}$$

Then, we can compute the hypervolume for both approximations. The difference between both volumes indicates how robust the Pareto front approximation is when going to test data, and we refer to it as the *approximation gap*. If it is zero, the Pareto set remains identical when moving from validation to test data. If it is greater than zero, then returned HPCs are dominated on the test data.

We can now compare two MHPO methods, A and B, based on their hypervolume using the following three criteria: (1) *hypervolume difference:* by checking if the optimistic estimate of the hypervolume of an MHPO method A is smaller than the pessimistic estimate of the hypervolume of an MHPO method B, (2) *dominance:* by using the notion of the optimistic and pessimistic Pareto set to check if pessimistic Pareto front approximation of A dominates the optimistic Pareto front approximation of B, following the popular idea of Pareto front dominance (Emmerich and Deutz, 2018), and (3) *approximation gap:* by comparing the gap between the optimistic and pessimistic hypervolume across MHPO methods whereas a smaller gap indicates a more robust approximation of the Pareto front.

5   Experimental Evaluation

In this section, we first show that the approximation gap appears in practice and second, experimentally check whether we can now compare algorithms again.

5.1   Demonstration of Approximation Gap

We first demonstrate the existence of the approximation gap by tuning the hyperparameters of a machine learning algorithm. Concretely, we tune the hyperparameters of a random forest model (Breiman, 2001) with 40 iterations of random search (Bergstra and Bengio, 2012) on the German credit dataset (Dua and Graff, 2017). We provide the configuration space and dataset description in Appendix A. We use precision and recall as objectives, motivated by the fact that both are often ad-hoc combined into the F1 score (Manning et al., 2008) despite this being an inherently multi-objective problem. Following Horn and Bischl, (2016), we tune class weights to account for the unbalanced targets. We split the dataset into 60% train, 20% valid, and 20% test data. For every HPC, we train a single model, record the precision and recall metrics on both the validation and test set and visualize the results in Fig. 3.

The plots are similarly structured as Figs. 1 and 2, and we depict validation performance (in orange) and test performance (in green) w.r.t. both objectives for all evaluated HPCs. The left-hand-side plot highlights the validation losses and the approximation of the Pareto front using the validation set. The middle plot shows how the performance changes when evaluating these HPCs on the test set. Furthermore, we show the hypothetical true Pareto-set $\tilde{\mathcal{P}}_{test}$ on the test data (which we cannot compute in practice; in grey). The right-hand-side plot shows the optimistic (in green) and pessimistic Pareto-set (in pink), which we described above. We observe $\tilde{\mathcal{P}}_{test} \prec \mathcal{P}_{optimistic} \prec \mathcal{P}_{pessimistic}$, while perfect generalization to the test set would give us $\tilde{\mathcal{P}}_{test} = \mathcal{P}_{optimistic} = \mathcal{P}_{pessimistic}$.

5.2   Can We Compare Two Algorithms Again?

Having seen the approximation gap, we now experimentally check whether the three criteria introduced in Sect. 4 enable comparisons between two different algorithms again. For this, we optimize the hyperparameters of a random forest and a linear classifier with random search for 50, 100, 200, and 500 iterations. We use the same experimental setup and configuration space for the random forest as above and display the configuration space for the linear model in Appendix A.

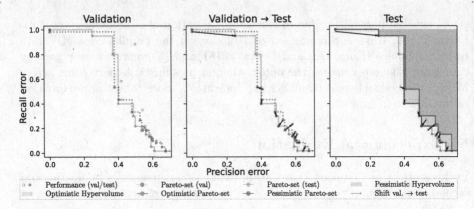

Fig. 3. Precision vs Recall. The left plot focuses on the validation error, the middle plot depicts the test error of points from the Pareto set on the validation set, and the right-hand-side plot depicts the approximation of the optimistic and the pessimistic Pareto sets. (Color figure online)

Table 1. Hypervolume indicators and approximation gap obtained by random search optimizing the hyperparameters of a random forest (top) and a linear model (bottom).

		50	100	200	500
Random Forest	Validation HV	0.5660	0.6357	0.6426	0.6563
	Pessimistic HV	0.5651	0.5972	0.6060	0.5721
	Optimistic HV	0.5833	0.6128	0.6272	0.6382
	Approximation Gap	0.0181	0.0156	0.0212	0.0661
Linear Model	Validation HV	0.5804	0.5943	0.6330	0.6340
	Pessimistic HV	0.5970	0.5628	0.5860	0.5641
	Optimistic HV	0.5989	0.5798	0.5994	0.5918
	Approximation Gap	0.0009	0.0170	0.0134	0.0277

First, we show the hypervolume indicator on the validation set, the pessimistic and optimistic hypervolume indicator, and the approximation gap in Table 1. As expected, we see that the validation hypervolume increases monotonically with more function evaluations, and after 100 function evaluations, the random forest has a larger validation hypervolume than the linear model, even with 500 function evaluations. Next, we look at the pessimistic and optimistic hypervolume. We can observe that there is no guarantee that they increase together with the validation hypervolume, which means that solutions obtained on the validation set do not generalize to the test set. This can be seen, for example, for the random forest, where the pessimistic hypervolume decreases when going from 200 to 500 function evaluations, while the optimistic hypervolume increases. For the linear model, we can even observe that both the optimistic

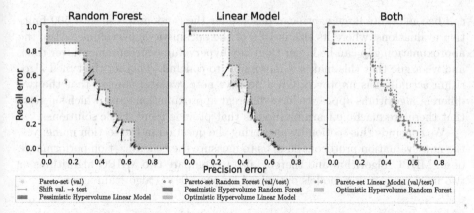

Fig. 4. Optimistic and pessimistic Pareto fronts for the random forest (left), the linear model (middle), and both (right) after 200 iterations of random search. For both models, we plot the pessimistic Pareto front in a darker color and using circle markers and the optimistic Pareto front in a lighter color and using star markers; and we use the same colors in the plot on the right-hand side. Furthermore, in the left and middle plots, we also give the validation Pareto front in light orange (similar to Fig. 3). (Color figure online)

and pessimistic hypervolume decrease, which can be seen when going from 50 to 100 and from 200 to 500 function evaluations. We can now also compare the two algorithms by comparing the hypervolume indicators (method (1) from Sect. 4), checking whether the pessimistic hypervolume indicator of one algorithm is larger than the optimistic hypervolume indicator of the other. Using this comparison method, we can conclude (1) that the linear model performs better than the random forest after 50 MHPO function evaluations, (2) that we cannot make a statement about 100 function evaluations, (3) that the random forest is better after 200 function evaluations, and (4) that we cannot make a statement at 500 function evaluations.

Second, we display the Pareto fronts of the two optimized algorithms in Fig. 4 to check whether one Pareto front dominates the other (method (2) from Sect. 4). We display results after 200 function evaluations, i.e., when the pessimistic hypervolume of the random forest is higher than the optimistic hypervolume of the linear model. This hypervolume dominance is a necessary but not a sufficient condition for Pareto front dominance. In this case, there are indeed solutions for the linear model (denoted as SGD) that are not dominated by the Pareto front of the random forest, making it impossible to state that one model is generally better than the other.

Finally, we examine the approximation gap of the two MHPO algorithms (method (3) from Sect. 4). The approximation gap is not a monotonic function. It can decrease when the number of function evaluations of the search algorithm increases (random forest from 50 to 100 function evaluations and linear model from 100 to 200 function evaluations). However, the approximation gap can become quite large as we can observe for the random forest with 500 function evaluations, where its size is 10% of the optimistic hypervolume. Also, the approximation gap can be larger than any hypervolume measurement over time, and we argue that this makes it impossible to conclude whether the tuned algorithm actually has improved. On a positive note, we also observe that the two different algorithms appear to have different approximation gaps, which suggests that there are machine learning models that provide more stable solutions.

We conclude this section by answering the question in the section name: yes, the new evaluation protocol allows us to measure the generalization performance of an MHPO algorithm and thereby also to compare two MHPO algorithms or two machine learning models optimized by one MHPO algorithm again.

6   Prior Evaluation Protocols for Multi-objective Optimization

This section reviews how prior works address the problem of measuring generalization performance in a multi-objective setting. We would first like to note that our setup differs from standard optimization problems under noise since we cannot recover the true function value by repeatedly evaluating the function of interest.[4] However, in our case, the performance of a model selected on the validation set suffers from a distribution shift on the test set. We are unaware of a method for describing such distribution shifts that happen when moving from the validation to the test set.[5]

To the best of our knowledge, no one has yet explicitly studied how to measure the generalization error of HPCs for machine learning models in a multi-objective setting. We found two works that evaluate multi-objective generalization: Horn et al., (2017) use the naive protocol we outline above, and Binder et al., (2020) solely compute, what we call, the optimistic Pareto front. Nonetheless, we would like to emphasize that these works employ these measures in an ad-hoc fashion without any discussion or justification. In contrast, we thoroughly introduce the approximation gap and the concepts behind it. In the field of MHPO, we found that researchers so far use scalarization to choose a final model to evaluate (Cruz et al., 2021), pick a model based on a single metric (Gardner et al., 2019), or use handcrafted heuristics to select a final model (Feffer et al., 2022).

A similar problem exists for constrained optimization: a solution that satisfies the constraints on the validation set can violate the constraint on the test set. Hernández-Lobato et al., (2016) found that "When the constraints are noisy,

---

[4] If the true function values of evaluated configurations cannot be recovered due to budget restrictions, our proposed evaluation protocol can be applied as well to deal with solutions that are no longer part of the Pareto front on the test set.

[5] Distributionally Robust Bayesian Optimization (Kirschner et al., 2020) is an algorithm that could be used in such a setting and the paper introducing it explicitly states AutoML as an application, but does neither demonstrate its applicability to AutoML nor elaborates on how to describe the distribution shift in a way the algorithm could handle it.

reporting the best observation is an overly optimistic metric because the best feasible observation might be infeasible in practice" and evaluate a "ground-truth score" by evaluating the final recommendation multiple times, treating a constrained violation as 100% classification error. They tuned a neural network on MNIST under inference time constraints and tuned Hamiltonian Monte Carlo under the constraint that the generated samples pass convergence diagnostic tests. In the field of noisy constrained Bayesian optimization, researchers have suggested an *identification step* to select the best point after optimization (Gelbart et al., 2014), and Letham et al., (2018) study the proportion of replicates in which the proposed method manages to find suitable solutions, but without scalarizing the final objective as done by Hernández-Lobato et al., (2016).

Moreover, for the problem of multi-objective ranking and selection (identification of the Pareto set from a finite set of choices), the F1 metric was proposed for judging the final result (Gonzalez et al., 2022). However, this does not quantify the solution quality in performance space. Last, the terminology of optimistic and pessimistic Pareto set has also been used in the context of approximating a Pareto front from the predictions of a probabilistic model (Iqbal et al., 2020).

7   Conclusions and Future Work

We have demonstrated that the standard evaluation protocol for single-objective HPO is inapplicable in the multi-objective setting and, as a remedy, introduced optimistic and pessimistic Pareto sets. Based on these, we can compare multi-objective algorithms using the *hypervolume difference*, *dominance*, or the new *approximation gap*. Furthermore, we can detect if the MHPO algorithm leads to an unstable solution, i.e., a large approximation gap, the analogue to over-fitting in single-objective optimization. In an experimental study, we have verified the existence of the approximation gap and demonstrated that we can now compare two machine learning models optimized for multiple metrics again.

In the future, we plan to (1) measure the effect of this problem over a large number of datasets and varying numbers of function evaluations, (2) extend our analysis to take measurement noise into account and (3) extend our protocol to multiple repetitions and cross-validation. Furthermore, we want to (4) evaluate additional multi-objective problems, e.g., trading off true-positive rates and false-positive rates (Horn and Bischl, 2016; Karl et al., 2022; Levesque et al., 2011) or fairness and predictive performance (Chakraborty et al., 2019; Cruz et al., 2021; Schmucker et al., 2021) and (5) study the related problem of distribution shifts in data streams.

Acknowledgements. Robert Bosch GmbH is acknowledged for financial support. Also, this research was partially supported by TAILOR, a project funded by EU Horizon 2020 research and innovation programme under GA No 952215. The authors of this work take full responsibility for its content.

A Experimental Details

Random Forest		Linear Model	
Hyperparameter name	Search space	Hyperparameter name	Search Space
criterion	[gini, entropy]	penalty	[l2, l1, elasticnet]
bootstrap	[True, False]	alpha	$[1e-6, 1e-2]$, log
max_features	[0.0, 1.0]	l1 ratio	[0.0, 1.0]
min_samples_split	[2, 20]	fit_intercept	[True, False]
min_samples_leaf	[1, 20]	eta0	$[1e-7, 1e-1]$
pos_class_weight exponent	[−7, 7]	pos_class_weight exp	[−7, 7]

We provide the random forest and linear model search spaces in Table A. We fit the linear model with stochastic gradient descent and use an *adaptive* learning rate and minimize the log loss (please see the scikit-learn (Pedregosa et al., 2011) documentation for a description of these). Because we are dealing with unbalanced data, we consider the class weights as a hyperparameter and tune the weight of the minority (positive) class in the range of $[2^{-7}, 2^7]$ on a log-scale (Horn and Bischl, 2011; Konen et al., 2016). To deal with categorical features, we use one hot encoding. We transform the features for the linear models using a quantile transformer with a normal output distribution.

We use the German credit dataset (Dua and Graff, 2017) because it is relatively small, leading to high variance in the algorithm performance, and unbalanced. We downloaded the dataset from OpenML (Vanschoren et al., 2014) using the OpenML-Python API (Feurer et al., 2021) as task ID 31, but conducted our own 60/20/20 split. It is a binary classification problem with 30% positive samples. The dataset has 1000 samples and 20 features. Out of the 20 features, 13 are categorical. The dataset contains no missing values.

References

Benmeziane, H., El Maghraoui, K., Ouarnoughi, H., Niar, S., Wistuba, M., Wang, N.: A comprehensive survey on Hardware-aware Neural Architecture Search. arXiv:2101.09336 [cs.LG] (2021)

Bergstra, J., Bengio, Y.: Random search for hyper-parameter optimization. J. Mach. Learn. Res. **13**, 281–305 (2012)

Binder, M., Moosbauer, J., Thomas, J., Bischl, B.: Multi-objective hyperparameter tuning and feature selection using filter ensembles. In: Ceberio, J. (ed.) Proceedings of the Genetic and Evolutionary Computation Conference (GECCO 2020), pp. 471–479. ACM Press (2020)

Breiman, L.: Random forests. Mach. Learn. J. **45**, 5–32 (2001)

Chakraborty, J., Xia, T., Fahid, F., Menzies, T.: Software engineering for fairness: a case study with Hyperparameter Optimization. In: Proceedings of the 34th IEEE/ACM International Conference on Automated Software Engineering (ASE). IEEE (2019)

Cruz, A., Saleiro, P., Belem, C., Soares, C., Bizarro, P.: Promoting fairness through hyperparameter optimization. In: Bailey, J., Miettinen, P., Koh, Y., Tao, D., Wu, X. (eds.) Proceedings of the IEEE International Conference on Data Mining (ICDM 2021), pp. 1036–1041. IEEE (2021)

Dua, D., Graff, C.: UCI machine learning repository (2017)

Elsken, T., Metzen, J., Hutter, F.: Efficient multi-objective Neural Architecture Search via Lamarckian evolution. In: Proceedings of the International Conference on Learning Representations (ICLR 2019) (2019a). Published online: https://iclr.cc/

Elsken, T., Metzen, J., Hutter, F.: Neural architecture search: a survey. J. Mach. Learn. Res. **20**(55), 1–21 (2019b)

Emmerich, M.T.M., Deutz, A.H.: A tutorial on multiobjective optimization: fundamentals and evolutionary methods. Nat. Comput. **17**(3), 585–609 (2018)

Feffer, M., Hirzel, M., Hoffman, S., Kate, K., Ram, P., Shinnar, A.: An empirical study of modular bias mitigators and ensembles. arXiv:2202.00751 [cs.LG] (2022)

Feurer, M., Hutter, F.: Hyperparameter optimization. In: Hutter et al. (2019), chap. 1, pp. 3–38, available for free at http://automl.org/book

Feurer, M., et al.: OpenML-Python: an extensible Python API for OpenML. J. Mach. Learn. Res. **22**(100), 1–5 (2021)

Gardner, S., et al.: Constrained multi-objective optimization for automated machine learning. In: Singh, L., De Veaux, R., Karypis, G., Bonchi, F., Hill, J. (eds.) Proceedings of the International Conference on Data Science and Advanced Analytics (DSAA 2019), pp. 364–373. ieeecis, IEEE (2019)

Gelbart, M., Snoek, J., Adams, R.: Bayesian optimization with unknown constraints. In: Zhang, N., Tian, J. (eds.) Proceedings of the 30th Conference on Uncertainty in Artificial Intelligence (UAI 2014), pp. 250–258. AUAI Press (2014)

Gonzalez, S., Branke, J., van Nieuwenhuyse, I.: Multiobjective ranking and selection using stochastic Kriging. arXiv:2209.03919 [stat.ML] (2022)

Hernández-Lobato, J., Gelbart, M., Adams, R., Hoffman, M., Ghahramani, Z.: A general framework for constrained Bayesian optimization using information-based search. J. Mach. Learn. Res. **17**(1), 5549–5601 (2016)

Horn, D., Bischl, B.: Multi-objective parameter configuration of machine learning algorithms using model-based optimization. In: Likas, A. (ed.) 2016 IEEE Symposium Series on Computational Intelligence (SSCI), pp. 1–8. IEEE (2016)

Horn, D., Dagge, M., Sun, X., Bischl, B.: First investigations on noisy model-based multi-objective optimization. In: Trautmann, H., et al. (eds.) EMO 2017. LNCS, vol. 10173, pp. 298–313. Springer, Cham (2017). https://doi.org/10.1007/978-3-319-54157-0_21

Hutter, F., Kotthoff, L., Vanschoren, J. (eds.): Automated Machine Learning: Methods, Systems, Challenges. Springer, Heidelberg (2019). Available for free at http://automl.org/book

Iqbal, M., Su, J., Kotthoff, L., Jamshidi, P.: Flexibo: Cost-aware multi-objective optimization of deep neural networks. arXiv:2001.06588 [cs.LG] (2020)

Karl, F., et al.: Multi-objective hyperparameter optimization - an overview. arXiv:2206.07438 [cs.LG] (2022)

Kirschner, J., Bogunovic, I., Jegelka, S., Krause, A.: Distributionally robust Bayesian optimization. In: Chiappa, S., Calandra, R. (eds.) Proceedings of the 23rd International Conference on Artificial Intelligence and Statistics (AISTATS 2020), pp. 2174–2184. Proceedings of Machine Learning Research (2020)

Konen, W., Koch, P., Flasch, O., Bartz-Beielstein, T., Friese, M., Naujoks, B.: Tuned data mining: a benchmark study on different tuners. In: Krasnogor, N. (ed.) Proceedings of the 13th Annual Conference on Genetic and Evolutionary Computation (GECCO 2011), pp. 1995–2002. ACM Press (2011)

Letham, B., Brian, K., Ottoni, G., Bakshy, E.: Constrained Bayesian optimization with noisy experiments. Bayesian Analysis (2018)

Levesque, J.C., Durand, A., Gagne, C., Sabourin, R.: Multi-objective evolutionary optimization for generating ensembles of classifiers in the roc space. In: Soule, T. (ed.) Proceedings of the 14th Annual Conference on Genetic and Evolutionary Computation (GECCO 2012), pp. 879–886. ACM Press (2011)

Manning, C., Raghavan, P., Schütze, H.: Introduction to Information Retrieval. Cambridge University Press, Cambridge (2008)

Molnar, C., Casalicchio, G., Bischl, B.: Quantifying model complexity via functional decomposition for better post-hoc interpretability. In: Cellier, P., Driessens, K. (eds.) ECML PKDD 2019. CCIS, vol. 1167, pp. 193–204. Springer, Cham (2020). https://doi.org/10.1007/978-3-030-43823-4_17

Morales-Hernández, A., Nieuwenhuyse, I.V., Gonzalez, S.: A survey on multi-objective hyperparameter optimization algorithms for machine learning. arXiv:2111.13755 [cs.LG] (2021)

Pedregosa, F., et al.: Scikit-learn: machine learning in Python. J. Mach. Learn. Res. **12**, 2825–2830 (2011)

Raschka, S.: Model evaluation, model selection, and algorithm selection in machine learning. arXiv:1811.12808 [stat.ML] (2018)

Schmucker, R., Donini, M., Zafar, M., Salinas, D., Archambeau, C.: Multi-objective asynchronous successive halving. arXiv:2106.12639 [stat.ML] (2021)

Vanschoren, J., van Rijn, J., Bischl, B., Torgo, L.: OpenML: networked science in machine learning. SIGKDD Explor. **15**(2), 49–60 (2014)

Zitzler, E., Deb, K., Thiele, L.: Comparison of multiobjective evolutionary algorithms: empirical results. Evol. Comput. **8**(2), 173–195 (2000)

Zitzler, E., Thiele, L., Laumanns, M., Fonseca, C., Fonseca, V.: Performance assessment of multiobjective optimizers: an analysis and review. IEEE Trans. Evol. Comput. **7**, 117–132 (2003)

Effects of Locality and Rule Language on Explanations for Knowledge Graph Embeddings

Luis Galárraga[✉][ID]

Inria, Univ. Rennes, CNRS, Irisa, Rennes, France
luis.galarraga@inria.fr

Abstract. Knowledge graphs (KGs) are key tools in many AI-related tasks such as reasoning or question answering. This has, in turn, propelled research in link prediction in KGs, the task of predicting missing relationships from the available knowledge. Solutions based on KG embeddings have shown promising results in this matter. On the downside, these approaches are usually unable to explain their predictions. While some works have proposed to compute post-hoc rule explanations for embedding-based link predictors, these efforts have mostly resorted to rules with unbounded atoms, e.g., $bornIn(x, y) \Rightarrow residence(x, y)$, learned on a global scope, i.e., the entire KG. None of these works has considered the impact of rules with bounded atoms such as $nationality(x, England) \Rightarrow speaks(x, English)$, or the impact of learning from regions of the KG, i.e., local scopes. We therefore study the effects of these factors on the quality of rule-based explanations for embedding-based link predictors. Our results suggest that more specific rules and local scopes can improve the accuracy of the explanations. Moreover, these rules can provide further insights about the inner-workings of KG embeddings for link prediction.

Keywords: knowledge graph embeddings · explainable AI

1 Introduction

The continuous advances in information extraction on the Web have given rise to large repositories of machine-friendly statements modeled as knowledge graphs (KGs). These are collections of facts of the form $p(s, o)$ that describe real-world entities, e.g., $capital(Italy, Rome)$. In this formalism, the predicate p in a statement $p(s, o)$ can be seen as a directed labeled edge that connects the subject s to the object o. KGs allow computers to "understand" the real world, and find applications in multiple AI-related tasks such as entity-centric IR, reasoning, question answering, smart assistants, etc. Since KGs usually suffer from incompleteness, a central task in KGs is *link prediction*, where the goal is to infer new facts from the available knowledge. Link prediction constitutes a fundamental step towards proper knowledge graph completion.

Approaches for link prediction in KGs abound and fall mainly into two paradigms. On the one hand, *symbolic methods* [12,16,19] mine explicit patterns on the graph, e.g., the rule $capital(x, y) \Rightarrow inCountry(y, x)$, and use those patterns to infer new relationships between entities. On the other hand, approaches based on latent factors [4,21,22,29,31,34] embed predicates p and entities s, o in a latent space driven by a score function that ranks true facts better than false ones. For example, TransE [4] learns d-dimensional embeddings (in bold) for predicates and entities such that $\mathbf{s}+\mathbf{p} \approx \mathbf{o}$, if $p(s, o)$ holds in reality. TransE's score function for facts is then $-\|\mathbf{s} + \mathbf{p} - \mathbf{o}\|_l$ ($l = \{1, 2\}$).

Embedding-based methods have exhibited promising performance for link prediction, however their main downside is that they operate as black boxes: one cannot obtain an explanation of the logic behind a predicted fact $p(s, o)$ from the latent representations of s, p, and o. This has therefore motivated some works on mining rule-based explanations for KG embeddings [7,9,26,27]. Those explanations can help us, for instance, verify if the embeddings meet expected reasoning guarantees such as transitivity, i.e., $p(x, z) \wedge p(z, y) \Rightarrow p(x, y)$, or detect biases in the data. It is known that redundancy in the form of inverse predicates, e.g., $hyponym(feline, cat)$, $hypernym(cat, feline)$ in benchmark datasets, led to overestimated accuracies for state-of-the-art embedding-based link predictors [3,18]. Had a mechanism to understand that the embeddings mainly captured patterns such as $hyponym(x, y) \Rightarrow hypernym(y, x)$, this issue could have been detected in advance.

A limitation of existing explanations for KG embeddings is that they only capture global inference patterns. This is tantamount to mining explanations in the language of unbounded atoms, i.e., rules with no constants in the arguments such as $bornIn(x, z) \wedge officialLang(z, y) \Rightarrow speaks(x, y)$, that hold *globally*, that is, on the entire KG. However, such rules cannot express specific entity associations such as $nationality(x, USA) \Rightarrow speaks(x, English)$, presumably captured by link predictors. On those grounds, Sect. 4 addresses the following research question (**RQ1**): **what is the impact of specific rules in the quality of the explanations for embedding-based predictors?**. Moreover, and in line with existing works in interpretable AI [23,24], we also study a second research question (**RQ2**): **how does learning explanation rules on specific regions of the KG, i.e., local explanations, impact the quality of the resulting rules?**. Before answering these questions, we discuss basic concepts and related work in Sect. 2, and explain how to compute rule-based explanations for link predictors in Sect. 3.

2   Preliminaries

2.1   Background Concepts

Knowledge Graphs. A knowledge graph $\mathcal{K} = (\mathcal{V}, \mathcal{E}, l_v, l_e)$ is a directed labeled graph with sets of vertices $\mathcal{V}$ and edges $\mathcal{E}$, where the injective functions $l_v : \mathcal{V} \rightarrow \mathcal{I}$, $l_e : \mathcal{E} \rightarrow \mathcal{P}$ assign labels to the vertices and edges. The sets $\mathcal{I}$ and $\mathcal{P}$ contain entity and predicate labels. An edge labeled *capital* departing from a

vertex labeled *France* to a vertex labeled *Paris* denotes the *statement* or *fact*
capital(France, Paris), i.e., France has capital Paris. Hence, a KG $\mathcal{K} \subset \mathcal{I} \times \mathcal{P} \times \mathcal{I}$
is also a set of facts $p(s, o)$ with subject s, predicate p, and object o. Usually,
standard KGs store only facts believed to be true.

We define the *potential set* $\Omega(\mathcal{K})$ of a KG as the universe of facts that could
be constructed from the entities and predicates in $\mathcal{K}$. More formally, $\Omega(\mathcal{K}) = \mathcal{D}_v(\mathcal{K}) \times \mathcal{D}_e(\mathcal{K}) \times \mathcal{D}_v(\mathcal{K})$ where

$$\mathcal{D}_v(\mathcal{K}) = \{l_v(v) : v \in \mathcal{V}\}, \qquad \mathcal{D}_e(\mathcal{K}) = \{l_e(e) : e \in \mathcal{E}\}$$

are the *entity and predicate domains* of $\mathcal{K}$. Furthermore, we define the *potential
set* of a predicate p as $\Omega(\mathcal{K}) \supseteq \Omega^p(\mathcal{K}) = \{p(s, o) : (s, o) \in \mathcal{D}^p(\mathcal{K}) \times \bar{\mathcal{D}}^p(\mathcal{K})\}$ with

$$\mathcal{D}^p(\mathcal{K}) = \{s : \exists o : p(s, o) \in \mathcal{K}\}, \qquad \bar{\mathcal{D}}^p(\mathcal{K}) = \{o : \exists s : p(s, o) \in \mathcal{K}\}.$$

$\Omega^p(\mathcal{K})$ therefore defines the set of all possible facts that could be constructed
with the known subjects and objects of predicate p.

Horn Rules. An *atom* A is a statement with constant predicate such that its
subject and object arguments can be variables $v \in \mathbb{V}$ with $\mathbb{V} \cap \mathcal{I} = \emptyset$. If A
has only variable arguments, we say A is *unbounded*, otherwise it is *bounded*.
A *Horn rule* R is a statement of the form $\boldsymbol{B} \Rightarrow H$ where the *body* $\boldsymbol{B}$ is
a conjunction of atoms $\bigwedge_{1 \leq i \leq n} A_i$, and H is the head atom. For instance,
the rule *parent(x, z)* $\wedge$ *nationality(z, y)* $\Rightarrow$ *nationality(x, y)* states that parents
and children have the same nationality. These rules usually come with scores
that quantify their precision. It is common to require atoms in rules to have
at least one variable, be transitively connected, and form *safe* rules, that is,
ensure that the head variables occur also in the body. This condition guarantees that the head variables are universally quantified, allowing for concrete
predictions via *substitutions*. A substitution $\sigma : \mathbb{V} \to \mathcal{I}$ is a partial mapping
from variables to constants, such that its application to atoms or rules replaces
each variable with its corresponding constant in the mapping. For example,
applying the substitution $\sigma = \{x \to$ *Marie Curie*$, y \to$ *France*$\}$ to the atom
$A : nationality(x, y)$, gives a new atom $\sigma(A) : nationality($*Marie Curie, France*$)$.
We say a rule $R : \boldsymbol{B} \Rightarrow H$ *predicts a fact* A' in a KG $\mathcal{K}$, denoted by $R \wedge \mathcal{K} \vdash A'$,
iff $\exists \sigma : (\forall B \in \boldsymbol{B} : \sigma(B) \in \mathcal{K}) \wedge \sigma(H) = A'$. Put differently a rule predicts a fact
A' if there exist a substitution σ that (i) maps each atom in the rule's body to a
known KG fact, and (ii) maps the head atom to A'. If R predicts a statement A'
and $A' \in \mathcal{K}$, we say that R *predicts* A' *correctly*, i.e., the prediction is a known
fact, and we use the notation $R \wedge \mathcal{K} \vDash A'$.

Link Predictors. A *link predictor* $f : \Omega(\mathcal{K}) \to \mathbb{R}$ is a function that scores
the facts in the potential set of a KG, usually assigning higher values to true
facts. Link predictors are mostly used to answer queries of the forms $p(s, ?)$ or
$p(?, o)$, in other words, queries that ask for the most likely subject or object of
a statement given the other two components. Embedding-based link predictors

operate on latent representations for entities, predicates, and facts in $\Omega(\mathcal{K})$. Hence, they actually have the form $f = \hat{f} \circ h$, where $\hat{f} : \mathbb{C}^k \to \mathbb{R}^1$ is a function defined on a k-dimensional representation for facts, and $h : \Omega(\mathcal{K}) \to \mathbb{C}^k$ maps facts to k-dimensional vectors. If the semantics of the vector components are not understandable to humans, we say that f is a black box. That is the case for pure embedding-based link predictors such as TransE [4] or ComplEx [31].

Explanations. An explanation $E = \langle \mathcal{R}, g \rangle$ for a black-box link predictor $f :$ $\Omega(\mathcal{K}) \to \mathbb{R}$ consists of a set $\mathcal{R}$ of Horn rules and a function $g : \mathcal{R} \to \mathbb{R}$ that attributes higher scores to rules that "agree" with f. A rule $R : B \Rightarrow H$ agrees with f, if R predicts a fact $A \in \Omega(\mathcal{K})$ also predicted by f. This definition assumes the existence of a threshold θ such that $f(A) \geq \theta$ is interpreted as the black box also "thinking" that A is true. Explanations can be of different scope, namely *global* when they are learned on the potential set $\Omega^p(\mathcal{K})$ of a predicate p, or *local* when they are learned on smaller regions of $\Omega^p(\mathcal{K})$ as explained in Sect. 3.

2.2 Related Work

Link Prediction. This problem has received a lot of attention in the last 10 years with approaches lying on a spectrum from symbolic methods to embedding-based techniques. We refer the reader to [14] for a comprehensive survey. Symbolic techniques learn explicit patterns, e.g., arbitrary subgraphs, paths, association rules, Horn rules, etc., from KGs and use those patterns as features to predict missing links between entities [12,16,19]. In contrast, the common principle of embedding-based methods is to model entities and predicates as elements in a latent space, where predicates characterize interactions between entity embeddings. Those interactions are modeled as geometrical operations, e.g., translation in TransE [4] where $s + p \approx o$ for true facts $p(s, o)$ ($s, p, o \in \mathbb{R}^d$), or rotation in RotatE [30]. More recent methods resort to neural architectures [20,28] that exploit the vicinity of entities in the graph to learn proper latent representations for both entities and predicates.

In all cases, a scoring function – implemented by minimizing a loss function – guides the training of the embeddings, which are learned to yield high scores for true facts and low scores for false facts. The latter are obtained by corrupting the true facts in the KG – a task of utter importance for the quality of the embeddings [13,36].

Other methods combine the strengths of symbolic patterns and embeddings [6]. In [17], the authors improve the accuracy of different state-of-the-art embedding-based link predictors by removing those predictions that are not backed up by any of the Horn rules learned on the data. This strategy is complemented with a combined ranking that takes into account the individual rankings given by the rules and the embeddings. Some approaches [1,13,35] propose iterative algorithms that use rules and embeddings to produce better

---

[1] Most methods embed the entities in real spaces, i.e., in $\mathbb{R}^k$, but a few, e.g.,[31] resort to vectors of complex numbers.

examples for subsequent training. In contrast, other methods [11,25] instruct the embeddings to comply to explicit reasoning patterns, e.g., transitivity, $p(x,z) \land p(z,y) \Rightarrow p(x,y)$.

Explaining the Black Box. Unlike symbolic approaches, link predictors based on embeddings are black boxes. Hence, there have been some efforts to explain their logic by mining explicit patterns [7,26,27] with attribution scores. Among these patterns, Horn rules are the most popular. The rules are extracted using state-of-the-art rule or path mining algorithms [2,8,15,16], whereas the attribution scores are learned via machine learning, e.g., linear or logistic regression in the spirit of standard explanation techniques such as LIME [24]. Nevertheless, none of these approaches exploits the power of Horn rules at its best. For instance, [7,27] mine rule explanations of up to two atoms, e.g., $bornIn(x,y) \Rightarrow livesIn(x,y)$, whereas DistMult [34] can only learn *pure* paths such as $bornIn(x,z) \land inCountry(z,y) \Rightarrow nationality(x,y)$. Hence, none of these methods can induce explanations in the language of bounded atoms such as $nationality(x,UK) \Rightarrow speaks(x,English)$. Furthermore, all these endeavors mine global explanations. Embedding-based models can, though, be very complex and therefore hard to approximate in the general sense. Thus we explore the effects of bounded atoms and locality in the quality of the explanations.

3   Explaining KG Embeddings for Link Prediction

Algorithm 1 describes a generic procedure to compute rule-based explanations for a black-box link predictor f trained on a KG $\mathcal{K}$, containing both true $(\mathcal{K}^+)$ and corrupted facts $(\mathcal{K}^-)$, in line with existing approaches [7,26,27]. The rules are learned on a context C consisting of facts of a given predicate p. We elaborate on the stages of the algorithm and the different ways to define the learning context.

Algorithm 1: Build Explanation

Input: link predictor $f : \Omega(\mathcal{K}) \to \mathbb{R}$ trained on $\mathcal{K} = \mathcal{K}^+ \cup \mathcal{K}^-$, context
$\quad\quad C \subset \Omega^p(\mathcal{K})$
Output: an explanation $E = \langle \mathcal{R}, g \rangle$ with set of rules $\mathcal{R}$, $g : \mathcal{R} \to \mathbb{R}$
1 $\hat{\mathcal{K}} := \emptyset$
2 **foreach** $A := p(s,o) \in C$ **do**
3 $\quad$ **if** $f(A) \geq \theta$ **then**
4 $\quad\quad$ $\hat{\mathcal{K}} := \hat{\mathcal{K}} \cup \{p^f(s,o)\}$
5 $\quad$ **else**
6 $\quad\quad$ $\hat{\mathcal{K}} := \hat{\mathcal{K}} \cup \{\neg p^f(s,o)\}$
7 $\mathcal{R} :=$ rule mining on $\mathcal{K} \cup \hat{\mathcal{K}}$ for predicates $p^f, \neg p^f$
8 **return** *build-rule-based-surrogate*$(\mathcal{R}, \mathcal{K}, \hat{\mathcal{K}}, f)$

Binarizing the Black Box. To learn Horn rules that mimic a black-box link predictor f, we need to convert f's scores for facts into true or false verdicts. To this end, lines 2–6 label each fact in the context C by computing f's score and then applying a threshold to decide whether the fact is deemed true or not by f. This set $\hat{\mathcal{K}}$ of annotated facts is represented by the surrogate predicates p^f, $\neg p^f$. For instance the fact $speaks^f(A.\ Einstein, English)$ means that f "thinks" that Einstein speaks English.

Rule Mining. Line 7 in Algorithm 1 learns a set $\mathcal{R}$ of Horn rules of the forms $B \Rightarrow p^f(s,o)$ and $B \Rightarrow \neg p^f(s,o)$ with confidence scores from the original KG $\mathcal{K}$ and the black-box annotated context $\hat{\mathcal{K}}$.

Learning the Explanation. Finally, line 8 uses the rules in $\mathcal{R}$ as features to learn a surrogate model $f_s : \mathbb{R}^{|\mathcal{R}|} \to \mathbb{R}$ that mimics the binarized f and provides importance scores for the mined rules. Given a statement $A = \bar{p}^f(s,o) \in \hat{\mathcal{K}}$ with $\bar{p}^f \in \{p^f, \neg p^f\}$, we encode A as a vector $x_A \in \mathbb{R}^{|\mathcal{R}|}$ such that its i-th entry is set as follows:

$$x_A[i] = \begin{cases} sgn(A) \times conf(R_i) & R_i \wedge (\mathcal{K} \cup \hat{\mathcal{K}}) \vDash A \\ -sgn(A) \times conf(R_i) & R_i \wedge (\mathcal{K} \cup \hat{\mathcal{K}}) \vdash A' \text{ with } A' \neq A \\ 0 & \text{otherwise} \end{cases}$$

Here $sgn(A) = 1$ if $A = \bar{p}^f(s,o)$, otherwise $sgn(A) = -1$. If a rule $R_i \in \mathcal{R}$ *predicts correctly* a statement $A \in \hat{\mathcal{K}}$, the i-th component of x_A holds a value equals the confidence of R_i (reported by the rule mining phase) with the same polarity of f's prediction. In that case, the rule R_i agrees with f and is a potential explanation for f's answer on A. If R_i is a potential explanation for some other fact A', we change the sign of confidence value. In any other case, we assign a score of 0 to the entry. We use the x_A vectors and the binarized labels – given by $sgn(A)$ – to train a surrogate logistic regression classifier f_s, whose coefficients define an attribution mapping $g : \mathcal{R} \to \mathbb{R}$ for rules – our explanation. The surrogate f_s can provide both binary labels and probability scores for facts, and its coefficients can be used to rank the rules predicting true and false verdicts $p^f(s,o), \neg p^f(s,o)$.

Explanation Context. Existing explanation approaches for KG embeddings [26,27] mine global explanations, where the context C given as input to Algorithm 1 contains a large sample of true and false statements. The latter are obtained by corrupting the true facts, so that for each fact $p(s,o)$ we also add $\{p(s',o), p(s,o')\}$ ($s \neq s', o \neq o'$). The resulting surrogate f_s approximates f's general logic when predicting p-labeled links.

A drawback of explanations based on global surrogates is that they assume that rules have always the same importance for all p-labeled predictions. Such a simplistic assumption can make explanation mining uninformative, if for example, the black box has a fine-grained behavior, i.e., it implements different logics for different regions of the KG. On those grounds, we propose to mine explanations within a *local* scope obtained by calling Algorithm 1 on different sub-contexts $C' \subseteq C$ with triples that are close to each other in the latent space.

These sub-contexts are obtained by applying agglomerative hierarchical clustering on $s \oplus o$, i.e., on the latent representation of pairs s, o for true facts $p(s, o) \in C^2$. We can also define *per-instance* contexts around a target fact $A = p(s, o)$ by calling Algorithm 1 on a sub-context $C' = \{A' = p(s', o) : A' \in C\} \cup \{A' = p(s, o') : A' \in C\} \cup \{A\}$, that is, on statements that share at least one argument with A.

4 Evaluation

To answer our research questions, we study the impact of bounded atoms (**RQ1**) and locality (**RQ2**) on the fidelity of rule explanations for embedding-based link predictors through a quantitative and an anecdotal evaluation.

4.1 Experimental Setup

Datasets and Link Predictors. We use the benchmark datasets fb15k-237, wn18rr, and yago3-10, on which we trained the bilinear methods ComplEx [31] and HolE [21], and the translational approach TransE [4]. We used the implementations and data offered by the Torch-KGE library [5].

Rule Mining. We mine Horn rules with AMIE [15], a state-of-the-art rule miner for large KGs. By default, AMIE mines closed Horn rules[3] of up to 3 unbounded atoms, but it can be instructed to mine longer rules, as well as to allow bounded atoms. Longer rules in combination with bounded atoms increase significantly the search space for rules, therefore we did not experiment with more than 3 atoms to avoid prohibitive runtimes [10]. AMIE does not support explicit counter-examples to estimate the precision of rules, as required by Algorithm 1, hence we extended the system to support explicit false facts in the precision computation. These counter-examples were generated through a variant of Bernoulli sampling that accounts for predicate domains [33]. We use all rules making at least 2 correct predictions with a precision of at least 10% to learn the surrogate model (see Sect. 3).

Explanations. We compute rule-based explanations for the studied link predictors using the test instances of the experimental datasets to construct contexts C of different scopes, i.e., global, local, and per-instance as explained in Sect. 3. For each call to Algorithm 1, we split C into train and test sets C_{train} and C_{test} (30%), so that we learn the explanations on C_{train} and evaluate them on C_{test}. Local clusters are computed using agglomerative hierarchical clustering instructed to return k clusters. We chose the most performing value of k between 2 and 6.

---

[2] $\oplus$ denotes concatenation; sub-contexts are corrupted to obtain counter-examples.
[3] These are safe rules where each variable occurs in at least 2 atoms.

Table 1. Fidelity on `fb15k-237`. Best performances are in bold; best locality results are underlined. The baseline **B** are global explanations with unbounded atoms. G, PI, and L stand for global, per-instance, and local explanations.

| | ROC-AUC | | | | | | S-MRR | | | | | | O-MRR | | | | | |
| | Unbounded | | | Bounded | | | Unbounded | | | Bounded | | | Unbounded | | | Bounded | | |
	B	L	PI	G	L	PI	B	L	PI	G	L	PI	B	L	PI	G	L	PI
complex	0.71	0.68	0.64	0.93	0.93	**0.95**	0.13	0.16	0.19	0.31	0.35	**0.44**	0.97	**1.00**	0.97	0.97	0.98	0.93
transe	0.72	0.70	0.64	**0.95**	0.92	**0.95**	0.12	0.20	0.19	0.22	**0.47**	0.45	0.97	**0.99**	0.98	0.98	0.93	0.91
hole	0.66	0.63	0.60	0.98	**0.99**	**0.99**	0.08	0.16	0.22	0.27	0.36	**0.50**	0.98	0.98	0.97	0.98	**1.00**	0.97

Table 2. Fidelity of rule-based explanations with bounded atoms.

| | ROC-AUC | | | S-MRR | | | O-MRR | | | ROC-AUC | | | S-MRR | | | O-MRR | | |
	G	L	PI	G	L	PI	G	L	PI	G	L	PI	G	L	PI	G	L	PI
complex	0.55	0.64	**0.68**	**0.93**	0.60	0.38	0.92	0.93	**1.00**	0.55	**0.75**	0.00	**0.94**	0.39	0.17	0.87	1.00	1.00
transe	0.51	0.55	**0.69**	0.71	**0.87**	0.32	0.93	0.92	**1.00**	**0.66**	0.63	0.63	**0.73**	0.50	0.50	0.94	0.97	**1.00**
hole	–	0.65	**0.73**	–	0.42	0.38	–	**1.00**	**1.00**	0.71	**0.81**	0.65	**0.90**	0.38	0.26	0.93	**1.00**	0.99
	(a) `wn18RR`									(b) `yago3-10`								

Link predictors are mainly used for two tasks: fact classification (true vs. false) and subject/object prediction for queries $p(?, o)$ and $p(s, ?)$ where potential candidates are ranked by their score. We quantify the fidelity of our surrogate models (their ability to approximate the link predictors) for these two tasks via standard metrics, namely the ROC-AUC score and the mean reciprocal rank (MRR). The threshold θ to binarize f's scores (line 6 in Algorithm 1) is chosen via logistic regression as follows: we learned a logistic regression classifier $f_c : C \to [0, 1]$ using f's scores on C as input features and the real truth value of the facts as label – corrupted facts are assumed false. In that spirit, f_c estimates the probability that a given statement is true. θ is then chosen so that $f_c(\theta) = 0.5$.

4.2   Results

Quantitative Evaluation. Tables 1 and 2 report the average ROC-AUC and MRR for the different explanation setups, namely unbounded vs. bounded rules learned on global (G), local (L), and per-instance (PI) scopes. The scores are computed by averaging the fidelity obtained for each call to Algorithm 1 weighted by the size of the corresponding test set, i.e., $|C_{test}|$. We disaggregate the MRR into S-MRR and O-MRR – when the task is to predict the subject or object given the other two components.

Our baseline setting (denoted by **B**) are global unbounded rules as mined by existing approaches [7, 26, 27]. We highlight that we could not mine explanations with such a setting for `wn18RR` and `yago3-10` on any of the studied link predictors – not even for local or per-instance scopes. This happens because unbounded rules can only be extracted when the training KG contains very prevalent and general regularities in the interactions between the predicates. The datasets wn18RR and yago3-10, however, have much fewer predicates than fb15k-237: 11 and 37 for the former versus 237 for the latter. Bounded atoms also increase the coverage explanation for fb15k-237. While the baseline provides explanations for 18 different predicates for ComplEx on fb15k-237, allowing bounded rules increases the coverage to 58 predicates (HolE and TransE exhibit comparable increases). Moreover the results in Table 1 suggest that bounded atoms in rules generally increase fidelity.

It is important to remark that allowing constants in the rule atoms comes at the expense of longer runtimes and many more, potentially noisy, rules. On fb15k-237 with global scopes, for example, the number of unique rules mined from TransE increases from 1k to 193k. That said, only 134k of those rules get non-zero coefficients during the attribution phase – implemented via logistic regression. This phase is indeed designed to identify the rules with actual explanation power w.r.t the veredicts of the link predictor.

We also observe that rule-based surrogates tend to be better at mimicking the link predictors for object prediction. This is explained by the nature of KG predicates, which are usually defined in a subject-oriented manner, e.g., *nationality*(*J. Biden, USA*) and not *hasCitizen*(*USA, J. Biden*). This makes subject prediction generally harder to mimic, because, e.g., it is easier to predict the nationalities of J. Biden than to predict all USA citizens. Besides, this phenomenon is corroborated by the actual performances of the link predictors. For instance, ComplEx exhibits an average S-MRR of 0.29 on wn18RR, whereas the average O-MRR reaches 0.46. That said, Table 2a suggests that S-MRR fidelity can still be high even in the presence of subject-oriented predicates.

When we look at the effects of locality on fidelity, we notice mixed effects. On fb15k-237, locality hurts ROC-AUC performance for unbounded rules and brings moderate performance gains for the MRR. This is probably because good general unbounded rules require more "diverse" data. The situation is different for bounded rules, for which locality boosts fidelity in most cases. These results suggest that locality and bounded rules are complementary. A similar behavior can also be observed for coverage. For example, local scopes combined with bounded rules allow mining 507k unique rules (with non-zero attribution) for 130 different predicates for ComplEx on fb15k-237 vs. 112k rules/58 predicates and 1505 rules/62 predicates when only one of the features is enabled (the baseline mines 730 predicates covering 18 predicates). For per-instance scopes we can compute rule explanations for up to 2782 individual facts (out of 20k) covering 83 predicates (HolE on fb15k-237).

Anecdotal Evaluation. Table 3 shows a few examples of rule-based explanations for our experimental link predictors. These correspond to some of the best ranked rules according to the coefficients of the surrogate classifiers. The rules illustrate regularities preserved by the link predictors, since the body of the rules defines conditions satisfied by the facts of the KG, in contrast to the head that matches statements predicted by our black boxes (see Algorithm 1).

Table 3. Some rule explanations. °, † denote local and per-instance explanations.

fb15k-237
(1) $place\_of\_birth(x, Chicago) \Rightarrow nationality(x, USA)$ [**TransE**]
(2) $has\_lived\_in(x, Brooklyn) \Rightarrow nationality(x, USA)$ [**ComplEx**]
(3) $profession(x, Author) \Rightarrow gender(x, M)$ [**ComplEx**]
(4) $impersonates(z, x) \land gender(z, y) \Rightarrow gender(x, y)$ [**ComplEx, HolE**]
(5) $country(z, y) \land birth\_place(x, z) \Rightarrow nationality(x, y)$
(6) $company(z, x) \land athlete:sport(z, y) \Rightarrow sport(x, y)°$ [**HolE**]
(7) $fwc:club(z, x) \land sport(z, y) \Rightarrow sport(x, y)°$ [**HolE**]

yago3-10
(8) $affiliation(x, Umeå\ IK) \Rightarrow gender(x, F)$† [**TransE, ComplEx**]
(9) $wonPrize(x, O.Orange - Nassau) \Rightarrow wonPrize(x, D.S.Medal)$ [**ComplEx**]

wn18rr
(10) $meronym\_mb(Insecta, x) \Rightarrow hypernym(x, Animal)$† [**ComplEx, HolE**]

Rules with bounded atoms offer legible insights about the information that the link predictors may be capturing to make predictions.

A key observation is that the different link predictors do not seem to rely on the same information – as suggested by rules (1) and (2) for ComplEx and TransE on fb15k-237. This is supported by the fact that among the 47 predicates for which ComplEx finds global explanations with bounded atoms, only 3 have common rules with TransE. We bring our attention to rule (3), which suggests that embeddings do reproduce the biases in the source data[4]. Recall that fb15k-237 was mainly extracted from Wikipedia, known to have gender biases [32]. Those biases are easier to spot with rules with bounded atoms, which are a complement to more general explanations such as (4) and (5). We also highlight that local contexts can illustrate the semantics captured by the embeddings. This is exemplified by rules (6) and (7) that were learned on the same predicate but on two fact clusters. As we can see, our mining routine learned semantically equivalent rules, defined on different thematic domains, namely *athletes* and *fwc*; the latter refers to the 2010 FIFA World Cup.

5 Conclusion

We have studied the effects of specific rules with bounded atoms and local scopes on the quality of explanations for embedding-based link predictors on knowledge graphs. Our results suggest a rather positive impact on the explanation fidelity and the coverage of the explanations. Moreover, specific rules and local scopes exhibit a symbiotic relationship.

---

[4] Rule (8), on the other hand, refers to a women's football team.

Even though rule-based explanations reflect regularities preserved by black-box link predictors, they do not shed light on causality. In this line of though, we envision to compute causal explanations that help us understand the role of the different entities, predicates, and latent components of KG embeddings in the resulting predictions. We have also planned to elaborate more on the relationship between link prediction performance and explanation fidelity, in particular at the level of the individual predicates. The source code and experimental data of our work is available at https://gitlab.inria.fr/glatour/geebis.

Acknowledgment. This research was supported by TAILOR, a project funded by EU Horizon 2020 research and innovation programme under GA No. 952215.

References

1. UniKER: a unified framework for combining embedding and horn rules for knowledge graph inference. In: ICML Workshop on Graph Representation Learning and Beyond (GRL+) (2020)
2. Ahmadi, N., Huynh, V.-P., Meduri, V., Ortona, S., Papotti, P.: Mining expressive rules in knowledge graphs. J. Data Inf. Qual. **1**(1) (2019)
3. Akrami, F., Saeef, M.S., Zhang, Q., Hu, W., Li, C.: Realistic re-evaluation of knowledge graph completion methods: an experimental study. In: ACM SIGMOD Conference (2020)
4. Bordes, A., Usunier, N., Garcia-Duran, A., Weston, J., Yakhnenko, O.: Translating embeddings for modeling multi-relational data. In: Advances in Neural Information Processing Systems, vol. 26 (2013)
5. Boschin, A: TorchKGE: knowledge graph embedding in Python and PyTorch. In: International Workshop on Knowledge Graphs (2020)
6. Boschin, A., Jain, N., Keretchashvili, G., Suchanek, F.: Combining embeddings and rules for fact prediction. In: International Research School in AI in Bergen, Schloss Dagstuhl-Leibniz-Zentrum für Informatik (2022)
7. Carmona, I.S., Riedel, S.: Extracting interpretable models from matrix factorization models. In: International Conference on Cognitive Computation (2015)
8. Chen, Y., Wang, D.Z., Goldberg, S.: ScaLeKB: scalable learning and inference over large knowledge bases. VLDB J. **25**(6) (2016)
9. Gad-Elrab, M.H., Stepanova, D., Tran, T.-K., Adel, H., Weikum, G.: ExCut: explainable embedding-based clustering over knowledge graphs. In: Pan, J.Z., et al. (eds.) ISWC 2020. LNCS, vol. 12506, pp. 218–237. Springer, Cham (2020). https://doi.org/10.1007/978-3-030-62419-4_13
10. Galárraga, L., Teflioudi, C., Hose, K., Suchanek, F.M.: Fast rule mining in ontological knowledge bases with AMIE+. VLDB J. **24**(6) (2015)
11. Guo, S., et al.: Knowledge graph embedding preserving soft logical regularity. In: International Conference on Knowledge Management (2020)
12. Hou, Z., Jin, X., Li, Z., Bai, L.: Rule-aware reinforcement learning for knowledge graph reasoning. In: ACL/IJCNLP (Findings) (2021)
13. Jain, N., Tran, T.-K., Gad-Elrab, M.H., Stepanova, D.: Improving knowledge graph embeddings with ontological reasoning. In: Hotho, A., et al. (eds.) ISWC 2021. LNCS, vol. 12922, pp. 410–426. Springer, Cham (2021). https://doi.org/10.1007/978-3-030-88361-4_24

14. Ji, S., Pan, S., Cambria, E., Marttinen, P., Yu, P.S.: A survey on knowledge graphs: representation, acquisition, and applications. IEEE Trans. Neural Netw. Learn. Syst. **33**(2) (2022)

15. Lajus, J., Galárraga, L., Suchanek, F.: Fast and exact rule mining with AMIE 3. In: Harth, A., et al. (eds.) ESWC 2020. LNCS, vol. 12123, pp. 36–52. Springer, Cham (2020). https://doi.org/10.1007/978-3-030-49461-2_3

16. Lao, N., Mitchell, T., Cohen, W.W.: Random walk inference and learning in a large scale knowledge base. In: Conference on Empirical Methods in Natural Language Processing (2011)

17. Meilicke, C., Betz, P., Stuckenschmidt, H.: Why a Naive way to combine symbolic and latent knowledge base completion works surprisingly well. In: 3rd Conference on Automated Knowledge Base Construction (2021)

18. Meilicke, C., Fink, M., Wang, Y., Ruffinelli, D., Gemulla, R., Stuckenschmidt, H.: Fine-grained evaluation of rule- and embedding-based systems for knowledge graph completion. In: Vrandečić, D., et al. (eds.) ISWC 2018. LNCS, vol. 11136, pp. 3–20. Springer, Cham (2018). https://doi.org/10.1007/978-3-030-00671-6_1

19. Meng, C., Cheng, R., Maniu, S., Senellart, P., Zhang, W.: Discovering meta-paths in large heterogeneous information networks. In: The Web Conference (2015)

20. Nguyen, D.Q., Nguyen, T.D., Nguyen, D.Q., Phung, D.: A novel embedding model for knowledge base completion based on convolutional neural network. In: Conference of the North American Chapter of the Association for Computational Linguistics (2018)

21. Nickel, M., Rosasco, L., Poggio, T.: Holographic embeddings of knowledge graphs. In: AAAI Conference on Artificial Intelligence (2016)

22. Nickel, M., Tresp, V.: Tensor factorization for multi-relational learning. In: Blockeel, H., Kersting, K., Nijssen, S., Železný, F. (eds.) ECML PKDD 2013. LNCS (LNAI), vol. 8190, pp. 617–621. Springer, Heidelberg (2013). https://doi.org/10.1007/978-3-642-40994-3_40

23. Peake, G., Wang, J.: Explanation mining: post hoc interpretability of latent factor models for recommendation systems. In: ACM SIGKDD International Conference on Knowledge Discovery and Data Mining (2018)

24. Ribeiro, M.T., Singh, S., Guestrin, C.: Why should I trust you?: explaining the predictions of any classifier. In: ACM SIGKDD International Conference on Knowledge Discovery and Data Mining (2016)

25. Rocktäschel, T., Riedel, S.: End-to-end differentiable proving. In: Conference on Neural Information Processing Systems (2017)

26. Ruschel, A., Gusmão, A.C., Polleti, G.P., Cozman, F.G.: Explaining completions produced by embeddings of knowledge graphs. In: Kern-Isberner, G., Ognjanović, Z. (eds.) ECSQARU 2019. LNCS (LNAI), vol. 11726, pp. 324–335. Springer, Cham (2019). https://doi.org/10.1007/978-3-030-29765-7_27

27. Sanchez, I., Rocktaschel, T., Riedel, S., Singh, S.: Towards extracting faithful and descriptive representations of latent variable models. In: AAAI Spring Symposium on Knowledge Representation and Reasoning (2015)

28. Shang, C., Tang, Y., Huang, J., Bi, J., He, X., Zhou, B.: End-to-end structure-aware convolutional networks for knowledge base completion. In: AAAI Conference on Artificial Intelligence (2019)

29. Socher, R., Chen, D., Manning, C.D., Ng, A.Y.: Reasoning with neural tensor networks for knowledge base completion. In: Conference on Neural Information Processing Systems (2013)

30. Sun, Z., Deng, Z.-H., Nie, J.-Y., Tang, J.: RotatE: knowledge graph embedding by relational rotation in complex space. In: International Conference on Learning Representations (2019)
31. Trouillon, T., Welbl, J., Riedel, S., Gaussier, É., Bouchard, G.: Complex embeddings for simple link prediction. In: International Conference on Machine Learning (2016)
32. Wagner, C., Graells-Garrido, E., Garcia, D., Menczer, F.: Women through the glass ceiling: gender asymmetries in Wikipedia. EPJ Data Sci. 5, 1–24 (2016)
33. Wang, Z., Zhang, J., Feng, J., Chen, Z.: Knowledge graph embedding by translating on hyperplanes. In: AAAI Conference on Artificial Intelligence, vol. 28, no. 1 (2014)
34. Yang, B., Yih, S.W., He, X., Gao, J., Deng, L.: Embedding entities and relations for learning and inference in knowledge bases. In: International Conference on Learning Representations (2015)
35. Zhang, W., et al.: Iteratively learning embeddings and rules for knowledge graph reasoning. In: The Web Conference (2019)
36. Zhang, Y., Yao, Q., Chen, L.: Efficient, Simple and Automated Negative Sampling for Knowledge Graph Embedding (2020)

Shapley Values with Uncertain Value Functions

Raoul Heese[1]($\boxtimes$) (iD), Sascha Mücke[2] (iD), Matthias Jakobs[2] (iD), Thore Gerlach[3] (iD), and Nico Piatkowski[3] (iD)

[1] Fraunhofer ITWM, Kaiserslautern, Germany
`raoul.heese@itwm.fraunhofer.de`
[2] TU Dortmund, Dortmund, Germany
[3] Fraunhofer IAIS, Sankt Augustin, Germany

Abstract. We propose a novel definition of Shapley values with uncertain value functions based on first principles using probability theory. Such uncertain value functions can arise in the context of explainable machine learning as a result of non-deterministic algorithms. We show that random effects can in fact be absorbed into a Shapley value with a noiseless but shifted value function. Hence, Shapley values with uncertain value functions can be used in analogy to regular Shapley values. However, their reliable evaluation typically requires more computational effort.

Keywords: Shapley Values · Uncertainty · Explainable Machine Learning · Game Theory

1 Introduction

The ability to interpret the predictions of machine learning (ML) models is an important requirement for many data-driven decision support applications, e. g., in computational biology [1], medicine [2], materials science [3], and finance [4], just to name a few. There are a variety of model-agnostic methods that enable explainability of otherwise black-box models [5]. The theory of Shapley values (SVs) [6], a concept from coalitional game theory, builds the foundation for a collection of such methods [7–12]. They all have in common that a SV—in the sense of an importance score—is attributed to each feature of an arbitrary predictive model. Based on these scores, explanations can be made about which features are responsible for which prediction. For a recent review of SVs in explainable ML (XAI), see, e. g., [13,14] and references therein.

SVs for XAI are based on the premise of a value function (VF) that quantifies the impact of each feature to a model's prediction by a real number. There are various approaches to define suitable VFs for such a task [12], but their evaluation typically involves a statistical or randomized (i. e., non-deterministic) ingredient. For example, a straightforward approach for a classifier is to retrain the model for all possible feature combinations and use its prediction of class

B. Crémilleux et al. (Eds.): IDA 2023, LNCS 13876, pp. 156–168, 2023.
https://doi.org/10.1007/978-3-031-30047-9_13

probabilities to specify the VF [15]. However, the training procedure os often a randomized algorithm, e. g., due to random initializations, data splitting, or randomized search heuristics—leading to different outcomes if repeated with different random seeds. As a result, the corresponding VF becomes a random variable that yields different outcomes if evaluated multiple times.

The effect of uncertain VFs for SVs in the context of XAI has not been extensively studied in the literature before. To our knowledge, [16], which is based on the *DeepSHAP* approximation [10,17], is the only work in this direction. Therein, a purely empirical perspective on sampling uncertainty is studied without any explicit sources of randomness. More generally, in [18,19], the authors define the uncertainty of coalition games as the standard deviation when considering the marginal contributions of each player entering a random coalition, which can also be understood as a measure of the strategic risk. An important insight is that whenever two players have the same SVs, they do not necessarily share the same risk. However, no randomness from the VF is considered in these works. In [20,21], the authors study uncertain coalition games [22] based on uncertainty theory [23] and introduce risk averse SVs. A brief review of related work can also be found in [21]. Since the calculus of uncertainty theory is based on special assumptions, the results are not straightforwardly applicable to the uncertainties that arise for SVs in the context of XAI. In conclusion, there is no self-consistent work on SVs with uncertain VFs based on probability theory.

We think, however, that the informative value of explainability via SVs can only be fairly judged if it incorporates a sound discussion of such uncertainties. As stated in [24], "it should be emphasized that the value of a game is an a priori measure—before the game is actually played." And indeed, SVs can only provide an expectation value of the feature importance. Especially in this light, it seems almost natural to ask the question what influence uncertainty has on the underlying probability density of this expected value and what happens to the higher moments. In this manuscript, we try to get a little closer to answering such questions.

Specifically, we present two major contributions:

1. In a novel approach, we define SVs with uncertain VFs based on first principles using probability theory.
2. We show that SVs with uncertain VFs correspond to regular SVs with shifted (deterministic) VFs, where the shift is determined by the mean bias of the marginal contributions. Our proposed definition consequently fulfills all properties of regular SVs. Furthermore, SVs with uncertain VFs can be used in analogy to regular SVs but with a higher computational effort that depends on the desired confidence.

The remaining part of this manuscript is structured as follows. In Sect. 2, we review regular SVs (without uncertainty) from a mostly general perspective. Subsequently, we introduce our definition of SVs with uncertain VFs in Sect. 3 and explain their properties. In Sect. 4, we briefly discuss the implications for XAI and show a numerical experiment. Finally, we conclude with a brief summary and outlook in Sect. 5. Throughout this paper, uncertain quantities are denoted in boldface. Moreover, lowercase letters denote a probability density that corresponds to its respective probability measure (e. g., p and $\mathbb{P}$).

2 Shapley Values

SVs, which originated in game theory [6,25], are based on the following premise: A coalition (or group) of players cooperates in what is called a coalition game and achieves a certain profit from this cooperation. The question that SVs try to answer is to evaluate the individual contribution of each player to the overall outcome. In other words, SVs represent an approach to attribute the total gains of a coalition game to each individual of a group of participating players. In this section, we start with a formal definition of SVs and then discuss some of their properties.

2.1 Definition

A coalition game is in this context defined by the VF (also called characteristic function), which specifies the payoff of the game as a real number under the presumption that only a subset of players participate:

Definition 1 (cf. [6]). *Let $S \subseteq \mathcal{S} := \{1, \ldots, N\}$ be a coalition of N players. A function*

$$v : S \to \mathbb{R} \tag{1}$$

that assigns coalitions to profit is called a value function (VF).

In total, there are 2^N possible coalitions including the empty set.

The individual contribution of each player to the game is represented by the associated SV:

Definition 2 (cf. [6]). *Given a player $i \in \{1, \ldots, N\}$ and a VF v, the corresponding Shapley value (SV)*

$$\Phi_i(v) \equiv \Phi_i := \sum_{S \subseteq \mathcal{S} \setminus \{i\}} w(S) \Delta_i v(S) \tag{2}$$

is a weighted sum over its marginal contributions

$$\Delta_i v(S) := v(S \cup \{i\}) - v(S) \tag{3}$$

when added to all coalitions of other players. The corresponding weights

$$w(S) \equiv w(|S|) := \frac{1}{N} \binom{N-1}{|S|}^{-1} = \frac{|S|! \, (N - |S| - 1)!}{N!} \tag{4}$$

depend on the cardinality of each coalition.

By default, the VF is typically normalized such that $v(\varnothing) := 0$, where $\varnothing$ denotes the empty set. This normalization can be achieved by a linear shift $v(S) \mapsto v(S) - v(\varnothing)$ in (1). However, since the VF enters twice in (2) in form of a subtraction, such a shift has no effect on Φ_i. Hence, we do not presume this kind of normalization in the following.

2.2 Properties

SVs exhibit many desirable properties, for example:

Efficiency $\sum_{i=1}^{N} \Phi_i = v(\mathcal{S}) - v(\varnothing)$.

Symmetry   If $\forall S \subseteq \mathcal{S}\backslash\{i,j\}$, $v(S \cup \{i\}) = v(S \cup \{j\})$, then $\Phi_i = \Phi_j$.

Linearity   For two VFs v and v', $\Phi_i(v) + \Phi_i(v') = \Phi_i(v + v') \forall i \in \mathcal{S}$. For $\alpha \in \mathbb{R}$, $\alpha\Phi_i(v) = \Phi_i(\alpha v) \forall i \in \mathcal{S}$.

Null Player   If $\forall S \subseteq \mathcal{S}\backslash\{i\}$, $v(S \cup \{i\}) = v(S)$, then $\Phi_i = 0$.

In fact, it can be shown that SVs represent the unique attribution method that satisfies these four properties simultaneously [26].

2.3 Probabilistic View

The sum of the weights from (4) over all possible coalitions is normalized according to $\sum_{S \subseteq \mathcal{S}\backslash\{i\}} w(S) = 1$, which allows us to view the coalitions in (2) as a random variable $\boldsymbol{S} \sim \mathbb{P}$ over the set $\mathcal{S}\backslash\{i\}$ with probability mass function $\mathbb{P}(S) := \mathbb{P}(\boldsymbol{S} = S) = w(S)$. This approach, in turn, also makes the marginal contributions from (3) a random variable, that is, $\boldsymbol{V}_i := \Delta_i v(\boldsymbol{S}) \sim \mathbb{P}_i$ with probability mass function

$$\mathbb{P}_i(u) := \mathbb{P}_i(\boldsymbol{V}_i = u) = \sum_{S \subseteq \mathcal{S}\backslash\{i\}} \mathbb{P}(S) \cdot \mathbb{I}\{\Delta_i v(S) = u\}, \tag{5}$$

where the indicator function $\mathbb{I}\{\Delta_i v(S) = u\} \in \{0,1\}$ is 1 if $\Delta_i v(S) = u$ and 0 otherwise. By construction, $\boldsymbol{V}_i$ has only finite support on the discrete set $\Delta_i \mathcal{V} := \{\Delta_i v(S) \mid S \subseteq \mathcal{S}\backslash\{i\}\}$. These random variables allow a different perspective on (2) in form of expectation values. Specifically,

$$\Phi_i = \mathbb{E}[\boldsymbol{V}_i] = \mathbb{E}[\Delta_i v(\boldsymbol{S})]. \tag{6}$$

Hence, the determination of Φ_i can also be understood as estimating one of the corresponding expectation values.

2.4 Higher Moments

Since we have full knowledge about the probability mass function $\mathbb{P}_i$, we can also evaluate higher moments. Specifically, the nth moment reads

$$\mathbb{E}[\boldsymbol{V}_i^n] = \sum_{S \subseteq \mathcal{S}\backslash\{i\}} w(S)\,[\Delta_i v(S)]^n = \mathbb{E}[\Delta_i v(\boldsymbol{S})^n] \tag{7}$$

for $n \in \mathbb{N}$. The first and second moments can be used to calculate the variance [18,19]

$$\sigma_i^2 := \mathbb{E}[\boldsymbol{V}_i^2] - \mathbb{E}[\boldsymbol{V}_i]^2 = \mathbb{E}[\Delta_i v(\boldsymbol{S})^2] - \Phi_i^2. \tag{8}$$

The variance (or the corresponding standard deviation σ_i) is a measure for the dispersion of the marginal contributions.

Therefore, we consider (8) as a measure of the *intrinsic* uncertainty of Φ_i. With "intrinsic" we refer to the fact that the VF $v(S)$ is (so far) presumed to be deterministic such that an uncertainty of Φ_i can only arise from the distribution of marginal contributions V_i that is determined by the choice of $v(S)$. In the next section, we discuss uncertain VFs and their implication for SVs.

3   Uncertain Shapley Values

The evaluation of a VF that is determined by a randomized procedure introduces uncertainty. In typical ML pipelines, randomization may occur as part of the learning algorithm (e. g., dropout [27]), and by the random choice of the training data. A plethora of methods for quantifying and mitigating uncertainty has been studied extensively in ML under the umbrella of Bayesian methods. Surprisingly, such methods are not well-studied in the context of SVs, as described in the introduction.

In this section, we present our definition of SVs with uncertain VFs, or *uncertain SVs* for short. First, we introduce uncertain VFs as random variables that lead to a probability distribution with an expectation value corresponding to the uncertain SVs. Next, we discuss various properties that arise from this definition.

3.1   Uncertain Value Functions

We presume that the VF $v(S)$ in (2) is replaced by an uncertain VF in form of a random variable:

Definition 3. *Let $S \subseteq \mathcal{S} := \{1, \dots, N\}$ be a coalition of N players. The random variable*

$$\boldsymbol{v}(S) \equiv \boldsymbol{v} := v(S) + \boldsymbol{\nu}(S) \tag{9}$$

that consists of the sum of a VF $v(S)$ as defined in (2) and a random variable $\boldsymbol{\nu}(S) \equiv \boldsymbol{\nu} \sim \mathbb{Q}(\cdot \mid S)$ that represents uncertainty or noise in the determination of $v(S)$ is called an uncertain VF. The associated (conditional) probability measure $\mathbb{Q}(\cdot \mid S)$ has real support but is not constrained by additional assumptions.

Our definition is very generic and covers both aleatoric uncertainty (from stochastic errors) and epistemic uncertainty (from systematic errors). Based on this new kind of VF, the marginal contributions $\Delta_i v(S)$, (2), can be replaced according to $\Delta_i v(S) \rightarrow \Delta_i \boldsymbol{v}(S)$ by the marginal contribution under uncertainty $\Delta_i \boldsymbol{v}(S) := \Delta_i v(S) + \boldsymbol{\epsilon}_i(S)$, where $\boldsymbol{\epsilon}_i(S) \equiv \boldsymbol{\epsilon}_i := \boldsymbol{\nu}(S \cup \{i\}) - \boldsymbol{\nu}(S) \sim \mathbb{H}_i(\cdot \mid S)$ denotes a random variable with density

$$h_i(\epsilon \mid S) := \int_{\mathbb{R}} q(\nu \mid S \cup \{i\}) \, q(\nu - \epsilon \mid S) \, \mathrm{d}\nu, \tag{10}$$

which corresponds to a convolution.

3.2   Probabilistic View

With these presumptions, we arrive at the random variable $\tilde{V}_i := V_i + \epsilon_i = \Delta v_i(S) + \nu(S \cup \{i\}) - \nu(S)$ with $S_i \sim \mathbb{P}$ and $\nu(S) \equiv \nu \sim \mathbb{Q}(\cdot \mid S)$, respectively. In particular, the discrete measure $\mathbb{P}_i$, (5), becomes continuous such that $\tilde{V}_i \sim \tilde{\mathbb{P}}_i$ with

$$\tilde{p}_i(u) = \sum_{S \subseteq \mathcal{S} \setminus \{i\}} \mathbb{P}(S) \, h_i(u - \Delta_i v(S) \mid S). \tag{11}$$

This expression can also be rewritten as $\tilde{p}_i(u) = \sum_{S \subseteq \mathcal{S} \setminus \{i\}} w(S) \int_{\mathbb{R}} h_i(\epsilon \mid S)$ $\delta(\Delta_i v(S) + \epsilon - u)\, \mathrm{d}\epsilon$ to highlight the similarity between $\tilde{p}_i$ and its noiseless analogon $\mathbb{P}_i$ from (5). Specifically, the indicator function in (5) translates to the Dirac delta function δ under an integral over the probability density h_i.

These preliminary considerations allow us to define uncertain SVs in analogy to (6):

Definition 4. *Given a player $i \in \{1, \ldots, N\}$ and an uncertain VF $v(S)$ as defined in (9), the corresponding uncertain SV*

$$\tilde{\Phi}_i(v) \equiv \tilde{\Phi}_i := \mathbb{E}\left[\tilde{V}_i\right] \tag{12a}$$

is the expectation value of a random variable $\tilde{V}_i$ as given by (11).

The expression (12a) can be rewritten as

$$\tilde{\Phi}_i = \sum_{S \subseteq \mathcal{S} \setminus \{i\}} w(S) \Delta_i^\epsilon v(S) \quad \text{with} \quad \Delta_i^\epsilon v(S) := \Delta_i v(S) + \mathbb{E}\left[\epsilon_i \mid S\right] \tag{12b}$$

by exploiting the linearity of the expectation value. These two forms of $\tilde{\Phi}_i$ highlight different perspectives, similar to (2) and (6). In (12a), two random variables V_i (from the sum over coalitions) and ϵ_i (from uncertain VFs) contribute to the expectation value. On the other hand, in (12b) only one random variable ϵ_i is considered. This form can also be found in [20]. A third representation

$$\tilde{\Phi}_i = \Phi_i + \Gamma_i \quad \text{with} \quad \Gamma_i(v) \equiv \Gamma_i(\nu) \equiv \Gamma_i := \sum_{S \subseteq \mathcal{S} \setminus \{i\}} \mathbb{P}(S) \mathbb{E}\left[\epsilon_i \mid S\right] = \mathbb{E}\left[\epsilon_i(S)\right] \tag{12c}$$

can be obtained from (12b) if we recall Φ_i from (2). This representation particularly highlights that the difference between uncertain SVs and regular (i.e., noiseless) SVs is given by Γ_i as an expectation value of $\epsilon_i(S) \sim \mathbb{G}_i$ with $g_i(\epsilon) = \sum_{S \subseteq \mathcal{S} \setminus \{i\}} \mathbb{P}(S) \, h_i(\epsilon \mid S)$.

By definition, the mean of the marginal contribution noise $\mathbb{E}\left[\epsilon_i \mid S\right] = \mathbb{E}\left[\nu(S \cup \{i\}) \mid S\right] - \mathbb{E}\left[\nu(S) \mid S\right]$ covers the noise-induced expectation value shift when adding a player i to the coalition S. Hence, Γ_i can be considered as the mean bias of the marginal contributions of player i. If the expectation value of the noise is independent of S (e.g., because h_i is independent of S), Γ_i vanishes.

Summarized, we have found three representations with uncertain SVs, (12a) to (12c). All of them are identical, but correspond to different interpretations as we have explained. In the borderline case of vanishing uncertainty (i.e., $\boldsymbol{\nu}(S) \equiv 0$), the VF becomes deterministic again since $q(\nu \mid S) = \delta(\nu)$, which leads to $h_i(\epsilon \mid S) = \delta(\epsilon)$ and therefore $\Gamma_i = 0$. Accordingly, the uncertain SV reduces to the regular SV, as expected.

3.3 Properties

Our previous considerations allow us to establish a direct connection between uncertain SVs and regular SVs:

Theorem 1. *Let $\mathbf{v}(S) = v(S) + \boldsymbol{\nu}(S)$ be an uncertain VF in the sense of (9) with a deterministic part $v(S)$ and a random part $\boldsymbol{\nu}(S)$. We denote the corresponding uncertain SV, (12), by $\tilde{\Phi}_i(\mathbf{v})$. Then, $\tilde{\Phi}_i(\mathbf{v}) = \Phi_i(v')$ for all $i \in \mathcal{S}$, where $\Phi_i(v')$ denotes the regular SV, (2), with respect to a deterministic VF $v'(S) := v(S) + \gamma(S)$, where $\gamma(S) := \sum_{j \in S} \Gamma_j$ is based on Γ_j from (12c).*

Proof. First, take note that for all $\Gamma_i \in \mathbb{R}$ and $i \in \mathcal{S}$, we can define the noiseless VF $v''(S) := \gamma(S) = \sum_{j \in S} \Gamma_j$ for all $S \subseteq \mathcal{S}$ with the effect that $\Phi_i(v'') = \Gamma_i$ for all $i \in \mathcal{S}$. Second, as a consequence of (12c) and the Linearity property of regular SVs, $\tilde{\Phi}_i(\mathbf{v}) = \Phi_i(v) + \Gamma_i = \Phi_i(v) + \Phi_i(v'') = \Phi_i(v + v'')$ for all $i \in \mathcal{S}$. Finally, define $v'(S) := v(S) + v''(S)$ for all $S \subseteq \mathcal{S}$. $\qquad\square$

Hence, uncertain SVs can in fact be considered as regular SVs with a suitably shifted VF. The shift $\gamma(S)$ corresponds to the cumulated mean bias of the marginal contributions for all players in S.

As a consequence, the properties of regular SVs (Efficiency, Symmetry, Linearity, Null Player) are also uniquely fulfilled with uncertain SVs. Moreover, using (12c), we can straightforwardly rewrite these properties in a form that highlights the uncertainty representation:

Uncertain Efficiency $\sum_{i=1}^{N} \tilde{\Phi}_i = v(\mathcal{S}) - v(\varnothing) + \gamma(\mathcal{S})$.

Uncertain Symmetry   If $\forall S \subseteq \mathcal{S}\backslash\{i,j\}$, $v(S \cup \{i\}) = v(S \cup \{j\})$, then $\tilde{\Phi}_i = \tilde{\Phi}_j + \Gamma_i - \Gamma_j$.

Uncertain Linearity   For two noisy VFs $\boldsymbol{v}$ and $\boldsymbol{v}'$, $\tilde{\Phi}_i(\boldsymbol{v}) + \tilde{\Phi}_i(\boldsymbol{v}') = \tilde{\Phi}_i(\boldsymbol{v} + \boldsymbol{v}') \forall i \in \mathcal{S}$. For $\alpha \in \mathbb{R}$, $\alpha\tilde{\Phi}_i(\boldsymbol{v}) = \tilde{\Phi}_i(\alpha\boldsymbol{v}) \forall i \in \mathcal{S}$.

Uncertain Null Player   If $\forall S \subseteq \mathcal{S}\backslash\{i\}$, $v(S \cup \{i\}) = v(S)$, then $\tilde{\Phi}_i = \Gamma_i$.

Due to the linearity of the expectation value, $\Gamma_i(\boldsymbol{v} + \boldsymbol{v}') = \Gamma_i(\boldsymbol{v}) + \Gamma_i(\boldsymbol{v}')$ and $\Gamma_i(\alpha\boldsymbol{v}) = \alpha\Gamma_i(\boldsymbol{v})$, respectively. We have also made use of $\gamma(\mathcal{S}) = \sum_{i=1}^{N} \Gamma_i$. In addition to the properties of regular SVs, the properties of uncertain SVs from [20] apply straightforwardly based on (9) and (12b). We refer to [20] for a derivation.

3.4 Higher Moments

Higher moments of the random variable $\tilde{V}_i$ can be determined similar to the noiseless case, (7). Specifically, the nth moment reads

$$\mathbb{E}\left[\tilde{V}_i^n\right] = \sum_{k=0}^{n} \binom{n}{k} \sum_{S \subseteq \mathcal{S}\backslash\{i\}} w(S) \left[\Delta_i v(S)\right]^k \mathbb{E}\left[\epsilon_i^{n-k} \mid S\right] \tag{13}$$

for $n \in \mathbb{N}$, where $\mathbb{E}\left[\epsilon_i^n \mid S\right] = \sum_{k=0}^n \binom{n}{k}(-1)^{n-k}\mathbb{E}\left[\nu^k \mid S \cup \{i\}\right]\mathbb{E}\left[\nu^{n-k} \mid S\right]$ follows from (10). Both equalities are direct implications of the binomial theorem. Consequently, any moment of $\tilde{V}_i$ can be explicitly calculated via (13) based on the corresponding moments of the random variable ν—knowledge about the underlying probability measures is not required.

For example, the variance reads

$$\tilde{\sigma}_i^2 := \mathbb{E}\left[\tilde{V}_i^2\right] - \mathbb{E}\left[\tilde{V}_i\right]^2 = \sigma_i^2 + \sigma_{\Gamma_i}^2 + \xi_i \tag{14}$$

with σ_i^2 from (8), the variance of the noise of the marginal contributions $\sigma_{\Gamma_i}^2 := \mathbb{E}\left[\epsilon_i^2\right] - \Gamma_i^2$, and the correlation $\xi_i := 2\sum_{S \subseteq \mathcal{S}\setminus\{i\}} w(S)\Delta_i v(S)\mathbb{E}\left[\epsilon_i \mid S\right] - 2\Phi_i\Gamma_i$. Hence, we find that the noise from the VF introduces a noise uncertainty $\sigma_{\Gamma_i}^2 \geq 0$ and a correlation term $\xi_i \in \mathbb{R}$ in addition to the intrinsic uncertainty $\sigma_i^2 \geq 0$.

4   Explainable Machine Learning

When the concept of SVs is used for XAI, features (or feature indices, to be more precise) take the role of players and the VF is determined by the model output. Suitable VFs can be realized in many different variants leading to different kinds of SVs [12]. For example, explanations can be performed for a single data point or an entire data set and might or might not require a retraining of the model. Uncertain SVs can be used in analogy to regular SVs to achieve explainability. Depending on the choice of the VF, different kinds of random effects will occur. A detailed discussion of such effects, however, is beyond the scope of this paper.

In the present section, we first briefly outline the practical challenge in the evaluation of uncertain SVs and highlight the fundamental difference to the evaluation of regular SVs. Subsequently, we present a simple numerical experiment to demonstrate the effects of randomness of a VF in the context of XAI.

4.1   Evaluation of Uncertain Shapley Values

The task of calculating regular SVs is NP-hard [28] and requires the evaluation of 2^N VFs according to (2). On the other hand, uncertain SVs can by design not be evaluated exactly if no a priori knowledge about the underlying probability distribution is available, i. e., (10) is unknown. Instead, they can only be estimated, which requires to evaluate VFs multiple times.

To compare the evaluation effort of regular SVs and uncertain SVs, we presume in the following that we explicitly sum over all coalitions in (2) and (12b), respectively. For regular SVs, this involves 2^N VF evaluations. For uncertain SVs, we can repeat each evaluation $n \in \mathbb{N}$ times, which in total requires $n2^N$ VF evaluations. An unbiased estimator for $\tilde{\Phi}_i$ is then given by the sample mean $\overline{\Phi}_i := \sum_{S \subseteq \mathcal{S}\setminus\{i\}} w(S)\left[\overline{v}(S \cup \{i\}) - \overline{v}(S)\right]$ with $\overline{v}(S) := \frac{1}{n}\sum_{k=1}^n v_k(S)$ based on the i.i.d. VF samples $v_1(S), \ldots, v_n(S)$ for all $S \subseteq \mathcal{S}$ and $i \in \mathcal{S}$. As is well-known, a confidence interval of $\overline{\Phi}_i$ for sufficiently many samples reads $\overline{\Phi}_i \pm \propto s_i/\sqrt{n}$ with the corresponding sample variance s_i^2.

The increased computational complexity in the evaluation of an uncertain SV in comparison with a regular SV ($n2^N$ instead of 2^N) is the price that has to be paid for an unknown randomness in the VF and its margin depends on the desired confidence of $\overline{\Phi}_i$. A straightforward approximation method for regular SVs is to treat the coalition as a random variable in the sense of (6) and estimate the expectation value. A similar approach can also be employed with uncertain SVs via (12a). For an overview over more advanced approximation methods of regular SVs, we refer to [29–31] and references therein.

4.2   Numerical Experiment

In this section, we study our theoretical results in the context of XAI. Specifically, we discuss the effects of two different kinds of noises, Bernoulli noise and Gaussian noise, on an exemplary VF (that can in principle be calculated noiselessly) and analyze the resulting uncertain SVs. To this end, we consider a synthetic data set created from the method `make_regression` of the Python library `scikit-learn` [32]. This method generates a random linear combination of normal distributions. We sampled $K = 10\,000$ data points $\mathcal{D} = \{(x^{(i)}, y_i)\}_{i=1,\ldots,K}$, each point of index i consisting of twelve features $x^{(i)} \in \mathbb{R}^{12}$ and one target value $y_i \in \mathbb{R}$. All features are informative. Additionally, we specified an additive Gaussian noise level of 0.1. The random seed we used is $97\,531$, which allows for reproducing the data.

Our goal is to study the influence of each feature with respect to the R^2 score $R^2(f, \mathcal{D}) := 1 - \sum_{i=1}^{K}(y_i - f(x^{(i)}))^2 / \sum_{i=1}^{K}(y_i - \overline{y})^2$ with $\overline{y} := \frac{1}{K}\sum_i y_i$, which can be used to quantify the quality of a regression model f, where $f(x^{(i)})$ describes the prediction of the model for the data point $x^{(i)}$. The R^2 score describes the proportional amount of variation in y that can be predicted from x and takes values in $(-\infty, 1]$, with 1 indicating a perfect match between model and data. Based on this score, we can define the VF $v(S) = v(S; f, \mathcal{D}) := R^2(f, \{(x_{S|0}^{(1)}, y_1), \ldots, (x_{S|0}^{(K)}, y_K)\})$, where feature values that are not in the coalition S are set to 0, denoted by $x_{S|0}$ [7]. As our model of choice, we fit a linear regression with intercept term to the data, i.e., $f_{\theta,b}(x) := \theta^\top x + b$.

To obtain two use cases, we define two noisy versions of our VF $v(S; f, \mathcal{D})$ by adding random variables that are independent of S. First, ν follows a Bernoulli distribution with $p = 0.33$ multiplied with the constant $c = 0.05$, which has the effect of randomly adding a constant offset to some values. Second, ν' follows a normal distribution with mean 0 and standard deviation 0.01. We arrive at three VFs: v (noiseless), $v := v + \nu$ (Bernoulli noise), and $v' := v + \nu'$ (Gaussian noise).

We plot in Fig. 1 the probability mass function $\mathbb{P}_i$ and densities $\tilde{p}_i$ according to (5) and (11), respectively, for all $i \in \mathcal{S} = \{1, \ldots, 12\}$. For the Bernoulli noise, the support is discrete and we therefore denote the corresponding probability mass function by $\tilde{\mathbb{P}}_i$. The probability mass function $\mathbb{P}_i$ shown in Fig. 1a is the noiseless distribution of $\Delta_i v(S)$ over all coalitions S, weighted with their respective $w(S)$ for each feature i as defined in (5). Its expectation value corresponds to Φ_i according to (6), which is listed in Table 1 for all $i \in \mathcal{S}$.

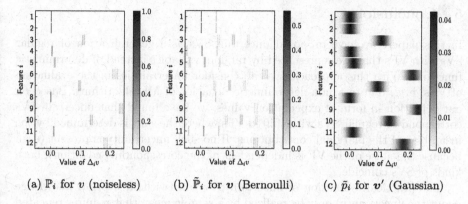

(a) $\mathbb{P}_i$ for v (noiseless)   (b) $\tilde{\mathbb{P}}_i$ for v (Bernoulli)   (c) $\tilde{p}_i$ for v' (Gaussian)

Fig. 1. Probability mass functions $\mathbb{P}_i$ and $\tilde{\mathbb{P}}_i$ and probability density $\tilde{p}_i$ according to (5) and (11), respectively, for all features $i \in \{1, \ldots, 12\}$. Darker colors indicate a higher probability. We investigate the effect of additive Bernoulli and Gaussian noise, respectively, on a noiseless VF.

Table 1. Regular SVs (without uncertainty) and their intrinsic variances w.r.t. $w(S)$, where i denotes the feature index.

i	Φ_i	σ_i^2	i	Φ_i	σ_i^2
1	-0.000050	3.081014×10^{-9}	7	0.095895	8.100267×10^{-6}
2	-0.000009	1.126325×10^{-9}	8	0.000009	2.346858×10^{-10}
3	0.108173	1.008858×10^{-5}	9	0.235202	1.988726×10^{-5}
4	0.003419	3.832217×10^{-7}	10	0.135854	9.707781×10^{-6}
5	0.012733	3.634072×10^{-6}	11	0.010408	1.179264×10^{-6}
6	0.214356	3.949685×10^{-5}	12	0.184340	4.631823×10^{-5}

From (14), we can derive the variance of all uncertain SVs: The correlation terms ξ_i are all equal to zero, as h_i is independent of S. The intrinsic variances σ_i^2 that result from the distribution of the marginal contributions $\Delta_i v$ according to (8) are listed in Table 1. The Bernoulli distribution of ν leads to a symmetrical noise function h_i with mean 0 and support on $\{-c, 0, +c\}$, where both non-zero values occur with probability $p(1-p)$. The distribution of ν', on the other hand, leads to a Gaussian h_i with twice the original variance. Thus, the variance of the noise of the marginal contributions is given by $\sigma_{\Gamma_i}^2 = c^2 \cdot 2p(1-p) = 1.1055 \times 10^{-3}$ for the Bernoulli noise and $2 \cdot 0.01^2 = 2 \times 10^{-4}$ for the Gaussian noise, respectively. According to (14), these values are added to the corresponding σ_i^2 from Table 1 to obtain the final variance. Notice that the variance incurred from the Gaussian or Bernoulli noise term is up to 6 orders of magnitude larger than the variances reported in the table. Hence, the final variance, and consequently the number of samples one has to draw to obtain a specific confidence region is dominated by the noise. Such insights cannot be derived with existing frameworks with uncertain SVs.

5 Conclusions

In this paper, we have proposed uncertain SVs as a generalization of regular SVs with VFs that are represented by random variables instead of deterministic functions. With this approach, we can consider uncertainties in the evaluation of VFs, e.g., due to a non-deterministic behavior of ML algorithms. Based on our definition in form of expectation values, we have found that uncertain SVs correspond to regular SVs with shifted VFs, where the shift is determined by the mean bias of the marginal contributions. If no such uncertainty is present (e.g., because the noise of the VF is independent of the corresponding coalition), both kinds of SVs coincide.

The practical evaluation of uncertain SVs can (without a priori knowledge about the uncertainty) only be realized by a sample mean that requires repeated VF evaluations. Thus, a key difference between uncertain SVs and regular SVs is the higher computational effort required to achieve a desired level of confidence due to the VF uncertainty.

We consider our work as a solid mathematical framework based on first principles that is highly general and can be used as a potential starting point for further research. For example, with respect to different kinds of noises, e.g., multiplicative noise or noise that is correlated for different coalitions. Another possible research direction is the investigation of uncertainty that arises from typical VFs in an XAI context. Furthermore, we think that a study of the effects of uncertain VFs on specialized SV approximations such as *KernelSHAP* [10], *TreeSHAP* [33] or *DeepSHAP* is potentially insightful. It also remains an open question how our uncertainty treatment can be translated to related concepts like Owen values [34] or Banzhaf-Owen values [35]. Finally, quantum machine learning [36] is a particularly promising field of application beyond XAI because randomness is an inherent property of quantum systems.

Acknowledgments. The authors would like to thank Sabine Müller and Moritz Wolter for helpful discussions and constructive feedback. Parts of this research have been funded by the Federal Ministry of Education and Research of Germany and the state of North-Rhine Westphalia as part of the Lamarr-Institute for Machine Learning and Artificial Intelligence (LAMARR22B), as well as by the Fraunhofer Cluster of Excellence Cognitive Internet Technologies (CCIT) and by the Fraunhofer Research Center Machine Learning (FZML).

References

1. Watson, D.S.: Interpretable machine learning for genomics. Hum. Genet. **141**(9), 1499–1513 (2022)
2. Amann, J., Blasimme, A., Vayena, E., Frey, D., Madai, V.I., Precise4Q Consortium.: Explainability for artificial intelligence in healthcare: a multidisciplinary perspective. BMC Med. Inform. Decis. Making, **20**(1), 310 (2020)
3. Zhong, X., Gallagher, B., Liu, S., Kailkhura, B., Hiszpanski, A., Han, T.Y.J.: Explainable machine learning in materials science. NPJ Comput. Mater. **8**(1), 204 (2022)

4. Carta, S., Podda, A.S., Reforgiato Recupero, D., Stanciu, M.M.: Explainable AI for financial forecasting. In: Nicosia, G., et al. (eds.) LOD 2021. LNCS, vol. 13164, pp. 51–69. Springer, Cham (2022). https://doi.org/10.1007/978-3-030-95470-3_5
5. Molnar, C.: Interpretable Machine Learning 2 ed (2022). github.io
6. Shapley, L.S.: A value for n-person games. Contrib. Theory Games **2**(28), 307–317 (1953)
7. Grömping, U.: Estimators of relative importance in linear regression based on variance decomposition. Am. Stat. **61**(2), 139–147 (2007)
8. Štrumbelj, E., Kononenko, I.: An efficient explanation of individual classifications using game theory. J. Mach. Learn. Res. **11**, 1–18 (2010)
9. Štrumbelj, E., Kononenko, I.: Explaining prediction models and individual predictions with feature contributions. Knowl. Inf. Syst. **41**(3), 647–665 (2014)
10. Lundberg, S.M., Lee, S.I.: A unified approach to interpreting model predictions. In: Advances in Neural Information Processing Systems (2017)
11. Merrick, L., Taly, A.: The explanation game: explaining machine learning models using Shapley values. In: Holzinger, A., Kieseberg, P., Tjoa, A.M., Weippl, E. (eds.) CD-MAKE 2020. LNCS, vol. 12279, pp. 17–38. Springer, Cham (2020). https://doi.org/10.1007/978-3-030-57321-8_2
12. Sundararajan, M., Najmi, A.: The many Shapley values for model explanation. In: Daumé III, H., Singh, A.,(eds.) International Conference on Machine Learning of Proceedings of Machine Learning Research, vol. 119, pp. 9269–9278. PMLR (2020)
13. Belle, V., Papantonis, I.: Principles and practice of explainable machine learning. Front. Big Data **4**, 39 (2021)
14. Rozemberczki, B., Watson, L., Bayer, P., Yang, H.T., Kiss, O., Nilsson, S., Sarkar, R.: The shapley value in machine learning. Olivér Kiss (2022)
15. Štrumbelj, E., Kononenko, I., Robnik Šikonja, M.: Explaining instance classifications with interactions of subsets of feature values. Data Knowl. Eng. **68**(10), 886–904 (2009)
16. Li, X., Zhou, Y., Dvornek, N.C., Gu, Y., Ventola, P., Duncan, J.S.: Efficient Shapley explanation for features importance estimation under uncertainty. In: Martel, A.L., et al. (eds.) MICCAI 2020. LNCS, vol. 12261, pp. 792–801. Springer, Cham (2020). https://doi.org/10.1007/978-3-030-59710-8_77
17. Shrikumar, A., Greenside, P., Kundaje, A.: Learning important features through propagating activation differences. In: International Conference on Machine Learning, Proceedings of Machine Learning Research, pp. 3145–3153. JMLR.org (2017)
18. Kargin, V.: Uncertainty of the Shapley value. Int. Game Theory Rev. **07**(04), 517–529 (2005)
19. Fatima, S.S., Wooldridge, M., Jennings, N.R.: An analysis of the Shapley value and its uncertainty for the voting game. In: La Poutré, H., Sadeh, N.M., Janson, S. (eds.) AMEC/TADA -2005. LNCS (LNAI), vol. 3937, pp. 85–98. Springer, Heidelberg (2006). https://doi.org/10.1007/11888727_7
20. Gao, J., Yang, X., Liu, D.: Uncertain Shapley value of coalitional game with application to supply chain alliance. Appl. Soft Comput. **56**, 551–556 (2017)
21. Dai, B., Yang, X., Liu, X.: Shapley value of uncertain coalitional game based on Hurwicz criterion with application to water resource allocation. Group Decis. Negot. **31**(1), 241–260 (2022)
22. Yang, X., Gao, J.: Uncertain differential games with application to capitalism. J. Uncertainty Anal. Appl. **1**(1), 17 (2013)
23. Liu, B.: Uncertainty Theory - A Branch of Mathematics for Modeling Human Uncertainty, vol. 300. Springer, Heidelberg (2010)

24. Hart, S.: Game Theory, chapter Shapley Value, pp. 210–216. Palgrave Macmillan UK, London (1989)
25. Aumann, R.J., Shapley, L.S.: Values of Non-Atomic Games. Princeton University Press, New Jersey (1974)
26. Dubey, P.: On the uniqueness of the Shapley value. Internet. J. Game Theory **4**, 131–139 (1975)
27. Srivastava, N., Hinton, G., Krizhevsky, A., Sutskever, I., Salakhutdinov, R.: Dropout: a simple way to prevent neural networks from overfitting. J. Mach. Learn. Res. **15**(56), 1929–1958 (2014)
28. Deng, X., Papadimitriou, C.H.: On the complexity of cooperative solution concepts. Math. Oper. Res. **19**(2), 257–266 (1994)
29. Aas, K., Jullum, M., Løland, A.: Explaining individual predictions when features are dependent: more accurate approximations to Shapley values. Artif. Intell. **298**, 103502 (2021)
30. Touati, S., Radjef, M.S., Sais, L.: A Bayesian Monte Carlo method for computing the Shapley value: application to weighted voting and bin packing games. Comput. Oper. Res. **125**, 105094 (2021)
31. Mitchell, R., Cooper, J., Frank, E., Holmes, G.: Sampling permutations for Shapley value estimation (2021)
32. Pedregosa, F., et al.: Scikit-learn: machine learning in Python. J. Mach. Learn. Res. **12**, 2825–2830 (2011)
33. Lundberg, S.M., et al.: From local explanations to global understanding with explainable AI for trees. Nat. Mach. Intell. **2**(1), 56–67 (2020)
34. López, S., Saboya, M.: On the relationship between Shapley and Owen values. Cent. Eur. J. Oper. Res. **17**(4), 415 (2009)
35. Saavedra-Nieves, A., Fiestras-Janeiro, M.G.: Sampling methods to estimate the Banzhaf-Owen value. Ann. Oper. Res. **301**(1), 199–223 (2021)
36. Cerezo, M., Verdon, G., Huang, H.Y., Cincio, L., Coles, P.J.: Challenges and opportunities in quantum machine learning. Nat. Comput. Sci. **2**(9), 567–576 (2022)

Revised Conditional t-SNE: Looking Beyond the Nearest Neighbors

Edith Heiter[1(✉)], Bo Kang[1], Ruth Seurinck[1,2], and Jefrey Lijffijt[1]

[1] Ghent University, Ghent, Belgium
{edith.heiter,bo.kang,ruth.seurinck,jefrey.lijffijt}@ugent.be
[2] VIB Center for Inflammation Research, Ghent, Belgium

Abstract. Conditional t-SNE (ct-SNE) is a recent extension to t-SNE that allows removal of known cluster information from the embedding, to obtain a visualization revealing structure beyond label information. This is useful, for example, when one wants to factor out unwanted differences between a set of classes. We show that ct-SNE fails in many realistic settings, namely if the data is well clustered over the labels in the original high-dimensional space. We introduce a revised method by conditioning the high-dimensional similarities instead of the low-dimensional similarities and storing within- and across-label nearest neighbors separately. This also enables the use of recently proposed speedups for t-SNE, improving the scalability. From experiments on synthetic data, we find that our proposed method resolves the considered problems and improves the embedding quality. On real data containing batch effects, the expected improvement is not always there. We argue revised ct-SNE is preferable overall, given its improved scalability. The results also highlight new open questions, such as how to handle distance variations between clusters.

1 Introduction

Motivation. t-distributed Stochastic Neighbor Embedding (t-SNE) is widely used to compute low-dimensional visualizations for high-dimensional data. Conditional t-SNE (ct-SNE) [3] is an extension of t-SNE that allows to factor out prior knowledge from the embedding. Providing discrete labels for all data points to ct-SNE allows same-labeled points to be embedded further apart than their distances would require—ideally revealing complementary structure present in the data.

We illustrate the idea of conditional t-SNE using a synthetic dataset ($n = 1500$, $d = 10$) in Fig. 1a. The data is generated such that each point belongs to one of two clusters in dim 1–4 (**blue**, orange) and one of three clusters in dim 5–6 (O, △, □). The remaining four dimensions are Gaussian noise. The t-SNE embedding in Fig. 1a shows a separate cluster for each class label combination. If we already know about the clustering in dim 1–4, we could provide the labels **blue**/orange to ct-SNE. Ideally, this would reveal the remaining structure in the data: Three clusters in dimension 5–6 (O, △, □). Figure 1b shows that ct-SNE wrongly merges the clusters, i.e., ● are merged with ▲, while revised ct-SNE does show the correct clusters (Fig. 1c).

B. Crémilleux et al. (Eds.): IDA 2023, LNCS 13876, pp. 169–181, 2023.
https://doi.org/10.1007/978-3-031-30047-9_14

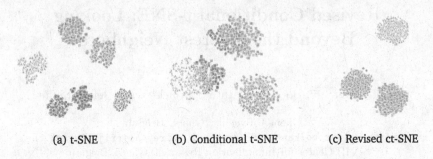

| (a) t-SNE | (b) Conditional t-SNE | (c) Revised ct-SNE |

Fig. 1. Illustration of label information used by ct-SNE and revised ct-SNE. (a) t-SNE shows the labeled data consists of several clusters. Provided with the class labels **blue** and orange, *ct-SNE (b) merges points from both colors–but the wrong shapes*, while revised ct-SNE (c) shows the three expected clusters. (Color figure online)

Use Cases of ct-SNE. (Revised) ct-SNE retains the unsupervised nature of t-SNE while adding supervision through labels to explicate what is *not* the target. This stands in contrast to supervised dimensionality reduction methods that incorporate label information to improve downstream prediction tasks, for example by increasing class separation in low-dimensional embeddings (see, e.g., [2,10]). As such, revised ct-SNE is useful in a situation where t-SNE is useful and when additionally there is known unwanted structure in the data. This may be in an iterative EDA setting, when clusters are identified, explored, and labeled, after which the user wants to explore further. Another setting is when label information about prominent structure in the data is available a priori, and this information acts as a confounder [3].

The presence of undesired and known class separation occurs for example with biological data containing single-cell RNA samples from various sources. The unwanted class separation is called the *batch effect* and can be seen as variation in the data that does not have a biological explanation. It often occurs when combining samples from different organisms, tissues, or when cells have been sequenced with different technologies. For this setting also other t-SNE variants have been proposed. Poličar et al. [8] suggest to embed one dataset (batch) using t-SNE and use this as a reference embedding. The other datasets are then embedded sample by sample on top of the initial embedding. This is different from our (revised) ct-SNE, in that it prevents any interaction of same-labeled samples by design. In addition, (revised) ct-SNE allows the user to tune the degree of class separation.

Contributions. In this paper we provide a thorough analysis of the root cause of ct-SNE's failures. We identify that the approximation of high-dimensional similarities discards essential structural information. In addition, the asymmetry of the KL-divergence hinders ct-SNE in achieving its goal. To overcome these limitations, we propose two modifications. First, we compute distances for same-labeled and differently-labeled neighbors separately. Second, we condition the high-dimensional instead of the low-dimensional similarities. Finally, we implemented revised ct-SNE into FIt-SNE [6] which leads to a considerable speed-up.

2 Background: Conditional t-SNE

In this section we review ct-SNE and point out details that might negatively affect the embedding quality. The objective of ct-SNE is to embed a dataset $X \in \mathbb{R}^{n \times d}$ to a lower dimension $Y \in \mathbb{R}^{n \times d'}$ with $d' \ll d$ by minimizing the Kullback-Leibler (KL) divergence between pairwise similarities in the high (HD) and low-dimensional (LD) space. The HD similarities

$$p_{j|i} = \frac{\exp(-\|x_i - x_j\|^2 / 2\sigma_i^2)}{\sum_{k \neq i} \exp(-\|x_i - x_k\|^2 / 2\sigma_i^2)}, \qquad p_{ij} = \frac{p_{i|j} + p_{j|i}}{2n}$$

are defined with a point-specific kernel bandwidth σ_i that depends on the density of the neighborhood around each point. It is computed by binary search such that each similarity distribution p_i has the same user-defined perplexity u. The LD similarities are based on a t-distribution

$$q_{ij} = \frac{\left(1 + \|y_i - y_j\|^2\right)^{-1}}{\sum_{k \neq l} \left(1 + \|y_k - y_l\|^2\right)^{-1}}.$$

In ct-SNE, the LD similarities are conditioned on the label matrix $\Delta \in \{0,1\}^{n \times n}$, based on the idea that $q_{ij|\Delta}$ should be higher for pairs of points with the same label ($\delta_{ij} = 1$) than for points with a different label ($\delta_{ij} = 0$). The conditional LD similarities are defined as

$$r_{ij} = q_{ij|\Delta} = \begin{cases} \alpha q_{ij}/U & \text{if } \delta_{ij} = 1 \\ \beta q_{ij}/U & \text{if } \delta_{ij} = 0 \end{cases}$$

and normalized with $U = \alpha \sum_{k \neq l: \delta_{kl}=1} q_{kl} + \beta \sum_{k \neq l: \delta_{kl}=0} q_{kl}$. We refer to the original publication for the detailed derivation and the exact relationship between the parameters $\alpha > 1 > \beta > 0$. Minimizing $\mathrm{KL}(p \,\|\, r)$ with $\alpha > \beta$ requires differently-labeled points to be embedded closer to each other to still match their pairwise HD similarity.

We illustrate the effect of ct-SNE on its gradient

$$\nabla_{y_i} \mathrm{KL}(p \,\|\, r) = 4 \sum_{j \neq i} \left(\underbrace{p_{ij} q_{ij} Z(y_i - y_j)}_{\text{attractive force}} - \underbrace{\frac{\delta_{ij}\alpha + (1 - \delta_{ij})\beta)}{U} q_{ij}^2 Z(y_i - y_j)}_{\text{repulsive force}} \right)$$

with $Z = \sum_{k \neq l} (1 + \|y_k - y_l\|^2)^{-1}$. The attractive part will pull neighboring points i and j closer together while the repulsive part pushes all points apart. First, we note that ct-SNE increases (decreases) the repulsive force between same-labeled (differently-labeled) points, while the attractive forces are the same as in t-SNE. To speed up the computation of the attractive forces van der Maaten [7] proposed to exploit the fast decay of the Gaussian kernel and retain only the similarities $p_{\cdot|i}$ for the set of $|\mathcal{N}_i| = 3u$ nearest neighbors, where u is the perplexity.

The Problem. The goal of ct-SNE is to bring secondary structure to the front by discounting certain points, i.e., increasing repulsive forces for points with

samples inside the
3u neighborhood

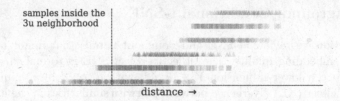

distance →

Fig. 2. High-dimensional distances to a random ● point aggregated per label. Only samples to the left of the dashed vertical line (all from cluster ● or ■) will exert an attractive force. (Color figure online)

the same label. With a fixed number of $3u$ nearest neighbors—that directly affect the placement of each point—ct-SNE can still only show new structure that reaches into this neighborhood. For the synthetic data, the cluster sizes are larger than $3u$, hence points with different labels are by definition outside the $3u$ neighborhood. In Fig. 2 we show the points that are part of $\mathcal{N}_\bullet$. Since the neighbors all have the same **blue** label, ct-SNE has no information on the similarity to orange labeled points. The overlap between the ● and ■ points in Fig. 1b occurs solely due to the decreased repulsive forces between differently-labeled samples. It is coincidental and *wrong*, in the sense points from ● are closer to ●, but this information is omitted and a wrong solution emerges (Fig. 1b).

Alternative Solutions. Increasing the neighborhood size does not lead to the desired results. We explain two ways that seem promising to circumvent the problem but do not work in practice. First, one could keep all pairwise HD similarities instead of approximating them. This results in non-zero attractive forces for differently-labeled points, but increases the complexity to compute these forces in every gradient update to $\mathcal{O}(n^2)$. In addition, the KL-divergence is asymmetric and weighs high p_{ij} to be more important to match with the LD similarity than small similarities. We implemented this method and see in Fig. 3a that the embedding on the synthetic data has barely changed. We presume the HD similarities are too small to have an effect on the embedding.

The second idea is to increase the perplexity. A higher perplexity will increase the neighborhood size by definition and differently-labeled neighbors might get assigned a higher similarity than with the first solution (and smaller perplexity). For large datasets with few class labels to be factored out, this might still be impractical, because it could be necessary to use a perplexity of $n/2$ to have sufficiently high attractive forces[1]. What is even more unfavorable is the loss of *locality* that goes hand in hand with a higher perplexity and stands in opposition with the original idea of t-SNE to preserve local neighborhoods. Figure 3b shows that on the synthetic data a high perplexity indeed leads to better preservation of the similarities between differently-labeled points, but locality is lost and instead two clusters emerge, instead of the expected three clusters.

---

[1] Assuming the distances between differently-labeled samples are larger than between same-labeled samples. A perplexity of $\frac{n}{2}$ would only assign same-labeled points a high similarity.

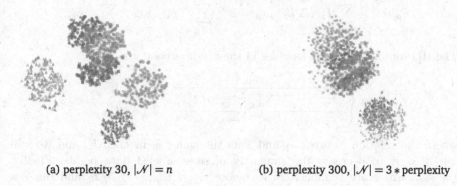

(a) perplexity 30, $|\mathcal{N}| = n$ (b) perplexity 300, $|\mathcal{N}| = 3 *$ perplexity

Fig. 3. Visualizations of ct-SNE embeddings ($\beta = 1\mathrm{e}{-}12$) with color label as prior knowledge. Not approximating the similarities in (a), the clusters still overlap arbitrarily as the attractive forces are too small. A higher perplexity (b) leads to correctly merged clusters that are not well separated. Changing β does not affect the results.

3 Revised Conditional t-SNE

In this section we argue how an adjusted approximation and a different formulation of ct-SNE might help retain important neighborhood information when factoring out prior knowledge. We propose two changes to ct-SNE to provide enough structural information about differently-labeled nearest neighbors and still discount the similarity to same-labeled points.

Expanding the Set of Nearest Neighbors. First, we search for nearest neighbors separately for same and differently-labeled points. We use $\mathcal{N}_{i,\delta_{ij}=1}$ as the set of $1.5u$ nearest neighbors with the same class label as i and $\mathcal{N}_{i,\delta_{ij}=0}$ denotes the set with $1.5u$ differently-labeled nearest neighbors. This does not add runtime to the gradient updates (when still using $3u$ neighbors in total) but requires to build and search in separate nearest-neighbor data structures (e.g., vantage-point trees [11], ANNOY [1]) for each label. As we saw in Fig. 3a, this change alone will not be sufficient.

Condition the High-Dimensional Similarities. The second change is to condition the HD instead of the LD similarities. This will affect the attractive forces, in contrast to the repulsive forces in ct-SNE. We define

$$r_{j|i} = P_{p_i}(j|\Delta) = \frac{P_{p_i}(\Delta|j) \cdot p_{j|i}}{P_{p_i}(\Delta)},$$

where $P_{p_i}(\Delta|j) = \frac{\prod_l n_l!}{n!}\beta^{\delta_{ij}}\alpha^{1-\delta_{ij}}$ is defined as in ct-SNE but we flipped the parameters[2]. The notation P_{p_i} denotes that we compute the similarity distribution for a fixed sample i and the corresponding values of $p_{<\cdot>|i}$. The marginal probability is also defined for each i separately as

---

[2] We use α and β instead of α' and β' as in the original paper [3], and thus need to normalize with the number of distinct label assignments.

$$P_{p_i}(\Delta) = \sum_{k \neq i} P_{p_i}(\Delta|k) \cdot p_{k|i} = \beta \sum_{k \neq i: \delta_{ik}=1} \cdot p_{k|i} + \alpha \sum_{k \neq i: \delta_{ik}=0} p_{k|i}.$$

The HD similarities for all samples in the neighborhood are

$$r_{j|i} = \begin{cases} \dfrac{\beta p_{j|i}}{\beta \sum_{k \neq i: \delta_{ik}=1} p_{k|i} + \alpha \sum_{k \neq i: \delta_{ik}=0} p_{k|i}} & \text{if } j \in \mathcal{N}_{i, \delta_{ij}=1} \\ \dfrac{\alpha p_{j|i}}{\beta \sum_{k \neq i: \delta_{ik}=1} p_{k|i} + \alpha \sum_{k \neq i: \delta_{ik}=0} p_{k|i}} & \text{if } j \in \mathcal{N}_{i, \delta_{ij}=0}, \end{cases}$$

where the relation between α and β is the same as in ct-SNE, and we will use $\beta < \alpha$ to decrease the similarity of same-labeled data points. Finally, the similarities are symmetrized as before $r_{ij} = (r_{j|i} + r_{i|j})/2n$ and the loss $KL(r \parallel q)$ measures the KL-divergence between the conditioned HD similarities and the LD similarities. Since the input similarities do not depend on the embeddings y_i, the gradient is the same as in t-SNE with r_{ij} instead of p_{ij}. This allows us to integrate our changes into FIt-SNE [6] offering a fast interpolation-based acceleration of the gradient computation. Our code is available at github.com/aida-ugent/revised-conditional-t-SNE.

Estimating the Point-Wise Variance. This new formulation of adjusting the HD similarities raises the question whether the point-wise variance of the Gaussian kernel should be computed using r_i or p_i. On r_i, the binary search for the variance satisfying the user-defined perplexity is not well-defined as the perplexity is not monotonously increasing with the variance. Thus, two or more possible solutions exist that can have opposite characteristics as shown in Fig. 4. The other option is to first estimate the variance on p_{ij} and then change the similarities with α or β. However, the effective perplexity of r_{ij} might differ from the specified perplexity defined by the user.

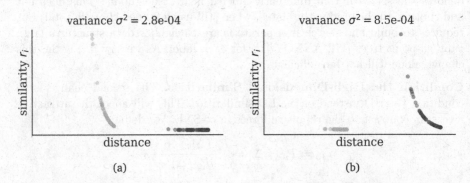

(a)   (b)

Fig. 4. Two similarity distributions r_i with $\beta = 1e-4$ where a different variance leads to the same perplexity of 50. The distances are computed for cell $i = 12823$ (indrop, beta) of the pancreas dataset and the colors correspond to the technology label. (Color figure online)

4   Evaluation

To compare revised ct-SNE with ct-SNE we provide experimental results on a synthetic and two biological datasets. We first describe the evaluation setup including the chosen quality measures and then describe the results. Embeddings of revised ct-SNE with variance estimation on p_i and experimental results on the second biological dataset can be found in the supplement at arxiv.org/abs/2302.03493.

4.1   Setup

We first provide the characteristics of the datasets and then define the evaluation measures. All experiments were run on a laptop with Intel® Core™ i7-10850H CPU @ 2.70 GHz with 16 GB RAM.

Datasets

Synthetic data Each point in this $n = 1500, d = 10$ dataset belongs to one of two clusters in dimensions 1–4 and one of three clusters in dimensions 5–6. The cluster centers are sampled from $\mathcal{N}(0, 25)$ and $\mathcal{N}(0, 1)$ respectively. For each point, we add noise from $\mathcal{N}(0, 0.01)$ to the cluster centers and append four dimensions of noise from $\mathcal{N}(0, 1)$. The clusters in dim 5–6 are of equal size, while 600 points belong to blue and 900 to the orange cluster. We provide the cluster labels of dim 1–4 as prior knowledge to ct-SNE and expect the embedding to show the structure implanted in dimensions 5–6.

Pancreas data [9] is a widely-used single cell RNAseq dataset ($n = 14890, d = 34363$) to benchmark data integration methods. It contains gene counts of human pancreatic islets cells from 8 sources sequenced with 5 different technologies—SMARTSeq2 (2394), Fluidigm C1 (638), CelSeq (1004), CelSeq2 (2285), and inDrops (8569). We provide the technology labels as prior knowledge to merge cells from different technologies together and expect a grouping according to the 13 celltypes. We followed the standard preprocessing steps for single-cell RNA datasets including the selection of 2000 highly-variable genes, normalization, standardization, and PCA to retain 50 principal components.

Evaluation Measures. To compare the embeddings quantitatively, we compute a normalized HD and LD neighborhood overlap score [4, 5] and the degree of label mixing with the *Laplacian score* that was also used to evaluate ct-SNE. For both measures we use a fixed neighborhood size equal to the perplexity which is 30 for the synthetic and 50 for the pancreas dataset. For the pancreas data, we compute both measures on a random subset of 5% of the data.

R_{NX} neighborhood preservation measures the normalized agreement of HD and LD neighborhoods as proposed by Lee and Verleysen [5]. Denoting the

k-sized HD and LD neighborhoods of data point i as v_i^k and n_i^k, the average neighborhood overlap rate is defined as

$$Q_{NX}(k) = \frac{1}{kn} \sum_{i=1}^{n} |v_i^k \cap n_i^k|.$$

Since a random embedding would yield a score of $\mathbb{E}[Q_{NX}(k)] = \frac{k}{n-1}$, these values are scaled to $R_{NX}(k) = \frac{(n-1)Q_{NX}(k)-k}{n-1-k} \in [0,1]$, measuring the improvement over a random embedding. We adjust this measure to reflect the idea of factoring out class label information. Given a set of LD neighbors n_i^k, we ensure that v_i^k contains equally many points with the same (and different) label, i.e., $|\{j \mid j \in v_i^k, \delta_{ij} = 1\}| = |\{j \mid j \in n_i^k, \delta_{ij} = 1\}|$. The distribution of same and differently labeled neighbors is determined by the embedding and differs per point.

Laplacian scores proposed by Kang et al. [3] measure the fraction of LD nearest neighbors with a different label. It can be compared to a baseline with random label assignment, where the expected Laplacian score is $\sum_{l \in L} \frac{n_l(n-n_l)}{n(n-1)}$ where label $l \in L$ has n_l samples. When factoring out structure encoded by a labeling of the data (e.g. dim 1–4 labels for the synthetic data or the technology feature for pancreas), we expect an increase of the Laplacian score evaluated on the same label. An increase of the Laplacian evaluated on a different class label is not necessarily desirable.

4.2   Results

We compare embeddings of t-SNE with embeddings by ct-SNE and revised ct-SNE for different values of β, and using variance estimation with either r_i or p_i.

Synthetic Data. To embed the synthetic dataset, we use a perplexity of $u = 30$, $\theta = 0.2$, and 750 epochs. This took about 10s for all methods. The R_{NX} and Laplacian scores are shown in Fig. 5 and in Fig. 1 we show embeddings of ct-SNE with $\beta = 1e{-}4$ and revised ct-SNE with $\beta = 1e{-}20$ as they score highest on the Laplacian (dim 1–4). The t-SNE embedding has a Laplacian score of 0 since the 30 nearest neighbors have the same labels (same color and shape) as we can visually confirm in Fig. 1a. Revised ct-SNE converges to a Laplacian score (dim 1–4) equivalent to a random embedding. We conclude that the structure captured by the labels in dimensions 1–4 has successfully been factored out in the embedding. The embedding by ct-SNE scores lower on the Laplacian (dim 1–4 labels) but higher when using the labels in dimensions 5–6 (O, △, □). This indicates that not only the imposed structure in dimensions 1–4 but also from dimensions 5–6 has been erroneously removed. The HD neighborhoods for a fixed level of the Laplacian (dim 1–4) are more accurately preserved by revised ct-SNE as shown in Fig. 5b.

Pancreas Data. To embed the pancreas data we use a perplexity of 50, $\theta = 0.5$, and 1000 iterations and show the t-SNE and conditional embeddings with

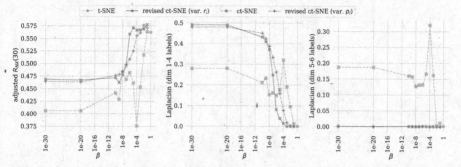

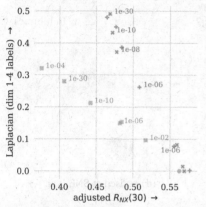

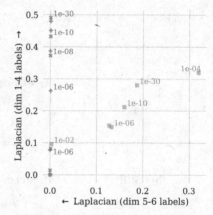

(a) R_{NX} and Laplacian scores for different values of β evaluated on neighborhoods of size $k = 30$. The t-SNE scores are depicted at $\beta = 1$.

(b) Trade-off between R_{NX} and Laplacian scores based on the dim1-4 labels.

(c) Trade-off between Laplacian scores.

Fig. 5. Evaluation results for the synthetic dataset. In (a) we show the neighborhood agreement and Laplacian for varying β, while (b) and (c) allow to compare two of the measures for a subset of all β values.

$\beta = 1e-30$ in Fig. 6. The runtime of ct-SNE for the pancreas dataset is 168 s compared to 28 s for revised ct-SNE. The evaluation results depicted in Fig. 6a show that the neighborhood agreement $R_{NX}(50)$ drops significantly from 0.44 for t-SNE to 0.30 (ct-SNE) and 0.24 (revised ct-SNE) with $\beta = 1e-30$. Conditional t-SNE and revised ct-SNE both converge to a Laplacian (technology) score that is lower than the score for a random mixing of labels (0.61). The Laplacian for the celltype labels however is high for ct-SNE, indicating a mix of different celltypes in the local neighborhoods. Indeed, visualization of the trade-offs (Figs. 7a and 7b) learns us that ct-SNE manages to retain R_{NX} better, while mixing the cells from different technologies, whereas revised ct-SNE leads to better trade-offs between the Laplacian scores (less mixing between celltypes, while mixing samples from different batches).

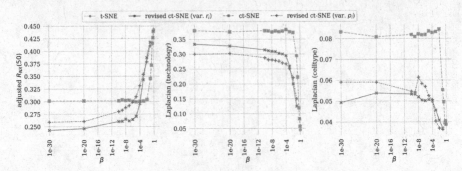

(a) R_{NX} and Laplacian scores for different values of β evaluated on neighborhoods of size $k = 50$. The t-SNE scores are depicted at $\beta = 1$.

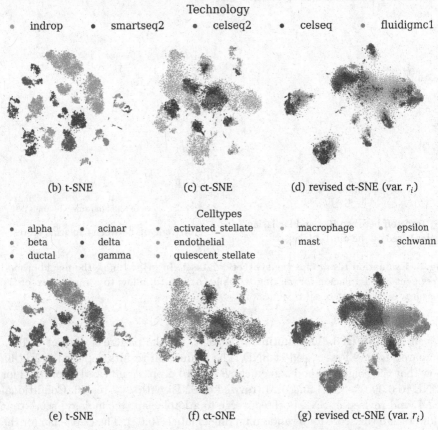

Fig. 6. Visualizations and Laplacian scores of pancreas data embeddings where the technology labels are provided as prior information to ct-SNE and revised ct-SNE with $\beta = 1\text{e}{-}30$. Cell coloring by technology (b)–(d) and cell type (e)–(g). (Color figure online)

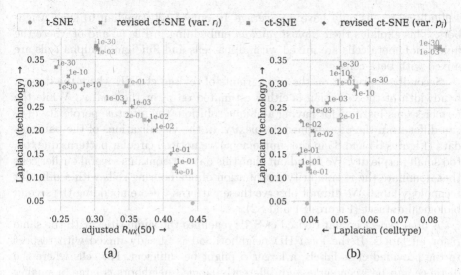

Fig. 7. Evaluation scores for embeddings of the pancreas data. Trade-offs between R_{NX} and Laplacian in (a) and Laplacian technology versus celltype in (b).

We also notice that $\beta > 1e-4$ is sufficient for ct-SNE while the Laplacian (technology) for revised ct-SNE only plateaus for smaller values of β. We speculate that revised ct-SNE requires smaller values as we change Gaussian HD similarities instead of values from a fat-tailed t-distribution. The trade-offs visualized in Figs. 7a and 7b suggest that $\beta = 0.1$ for ct-SNE and $\beta = 1e-4$ for revised ct-SNE might be suitable starting points that can be adjusted in both directions. Finally, the differences in evaluation scores between point-wise variance estimation using r_i and p_i are small and inconclusive.

5   Discussion

The experiments showed that revised ct-SNE outperformed ct-SNE on the synthetic dataset. On the pancreas dataset, ct-SNE achieved a better trade-off between neighborhood preservation and mixing cells from different technologies, but at the same time in a higher fraction of neighbors with a different celltype. Revised ct-SNE benefits from the FIt-SNE implementation, leading to a faster runtime.

First, we explain the different results of ct-SNE on the two datasets. The original implementation of ct-SNE can suppress a grouping of data points according to given labels, but its ability to reveal larger structures is limited. We showed that the revised formulation avoids *random* placement of points in both the synthetic and the pancreas dataset. The original ct-SNE performs better on the pancreas data than on the synthetic data, because most local neighborhoods also contain cells with a different technology label—which is not the case in the synthetic data. For cells sequenced with ● celseq, there are on average 68 out of 150 nearest neighbors with a different technology label (median 63). However,

● fluidigmc1 cells only have on average 10 out of 150 (median 0) such neighbors. This explains their *almost random* embedding in Fig. 6c and 6f where the fluidigmc1, beta cells are mixed with alpha cells and fluidigmc1, alpha cells are mixed with beta cells.

Secondly, we reflect on the two variants of revised ct-SNE where the desired bandwidth of the Gaussian is either estimated on p_i or the final r_i. While the former seems justified for having a unique solution, the *effective* perplexity of r_i can differ from the user-defined perplexity. In the visualizations of the pancreas data (Figures S4 and S5 in the supplement) we noticed circular patterns due to a too small perplexity. We found out that this dataset contains several outlier cells that dominate the r_i similarity distribution of neighboring cells with a different technology label. We did not observe these patterns when embedding the second biological dataset (Figures S1 and S2).

A final aspect is that revised ct-SNE redefines the similarities with the same β for all labels. If the local HD neighborhood is already mixed with respect to the provided class labels, a larger β might be sufficient. For cells where the distance gap between same and differently-labeled neighbors is large, a smaller β is necessary to reweigh the similarities. In Fig. 6 we show the embeddings with $\beta = 1e-30$ which might overshoot the goal for some cells. That means, a too small β can remove the neighborhood information of same-labeled points completely. In summary, neither computing the variance on p_i nor on r_i ensures a stable balance between same and differently-labeled neighbors.

Future Work. Based on the result that revised ct-SNE addresses some shortcomings of ct-SNE but brings a different set of drawbacks, we see various avenues for future work. Firstly we assume that the R_{NX} scores for revised ct-SNE embeddings would increase when the same-labeled similarities do not vanish for small β. Additionally, one could allow for separate *merging strengths* for every label, based on the assumption that all labels should be mixed in the resulting embedding. To ensure a certain trade-off between effective same and differently-labeled neighbors one could compute two separate similarity distributions with two different perplexities. This is a similar idea as implemented in Class-aware t-SNE [2] where the perplexity is adjusted to reach a certain ratio of same-labeled points in the neighborhood. Finally, the embeddings by ct-SNE and revised ct-SNE could be compared to the reference embeddings by Poličar et al. [8] or a combination of data integration methods for biological datasets and t-SNE.

6   Conclusion

We presented revised ct-SNE to find low-dimensional embeddings that show structure beyond a previously known clustering. Conditional t-SNE can fail to reveal this structure when focusing only on the local neighborhoods using small perplexities. To resolve this limitation, we reformulate the original idea to condition the high-dimensional similarities and explicitly include nearest neighbors with different labels. Our experiments on synthetic data confirmed that the revised ct-SNE improved on the criticized aspects of the original method, but ct-SNE performed better in terms of label mixing and neighborhood preservation on real-world single-cell data. Finally, we investigated limitations of revised ct-SNE and proposed to control the number of effective same and differently-labeled neighbors more explicitly.

Acknowledgements. The authors would like to thank Tijl De Bie and Yvan Saeys for the discussions about conditional t-SNE. This research was funded by the ERC under the EU's 7th Framework and H2020 Programmes (ERC Grant Agreement no. 615517 and 963924), the Flemish Government (AI Research Program), and the FWO (project no. 11J2322N, G0F9816N, 3G042220).

References

1. Bernhardsson, E.: Annoy: approximate nearest neighbors in C++/Python (2013). https://github.com/spotify/annoy
2. de Bodt, C., Mulders, D., Sánchez, D.L., Verleysen, M., Lee, J.A.: Class-aware t-SNE: cat-SNE. In: ESANN (2019)
3. Kang, B., García García, D., Lijffijt, J., Santos-Rodríguez, R., De Bie, T.: Conditional t-SNE: more informative t-SNE embeddings. Mach. Learn. **110**(10), 2905–2940 (2021)
4. Lee, J.A., Peluffo-Ordóñez, D.H., Verleysen, M.: Multi-scale similarities in stochastic neighbour embedding: reducing dimensionality while preserving both local and global structure. Neurocomputing **169**, 246–261 (2015)
5. Lee, J.A., Verleysen, M.: Quality assessment of dimensionality reduction: rank-based criteria. Neurocomputing **72**(7–9), 1431–1443 (2009)
6. Linderman, G.C., Rachh, M., Hoskins, J.G., Steinerberger, S., Kluger, Y.: Fast interpolation-based t-SNE for improved visualization of single-cell RNA-seq data. Nat. Methods **16**(3), 243–245 (2019)
7. van der Maaten, L.: Accelerating t-SNE using tree-based algorithms. JMLR **15**(1), 3221–3245 (2014)
8. Poličar, P.G., Stražar, M., Zupan, B.: Embedding to reference t-SNE space addresses batch effects in single-cell classification. Mach. Learn. **112**, 721–740 (2021)
9. Satija Lab: panc8.SeuratData: Eight Pancreas Datasets Across Five Technologies (2019). R package version 3.0.2
10. Vu, V.M., Bibal, A., Frénay, B.: HCt-SNE: hierarchical constraints with t-SNE. In: International Joint Conference on Neural Networks (IJCNN), pp. 1–8. IEEE (2021)
11. Yianilos, P.N.: Data structures and algorithms for nearest neighbor search in general metric spaces. In: Proceedings of ACM-SIAM Symposium on Discrete algorithms, vol. 66, pp. 311–321 (1993)

On the Change of Decision Boundary and Loss in Learning with Concept Drift

Fabian Hinder[✉][iD], Valerie Vaquet[iD], Johannes Brinkrolf[iD],
and Barbara Hammer[iD]

CITEC, Bielefeld University, Bielefeld, Germany
{fhinder,vvaquet,jbrinkro,bhammer}@techfak.uni-bielefeld.de

Abstract. Concept drift, i.e., the change of the data generating distribution, can render machine learning models inaccurate. Many technologies for learning with drift rely on the interleaved test-train error (ITTE) as a quantity to evaluate model performance and trigger drift detection and model updates. Online learning theory mainly focuses on providing generalization bounds for future loss. Usually, these bounds are too loose to be of practical use. Improving them further is not easily possible as they are tight in many cases. In this work, a new theoretical framework focusing on more practical questions is presented: change of training result, optimal models, and ITTE in the presence (and type) of drift. We support our theoretical findings with empirical evidence for several learning algorithms, models, and datasets.

Keywords: Concept Drift · Stream Learning · Learning Theory · Error Based Drift Detection

1 Introduction

The world that surrounds us is subject to constant change, which also affects the increasing amount of data collected over time, in social media, sensor networks, IoT devices, etc. Those changes, referred to as concept drift, can be caused by seasonal changes, changing demands of individual customers, aging or failing sensors, and many more. As drift constitutes a major issue in many applications, considerable research is focusing on this setting [5]. Depending on the domain of data and application, different drift scenarios might occur: For example, covariate shift refers to the situation that training and test sets have different marginal distributions [9].

In recent years, a large variety of methods for learning in presence of drift has been proposed [5] where a majority of the approaches targets supervised learning scenarios. Here, one distinguishes between virtual and real drift, i.e., non-stationarity of the marginal distribution only or also the posterior. Learning technologies often rely on windowing techniques and adapt the model based

We gratefully acknowledge funding by the BMBF TiM, grant number 05M20PBA.

on the characteristics of the data in an observed time window. Here, many approaches use non-parametric methods or ensemble technologies [7]. Active methods explicitly detect drift, usually referring to drift of the classification error, and trigger model adaptation this way, while passive methods continuously adjust the model [5]. Hybrid approaches combine both methods by continuously adjusting the model unless drift is detected and a new model is trained.

In most techniques, evaluation takes place by means of the so-called interleaved test-train error (ITTE), which evaluates the current model on a new data point before using it for training. This error is used to evaluate the overall performance of the algorithm, as well as to detect drifts in case of significant changes in the error or to control important parameters such as the window size [18]. Thereby, these techniques often rely on strong assumptions regarding the underlying process, e.g., they detect a drift when the classification accuracy drops below a predefined threshold during a predefined time. Such methods face problems if the underlying drift characteristics do not align with these assumptions.

Here, we want to shed some light on the suitability of such choices and investigate the mathematical properties of the ITTE when used as an evaluation scheme. As the phenomenon of concept drift is widespread, a theoretical understanding of the relation between drift and the adaption behavior of learning models becomes crucial. Currently, the majority of theoretical work for drift learning focuses on learning guarantees which are similar in nature to the work of Vapnik in the batch case [11,12,19]. Although those results provide interesting insights into the validity of learning models in the streaming setup, they focus on worst-case scenarios and hence provide very loose bounds on average only. In contrast, in this work, we focus on theoretical aspects of the learning algorithm itself in non-stationary environments, targeting general learning models including unsupervised ones. In contrast to the existing literature, we focus on alterations of models. This perspective is closely connected to the actual change of decision boundaries and average cases. In particular, we provide a mathematical substantiation of the suitability of the ITTE to evaluate model drift.

This paper is organized as follows: First (Sect. 2) we recall the basic notions of statistical learning theory and concept drift followed by reviewing the existing literature, positioning this work with respect to it, and concretizing the research questions (Sect. 2.3). We proceed with a theoretical analysis focusing on (1) changes of the decision boundary in presence of drift (Sect. 3.1), (2) changes in the training result (Sect. 3.2), and (3) the connection of ITTE, drift, and the change of the optimal model (Sect. 3.3). Afterward, we empirically quantify the theoretical findings (Sect. 4) and conclude with a summary (Sect. 5).

2   Problem Setup, Notation, and Related Work

We make use of the formal framework for concept drift as introduced in [14,15] as well as classical statistical learning theory, e.g., as presented in [24]. In this section, we recall the basic notions of both subjects followed by a summary of the related work on learning theory in the context of concept drift.

2.1   Basic Notions of Statistical Learning Theory

In classical learning theory, one considers a hypothesis class $\mathcal{H}$, e.g., a set of functions from $\mathbb{R}^d$ to $\mathbb{R}$, together with a non-negative loss function $\ell : \mathcal{H} \times (\mathcal{X} \times \mathcal{Y}) \to \mathbb{R}_{\geq 0}$ that is used to evaluate how well a model h matches an observation $(x, y) \in \mathcal{X} \times \mathcal{Y}$ by assigning an error $\ell(h, (x, y))$. We will refer to $\mathcal{X}$ as the data space and $\mathcal{Y}$ as the label space. For a given distribution $\mathcal{D}$ on $\mathcal{X} \times \mathcal{Y}$ we consider $\mathcal{X}$- and $\mathcal{Y}$-valued random variables X and Y, $(X, Y) \sim \mathcal{D}$, and assign the loss $\mathcal{L}_{\mathcal{D}}(h) = \mathbb{E}[\ell(h, (X, Y))]$ to a model $h \in \mathcal{H}$. Using a data sample $S \in \cup_{N \in \mathbb{N}} (\mathcal{X} \times \mathcal{Y})^N$ consisting of i.i.d. random variables $S = ((X_1, Y_1), \ldots, (X_n, Y_n))$ distributed according to $\mathcal{D}$, we can approximate $\mathcal{L}_{\mathcal{D}}(h)$ using the empirical loss $\mathcal{L}_S(h) = \frac{1}{n} \sum_{i=1}^{n} \ell(h, (X_i, Y_i))$, which converges to $\mathcal{L}_{\mathcal{D}}(h)$ almost surely. Popular loss functions are the mean squared error $\ell(h, (x, y)) = (h(x) - y)^2$, cross-entropy $\ell(h, (x, y)) = \sum_{i=1}^{n} \mathbf{1}[y = i] \log h(x)_i$, or the 0–1-loss $\ell(h, (x, y)) = \mathbf{1}[h(x) \neq y]$. Notice that this setup also covers some unsupervised learning problems.

In machine learning, training a model often refers to minimizing the loss $\mathcal{L}_{\mathcal{D}}(h)$ using the empirical loss $\mathcal{L}_S(h)$ as a proxy. A learning algorithm A, such as gradient descent schemes, selects a model h given a sample S, i.e., $A : \cup_N (\mathcal{X} \times \mathcal{Y})^N \to \mathcal{H}$. Classical learning theory investigates under which circumstances A is consistent, that is, it selects a good model with high probability: $\mathcal{L}_{\mathcal{D}}(A(S)) \to \inf_{h^* \in \mathcal{H}} \mathcal{L}_{\mathcal{D}}(h^*)$ as $|S| \to \infty$ in probability. Since the model $A(S)$ is biased towards the loss $\mathcal{L}_S$ due to training, classical approaches aim for uniform bounds $\sup_{h \in \mathcal{H}} |\mathcal{L}_S(h) - \mathcal{L}_{\mathcal{D}}(h)| \to 0$ as $|S| \to \infty$ in probability.

2.2   A Statistical Framework for Concept Drift

The classical setup of learning theory assumes a time-invariant distribution $\mathcal{D}$ for all (X_i, Y_i). This assumption is violated in many real-world applications, in particular, when learning on data streams. Therefore, we incorporate time into our considerations by means of an index set $\mathcal{T}$, representing time, and a collection of (possibly different) distributions $\mathcal{D}_t$ on $\mathcal{X} \times \mathcal{Y}$, indexed over $\mathcal{T}$ [7]. In particular, the model h and its loss also become time-dependent. It is possible to extend this setup to a general statistical interdependence of data and time via a distribution $\mathcal{D}$ on $\mathcal{T} \times (\mathcal{X} \times \mathcal{Y})$ which decomposes into a distribution $\mathbb{P}_T$ on $\mathcal{T}$ and the conditional distributions $\mathcal{D}_t$ on $\mathcal{X} \times \mathcal{Y}$ [14,15]. Notice that this setup [15] is very general and can therefore be applied in different scenarios (see Sect. 5), albeit our main example is binary classification on a time interval, i.e., $\mathcal{X} = \mathbb{R}^d$, $\mathcal{Y} = \{0, 1\}$, and $\mathcal{T} = [0, 1]$.

Drift refers to the fact that $\mathcal{D}_t$ varies for different time points, i.e., $\{(t_0, t_1) \in \mathcal{T}^2 : \mathcal{D}_{t_0} \neq \mathcal{D}_{t_1}\}$ has measure larger zero w.r.t $\mathbb{P}_T^2$ [14]. One further distinguishes a change of the posterior $\mathcal{D}_t(Y|X)$, referred to as real drift, and of the marginal $\mathcal{D}_t(X)$, referred to as virtual drift. One of the key findings of [14] is a unique characterization of the presence of drift by the property of statistical dependency of time T and data (X, Y) if a time-enriched representation of the data $(T, X, Y) \sim \mathcal{D}$ is considered. Determining whether or not there is drift during a time period is referred to as drift detection.

Since the distribution $\mathcal{D}_t$ can shift too rapidly to enable a faithful estimation of quantities thereof, we propose to address time windows $W \subset \mathcal{T}$ and to consider all data points, that are observed during W, analogous to an observation in classical learning theory. This leads to the following formalization [17]:

Definition 1. *Let $\mathcal{X}, \mathcal{Y}, \mathcal{T}$ be measurable spaces. Let $(\mathcal{D}_t, \mathbb{P}_T)$ be a drift process [14, 15] on $\mathcal{X} \times \mathcal{Y}$ and $\mathcal{T}$, i.e., a distribution $\mathbb{P}_T$ on $\mathcal{T}$ and Markov kernels $\mathcal{D}_t$ from $\mathcal{T}$ to $\mathcal{X} \times \mathcal{Y}$. A time window $W \subset \mathcal{T}$ is a $\mathbb{P}_T$ non-null set. A sample (of size n) observed during W is a tuple $S = ((X_1, Y_1), \ldots, (X_n, Y_n))$ drawn i.i.d. from the mean distribution on W, that is $\mathcal{D}_W := \mathcal{D}(X, Y \mid T \in W)$.*

This resembles the practical procedure, where one obtains sample S_1 during W_1 from another sample S_2 during W_2, with $W_1 \subset W_2$, by selecting those entries of S_2 that are observed during W_1. In particular, if $W_1 = \{1, \ldots, t\}$, $W_2 = \{1, \ldots, t, t+1\}$ this corresponds to an incremental update. Other windowing strategies like sliding windows, removal of old samples, passive updates, etc. are also possible. Indeed, using this idea many questions in the context of drift can be implemented by choosing the right windowing scheme. For our research question we considers the general setup, and thus not focus on a specific windowing scheme.

Notice that by applying windowing different possible drift dynamics present in the drift process $\mathcal{D}_t$, e.g., abrupt, incremental, periodic, are simplified to a single abrupt drift. In particular, the continuous analysis of drift behavior is reduced to a binary problem, e.g., we can speak of a distribution before ($\mathcal{D}_{W_1}$) and after ($\mathcal{D}_{W_2}$) the drift.

In this work, we will consider data drawn from a single drift process, thus we will make use of the following short-hand notation $\mathcal{L}_t(h) := \mathcal{L}_{\mathcal{D}_t}(h)$ for a time point $t \in \mathcal{T}$ and $\mathcal{L}_W(h) = \mathcal{L}_{\mathcal{D}_W}(h)$ for a time window $W \subset \mathcal{T}$, where $\mathcal{D}_W = \mathbb{E}[\mathcal{D}_T \mid T \in W]$ denotes the mean of $\mathcal{D}_t$ during W and $\mathcal{L}(h) := \mathcal{L}_{\mathcal{T}}(h)$ is the loss on the entire stream. Notice that this is well defined, i.e., $\mathcal{L}_W(h) = \mathbb{E}[\ell(h, (X, Y)) \mid T \in W] = \mathbb{E}[\mathcal{L}_T(h) \mid T \in W]$ assuming $\mathcal{L}(h) < \infty$. In stream learning, some algorithms put more weight on newer observations, e.g., by continuously updating the model. Such considerations can be easily integrated into our framework, but we omit them for simplicity.

2.3   Related Work, Existing Methods, and Research Questions

Stream learning algorithms can be split into two categories [5]: passive methods, which adapt the model slightly in every iteration (Line 6 in Algorithm 1), and active methods, which train a new model once drift is detected (Line 8 in Algorithm 1). There also exist hybrid methods that integrate both characteristics as outlined in Algorithm 1.

Most existing theoretical work on stream learning in the context of drift derives learning guarantees as inequalities of the following form: the risk at a future time point $W_2 = \{t + 1\}$ is bounded using the risk on the current

---

Algorithm 1. Schematic Description of Typical Stream Learning Algorithm

---

1: **Input:** S data stream, A training algorithm, h_0 initial model, ℓ loss function, D
   drift detector
2: Initialize model $h \leftarrow h_0, L \leftarrow \emptyset$
3: **while** Not at end of stream S **do**
4:     Receive new sample (x, y) from stream S
5:     Compute ITTE $L \leftarrow \textsc{Update}(L, \ell(h, (x, y)))$
6:     Update model $h \leftarrow A(h, (x, y))$ ▷ Passive Adaption
7:     **if** Detect drift $D(L)$ **then**
8:         Reset model $h \leftarrow h_0$ OR Retrain on next samples   ▷ Active Adaption
9:     **end if**
10: **end while**

---

time window $W_1 = \{1, \ldots, t\}$ and a distributional difference in between those
windows [12,19,22]:

$$\underbrace{\mathcal{L}_{W_2}(h)}_{\text{application time risk}} \leq \underbrace{\mathcal{L}_{W_1}(h)}_{\text{train time risk}} + \underbrace{\sup_{h' \in \mathcal{H}} |\mathcal{L}_{W_2}(h') - \mathcal{L}_{W_1}(h')|}_{\text{distributional discrepancy}}. \tag{1}$$

Other similar ways to quantify the distributional discrepancy exist [3,10,26].
Most approaches aim for a good upper bound of the train time risk [11,19,22].
Eq. (1) then gives rise to convergence guarantees usually applied by splitting
the so far observed stream into several chunks and training a model on each of
them [11,12,26]. Notice that due to the windowing, the problem is closely linked
to domain adaption which is analyzed using a very similar theory [3,23].

A crucial aspect of the inequality is the distributional discrepancy. Notice
that it is closely related to other statistical quantities like the total variation
norm [12,14,19] or the Wasserstein distance. It provides a bound that refers
to the worst possible outcome regarding the drift. Although this scenario can
theoretically occur (see examples given in [11, Theorem 2]), it is not likely in
practice. For example, for binary classification with 0–1-loss, i.e., $\mathcal{Y} = \{-1, 1\}$
and $\ell(h, (x, y)) = (1 - y \cdot h(x))/2$, we can turn the difference into a label swap:

$$\sup_{h \in \mathcal{H}} |\mathcal{L}_{W_1}(h) - \mathcal{L}_{W_2}(h)| = 2 \sup_{h \in \mathcal{H}} \mathcal{L}_{(\mathcal{D}_{W_1} + \overline{\mathcal{D}_{W_2}})/2}(h),$$

where $\overline{\mathcal{D}_{W_2}}$ denotes the distribution during W_2 with swapped label. Thus, we
obtain large discrepancies even if the decision boundary is not affected by drift.
Comparable statistics are used for unsupervised drift detection [17]. These points
imply that error bounds are not necessarily a suitable tool to study the effect of
concept drift on learning algorithms, models, and losses.

To overcome this problem, in practice, drift learning algorithms rely on a
comparison of the current and historical loss estimated using the ITTE scheme
instead. However, this procedure has its own deficiencies. As can be seen in
Fig. 1, the loss of a fixed model can change without a change of the optimal
model and vice versa. Based on these insights, in this contribution we aim to

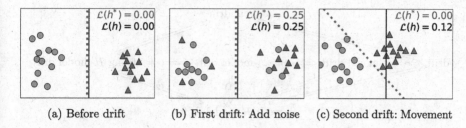

(a) Before drift   (b) First drift: Add noise   (c) Second drift: Movement

Fig. 1. Effect of drift on model loss of a fixed and optimal model. Graphic shows fixed model h (black line), optimal model h^* (red dotted line), and model losses. (Color figure online)

provide a better understanding of the usage of stream learning algorithms in the context of drift and novel techniques derived thereof, answering the following questions:

1. How are model changes related to different types (real/virtual) of drift?
2. What is the relation between optimal models and the output of learning algorithms on different time windows? When to retrain the model?
3. How are changes in the optimal model mirrored in changes in the ITTE?

3 Theoretical Analysis

To answer these research questions we propose three formal definitions, each reflecting a different aspect and perspective of drift – each designed to answer one of our research questions. We then compare those definitions, show formal implications, and provide counterexamples in case of differences, in order to provide the desired answers. We summarize our findings in Fig. 2, displaying different types of drift definitions and their implications.

We will refer to the types of drift that affect models as *model drift*. It is a generalization of the notion of model drift in the work [14], which is based on the comparison of the distribution for two different time windows, i.e., $\mathcal{D}_{W_1} \neq \mathcal{D}_{W_2}$. We extend this idea to incorporate model and loss-specific properties.

3.1 Model Drift as Inconsistency of Optimal Models

The concept of model drift can be considered from two points of view: different training results (see Sect. 3.2) and inconsistency of optimal models. We deal with the latter notion first. Using loss as a proxy for performance, we consider that a model performs well if it has a loss comparable to the minimal achievable loss. We refer to this as *hypothetical-* or *$\mathcal{H}$-model drift*, which is defined as follows:

Definition 2. *Let $\mathcal{H}$ be a hypothesis class, ℓ a loss function on $\mathcal{H}$, and $\mathcal{D}_t$ be a drift process. We say that $\mathcal{D}_t$ has strong $\mathcal{H}$-model drift iff there exist time windows without a common well-performing model, i.e., there exist measurable*

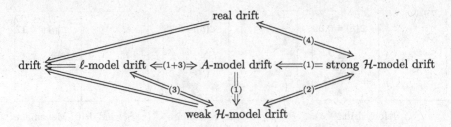

Fig. 2. Definitions and implications. Numbers indicate needed assumptions: (1) A is consistent, (2) loss uniquely determines model, (3) optimal loss is unchanged, (4) universal hypothesis class of probabilistic models with non-regularized loss.

$\mathbb{P}_T$ non-null sets $W_1, W_2 \subset T$ and $C > 0$ such that for every $h \in \mathcal{H}$ we have $\mathcal{L}_{W_1}(h) > \inf_{h^* \in \mathcal{H}} \mathcal{L}_{W_1}(h^*) + C$ or $\mathcal{L}_{W_2}(h) > \inf_{h^* \in \mathcal{H}} \mathcal{L}_{W_2}(h^*) + C$. We say that $\mathcal{D}_t$ has weak $\mathcal{H}$-model drift iff some model stops being optimal, i.e., for some time windows W_1, W_2 there exists a $C > 0$ such that for all $\varepsilon < C$ there is some $h \in \mathcal{H}$ such that $\mathcal{L}_{W_1}(h) \leq \inf_{h^* \in \mathcal{H}} \mathcal{L}_{W_1}(h^*) + \varepsilon$ and $\mathcal{L}_{W_2}(h) > \inf_{h^* \in \mathcal{H}} \mathcal{L}_{W_2}(h^*) + C$.

$\mathcal{H}$-model drift refers to the fact that models that perform well during one time window perform poorly during the other. Thus, it is used to study the change in optimal models. The difference between strong and weak $\mathcal{H}$-model drift is only relevant if there are multiple optimal models: Weak $\mathcal{H}$-model drift refers to the case that there exists at least one model which is only optimal for one time window. Strong $\mathcal{H}$-model drift requires this to hold for all optimal models. Thus, strong $\mathcal{H}$-model drift implies that model adaptation is strictly necessary for optimum results, whereas the necessity of model adaptation for weak $\mathcal{H}$-model drift might depend on the specific choice of the model. Strong $\mathcal{H}$-model drift implies weak $\mathcal{H}$-model drift. This raises the question under which circumstances the converse is also true. It turns out that loss functions inducing unique optima are sufficient:

Lemma 1. *If $\mathcal{D}_t$ has strong $\mathcal{H}$-model drift for windows W_1, W_2, then it has weak $\mathcal{H}$-model drift on the same windows. If the optimal model is uniquely determined by the loss, i.e., for all $h_i, h_i' \subset \mathcal{H}$ with $\mathcal{L}_{W_1}(h_i), \mathcal{L}_{W_1}(h_i') \xrightarrow{i \to \infty} \inf_{h^* \in \mathcal{H}} \mathcal{L}_{W_1}(h^*)$ we have $\limsup_{i \to \infty} |\ell(h_i, (x, y)) - \ell(h_i', (x, y))| = 0$ for all $(x, y) \in \mathcal{X} \times \mathcal{Y}$ and ℓ bounded, then the converse is also true. The additional assumption is necessary.*

Proof. Due to space restrictions, all proofs can be found in the ArXiv version [16].

Notice that the uniqueness criterion becomes particular intuitive for functions $h : \mathcal{X} \to \mathcal{Y}$ and losses induced by a metric, i.e., $\ell(h, (x, y)) = d(h(x), y)$, in which case we can bound $|\ell(h, (x, y)) - \ell(h', (x, y))| \leq d(h(x), h'(x))$. Thus, the criterion requires models with little variance to ensure that the notions of strong and weak $\mathcal{H}$-model drift coincide. This can be achieved by a regularization term such as limiting the weight norm. As an immediate consequence we have:

Corollary 1. *For k-nearest neighbor, RBF-networks, and decision tree virtual drift cannot cause strong $\mathcal{H}$-model drift, i.e., we can keep the training window. For SVMs and linear regression based on the mean squared error virtual drift can cause strong $\mathcal{H}$-model drift, i.e., we may have to reset the training window.*

Obviously, (weak) $\mathcal{H}$-model drift implies drift because if there is no change of the loss, i.e., $\mathcal{L}_t(h) = \mathcal{L}_s(h)$ for all $h \in \mathcal{H}$, $s, t \in \mathcal{T}$, there cannot be (weak) $\mathcal{H}$-model drift. The converse is not so clear. We address this question in the following, targeting real drift.

Theorem 1. *Let $\mathcal{Y} = \{0, 1\}$, $\mathcal{T} = [0, 1]$, and $\mathcal{X} = \mathbb{R}^d$. Let $\mathcal{D}_t$ be a drift process, $\mathcal{H}$ be a hypothesis class of probabilistic, binary classifiers, i.e., maps $h : \mathcal{X} \to [0, 1]$, with MSE-loss, i.e., $\ell(h, (x, y)) = (h(x) - y)^2$, and assume that $\mathcal{H}$ is universal, i.e., dense span in the compactly supported continuous functions $C_c(\mathcal{X})$. Then, $\mathcal{D}_t$ has real drift if and only if $\mathcal{D}_t$ has strong $\mathcal{H}$-model drift.*

This theorem includes crucial ingredients which are necessary to guarantee the result. As an example, the model class has to be very flexible, i.e., universal, to adapt to arbitrary drift, and the loss function must enable such adaptation.

Together with Corollary 1 this completes our study of Question 1, whether or not concept drift affects the optimal model depends on the drift and the model class. Real drift usually affects all model classes while virtual drift poses only a problem if the model class is not flexible enough to adapt to the change.

So far we considered the change of decision boundaries through the lens of models, disregarding how they are achieved. We will take on a more practical point of view by considering models as an output of training algorithms applied to windows in the next section.

3.2 Model Drift as Time Dependent Training Result

Another way to consider the problem of model drift is to consider the output of a training algorithm. This idea leads to the second point of view: drift manifests itself as the fact that the model obtained by training on data from one time point differs significantly from the model trained on data of another time point. We will refer to this notion as *algorithmic-* or *A-model drift*. It answers the question of whether replacing a model trained on past data (drawn during W_1) with a model trained on new data (drawn during W_2) improves performance. Using loss as a proxy we obtain the following definition:

Definition 3. *Let $\mathcal{H}$ be a hypothesis class, ℓ a loss function on $\mathcal{H}$, and $\mathcal{D}_t$ be a drift process. For a training algorithm A we say that $\mathcal{D}_t$ has A-model drift iff model adaptation yields a significant increase in performance with a high probability, i.e., there exist time windows W_1, W_2 such that for all $\delta > 0$ there exists a $C > 0$ and numbers N_1 and N_2 such that with probability at least $1 - \delta$ over all samples S_1 and S_2 drawn from $\mathcal{D}_{W_1}$ and $\mathcal{D}_{W_2}$ of size at least N_1 and N_2, respectively, it holds $\mathcal{L}_{W_2}(A(S_1)) > \mathcal{L}_{W_2}(A(S_2)) + C$.*

Note that we do not specify how the algorithm processes the data, thus we also capture updating procedures. Removal of old data points, e.g., $W_1 = \{1, \ldots, t_1, \ldots, t_2\}$, $W_2 = \{t_1, \ldots, t_2\}$, is also a relevant instantiation of this setup. Unlike $\mathcal{H}$-model drift which is concerned with consistency, it focuses on changes of the training result. The following theorem connects both notions:

Theorem 2. *Let $\mathcal{D}_t$ be a drift process, $\mathcal{H}$ a hypothesis class with loss ℓ, and learning algorithm A. Consider the following statement with respect to the same time windows W_1 and W_2: (i) $\mathcal{D}_t$ has strong $\mathcal{H}$-model drift for windows W_1, W_2. (ii) $\mathcal{D}_t$ has A-model drift for windows W_1, W_2. (iii) $\mathcal{D}_t$ has weak $\mathcal{H}$-model drift for windows W_1, W_2. If A is a consistent training algorithm, i.e., for sufficiently large sample sizes we obtain arbitrarily good approximations of the optimal model [24, Definition 7.8], then $(i) \Rightarrow (ii) \Rightarrow (iii)$ holds. In particular, if we additionally assume that the optimal model is uniquely determined by the loss (see Lemma 1) then all three statements are equivalent. If A is not consistent, then none of the implications hold.*

The relevance of this result follows from the fact that it connects theoretically optimal models to those obtained from training data when learning with drift. The result implies that model adaption does not increase performance if there is no drift. Further, if the model is uniquely determined any consistent algorithm will suffer from drift in the same situations. Formally, the following holds:

Corollary 2. *Let $\mathcal{D}_t$ be a drift process, $\mathcal{H}$ a hypothesis class with consistent learning algorithms A and B. Assume that the optimal model is uniquely determined by the loss, then for windows W_1 and W_2, A-model drift is present if and only if B-model drift is present.*

This concludes our consideration of Question 2: For reasonable, i.e., consistent, learning algorithms the question whether or not the resulting model is affected by the drift mainly depends on the number of possible, well-performing models. If this number is small, then any kind of drift is likely to have an effect on the training result. In this case, the choice of learning algorithm is negligible.

Although the results regarding A-model drift give us relevant insight, they do not yet include one important aspect of practical settings: A-model drift compares already trained models, yet training a new model for every possible time window is usually unfeasible. Due to this fact, many algorithms investigate incremental updates and refer to the ITTE as an indicator of model accuracy and concept drift [7]. Next, we will investigate the validity of this approach.

3.3   Interleaved Test-Train Error as Indicator for Model Drift

A common technique to detect concept drift is to relate it to the performance of a fixed model. In this setup a decrease in performance indicates drift. Using loss as a proxy for performance we obtain the notion of *loss-* or *ℓ-model drift*:

Definition 4. *Let $\mathcal{H}$ be a hypothesis class, ℓ a loss function on $\mathcal{H}$, and $\mathcal{D}_t$ be a drift process. We say that $\mathcal{D}_t$ has ℓ-model drift iff the loss of an optimal model* *changes, i.e., for time windows W_1, W_2 there exists a $C > 0$ such that for all $\varepsilon < C$ there is some $h \in \mathcal{H}$ such that $\mathcal{L}_{W_1}(h) \leq \inf_{h^* \in \mathcal{H}} \mathcal{L}_{W_1}(h^*) + \varepsilon$ and $\mathcal{L}_{W_2}(h) > \mathcal{L}_{W_1}(h) + C$. We say that the optimal loss is non-decreasing/non-increasing/constant iff $\inf_{h^* \in \mathcal{H}} \mathcal{L}_{W_1}(h^*) \leq / \geq / = \inf_{h^* \in \mathcal{H}} \mathcal{L}_{W_2}(h^*)$ holds.*

The common instantiation of ITTE is obtained by considering $W_1 = \{1, \ldots, t\}$, $W_2 = \{1, \ldots, t, t+1\}$. It is easy to see that ℓ-model drift implies drift, the connection to the other notions of model drift is not so obvious as a change in the difficulty of the learning problem does not imply a change of the optimal model or vice versa: an example is the setup of binary classification and drift induced change of noise level (Fig. 1). Assumptions regarding the minimal loss lead to the following result:

Lemma 2. *Assume the situation of Definition 4. For time windows W_1, W_2 it holds: (i) For non-decreasing optimal loss, weak $\mathcal{H}$-model drift implies ℓ-model drift. (ii) For non-increasing optimal loss, ℓ-model drift implies weak $\mathcal{H}$-model drift. The additional assumption is necessary.*

As a direct consequence of this lemma and Theorem 2, we obtain a criterion that characterizes in which cases active methods based on the ITTE are optimal. Here, we do not require that the loss uniquely determines the model:

Theorem 3. *Let $\mathcal{D}_t$ be a drift process and $\mathcal{H}$ be a hypothesis class with loss ℓ. Assume the optimal loss is constant. Then for time windows W_1, W_2 and any consistent learning algorithm A it holds: $\mathcal{D}_t$ has A-model drift if and only if it has ℓ-model drift with respect to $h = A(S_1)$, i.e., $\forall \delta > 0 \exists N > 0 \forall n > N : \mathbb{P}_{S_1 \sim \mathcal{D}_{W_1}^n}[\mathcal{L}_{W_2}(A(S_1)) > \mathcal{L}_{W_1}(A(S_1)) + C] > 1 - \delta$.*

Notice that this result provides a theoretical justification for the common practice in active methods, to use drift detectors on the ITTE to determine whether or not to retrain the model. The statement only holds if the optimal loss is constant – otherwise, the ITTE is misleading and can result in both false positive and false negative implications (see Fig. 1). This answers Question 3.

4 Empirical Evaluation

In the following, we demonstrate our theoretical insights in experiments and quantify their effects. All results which are reported in the following are statistically significant (based on a t-test, $p < 0.001$). All experiments are performed on the following standard synthetic benchmark datasets AGRAWAL [1], LED [2], MIXED [8], RandomRBF [20], RandomTree [20], SEA [25], Sine [8], STAGGER [8] and the following real-world benchmark datasets "Electricity market prices" (Elec) [13], "Forest Covertype" (Forest) [4], and "Nebraska Weather" (Weather) [6]. To remove effects due to unknown drift in the real-world datasets, we apply a permutation scheme [17], and we induce real drift by swapping labels of two randomly chosen classes. As a result, all datasets have controlled real

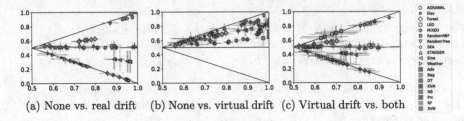

(a) None vs. real drift   (b) None vs. virtual drift   (c) Virtual drift vs. both

Fig. 3. Comparison of types of drift for different datasets (marker) and models (color) by means of accuracy. x-axis shows accuracy on original distribution (test-set), y-axis shows accuracy on drifted distribution, black lines mark $y = \pm x, 50\%$. For the sake of clarity, error bars show $1/2$ of standard deviation.

drift and no virtual drift. We induce virtual drift by segmenting the data space (leaves of decision trees on random linear transformations fitted to uniform random noise) and randomly associate every segment with one time window, i.e., the samples in a segment are either used or dropped completely. For comparability, all problems are turned into binary classification tasks with class imbalance below 25%. This way we obtained 2×2 distributions with controlled drifting behavior, i.e., $\mathcal{D}_{ij}(X, Y) = \mathcal{D}_i(X)\mathcal{D}_j(Y|X)$, $i, j \in \{0, 1\}$. Notice that this corresponds to windowing and thus does not reduce the generality as already discussed in Sect. 2.2, i.e., phenomena like incremental or periodic drift are also captured.

To show the effect of real and virtual drift on classification accuracy, we draw train and test samples from those distributions which correspond to the time windows in Sect. 3 and compute the train-test error of the following models: Decision Tree (DT), Random Forest (RF), k-Nearest Neighbour (k-NN), Bagging (Bag; with DT), AdaBoost (Ada; with DT), Gaussian Naïve Bayes (NB), Perceptron (Prc), and linear SVM (SVM) [21]. We repeated the experiment 1,000 times. The results are shown in Fig. 3. We found that real and virtual drift causes a significant decrease in accuracy compared to the non-drifting baseline for all models and datasets (except for Prc and SVM on AGRAWAL on virtual drift where the results are inconclusive). A combination of real and virtual drift decreased the accuracy even further if compared to the non-drifting baseline and virtual drift only. These findings are in strong agreement with Theorem 3 and show that virtual drift can cause a significant decrease in accuracy although it is usually considered less relevant for the performance of a model.

To evaluate the necessity to reset the training window after drift we combined two windows that differ in one drift type, i.e., virtual or real drift, and proceed as before. An overview of the results is presented in Fig. 4a and 4b. As expected, the models trained on the composed windows outperform the ones trained on the non-composed samples (except for SVM on AGRAWAL with virtual drift where the results are inconclusive). In comparison to the non-drifting baseline, the composed real drift models are outperformed, and the composed virtual drift model are mainly inconclusive. An analysis of the usage of additional information in the latter scenario is presented in Fig. 4c: For c composed virtual, v virtual,

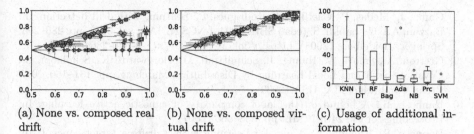

(a) None vs. composed real drift

(b) None vs. composed virtual drift

(c) Usage of additional information

Fig. 4. Evaluation of composed windows. Plots (a) and (b) use the same color/marker scheme as Fig. 3. (Color figure online)

and n no drift accuracy, we normalize the increase of accuracy beyond generalization $(c - v)$ by the decrease of accuracy due to limited flexibility $(n - c)$, i.e., $|(c - v)/(n - c)|$. As can be seen, NB, Prc, and SVM do not profit, DT and k-NN profit most, RF, Bag, and Ada profit moderately. These findings are in strong agreement with Corollary 1 and Lemma 1 as they quantitatively show that more flexible models are better at handling virtual drift in the training window.

5 Discussion and Conclusion

In this work, we considered the problem of online and stream learning with drift from a theoretical point of view. Our main results aim at the application of active methods that adapt to drift in data streams by mainly considering the ITTE. In contrast to many other works in this area, we focused on consistency and/or change of the decision boundary as indicated by models and loss functions. Furthermore, out approach applies to semi- and unsupervised setups assuming a suitable loss function characterizes the model, e.g., k-means, PCA, density estimation, etc. More general notions of time, e.g., computational nodes as in federated learning or domains as in domain adaption, are also covered.

References

1. Agrawal, R., Imielinski, T., Swami, A.N.: Database mining: a performance perspective. IEEE Trans. Knowl. Data Eng. **5**, 914–925 (1993)
2. Asuncion, A., Newman, D.: UCI machine learning repository (2007)
3. Ben-David, S., Blitzer, J., Crammer, K., Kulesza, A., Pereira, F., Vaughan, J.W.: A theory of learning from different domains. Mach. Learn. **79**, 151–175 (2010)
4. Blackard, J.A., Dean, D.J., Anderson, C.W.: Covertype data set (1998). http://archive.ics.uci.edu/ml/datasets/Covertype
5. Ditzler, G., Roveri, M., Alippi, C., Polikar, R.: Learning in nonstationary environments: a survey. IEEE Comp. Int. Mag. **10**(4), 12–25 (2015)
6. Elwell, R., Polikar, R.: Incremental learning of concept drift in nonstationary environments. IEEE Trans. Neural Netw. **22**(10), 1517–1531 (2011)
7. Gama, J.a., Žliobaitė, I., Bifet, A., Pechenizkiy, M., Bouchachia, A.: A survey on concept drift adaptation. ACM Comput. Surv. **46**(4), 44:1–44:37 (2014)

8. Gama, J., Medas, P., Castillo, G., Rodrigues, P.: Learning with drift detection. In: Bazzan, A.L.C., Labidi, S. (eds.) SBIA 2004. LNCS (LNAI), vol. 3171, pp. 286–295. Springer, Heidelberg (2004). https://doi.org/10.1007/978-3-540-28645-5_29
9. Gretton, A., Smola, A., Huang, J., Schmittfull, M., Borgwardt, K., Schölkopf, B.: Covariate Shift and Local Learning by Distribution Matching, pp. 131–160. MIT Press, Cambridge, MA, USA (2009)
10. Hanneke, S.: A bound on the label complexity of agnostic active learning. In: Proceedings of the 24th ICML, pp. 353–360 (2007)
11. Hanneke, S., Kanade, V., Yang, L.: Learning with a drifting target concept. In: Chaudhuri, K., Gentile, C., Zilles, S. (eds.) ALT 2015. LNCS (LNAI), vol. 9355, pp. 149–164. Springer, Cham (2015). https://doi.org/10.1007/978-3-319-24486-0_10
12. Hanneke, S., Yang, L.: Statistical learning under nonstationary mixing processes. In: The 22nd International Conference on Artificial Intelligence and Statistics, pp. 1678–1686. PMLR (2019)
13. Harries, M., cse tr, U.N., Wales, N.S.: Splice-2 comparative evaluation: Electricity pricing. Technical report (1999)
14. Hinder, F., Artelt, A., Hammer, B.: Towards non-parametric drift detection via dynamic adapting window independence drift detection (dawidd). In: ICML (2020)
15. Hinder, F., Artelt, A., Hammer, B.: A probability theoretic approach to drifting data in continuous time domains. arXiv preprint arXiv:1912.01969 (2019)
16. Hinder, F., Vaquet, V., Brinkrolf, J., Hammer, B.: On the change of decision boundaries and loss in learning with concept drift. arXiv preprint arXiv:2212.01223 (2022)
17. Hinder, F., Vaquet, V., Hammer, B.: Suitability of different metric choices for concept drift detection. In: Bouadi, T., Fromont, E., Hüllermeier, E. (eds.) IDA 2022. LNCS, vol. 13205, pp. 157–170. Springer, Cham (2022). https://doi.org/10.1007/978-3-031-01333-1_13
18. Losing, V., Hammer, B., Wersing, H.: Incremental on-line learning: a review and comparison of state of the art algorithms. Neurocomputing **275**, 1261–1274 (2018)
19. Mohri, M., Muñoz Medina, A.: New analysis and algorithm for learning with drifting distributions. In: Bshouty, N.H., Stoltz, G., Vayatis, N., Zeugmann, T. (eds.) ALT 2012. LNCS (LNAI), vol. 7568, pp. 124–138. Springer, Heidelberg (2012). https://doi.org/10.1007/978-3-642-34106-9_13
20. Montiel, J., Read, J., Bifet, A., Abdessalem, T.: Scikit-multiflow: a multi-output streaming framework. J. Mach. Learn. Res. **19**(72), 1–5 (2018)
21. Pedregosa, F., et al.: Scikit-learn: machine learning in Python. J. Mach. Learn. Res. **12**, 2825–2830 (2011)
22. Rakhlin, A., Sridharan, K., Tewari, A.: Online learning via sequential complexities. J. Mach. Learn. Res. **16**(1), 155–186 (2015)
23. Redko, I., Morvant, E., Habrard, A., Sebban, M., Bennani, Y.: A survey on domain adaptation theory: learning bounds and theoretical guarantees. arXiv preprint arXiv:2004.11829 (2020)
24. Shalev-Shwartz, S., Ben-David, S.: Understanding Machine Learning: From Theory to Algorithms. Cambridge University Press, Cambridge (2014)
25. Street, W.N., Kim, Y.: A streaming ensemble algorithm (SEA) for large-scale classification. In: Proceedings of the Seventh ACM SIGKDD International Conference on Knowledge Discovery and Data Mining, San Francisco, CA, USA, 26–29 August 2001, pp. 377–382 (2001)
26. Yang, L.: Active learning with a drifting distribution. In: Advances in Neural Information Processing Systems, vol. 24 (2011)

AID4HAI: Automatic Idea Detection for Healthcare-Associated Infections from Twitter, a Framework Based on Active Learning and Transfer Learning

Zahra Kharazian[1,3]($\boxtimes$), Mahmoud Rahat[1], Fábio Gama[2],
Peyman Sheikholharam Mashhadi[1], Sławomir Nowaczyk[1], Tony Lindgren[3],
and Sindri Magnússon[3]

[1] Center for Applied Intelligent Systems Research (CAISR), Halmstad University,
Halmstad, Sweden
{mahmoud.rahat,peyman.mashhadi,slawomir.nowaczyk}@hh.se
[2] Department of Innovation Management, Sweden, Halmstad University,
Halmstad, Sweden
fabio.gama@hh.se
[3] Department of Computer and System Science (DSV), Stockholm University,
Stockholm, Sweden
{zahra.kharazian,tony,sindri.magnusson}@dsv.su.se

Abstract. This research is an interdisciplinary work between data scientists, innovation management researchers and experts from Swedish academia and a hygiene and health company. Based on this collaboration, we have developed a novel package for automatic idea detection with the motivation of controlling and preventing healthcare-associated infections (HAI). The principal idea of this study is to use machine learning methods to extract informative ideas from social media to assist healthcare professionals in reducing the rate of HAI. Therefore, the proposed package offers a corpus of data collected from Twitter, associated expert-created labels, and software implementation of an annotation framework based on the Active Learning paradigm. We employed Transfer Learning and built a two-step deep neural network model that incrementally extracts the semantic representation of the collected text data using the BERTweet language model in the first step and classifies these representations as informative or non-informative using a multi-layer perception (MLP) in the second step. The package is called AID4HAI (Automatic Idea Detection for controlling and preventing Healthcare-Associated Infections) and is made fully available (software code and the collected data) through a public GitHub repository (https://github.com/XaraKar/AID4HAI). We believe that sharing our ideas and releasing these ready-to-use tools contributes to the development of the field and inspires future research.

Keywords: automatic idea detection · healthcare-associated infections · human-in-the-loop · active learning · feedback loops · supervised machine learning · natural language processing

© The Author(s), under exclusive license to Springer Nature Switzerland AG 2023
B. Crémilleux et al. (Eds.): IDA 2023, LNCS 13876, pp. 195–207, 2023.
https://doi.org/10.1007/978-3-031-30047-9_16

1   Introduction

Healthcare-Associated Infections (HAIs) are among the most prevalent contamination events in healthcare settings, posing significant challenges to patient care. Multimodal interventions have been advocated, implemented, and studied to control and prevent HAIs, and cross-transmission of multidrug-resistant organisms worldwide [14]. Despite numerous efforts, the compliance of interventions among healthcare professionals remains below the WHO recommendations, hampering patient safety [8,10]. Consequently, healthcare professionals, firms, and policymakers seek innovative ideas to increase interventions compliances and improve patient care. They now ask questions such as "what can healthcare professionals do in the real world to reduce HAIs?" and "how to prevent HAIs in different settings effectively?" Moreover, new knowledge is needed to promote behavior changes and education, monitor performance feedback, and create a safe climate.

Notably, the literature in idea identification has primarily investigated traditional methods involving interviews, ethnographic market research, repertory grid technique, and lead user workshops to identify novel ideas in healthcare settings [21]. Although prior research has highlighted important aspects to identify ideas and established some principles for best practices, it needs more details regarding how to scale the identification methods across different continents, reduce the operationalization costs, and ultimately seek local ideas discussed in real-time [22]. For example, Kesselheim et al. [11] conducted interviews to investigate the idea generation processes and clinical doctors' involvement in coronary artery stents but has been restricted to local settings. Likewise, Smith et al. [20] used text-matching algorithms on patents to investigate premarket approval applications in four medical device firms (Medtronic, Johnson & Johnson, Boston Scientific, and Guidant). Nevertheless, the study is restricted to secondary data, which undermines the possibility of identifying unpublished and potentially disruptive ideas.

A new and efficient way of identifying ideas is to use classification algorithms that can screen large amount of text and identify those bits of information that are more likely to contain ideas [4]. One way to access such a large pool of information is to use social media platforms such as Twitter and analyze the human-generated text using Natural Language Processing (NLP) [7]. The literature on NLP emphasizes using Transfer Learning to extract semantic representation in social media [15]. The *BERTweet* language model has been primarily investigated and provided valuable insights for various downstream tasks [17]. However, the potential of transfer learning and domain adaptation is not limited to text processing. We have explored this perspective in our previous studies in other domains and received outstanding results [12,18].

Patients, healthcare professionals, scholars, and industry representatives are constantly using social media to communicate their needs and promote new healthcare practices [16]. This study conducts a retrospective observational analysis of Twitter user's posting related to HAIs. Ideas are identified using supervised machine learning, demonstrating how technologies such as artificial intelligence can advance HAI interventions. The idea is to analyze a set of tweets and rank them based on their probability of conveying an idea or a problem (aka informative tweets). The informative tweets form the minority class, which is also referred to as positive samples. On the other hand, the majority class is non-informative tweets and corresponds to the negative class. The proposed framework (AID4HAI) analyzes the collected HAIs ideas and validates the theoretical and practical implications of the approach with the help of a Swedish hygiene and health company. We employ Active Learning (AL) at the core of our framework to incrementally improve a discriminative model for finding as many potential ideas as possible.

In the Active Learning setup, we usually have a small labeled and a large unlabeled data pool. The goal is to pick samples from the pool of unlabeled data that produce the most significant improvement in the model's performance and then present the selected samples to the annotators for labeling, eventually adding them to the set of labeled data. This can be done in different ways. A popular approach is the *least confidence* [13] query strategy which chooses samples to query by considering the uncertainty of the classifier prediction. According to Chen et al. [2], uncertainty sampling methods are not a perfect solution for imbalanced scenarios since the majority class size is much larger than the minority one, and they will presumably query too many samples from the majority class. A more recent query strategy for handling imbalanced classification aims to find samples that are under-represented in the labeled data distribution. To this end, they either train an auxiliary binary classifier [6] to distinguish between labeled and unlabeled data or train an outlier detection algorithm [1] on the labeled data to score samples of the unlabeled pool.

However, these approaches do not promote (enforce) the selection of samples from the minority class in a manner that would be sufficient for our case. Despite improving the decision boundary, the ratio of the majority to the minority class is still preserved, and there is no mechanism to make the proportion more balanced. This drawback can potentially lead to the shortage of samples from the minority class in highly imbalanced datasets and ultimately degrades the performance of discriminative models. The proposed framework prioritizes the minority class by selecting samples that are predicted as more informative according to the feedback from the trained model. This helps to make the training dataset more and more balanced over the iterations.

The rest of this paper is structured as follows: Sect. 2 discusses how the dataset is collected, pre-processed, and labeled in detail. Section 3 brings up the proposed iterative method. Experiments and results are demonstrated in Sect. 4. Finally, the study's conclusion and future works are mentioned in Sect. 5.

2   Data Collection and Labeling

2.1   Data Collection

The Twitter platform has been chosen as a data source in this study. For extracting data, we selected a list of 78 HAI-related keywords and accounts with the help of experts in business and specialists in the healthcare domain. This list contains 21 personal accounts from famous Infection Prevention (IP) specialists with a high number of followers on the Twitter platform, 6 HAI-related journals, 15 public health organizations, 11 health and hygiene companies, and 25 HAI-related keywords. The following keywords and accounts were used:

Infection preventionists: Tom Frieden, Jason Gallagher, Debbie Goff, Marc Mendelson, Jon Otter, Eli Perencevich, Kevin Pho, Laura Piddock, Didier Pittet, Daniel Uslan, Marion Koopmans, Debbie Xuereb, Carole Hallam, Heather Loveday, Pat Cattini, Ermira Tartari, Karen Wares, Hannah, Evonne T Curran, Martin Kiernan, and Helen Dunn.

Journals: Infection Control & Hospital Epidemiology, Lancet Infectious Diseases, Journal American Medical Association, New England Journal Medicine, Journal of Infection and Prevention, and Journal of Hospital Infection

Organizations: CDCFlu, CDCGov, CDC, WHO NIAIDNews, ECDC_EU, SHEA_Epi, IDSAInfo, APIC, HIS_infection, IPS_Infection, IPSRnD, NHInfectPrevent, ips_epdc, IFH_HomeHygiene, and ESCMID

Companies: Purell, Clean hands safe hands, DEB, GWA, Hygiene, EcoLab, Georgia-Pacific, Ophardt, SaniNudge, Essity, and Tork.

Keywords: Cross Infection, Health acquired infection, Hospitalacquired infection, #hospitalacquiredinfection, Healthcare acquired infection, #Health acquired infection, Cross contamination, Nosocomial Infection, Healthcare-Associated Infection, Healthcare Associated Infection, Hand hygiene, Hospital Infection, Hand disinfection, Hand washing, Hand sanitizer, Infection control, Disinfection, Infection prevention, Decontaminate hands, Surgical site infection, Central line-associated bloodstream infections, Catheter associated urinary tract infections, Ventilator associated pneumonia, #HAI, and #HCAI

We searched Twitter by each of these queries and collected about 4.5 million tweets using the Twitter API v2. It resulted in a dataset containing selected HAI-related tweets posted from 2019 till the beginning of 2022. The dataset encompasses the tweet id, text, user id, time, and key metadata (i.e., number of likes, replies, and retweets) for each tweet. The collected tweets are in English, and each tweet has up to 280 characters.

2.2   Data Pre-processing

All the tweets were collected based on HAI-related keywords and also from the personal accounts belonging to infection prevention specialists whose goal is to teach and inform people about infection prevention. Although these accounts are all related to our goal, there is still no guarantee that all the tweets are on topic for our study. Therefore, we filtered out the collected tweets and kept only those that have at least one of the following terms: infection, health, contamination, nosocomial, healthcare, hand, hygiene, disinfection, prevention, decontaminate, surgical, bloodstream, catheter, urinary, ventilator, pneumonia, sanitizer, rub, hospital, disease, wash, control

Furthermore, inspired by Christensen et al. [3], we filtered the collected tweets once more with some ideation and problem terms to increase the chance of finding more ideas and problems in HAI. The terms are: need, problem, been, still, difficult, puzzle, can't, would, headache, would be, they would, i think, idea, and could be. This filtering process narrowed down the dataset to 692616 tweets (the unlabeled data pool).

2.3   Data Labeling

The collected data is unlabeled, and determining whether each tweet belongs to the "informative" or "non-informative" classes requires data annotation. For this purpose, three annotators with healthcare background were recruited and instructed to label the data independently. They were given instructions through some educational sessions by experts from a hygiene company and physicians from a hospital in Sweden. Also, some examples of informative and non-informative tweets are shown to annotators to familiarize them with the concept of controlling HAI. Then they were asked to read each tweet and label it based on this question: "Does the text below contain any information that can improve or create products or services related to HAI? (Mark 1 for Yes (informative) and 0 for No (non-informative))"

The annotation has been done in four iterations. In each iteration, a new batch of data is labeled by three annotators individually. This corpus of the labeled dataset is then published on a public GitHub repository and can be used for further studies.

After the annotation, we used our annotator's labels to identify each tweet's importance and label them. We assigned label 1 to those tweets that a majority of the annotators (2 out of 3, or unanimously) perceived as belonging to the "informative" class (referred to as 2- and 3-stars). Samples with no vote of informativeness (0-star) belonged to the "non-informative" class. Samples with only one vote (1-star) are ambiguous; in a sense, their label is unclear, making it difficult to judge to which group they belong. On the one hand, they are not "informative" since they failed to receive the majority; on the other hand, they should not be considered "non-informative" since one of the annotators voted them as informative. These ambiguous samples are removed from the data and neither used for training nor evaluating the models.

3   Methodology

3.1   Model

In this project, we used Transfer Learning and built a two-step deep neural network model that identifies the HAI-related informative tweets from non-informative ones. Transfer Learning is an ML technique for transferring the knowledge learned from one domain to another domain. More specifically, we took advantage of the transformer layers of the *BERTweet* language model [17], which has been previously trained on 850 million general English tweets, as a first step to extract the semantic representation of our HAI-related English tweets. Then, we use these representations as input for the next layer of the model, which is a multi-layer perceptron (MLP) chosen for classifying tweets. The portion of data used for training, validation, and testing is 60%, 20%, and 20%, respectively. The training process stops when the validation loss does not improve after ten epochs using the early stopping method. The structure of the model can be seen in Fig. 1.

BERTweet has the same architecture as the BERT (Bidirectional Encoder Representations from Transformers) [5] and consists of transformer layers and self-attention heads. On the other hand, the MLP we use as the classification head is structured from max-pooling, batch normalization, a dense layer with a ReLU activation function, a dropout layer, and another dense layer with a softmax activation function. Overall, this large model contains approximately 135 million parameters.

3.2   Active Learning in Data Labeling

One of the challenges of this project is how to train an ML model with unlabelled and highly imbalanced data. In this case, since we are working with a highly skewed dataset, most ML models will fail to distinguish samples from the minority class, and their decision will be biased toward the majority class.

Moreover, manually labeling a large-scale set of tweets (by humans) is laborious, time-consuming, and costly. This challenge, together with dealing with highly imbalanced data, motivated us to employ an iterative method based on Active Learning and feedback loops to boost the chance of finding more positive samples among the pool of unlabeled and imbalanced datasets. According to a survey on active learning [19], the main advantage of this learning approach is that it enables the model to achieve good performance even by training on relatively small number of labeled samples.

Active learning is a method that can be used for optimizing the labeling process by prioritizing samples to query an expert/oracle for labeling or correcting the labels. In this research, based on our need to find more informative samples within the pool of imbalanced and unlabeled data, we favor a query strategy that allows the active learner to select those samples with a higher probability of having ideas in the minority class. We called it the *"Richest Minority"* query strategy. These probabilities are the predictions produced by the trained model in each iteration.

Our work uses a two-step deep neural network model including *BERTweet* language model and a multi-layer perception (MLP) to classify and evaluate the tweets. The output of this model is a vector of probabilities produced by the softmax activation function that indicates how much each tweet is assumed to belong to the informative class. By sorting these probability scores from high to low and following the *Richest Minority* query strategy, we can select the most informative set of tweets to query the oracle. We employed this query strategy in our iterative algorithm and managed to detect a number of informative ideas and problems for controlling and preventing healthcare-associated infections.

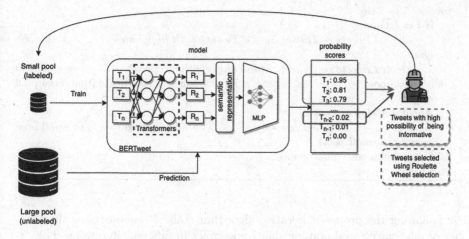

Fig. 1. Proposed algorithm based on Active Learning and Transfer Learning

Figure 1 and Algortihm 1 demonstrate the proposed iterative method. The algorithm follows the below steps:

1. A small subset of data is selected based on each tweet's metadata score, including the number of likes, replies, retweets, and quotes. To do so, we normalized each of these features using the MinMax scaler method, summed up these four normalized values, and considered it a new feature called "rate." Afterward, we sorted the rate values for each set of tweets grouped by their keywords/accounts and selected the first three tweets plus a random tweet among the rest. This procedure yielded a diversified "small pool," which contains 586 tweets. The rest of the unlabeled data is stored in the "large pool."
2. The small pool is labeled by human annotators.
3. Labeled samples (if any) from the previous iteration(s) are added to the small labeled pool.
4. The two-step model is used to extract the semantic representation of each tweet and train a classifier on them.
5. The trained model is used to predict the probability class score of the remaining unlabeled data.
6. To find more positive samples, using the *richest minority* query strategy, the 700 most informative samples (i.e., ones with the highest probability scores), as well as 300 random samples (using the *roulette wheel selection* method) are selected as the next small batch of data to query for annotation. The reason for adding random samples selected using the roulette wheel method is to avoid the Echo Chamber phenomenon [9], where the same or similar ideas are repeatedly discovered in the dataset.
7. Go back to step 2

Algorithm 1. Proposed algorithm

$large\_pool \leftarrow$ All filtered Tweets   ▷ size:692616
$small\_pool \leftarrow []$
for $i \leftarrow 1$ *to* 4 **do**
   if $i = 1$ **then**
      $selected\_tweets \leftarrow$ *Initial Seeds. Tweets with high rates*   ▷ size:586
   else
      $selected\_tweets \leftarrow richest\_minority\_query(large\_pool, ^y)$ ▷ size:1000
      large_pool $-= selected\_tweets$ ▷ remove the selected tweets from large pool
   end if
   Annotate(selected_tweets)
   $small\_pool += selected\_tweets$ ▷ add selected tweets to small pool
   $model.fit(small\_pool)$
   $^y \leftarrow$ model.predict(large_pool)
end for

Following the proposed iterative algorithm, Table 1 demonstrates the number of informative and non-informative samples in different iterations. The "3-star", "2-star", "1-star", and "0-star" columns show the score of tweets. The "Informative" column shows the summation of 3 and 2-star tweets, while the "non-informative" column shows the number of 0-star tweets in each iteration. "Aggregated informative" is the number of accumulated informative samples from the current and previous iterations. This growing data is used to retrain the model in further iterations.

4   Experiments and Results

We have designed an experiment to evaluate the performance of the proposed framework based on two data split configurations. Figure 2a shows the portions of data used to train models in each iteration. For instance, the first model is trained, validated, and tested on 60, 20, and 20% of the first batch of data. This proportion is kept for all other iterations but with the difference that the data for the rest of the iterations are being aggregated (from that iteration and the previous ones). The second configuration uses a similar setting (Fig. 2b). The only difference is that the test set here is the collection of test sets from

Table 1. Statistics of the labeled data in all iterations

iteration	3-star	2-star	1-star	0-star	total	informative	non-info	agg informative	agg non-info
1st	15	42	122	407	586	57	407	57	407
2nd	19	85	152	731	987	104	731	161	1138
3rd	26	90	197	664	974	116	664	277	1802
4th	17	81	196	676	970	98	676	375	2478
total	77	298	667	2478	3517	375	2478		

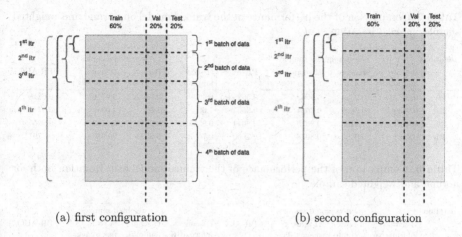

(a) first configuration   (b) second configuration

Fig. 2. The data split used for the experiment. Note that the second configuration uses the same test set for all iterations

all iterations. The purpose is to evaluate the model's performance without the influence of various test sets.

In each iteration, the performance of the trained model is evaluated and reported using the f1-score measure for each class and the macro average of both classes. It is known that the f1-score evaluation metric is less sensitive to class imbalance. Furthermore, another evaluation metric, the Area under the Precision-Recall Curve (PR-AUC), is calculated to assess the performance of the models.

One common practice for training models with imbalanced datasets is to add weight to the samples from the minority class. To evaluate the effect of adding sample weights, we trained the same model twice, once in the presence and once in the absence of sample weights. We assign weights equal to one to the samples from the majority class, while the samples from the minority class receive weights of 2 or 3 according to their respective number of votes. In other words, positive samples with two votes get weight 2, and positive samples with three votes get weight 3. The samples with only one vote are being ignored, as mentioned earlier. Tables 2 and 3, respectively, represent the results for the first and second data splits.

Comparing the values from "f1-score macro avg" of "non-weighted" and "weighted samples" in Table 2, one can conclude that assigning weight to the samples, improves the performance of the model, as expected. On average, the "f1-score macro avg" has increased by 7% over four iterations. The amount of improvement decreases gradually from the first to the last iteration. In the first iteration, sample weighting helped the model by 20.79%. This value decreased to 0.05% in the last iteration. Our hypothesis to explain this phenomenon is that as we go through the iterations, the training set's size increases, reducing the need for adding sample weights. Also, by comparing the "PR-AUC" of "non-weighted" versus "weighted samples" in each iteration, we can see this number

Table 2. Comparison of the performance of the trained model on normal and weighted samples in each iteration

iteration	non-weighted samples				weighted samples			
	f1-score informative	f1-score non-info	f1-score macro avg	PR-AUC	f1-score informative	f1-score non-info	f1-score macro avg	PR-AUC
1st	0.00	0.9333	0.4667	0.5625	0.4000	0.9492	0.6746	0.6718
2nd	0.9459	0.9909	0.9684	0.9565	1.00	1.00	1.00	0.9782
3rd	0.8872	0.9785	0.9329	0.8644	0.9552	0.9914	0.9733	0.9055
4th	0.7627	0.9729	0.8678	0.7960	0.7692	0.9673	0.8683	0.7791

Table 3. Comparison of the performance of the trained model over iteration both for normal and weighted samples

iteration	non-weighted samples				weighted samples			
	f1-score informative	f1-score non-info	f1-score macro avg	PR-AUC	f1-score informative	f1-score non-info	f1-score macro avg	PR-AUC
1st	0.00	0.9392	0.4696	0.5572	0.2655	0.9201	0.5928	0.3174
2nd	0.3974	0.9091	0.6532	0.4349	0.3638	0.8108	0.5872	0.5004
3rd	0.4444	0.9529	0.6987	0.6081	0.6032	0.9513	0.7772	0.6288
4th	0.6435	0.9605	0.8020	0.6830	0.6719	0.9596	0.8154	0.6924

has increased chiefly when the model has trained on weighted samples. A similar pattern for the effect of sample weight can be concluded from the results of Table 3.

By comparing the macro average f1-score of the trained models in consecutive iterations (see Table 3), we can see that the ability of the model to distinguish between informative and non-informative tweets gradually increases both for weighted and non-weighted scenarios. The performance starts from 0.46 and increases all the way to 0.80 for the non-weighted samples. The corresponding numbers for the weighted samples show an increase from 0.59 to 0.81.

Moreover, by subtracting the value of the f1-score of the informative class from the f1-score of the non-informative class in consecutive iterations, we can see this value is relatively high in the first iteration (0.93 for non-weighted and 0.65 for weighted). This number gradually decreased to the last iteration (0.31 for non-weighted and 0.28 for weighted). Also, by comparing the "PR-AUC" of the trained model on "non-weighted" versus "weighted samples" in consecutive iterations in Table 3, we can see this value has increased from 0.55 to 0.68 for "non-weighted" and has increased from 0.31 to 0.69 for "weighted samples." These patterns show the performance improvement of the model on classifying the imbalance dataset over iterations.

5   Conclusions and Future Works

The main contribution of this paper is to introduce a full framework capable of discovering ideas and problems to control and prevent Healthcare-Associated Infections (HAI). This framework contains a corpus of 4.5 million HAI-related tweets posted from 2019 till the beginning of 2022 using the Twitter API v2 from the Twitter platform. Moreover, our work introduces an iterative machine learning method based on active learning and feedback from the model's decision. It selects the informative tweets based on the novel *richest minority* query strategy. The collected and labeled dataset, as well as the algorithm's code, are published in a GitHub repository called AID4HAI.

In our experiments, the proposed framework managed to discover 375 informative HAI-related ideas and problems, within the four iterations. The ideas and problems concern a number of various topics and directions. Figure 3 plots a handful of automatically extracted ideas/problems from Twitter. Our innovation team helped us visualize the core idea of each tweet across a two-dimensional chart. The x-axis represents the spectrum of ideas suggesting products to services. The y-axis spreads ideas based on behavior or technological-driven.

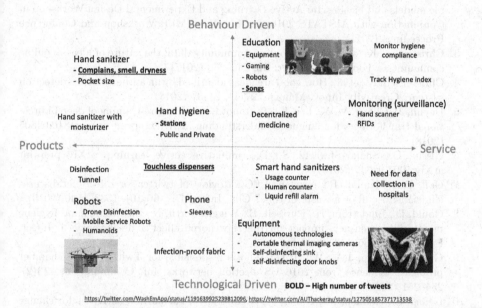

Fig. 3. Examples of extracted ideas plotted on a chart with an x-axis (service/product spectrum) and a y-axis (behavior/technology-driven).

The deep neural network model used in this study categorizes a tweet based on its informativeness. As a future work, it would be interesting to evaluate the tweets across additional dimensions of interest. Moreover, one could visualize the vectors from the model's attention layers and validate if they are focusing on the sensible tokens of the text.

Acknowledgment. The work was supported by research grants from KK-Foundation, Scania CV AB and the Vinnova program for Strategic Vehicle Research and Innovation (FFI).

We would like to express our utmost gratitude to Håkan Lindström and Peter Blomström and their team for their valuable help and professional support during this project. Also, we would like to thank our annotation team for helping us to provide this labeled data.

References

1. Barata, R., Leite, M., Pacheco, R., Sampaio, M.O., Ascensão, J.T., Bizarro, P.: Active learning for imbalanced data under cold start. In: Proceedings of the Second ACM International Conference on AI in Finance, pp. 1–9 (2021)
2. Chen, Y., Mani, S.: Active learning for unbalanced data in the challenge with multiple models and biasing. In: Active Learning and Experimental Design Workshop in Conjunction with AISTATS 2010, pp. 113–126. JMLR Workshop and Conference Proceedings (2011)
3. Christensen, K., et al.: Mining online community data: the nature of ideas in online communities. Food Qual. Prefer. **62**, 246–256 (2017)
4. Christensen, K., et al.: How good are ideas identified by an automatic idea detection system? Creativity Innov. Manage. **27**(1), 23–31 (2018)
5. Devlin, J., Chang, M.W., Lee, K., Toutanova, K.: Bert: pre-training of deep bidirectional transformers for language understanding. arXiv preprint arXiv:1810.04805 (2018)
6. Gissin, D., Shalev-Shwartz, S.: Discriminative active learning. arXiv preprint arXiv:1907.06347 (2019)
7. Goff, D.A., Kullar, R., Newland, J.G.: Review of twitter for infectious diseases clinicians: useful or a waste of time? Clin. Infect. Dis. **60**(10), 1533–1540 (2015)
8. Gould, D., Lindström, H., Purssell, E., Wigglesworth, N.: Electronic hand hygiene monitoring: accuracy, impact on the Hawthorne effect and efficiency. J. Infect. Prev. **21**(4), 136–143 (2020)
9. Guo, L., Rohde, J.A., Wu, H.D.: Who is responsible for Twitter's echo chamber problem? evidence from 2016 US election networks. Inf. Commun. Soc. **23**(2), 234–251 (2020)
10. Irgang, L., Holmén, M., Gama, F., Svedberg, P.: Facilitation activities for change response: a qualitative study on infection prevention and control professionals during a pandemic in Brazil. J. Health Organ. Manage. **35**, 886–903 (2021)
11. Kesselheim, A.S., Xu, S., Avorn, J.: Clinicians' contributions to the development of coronary artery stents: a qualitative study of transformative device innovation. PLoS ONE **9**(2), e88664 (2014)
12. Kharazian, Z., Rahat, M., Fatemizadeh, E., Nasrabadi, A.M.: Increasing safety at smart elderly homes by human fall detection from video using transfer learning approaches. In: 30th European Safety and Reliability Conference (ESREL2020) & 15th Probabilistic Safety Assessment and Management Conference (PSAM15), Venice, Italy, 1–5 November 2020 (2020)
13. Lewis, D.D.: A sequential algorithm for training text classifiers: corrigendum and additional data. In: ACM Sigir Forum, vol. 29, pp. 13–19. ACM New York, NY, USA (1995)

14. Lotfinejad, N., Peters, A., Tartari, E., Fankhauser-Rodriguez, C., Pires, D., Pittet, D.: Hand hygiene in health care: 20 years of ongoing advances and perspectives. Lancet. Infect. Dis **21**(8), e209–e221 (2021)

15. Malte, A., Ratadiya, P.: Evolution of transfer learning in natural language processing. arXiv preprint arXiv:1910.07370 (2019)

16. Martischang, R., et al.: Enhancing engagement beyond the conference walls: analysis of twitter use at# icpic2019 infection prevention and control conference. Antimicrob. Resist. Infect. Control **10**(1), 1–10 (2021)

17. Nguyen, D.Q., Vu, T., Nguyen, A.T.: Bertweet: a pre-trained language model for English tweets. arXiv preprint arXiv:2005.10200 (2020)

18. Rahat, M., Mashhadi, P.S., Nowaczyk, S., Rognvaldsson, T., Taheri, A., Abbasi, A.: Domain adaptation in predicting turbocharger failures using vehicle's sensor measurements. In: PHM Society European Conference, vol. 7, pp. 432–439 (2022)

19. Settles, B.: Active learning literature survey (2009)

20. Smith, S.W., Sfekas, A.: How much do physician-entrepreneurs contribute to new medical devices? Med. Care **51**, 461–467 (2013)

21. Thune, T., Mina, A.: Hospitals as innovators in the health-care system: a literature review and research agenda. Res. Policy **45**(8), 1545–1557 (2016)

22. Weigel, T., Goffin, K.: Creating innovation capabilities: mölnlycke health care's journey. Res. Technol. Manage. **58**(4), 28–35 (2015)

Explanations for Itemset Mining by Constraint Programming: A Case Study Using ChEMBL Data

Maksim Koptelov[1]([⊠]) [iD], Albrecht Zimmermann[1], Patrice Boizumault[1], Ronan Bureau[2], and Jean-Luc Lamotte[1]

[1] UNICAEN, ENSICAEN, CNRS – UMR GREYC, 14000 Caen, France
{maksim.koptelov,albrecht.zimmermann,patrice.boizumault,
jean-luc.lamotte}@unicaen.fr
[2] CERMN/UNICAEN, 14200 Caen, France
ronan.bureau@unicaen.fr

Abstract. In sensitive applications, such as drug development, offering experts an explanation for why data mining operations arrive at certain results adds a very valuable facet. In this work we benefit from modelling the task as a Constraint Satisfaction Problem (CSP) twice: by adding multiple constraints to the mining process and by deriving pattern failure explanations. We illustrate experimentally how to apply our method on data originally retrieved from the ChEMBL database [14]. We also report some interesting dependencies discovered by our method which are not easy to observe when analysing data manually.

Keywords: Itemset mining · Constraint programming · Explainable AI

1 Introduction

With the recent surge in applications of machine learning, mainly deep learning, techniques to a variety of fields, the need for explanations for those techniques has also increased. Most of the techniques explaining machine learning models exploit the supervised nature of the problem setting, solving problems such as:

- Can we learn a symbolic model giving the same predictions?
- What are the minimal changes that need to be done to a data instance to change its predicted label?
- Can we identify features or image regions that contribute strongly to the prediction result?

In unsupervised data mining, however, especially in constraint-based pattern mining, labeled examples are typically not available, increasing the challenge. As a result, there are arguably more workshops for (interesting) work-in-progress papers on explainable data mining than there are publications that were accepted for conference proceedings or journals on the subject. The ones that do exist ignore explanations of itemset mining, a classical data mining task.

In addition, the questions change: since mined patterns are often starting points for further development, for instance in drug development, or "food for thought" that help formulate research hypothesis, their plausability and persuasiveness need to be supported.

Finally, this is clearly an application dependent subject. There are a number of applications in sensitive areas such as pharmaceutical or medical domains for which explanations are obligatory. As a common example, a chemical compound selected by a black-box classifier from a database of molecules cannot be approved by a pharmacist as a drug candidate, because of the high risks associated with the following production process costs [6].

Indeed, due to such considerations, our partner researchers at CERMN[1] are in need of explanations for their itemset mining and motivated this study. Their task, requires finding molecular sub-structures which are able to discriminate between active and inactive molecules. In addition, they are obliged to use multiple constraints regarding the structure and properties of the resulting patterns, and they would like to have explanations on top of that.

More precisely, the answers to the following questions are desired:

- Why is this pattern not frequent/closed/emerging?
- Why could this constraint not be satisfied?
- How did the mining algorithm arrive at this particular solution satisfying the constraint(s)?

Straight-forward answers to these questions, e.g. "the pattern doesn't have enough overall support" or "the pattern has too much support in the class that was not targeted" are tautological and not very satisfying. Instead, a practitioner would be interested in knowing what element of the pattern or which other constraint forced the support below a given threshold or lead to the inclusion of transactions that reduce the growth rate.

These are questions that have already been asked in similar form in the constraint programming (CP) community [12,18]. We formulate our problem setting as one of constraint-based itemset mining, for which CSP solutions have been proposed [7,15]. In addition, past work has added explanations to CSP solvers [3,13]. We therefore base our work in part on proposals made to answer explanatory questions in CP. In this work, we develop an approach for pattern failure explanations, which is our main contribution. We demonstrate the application of it on data derived from the ChEMBL database.

The rest of the paper is organised as follows. Section 2 highlights important works related to the topic. Section 3 outlines the problem setting and used formalisms, including how to derive explanations for itemset mining based on constraint failure. Section 4 shows and discusses the results of our case study on the ChEMBL data. Finally, we review and discuss future improvements in Conclusion.

---

[1] Centre d'Etude et de Recherche du Médicament de Normandie: https://cermn. unicaen.fr.

2   Related Work

Following [9], we define *data mining* as the search for valid, novel, potentially useful, and ultimately understandable patterns in the data. One of the seminal tasks in data mining is itemset mining, for which a number of more specialized problems, such as frequent itemset mining, frequent closed itemset mining, discriminative or emerging itemset mining and others have been defined [11]. In this paper, we mainly focus on those first three tasks.

Traditional approaches for itemset mining take the form of specialized breadth-first [1,2,31] and depth-first algorithms [16,32]. An alternative approach involves using CP [7,15], a general declarative methodology for solving constraint satisfaction problems. Constraint programs specify the problem and a general solver tries to find a solution. A clear advantage lies in the universality of this approach. The new task could be modelled by adding new constraints while in traditional approaches the algorithm must be redesigned from scratch each time. Another advantage of CP systems is a possibility of result explanation [3,13]. In this work we benefit from the latter by modelling the itemset mining task with CP.

Most of the works on explainable AI are focused on explanations for machine learning [25,27]. While work on directly explainable data mining are rare, interactive data mining has been proposed as a first approximation of interpretable data mining involving both the miner and the domain expert, as well as the data itself [17,22]. The ultimate goal of such a process is to make pattern mining more practically useful by making the end user understand *during* the mining process how mining results come to pass. Discrimination-aware data mining exists for more than a decade now [19,26]. It mainly focuses on developing methods for protecting from unfair classification models, especially when they might affect somebody's life. Work on visual data mining [4,10,30] attempt to make the data mining process understandable through visualization. Some of them offer explanations for clustering or binary classification tasks [5,29]. Finally, there are few works which use explanations for improving the data mining results. For instance, [21] tries to mix data mining with domain expert knowledge in order to improve the quality of discovered patterns in the medical domain. Likewise, [20] developed an approach for mining surprising patterns and generating explanations. Based on association rule mining, the approach that they proposed uses expert knowledge to improve the search and provide explanations.

In this paper, we take a step towards explaining itemset mining, one of the core tasks of data mining. This is the first work in this direction to the best of our knowledge.

3   Preliminaries

As described above, the result of a constraint-based pattern mining operation is a set of patterns. A user might want to know now why certain patterns were included and others were not. The straight-forward answer is simple: the patterns satisfied the specified constraints (or not). This might not be sufficient information, however: specifying constraints and deciding on threshold values is not an easy task, and a small change may lead to a large change in results. In addition, especially when a number of complex constraints are combined, their interplay can lead to the inclusion or exclusion of patterns in unexpected manners, which are not easy to understand without additional explanation. Gaining such understanding will help in formulating future constraints. Our proposal for fournishing such explanations is to exploit *pattern failure* explanations in CSPs. In this section, we lay out the itemset mining problem, the CSP framework, and how to model itemset mining, as well as how to derive explanations.

3.1   Itemset Mining

The pattern mining task we address in this paper is the classical itemset mining one: *given* a set of *items* $\mathcal{I} = \{i_1, \ldots, i_m\}$, a transaction set $\mathcal{T} = \{t_1, \ldots, t_n \mid t_i \subseteq \mathcal{I}\}$, and a (combination of) constraint(s) $\mathcal{C} : \mathcal{I} \times \mathcal{T} \mapsto \{\text{true,false}\}$, *find* $Th(\mathcal{I}, \mathcal{T}, \mathcal{C}) = \{p \subseteq \mathcal{I} \mid \mathcal{C}(p, \mathcal{T}) = \text{true}\}$.

The *support* of an itemset is the cardinality of the set of transactions in which it is contained: $supp(p, \mathcal{T}) = |\{t \in \mathcal{T} \mid p \subseteq t\}|$. Given a threshold θ_f the minimum support (frequency) constraint is defined as $freq(p, \mathcal{T}) = \text{true} \Leftrightarrow supp(p, \mathcal{T}) \geq \theta_f$. An itemset is *closed* if none of its strict specializations has the same support: $closed(p, \mathcal{T}) = true \Leftrightarrow \forall p' \supset p : supp(p', \mathcal{T}) < supp(p, \mathcal{T})$.

Finally, given a labeling $l : \mathcal{I} \mapsto \{+, -\}$, $\mathcal{T}^+ = \{t \in \mathcal{T} \mid l(t) = +\}$, $\mathcal{T}^- = \mathcal{T} \setminus \mathcal{T}^+$, a quality measure $\sigma : \mathcal{I} \times \mathcal{T}^+ \times \mathcal{T}^- \mapsto \mathbb{R}$, a threshold θ_d, an itemset is *emerging/discriminative*: $disc(p, \mathcal{T}) = \text{true} \Leftrightarrow \sigma(p, \mathcal{T}^+, \mathcal{T}^-) \geq \theta_d$.

3.2   Constraint Programming

General CSP Context. A classical CSP is defined by a triplet (V, D, C) in which $V = \{X_1, X_2, ..., X_n\}$ is a set of variables, $D = \{D_1, D_2, ..., D_n\}$ the set of domains of variables, with D_i a finite set containing the possible values for the variable X_i, and $C = \{c_1, c_2, ..., c_k\}$ a set of constraints. A solution of the CSP is a complete instantiation S such that all the constraints C are satisfied by S.

Consider an example with $V = \{X_1, X_2, X_3\}$, $D_1 = \{1, 3, 5\}$, $D_2 = \{2, 3, 4\}$, $D_3 = \{2, 3, 7\}$ and $c_1 : X_1 < X_2$, $c_2 : X_2 = X_3$ (Fig. 1). There are two possible solutions for this problem: $X_1 = 1$, $X_2 = 2$, $X_3 = 2$ or $X_1 = 1$, $X_2 = 3$, $X_3 = 3$.

Explanations for CSPs. The CSP framework is not only a powerful tool for modelling different type of constraints, but also for providing explanations (Sect. 2). In this work, we deal with explanations for value removal as the simplest to implement and interpret.

An *explanation for value removal* is a subset of the set of constraints C such that the conjunction of these constraints leads to the removal of the value a from the domain of the variable X_i. In case of multiple explanations, this expression becomes a disjunction of conjunctions:

$$Expl(X_i \neq a) = \bigvee \left(\bigwedge_{i \in [1..k]} c_i \implies X_i \neq a \right).$$

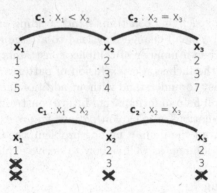

Fig. 1. An example of a CSP (top) and the result of its filtering (bottom)

An example of such explanations for the CSP in Fig. 1: $Expl(X_1 \neq 5) = c_1$ (there is no value > 5 in the domain of X_2), $Expl(X_2 \neq 4) = c_2$ (there is no value $= 4$ in the domain of X_3), $Expl(X_1 \neq 3) = c_1 \wedge Expl(X_2 \neq 4) = c_1 \wedge c_2$ (c_2 removes the value 4 from X_2, and c_1 in turn removes 3 from X_1).

Modeling Itemset Mining as a CSP. To model the itemset mining problem with CP, we follow [15]: the CSP must be defined by a triplet (V,D,C), in which $V = I \cup T$ a set of variables s.t.: $I = \{I_1, I_2, ..., I_m\}$ a set of items, $T = \{T_1, T_2, ..., T_n\}$ a set of transactions, $D = \{D_{I_1}, ..., D_{I_m}, D_{T_1}, ..., D_{T_n}\}$ a set of domains of variables with $D_i = \{0, 1\}$, $C = \{c_1, c_2, ..., c_k\}$ a set of constraints. As for the latter refined constraints proposed by [15] can be used according to the task.

Consider a toy example. Given a set of transactions $T = \{ACD, ABD, CD\}$ and minimum frequency $\theta_s = 2$, we would like to find all frequent closed patterns. To model the problem as a CSP, we define $DB = \{\{1, 0, 1, 1\}, \{1, 1, 0, 1\}, \{0, 0, 1, 1\}\}$, $V = \{I_1, I_2, I_3, I_4, T_1, T_2, T_3\}$, $D = \{D_{I_1}, D_{I_2}, D_{I_3}, D_{I_4}, D_{T_1}, D_{T_2}, D_{T_3}\}$ with $D_{X_i} = \{0, 1\}$, $C = \{c_1, c_2, ..., c_{11}\}$ with the constraints defined as in Fig. 2.

There are three solutions to this problem: AD, CD, D. Figure 2 also demonstrates the search process. Branching of the search tree usually stops when a solution is found, then the search backtracks to another branch until all the solutions are retrieved. In our setting, however, we continue the search until all the failures are found (Failure 1–7 in Fig. 2). We use them later to explain a pattern failure which we define as follows. A *pattern failure* is a state of the CSP in which one of the itemset domains is empty:

$$I_1 = [\,] \vee I_2 = [\,] \vee ... \vee I_m = [\,] \implies CSP \rightarrow Fail.$$

3.3   Explanations for Itemset Mining

As explained above, CSPs allow to derive explanations. The default approach does not allow to explain a success (a solution, specific pattern or presence of an

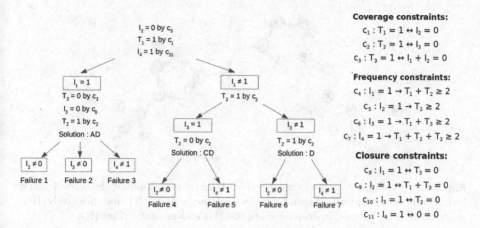

Fig. 2. Constraints and search tree for a toy example of itemset mining by CSP

item in the solution) in an effective way: it can only say that we have this as a solution because it satisfied all the constraints. However, it *is* possible to explain a failure (no solution at all, a particular pattern does not belong to the solution etc.) more effectively by interpreting constraints which led to that failure. In addition, there could be an exponential number of explanations. We therefore choose to keep only one: the shortest one. Here, we present an approach for that.

Our approach for finding explanations for pattern failure is a 4-step process:

S1 Initialize domains with the elements of a pattern whose failure (i.e. absence from the solution) needs to be explained. The pattern needs to be precisely specified by the user/chemical expert

S2 Obtain different explanations for pattern failure in the form of conjunctions and/or disjunctions of constraints which led to emptying one of the itemset domains

S3 Select the shortest explanation w.r.t. the number of constraints

S4 Interpret the constraints in that explanation using logical inference and/or analysing them manually

Following our example in Fig. 2, we can explain, for instance, why pattern AB is not in the solution. The shortest explanation will be:

$$Expl(AB \rightarrow \text{Fail}) = c_5 \wedge c_{11}.$$

We can interpret c_5 (the frequency constraint) as "if B is in the itemset ($I_2 = 1$), the itemset must be frequent ($T \geq 2$)". Since $T \geq 2$ is False, B must be removed from the pattern, which can be rephrased as "the pattern cannot be frequent if B is present". Closed itemset mining aims at avoiding redundant itemsets and the closure constraint checks if all transactions contain the same element as without it the itemset cannot be closed. We can thus interpret c_{11} (the closure constraint) as "there must be D in the itemset ($I_4 = 1$), otherwise it

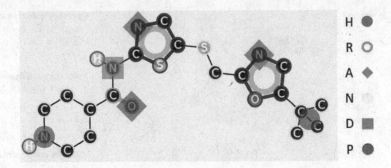

Fig. 3. An example of a molecule (left) and its pharmacophoric features (right): hydrogen-bond acceptors (A) and donors (D), negatively (N) and positively (P), charged ionizable groups, hydrophobic regions (H) and aromatic rings (R)

cannot be a closed pattern" ($I_4 = 1$ if and only if *True*, where *True* corresponds to $0 = 0$).

4   Case Study

We illustrate our approach on a set of molecular data, from which we aim to mine combinations of chemically meaningful subgraph patterns.

4.1   Data and Representation

Our data originally is a set of BCR-ABL inhibitors (target ID 1862) that have been extracted from the ChEMBL[2] database, a widely used database in computational drug discovery [14]. In this study, we would like to understand the mechanism of action on the BCR-ABL target.

After several steps of preprocessing, our set is composed of 739 molecules, 387 of which are labeled as active and 352 as inactive. A molecule is called *active* if it causes the target to react. If a molecule does not generate a sufficient reaction at the level of the target, it is considered to be *inactive*. Each molecule is represented as the 2D/3D arrangement of molecular features that are necessary for a drug candidate to interact with a biological target in a specific binding site [8]. In total there are 6 features in our data (Fig. 3). Graphs in this representation are also referred to as *pharmacophores*, with its *order* O_n equal to its number of vertices (Fig. 4). For example, the molecule in Fig. 3 includes the following pharmacophores: |P|D||5|, |P|A||5|, |P|A||7|, |P|R||6|, |P|A||12|, |R|R||3|, |R|A||0|, |R|H||1|, |R|H||6|, |A|A||6| etc. (28 in total).

From our data, we mined 258 distinct 2D pharmacophores of O_2 having minimum support 10, using Norns [24]. The objective of the study is to explain why a molecule is active by identifying the pharmacophores which cause activity.

---

[2] A manually curated database of bioactive molecules with drug-like properties.

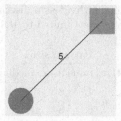

Fig. 4. Example of an O_2 pharmacophore $|P|D||5|$ with a positively charged ionizable group (P) and a hydrogen-bond donor features (D) with the distance 5 between them

4.2 Mining Task

We want to identify combinations of at most 8 such pharmacophores that are shared by a significant number of molecules (at least 12%–15% of the data) and appears more often in one of the two classes than in the other, and that are not subsets of each other.

4.3 Experimental Setting

Using mined pharmacophores, we can represent each molecule as a transaction encoding whether pharmacophores are present or not, giving us the classical itemset mining setting. We implemented refined constraints from [15] and a CSP solver in Python[3]. We adopted the AC3 algorithm [23] for the latter and implemented the MAC algorithm [28] for backtracking the search.

Following preliminary experiments, we use χ^2 as the discriminatory measure. We implemented χ^2 as in [15]. Constraint thresholds were set to $\theta_{size} = 8$ for size, $\theta_{supp} = 100$ for minimum support, and minimum thresholds for χ^2 ($\theta_{\chi^2} \in \{48, 64, 96, 128\}$). In addition, we experimented with adding a *purity* constraint, i.e. patterns present in one class only, which we defined as follows:

$$I_i = 1 \rightarrow min\left(\sum_{t^+} DB_{ti} \cdot T_t, \sum_{t^-} DB_{ti} \cdot T_t \right) = 0 \qquad (1)$$

Finally, we try to answer why changing one *item* in a pattern and adding another one changes the class of solution from *pure* (i.e. covering only active or inactive molecules) to *not pure* (covering both active and inactive molecules).

4.4 Experimental Results

As can be seen from Table 1, the pure solution constraint reduces the number of results dramatically – on average by three order of magnitude. Moreover, the results corresponding to the pure solution and θ at 48, 64 and 96 remain the

---

[3] https://github.com/koptelovmax/dmbycsp.

Table 1. Emerging pattern mining with χ^2 as a discriminative measure, pattern size not exceeding 8 and minimum frequency limited to 100

θ_{χ^2}	Pure	Found solutions			Pattern size			Coverage	Frequency		
		Total	Active	Inactive	min	max	median		min	max	mean
48	✗	85037	84530	507	1	8	7	100,00%	100	682	157,7
	✓	47	47	0	6	8	8	30,85%	101	175	126,7
64	✗	69060	68995	65	1	8	7	100,00%	100	656	164,6
	✓	47	47	0	6	8	8	30,85%	101	175	126,7
96	✗	44013	44013	0	1	8	7	99,32%	100	624	179,5
	✓	47	47	0	6	8	8	30,85%	101	175	126,7
128	✗	24645	24645	0	1	8	7	97,56%	119	547	198,3
	✓	27	27	0	6	8	8	30,04%	119	175	136,9

same. In addition, there are no inactive solutions in case of pure patterns or with θ at 96 and 128, which is not a problem per se since our aim is to explain active solutions. On one hand, a smaller number of patterns is easier to evaluate manually. The main drawback of this modeling that the pure solutions found cover only 30% of molecules. This is not really desirable for a chemical expert, and solutions combining to cover most of the molecules are required. We thus go to the next step of our study where we will try to understand the interior mechanics behind our mining process.

Towards Explaining Pattern Failure. After discussing with the chemical experts we collaborate with, they asked for an explanations for why changing one item (pharmacophore in our case) leads to changing the class of solution from pure to not pure:

$$ABC \text{ (pure)} \leftrightarrow AEC \text{ (not pure)}.$$

While studying this phenomena in more detail, we realised that the actual change of the class happens when one element is *removed* from the pattern (Fig. 5). In other words:

$$ABC \text{ (pure)} \rightarrow AC \text{ (not pure)} \rightarrow AEC \text{ (not pure)}.$$

We would like to explain the first part: why removing an item makes the pattern not pure. Consider, for instance, the first two lines in the example in Fig. 5, where solution 17863 is pure, and 17902 is not. Our methodology for answering that is the following:

1. Model the problem using the purity constraint (Eq. 1)
2. Explain using our method from Sect. 3.3 why the combination of molecule features |D|R||1| |D|R||3| |A|H||11| |R|R||1| |R|H||5| is a failure

17863	$	D	R		1	$ $	D	R		3	$ $	A	R		0	$ $	A	H		11	$ $	R	R		1	$ $	R	H		5	$	(171: +171 -0)	202.36
17902	$	D	R		1	$ $	D	R		3	$ $	A	H		11	$ $	R	R		1	$ $	R	H		5	$	(172 : +171 -1)	198.95
17899	$	D	R		1	$ $	D	R		3	$ $	A	R		2	$ $	A	H		11	$ $	R	R		1	$ $	R	H		5	$	(171: +170 -1)	197.42

Fig. 5. Pattern failure example. Columns are: id of solution, pattern, frequency and χ^2 value. Red colour represents removal of an item and blue is adding an item (Color figure online)

3. Verify why adding $|A|R||0|$ to the pattern gives a solution. For that:
   (a) Find its purity constraint
   (b) Explain why it became *true*

After an initialization step we move directly to S2 of our approach from Sect. 3.3, which will generate the following explanations:

$$Expl(|D|R||1| \; |D|R||3| \; |A|H||11| \; |R|R||1| \; |R|H||5| \rightarrow Fail) =$$

$$= Expl(|D|R||1| \neq 1) \lor Expl(|D|R||3| \neq 1) \lor Expl(|A|H||11| \neq 1) \lor$$

$$\lor Expl(|R|R||1| \neq 1) \lor Expl(|R|H||5| \neq 1) = c_{2361} \lor c_{2372} \lor c_{2440} \lor c_{2461} \lor c_{2488}.$$

According to S3, the shortest explanation is one of those constraints, for instance:

c_{2361} – purity constraint:
   before filtering: $|D|R||1| = 1 \rightarrow min(384, 298) = 0$ *False*
   after filtering: $|D|R||1| = 1 \rightarrow min(171, 1) = 0$ *False*

Finally, we try to interpret this following S4. After filtering, the CSP this constraint remains *false*, but its coverage changes – each of the items in the pattern covers 171 active molecules and 1 inactive one:

$$c_{2361}: |D|R||1| = 1 \rightarrow min(T_1 + T_2 + ... + T_{356} + T_{379}, \; T_{429}) = 0 \; \textbf{False}$$

We also know that the inactive molecule is represented by the variable T_{429} (or by ChEMBL ID 1984038).

Next, we would like to explain why adding $|A|R||0|$ to the solution makes the pattern pure. For that, one needs to instantiate the CSP with a new pattern including $|A|R||0|$ and check its purity constraint after filtering:

$$c_{2417} : |A|R||0| = 1 \rightarrow min(171, 0) = 0 \; True$$

As can be seen from c_{2417}, our pattern is included only in active molecules. To explain for a user who is not a data mining expert why removing $|A|R||0|$ from the pattern affects its purity, one can draw the Euler diagram (Fig. 6). In that diagram, the pattern containing all pharmacophores including $|A|R||0|$ will be present only in active molecules. This is the type of information which is laborious to observe manually, but can be easily derived using a CSP.

Finally, we would like to explain why the purity constraint associated with $|A|R||0|$ becomes *true*, especially given that before filtering it was *false*:

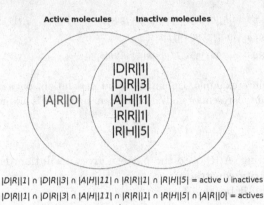

$|D|R||1| \cap |D|R||3| \cap |A|H||11| \cap |R|R||1| \cap |R|H||5| = $ active $\cup$ inactives

$|D|R||1| \cap |D|R||3| \cap |A|H||11| \cap |R|R||1| \cap |R|H||5| \cap |A|R||0| = $ actives

Fig. 6. Active, inactive molecules and their intersection

$$c_{2417} : |A|R||0| = 1 \rightarrow min(T_1+T_2+...+T_{787}, T_{388}+T_{389}+...+T_{739}) = 0 \; False$$

To do that we need to explain why $T_{388} \neq 1, ..., T_{739} \neq 1$:

$Expl(T_{388} \neq 1) = c_{388}$, where c_{388} – coverage constraint:
$$T_{388} = 1 \leftrightarrow |D|R||3| + |A|H||11| + |R|R||1| = 0 \; False$$

...

$Expl(T_{739} \neq 1) = c_{739}$, where c_{739} – coverage constraint:
$$T_{739} = 1 \leftrightarrow |D|R||1| + |A|H||11| + |R|H||5| = 0 \; False$$

These constraints can be interpreted as follows: the combination of molecular features $|D|R||3|$ $|A|H||11|$ $|R|R||1|$ must cover molecule T_{388} (ChEMBL ID 1836675), ..., and $|D|R||1|$ $|A|H||11|$ $|R|H||5|$ must cover T_{739} (ChEMBL ID 281470), otherwise the coverage condition fails.

This is the type of information which can be easily retrieved with our method, and which can be useful for chemical experts.

Towards Explaining Constant Constraints Outcomes. We noticed that certain constraints are always *true* or *false*. For instance, in our example in Fig. 2, there are two constraints which always remain constant: c_5 (always *false*) and c_{11} (always *true*). In that toy example they can be interpreted as follows: if there is B in the pattern it is always not frequent (c_5); there must be D in the solution to be closed (c_{11}). Both of these conditions hold for our simple CSP since each solution contains item D and non of them has B.

Now if we verify which constraints remain constant for our ChEMBL set with the constraint thresholds $\theta_{\chi^2} = 128$, $\theta_{size} = 8$, $\theta_{supp} = 100$, allowing pure solutions only, we find that 363 constraints (out of 2510 used to model the CSP) remain constant:

- 159 frequency constraints – always *false*
- 194 discriminative constraints – always *false*
- 2 size constraints – always *true*
- 9 purity constraints – always *true*

If we interpret them, we get the following information:

- frequency constraints – if there is $|P|P||3|$, $|P|D||10|$, $|P|D||11|$, ..., $|H|H||9|$ (159 in total) in the pattern, it is always not frequent
- discriminative constraints – if there is $|P|P||3|$, $|P|D||10|$, $|P|D||11|$, ..., $|H|H||9|$ (194 in total) in the pattern, it is always not discriminating
- size constraints – if the pattern is included in molecule T_{671} (ChEMBL ID 1241863) or T_{696} (ChEMBL ID 1241772) its size is always less than 8
- purity constraints – $|P|D||17|$, $|N|D||8|$, $|N|D||9|$, ..., $|R|H||19|$ (9 in total) are covered only by pure molecules:
    - $|P|D||17|$, ..., $|R|H||19|$ (7 in total) are included in active molecules only
    - $|N|D||8|$, $|N|D||9|$ – in inactive molecules only

This information can be read off without rerunning the mining operation. This can be useful for chemical experts to get quick-shot statistics on the data, explain why particular patterns in the solution do not include particular elements, modify the data set, or adjust constraint settings before repeating mining.

5   Conclusion

In this paper, we have explained how one can use constraint failure explanations in CSPs to explain why certain patterns do not appear in a solution set. These explanations can then be used to identify problematic data instances, or to modify constraint parameters. In a chemoinformatics use case, we have shown how such explanations and the identification of particular phenomena can look in practice.

A drawback of our method is that patterns to be explained need to be specified manually, and explanations need to be interpreted to arrive at statements about the data. In future work, we will therefore look at how generate patterns automatically, e.g. by looking at syntactically similar patterns, and how to post-process explanations to highlight interesting data. We would also think about how we could improve the explanations which we already generated. For the last we first need to get a detailed feedback from the experts.

References

1. Agrawal, R., Srikant, R.: Fast algorithms for mining association rules. In: VLDB, vol. 1215, pp. 487–499 (1994)
2. Bodon, F.: A fast apriori implementation. In: FIMI, vol. 3, p. 63 (2003)

3. Bogaerts, B., Gamba, E., Guns, T.: A framework for step-wise explaining how to solve constraint satisfaction problems. Artif. Intell. **300**, 103550 (2021)
4. Bouali, F., Guettala, A., Venturini, G.: Vizassist: an interactive user assistant for visual data mining. Vis. Comput. **32**(11), 1447–1463 (2016)
5. Cortez, P., Embrechts, M.: Using sensitivity analysis and visualization techniques to open black box data mining models. Inf. Sci. **225**, 1–17 (2013)
6. Couronne, C., Koptelov, M., Zimmermann, A.: PrePeP: a light-weight, extensible tool for predicting frequent hitters. In: Dong, Y., Ifrim, G., Mladenić, D., Saunders, C., Van Hoecke, S. (eds.) ECML PKDD 2020. LNCS (LNAI), vol. 12461, pp. 570–573. Springer, Cham (2021). https://doi.org/10.1007/978-3-030-67670-4_41
7. De Raedt, L., Guns, T., Nijssen, S.: Constraint programming for itemset mining. In: KDD, pp. 204–212 (2008)
8. Dror, O., et al.: Novel approach for efficient pharmacophore-based virtual screening: method and applications. J. Chem. Inf. Model. **49**(10), 2333–2343 (2009)
9. Fayyad, U., Piatetsky-Shapiro, G., Smyth, P.: From data mining to knowledge discovery in databases. AI Mag. **17**(3), 37–37 (1996)
10. Ferreira, M., Levkowitz, H.: From visual data exploration to visual data mining: a survey. IEEE Trans. Visual. Comput. Graph. **9**(3), 378–394 (2003)
11. Fournier-Viger, P., Lin, J.C.W., Vo, B., Chi, T., Zhang, J., Le, H.: A survey of itemset mining. Data Min. Knowl. Disc. **7**(4), e1207 (2017)
12. Freuder, E.: Explaining ourselves: human-aware constraint reasoning. In: Proceedings of the AAAI Conference on Artificial Intelligence, vol. 31 (2017)
13. Gamba, E., Bogaerts, B., Guns, T.: Efficiently explaining CSPs with unsatisfiable subset optimization. In: (IJCAI), pp. 1381–1388 (2021)
14. Gaulton, A., et al.: The chEMBL database in 2017. Nucleic Acids Res. **45**(D1), D945–D954 (2017)
15. Guns, T., Nijssen, S., De Raedt, L.: Itemset mining: a constraint programming perspective. Artif. Intell. **175**(12–13), 1951–1983 (2011)
16. Han, J., Pei, J., Yin, Y., Mao, R.: Mining frequent patterns without candidate generation: a frequent-pattern tree approach. Data Min. Knowl. Discov. **8**(1), 53–87 (2004). https://doi.org/10.1023/B:DAMI.0000005258.31418.83
17. Holzinger, A., Dehmer, M., Jurisica, I.: Knowledge discovery and interactive data mining in bioinformatics-state-of-the-art, future challenges and research directions. BMC Bioinform. **15**(6), 1–9 (2014)
18. Jussien, N., Ouis, S.: User-friendly explanations for constraint programming. In: International Conference on Principles and Practice of CP (2001)
19. Kashid, A., Kulkarni, V., Patankar, R.: Discrimination-aware data mining: a survey. Int. J. Data Sci. **2**(1), 70–84 (2017)
20. Kuo, Y.T., Lonie, A., Pearce, A.R., Sonenberg, L.: Mining surprising patterns and their explanations in clinical data. Appl. AI **28**(2), 111–138 (2014)
21. Kuo, Y.T., et al.: Domain ontology driven data mining: a medical case study. In: 2007 International Workshop on Domain Driven Data Mining, pp. 11–17 (2007)
22. Leeuwen, M.: Interactive data exploration using pattern mining. In: Interactive Knowledge Discovery and Data Mining in Biomedical Informatics, pp. 169–182 (2014)
23. Mackworth, A.K.: Consistency in networks of relations. AI **8**(1), 99–118 (1977)
24. Métivier, J.P., et al.: The pharmacophore network: a computational method for exploring structure-activity relationships from a large chemical data set. J. Med. Chem. **61**(8), 3551–3564 (2018)
25. Pedreschi, D., et al.: Meaningful explanations of black box AI decision systems. In: AAAI, vol. 33, pp. 9780–9784 (2019)

26. Pedreshi, D., et al.: Discrimination-aware data mining. In: KDD, pp. 560–568 (2008)
27. Ribeiro, M.T., Singh, S., Guestrin, C.: Why should i trust you? explaining the predictions of any classifier. In: KDD, pp. 1135–1144 (2016)
28. Sabin, D., Freuder, E.C.: Contradicting conventional wisdom in constraint satisfaction. In: Borning, A. (ed.) PPCP 1994. LNCS, vol. 874, pp. 10–20. Springer, Heidelberg (1994). https://doi.org/10.1007/3-540-58601-6_86
29. Soukup, T., Davidson, I.: Visual Data Mining: Techniques and Tools for Data Visualization and Mining. John Wiley & Sons, Hoboken (2002)
30. Velu, C., Kashwan, K.: Visual data mining techniques for classification of diabetic patients. In: IACC, pp. 1070–1075. IEEE (2013)
31. Wu, H., Lu, Z., Pan, L., Xu, R., Jiang, W.: An improved apriori-based algorithm for association rules mining. In: 6th FSKD, vol. 2, pp. 51–55. IEEE (2009)
32. Zaki, M.J.: Scalable algorithms for association mining. IEEE Trans. Knowl. Data Eng. 12(3), 372–390 (2000)

Translated Texts Under the Lens: From Machine Translation Detection to Source Language Identification

Massimo La Morgia(✉)(ID), Alessandro Mei(✉)(ID), Eugenio Nerio Nemmi(✉)(ID), Luca Sabatini, and Francesco Sassi(✉)(ID)

Sapienza University of Rome, Rome, Italy
{lamorgia,mei,nemmi,sassi}@di.uniroma1.it

Abstract. Machine Translation Systems are today used to break down linguistic barriers. People from different countries and languages can now interact with each other thanks to state-of-the-art translators from prominent software companies like Google and Microsoft. However, these tools are also used to expand the audience for phishing attacks, scam emails or to generate fake reviews to promote a product on different e-commerce platforms. In all these cases, detecting whether a text has been translated can be crucial information. In this work, we tackle the problem of the detection of translated texts from different angles. On top of addressing the classic task of machine translation detection, we investigate and find common patterns across different machine translation systems unrelated to the original text's source language. Then, we show that it is possible to identify the machine translation system used to generate a translated text with high performances (F1-score 88.5%) and that it is also possible to identify the source language of the original text. We perform our tasks over two datasets that we use to evaluate our models: Books, a new dataset we built from scratch based on excerpts of novels, and the well-known Europarl dataset, based on proceedings of the European Parliament.

Keywords: Machine Translation Systems · Machine Learning · Natural Language Processing

1 Introduction

Today, hundreds of thousands of people use commercial machine translation systems (MTSs) worldwide for personal or working purposes. They help bridge the gap in language barriers, especially on the Web, by facilitating communication between people. However, bad actors use these systems to target potential victims of email-phishing [32] massively or generate fake reviews of products to trick recommendation systems [16] and people into buying or choosing a specific product. For all these reasons, machine translation detectors are actively used to infer spam emails or to detect poor quality web pages [13].

In this work, we put automatically translated texts under the lens. We study the impact of the MTSs and the source language of the translated text on the

B. Crémilleux et al. (Eds.): IDA 2023, LNCS 13876, pp. 222–235, 2023.
https://doi.org/10.1007/978-3-031-30047-9_18

Machine Translation Detection (MTD) task leveraging Books, a novel dataset built from excerpts of novels. We find that MTSs have common patterns that can be learned by training on a single MTS; thus, we are able to identify translated text regardless of the MTS used for the translation, suggesting that the automatic translation process introduces recognizable patterns in the translation. Similarly, we discover that we can learn these patterns regardless of the source language of the translated text. We can train on a single MTS using text from a single source language and still detect the translated text on multiple MTSs and source languages with comparable performances.

We then investigate the possibility of identifying the MTS used to produce the translation and the source language of the original text. To explore these questions, we introduce, to the best of our knowledge, two new tasks: Machine Translation Identification (MTI) and Source Language Identification (SLI). In the former (MTI), we want to identify which MTS has been used to generate a translation, while in the latter (SLI), we want to identify the source language of a translated text. For the first task, MTI, we built a classifier that shows promising results, with an average F1-score of 88.5%. In the second task, SLI, we propose a stacked classifier able to identify the source language of a machine-translated text with an average F1-score of 78% among 4 languages. We believe that these tasks could be helpful in forensic analysis, where malicious actors attempt to obfuscate their writing style using MTSs [17,25]. In particular, in this paper, we try to answer the following research questions:

Q1. Is it possible to identify a translated text regardless of the MTS used or the source language of the text?
Q2. Is it possible to identify which translator has been used to translate the text?
Q3. Is it possible to recognize the source language of the translated text?

2   Datasets

Since we need specific information to explore our questions, we build new datasets. Indeed for Q1 and Q2, we need the translation of an *original* sample both by a human and an MTS, while for Q3, we need to know the source language. In particular, to assess our experiments over different settings and topic domains, we perform our study using two datasets: one extracted from novels and the other based on speech transcriptions. The first dataset we use is *Books*, a novel dataset we introduce. To build Books, we collect 100 books originally written in 4 different languages by 100 different established writers of the XX/XXI century [37]. In particular, we select 25 books for each of the following source languages: Italian, French, Spanish, and German. The selected books belong to several different domains and authors. Thus, they have very different writing styles. From each book, we select an excerpt of approximately 10,000 characters (on average 1642.67 words per novel) and their corresponding translation from the English edition. Finally, we produce 3 more English translations for each original excerpt using the APIs of 3 state-of-the-art commercial

Machine Translation Systems: Google Translate [12] (*GT*), Microsoft Translation [26] (*MT*), and DeepL [7] (*DL*). At the end of the process, the Books dataset is made of 400 different samples.

The second dataset we use for our experiments is Europarl [18]. It is a parallel corpus extracted from the proceedings of the European Parliament containing *speech transcripts* of European parliamentarians and the corresponding professional translations into each of the 20 European languages. The texts on this dataset include many speech-distinctive elements such as hesitations, broken sentences, and repetition [5]. Consistently with Books, we obtain 100 seed samples by extracting from Europarl 25 samples for each of the 4 languages we consider. Every sample is made using transcripts of speakers of the same source language and contains about 10,000 characters (on average 1512.81 words per sample). We pre-process the dataset using Moses [19], a statistical machine translation system that includes different tools and utilities to parse and parallelize the Europarl dataset. Then, we collect the parallel English translation of each seed sample. Finally, we translate each seed sample using the selected MTSs. Figure 1 summarizes the process of building the Books and Europarl dataset. Both datasets at the end contain 400 samples in English, which are produced starting from 100 seeds (25 for each language), of which 100 were made by translating the original seed by professional human translators and 300 using machine-translation systems (100 for each MTS).

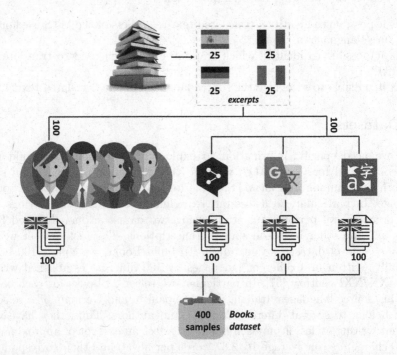

Fig. 1. Step by step representation of the process used to build Books dataset. The same pipeline was applied to build the Europarl dataset.

3   Experimental Settings

In this section, we describe the experimental settings of the tasks in terms of train/test splits, the pre-processing we apply, and the features we use. For all the experiments, we use 60% of the dataset as train and 40% as test. We use Python Scikit-learn [33] to implement all the models and the feature selection techniques. Whenever the model parameters are not specified, we use the default values.

3.1   Pre-processing and Feature Description

We apply three pre-processing techniques to extract our features. Firstly, we tokenize the texts. Tokenization is the process of separating a piece of text into smaller units called tokens (*e.g.,* words, char). We then apply the following processes:

- **Stemming.** It is the process of reducing inflected words to their root (words stem). (*e.g.,* writing → write; eating → eat)
- **Part-of-Speech (POS) tagging.** It is the process of identifying a word's appropriate part of speech in a text based on its definition and context.
- **Distortion Text.** It is a process where ASCII characters are replaced with a special character [36]. Table 1 shows an example of this pre-processing step.

Table 1. Example of text distortion.

Original Text	Distorted Text
I don't know. Just making conversation with you, Morty. What do you think, I-I-I... know everything about everything?	* ***'* **** **** ****** ************ **** *** ***** . **** ** *** ***** , *-*-*... **** ********** ***** **********?

Most of the features are based on *n-grams*, that are a sequence of N contiguous elements, in our case, character (char-gram) or words (word-gram). We use the notation *Char-gram (i-k)* (resp. *Word-gram (i-k)*) to denote all the char n-grams (resp. word n-grams) with $n \in \{i, \ldots, k\}$. Table 2 shows the features we use for our tasks and the feature number for the different tasks and datasets. Below a description of each feature:

- **Char-gram** is a sequence of N contiguous characters.
- **Sentence Length** is the average length of the sentences for each text based on the number of characters.
- **Words avg** is the average number of words for each sentence of the text.
- **Adjectives avg** is the average number of adjectives for each text.
- **Dist Char-gram (i-k)** are char-grams computed over the distortion text.
- **POS Word-gram(i-k)** are word-grams computed over Part of Speech (POS) tagged text.

– **Type Token Ratio** (*TTR*) is the ratio between the number of unique words and the total number of words for a given text. The idea behind this feature is to measure the vocabulary variety (in terms of words) of a text.

We use the Bag of Words to weigh the char-grams and word-grams, while we use Tf-idf (term frequency-inverse document frequency) to weigh the distortion text. The Bags of Words is a representation that creates vectors with the number of occurrences of a specified element in the text (*e.g.*, words), while the Tf-idf is an weighting schema that gives a larger value (weight) to elements that are less frequent in the document corpus.

Table 2. Features types and numbers of features for the MTI and SLI tasks on both datasets.

Feature Type	MTI		SLI	
	Books	Euro	Books	Euro
Char-gram (1–6)	318, 250	220, 593	261, 895	175, 247
Sentence Length	1	1	1	1
Words avg	1	1	1	1
Adjectives avg	1	1	1	1
Dist. Char-gram (5–8)	15, 134	12, 080	–	–
Dist. Char-gram (2–8)	–	–	13, 897	9, 522
POS Word-gram (1–6)	–	–	187, 481	145, 473
TTR	1	1	–	–
All	333, 388	232, 677	463, 276	330, 245

4 Machine Translation Detection

The goal of the *Machine Translation Detection* (*MTD*) task is to automatically detect whether a text has been translated by a machine translation system or is human-generated. This task was broadly studied in the literature with different approaches such as using fixed features [1,23], n-gram [2,34], coherence score [27] and similarity with round-trip translation [28]. In this section, we first want to replicate similar results to the state-of-the-art on our datasets Books, to verify that it is suitable for our purposes. Then, we design two experiments to explore further the underlying patterns of machine-translated texts.

For all the experiments in this section, we use the following model. We train a Multilayer Perceptron [15] with a single hidden layer made of 10 neurons and a BFGS optimizer [3] for weights optimization. Regarding the features, we compute all the char n-grams with $n \in \{1, \ldots, 6\}$ and then select the 2,500 more relevant n-grams according to the chi-square metric [9]. We finally normalize the features with the SkLearn StandardScaler. Figure 2 shows the results on Books and Europarl datasets.

We obtain a high F1-score on both corpora (0.9 on Books and 0.97 on Europarl), showing that our model can achieve excellent results in distinguishing machine-translated and human-translated texts.

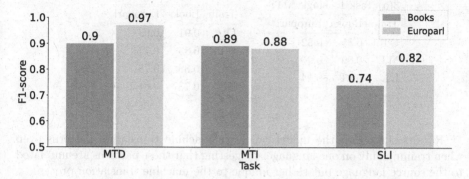

Fig. 2. F1-score for the Machine Translation Detection (MTD), Machine Translation Identification (MTI), and Source Language Identification (SLI) tasks on the Books and Europarl datasets.

4.1 Learning from a Single MTS

The next interesting point to explore is if there are any common patterns among the different MTSs that could be learned to identify a translated text, even if it is generated by an MTS that was not included in the training set. To verify this idea, we train our model using only samples translated by a single MTS and human-translated samples. We repeat the experiment 3 times, training the model at each iteration with samples produced by a different MTS and testing it only on the samples of the remaining MTSs.

Table 3(a) shows the results of this experiment for the different combinations. As we can see, the model is able to achieve good results (on average 88% of F1-score) when tested on samples generated by machine translators that are not represented in the training set. Interestingly, the model trained on MT achieves similar (average delta 0.015) results to those obtained by training the model using the whole dataset (*i.e.*, training on all the MTSs).

These results suggest that there are some common patterns among the MTSs that the model can learn from a single MTS.

4.2 Learning from a Single Language

Since we have 4 different source languages in our dataset, we want to understand the impact they might have on the MTD task. In this experiment, we train our model using only the translation from one source language and test it against the sample produced by the translation from other source languages and the human-translated samples. Table 3(b) shows the F1-score using the different source languages.

Table 3. 3(a): F1-score for Task 1 training on a single MTS' samples and testing on the others. 3(b): F1-score for Task 1 training on a single language and testing on the others.

3(a) Task 1 - single MTS

Train	Books	Europarl
GT	0.85	0.82
MT	0.89	0.95
DL	0.84	0.94

3(b) Task 1 - single language

Train	Books	Europarl
IT	0.91	0.93
FR	0.85	0.74
ES	0.88	0.78
DE	0.73	0.81

Results show that the model can learn machine translation patterns even when training only on one language, suggesting that these patterns are unrelated to the source language but rather unique to the machine translation process.

5   Machine Translator Identification

Results from the previous section suggest common patterns exist among the different MTSs that allow us to differentiate machine-translated texts from human-translated ones. In this section, we investigate if MTSs translations differ enough from each other to be able to identify which one has been used to translate a sample (Question Q2). Thus, given a machine-translated text T', our goal is to identify the MTS M that generated the text T'. We call this task *Machine Translator Identification* (*MTI*). In particular, we focus on identifying the 3 MTSs used to build the Books and Europarl datasets: *Google Translate, Microsoft Translation,* and *DeepL*. Given the task's goal, we use a sub-set of Europarl and Books datasets for the following experiments, removing the 100 samples representing the class of human translations from each dataset.

For this task, we build an ensemble classifier. The first level comprises three different classifiers: a Support Vector Machine, a Logistic Regression, and a Random Tree. Then, the outputs of the classifiers are used as input to feed a hard voting layer (SkLearn VotingClassifier) for the final prediction. Table 2 shows the type and the number of features we use to train the three classifiers at the first level of our architecture. For all the n-gram type features, we select only the 85% most significant ones using SelectPercentile of SkLearn, and we standardize them with the SkLearn StandardScaler. Figure 2 reports the F1-score for the two datasets. As we can notice, our classifier performs similarly on both datasets, with an F1-score of 0.89 and 0.88 for Books and Europarl, respectively. To better understand the results, we analyze the confusion matrices of the two classifications. The confusion matrix of Books (Table 4) shows that GT is the hardest MTS to identify, and its misclassified samples are mostly assigned to the MT class.

We found a possible explanation for these errors by analyzing the BLUE score [31]. The BLEU Score is an algorithm for evaluating the quality of a trans-

Table 4. Confusion Matrix on Books for the MTI task.

Predicted

		GT	MT	DL
Label	GT	30	6	4
	MT	1	39	0
	DL	0	2	38

lation. It measures the similarity of the translation to a reference one. For each pair of the MTSs, we measure the BLEU score, obtaining a value of 69 for the pair GT-MT, 63 for GT-DL, and 62.4 for DL-MT (Table 5).

Table 5. BLEU score for the MTS pairs.

	MT	DL	GT
GT	69	63	–
DL	62.4	–	–

The high BLEU score between GT and MT shows that they have similar translations, which could be the reason for the incorrect classification of the GT samples. Conversely, the low similarity between the MT and DL classes could lead to the high accuracy we observe in our experiment. Finally, we obtain similar results by analyzing the confusion matrix and the BLUE score for the Europarl dataset.

6 Source Language Identification

As a final task, we propose the **Source Language Identification** (SLI). The goal of the task is to identify the source language of a given machine-translated text. Thus, given a machine-translated text T' in a language $L2$ (in our case English), the goal of the task is to identify the language $L1$ of the text T. This task could be considered a variation of other tasks already studied in the literature, such as Native Language Identification (NLI), where the goal is to identify the native language (L1) of a person who writes in another language (L2) or determining the source language of a human-translated text (see Sect. 7), where the goal is to identify the source language of a text that has been human-translated. However, unlike the previous studies, our task focuses on identifying the source language of a text that is translated by a Machine Translation System and not by a human. For our experiments, we consider English as $L2$, and the possible $L1$ languages are: Italian, French, Spanish or German. Since we only care about translations of MTS (*i.e.*, text not translated by human), we modify our dataset in the same way as we did for the MTI task (Sect.5).

For this task, we use the stacking ensemble technique. In particular, we stacked an AdaBoost [10] model with 50 LinearSVC [6] and a Logistic Regression [39] model as base estimators. Table 2 shows the type and the number of features we use to train the stacking classifier. For all the n-gram features, we select the top 70% according to their F-value, computed with the variance analysis (ANOVA) [35]. Then, we standardize them with a StandardScaler. Figure 2 shows the F1-score of the model trained and tested on both our datasets. The results suggest that identifying the source language is easier in Europarl than in Books. As noted in [14], a possible reason could be that the Europarl dataset may contain some distinctive patterns for the source language of the speaker since it is a transcription of a talk. Instead, the Books dataset covers a wide area of topics and contains fewer clues about the author's source language. Table 6 shows the confusion matrices on the Books and Europarl dataset.

Table 6. Confusion Matrices on Books and Europarl for the SLI task.

Predicted

Label	DE	ES	FR	IT
DE	25	0	5	0
ES	2	23	4	1
FR	0	2	27	1
IT	1	5	9	15

Books

Predicted

Label	DE	ES	FR	IT
DE	27	3	0	0
ES	0	24	3	3
FR	0	3	24	3
IT	0	9	1	20

Europarl

The most challenging source language to detect on both datasets is Italian, frequently misclassified as Spanish or French. German is generally better identified than the other languages except for French on the Books dataset, with five classification errors. Indeed, German has the highest F1-score among all the classes, with a value of 0.86 in Books and 0.94 in Europarl. This is intuitive and expected, since German is a West Germanic language while the other 3 are Romance languages and have more features in common [30].

7 Related Work

Machine Translation Detection: The detection of automatic translations has been investigated in the past using multiple techniques. Both Aharoni et al. [1] and Li et al. [23] use fixed features taken from the English language that may be used regardless of the language in which the content was originally written (*i.e.*, source language). They respectively achieve an accuracy of 90% and 83%. Arase et al. [2] and Popescu et al. [34] use an n-gram based approach to perform the task, reaching a high accuracy of 96% and 99%. Other works used words distribution [22,29], that lead to a max accuracy of 98%, or coherence score [27], with an accuracy of 73%. More recently, Nguyen et al. [28] propose a method to detect translated texts using text similarity with round-trip translation that appears to be resistant for different translators and languages. In this case, they were able to achieve an accuracy of 94%.

Machine Translator Identification: Bhardwaj et al. [4] test 18 classifiers to detect translated text using commercial as well as in-house MTS. Looking at the identification of the MTS, previous works [1] show that testing machine translation detection over different MTSs produces different results. This suggests that these MTSs have different qualities of translation and that there are differences between them. In the same way, Bizzoni et al. [5] found similar results studying translationese ([11]) over different architectures. These studies show that there could be enough differences in MTS systems to be able to identify which translator has been used for a given translation.

Source Language Identification: We can have three slightly different settings for the source language identification task. The first setting is the well-known NLI task [38] where the goal is to identify the native language $L1$ of a person who writes a text in a second language $L2$ [21]. The second setting is when the translation has been performed by a person that is different from the one that wrote the original text. In [14], the author shows that it is possible to identify the source language of the translation of speeches in the Europarl corpus with an accuracy of more than 87%, without testing if these results hold for translated text (i.e. it is possible to detect the source language of an automatically translated text). Using human translation, also Lynch et al. [24] and Koppel et al. [20] perform the same task showing that it is possible to determine the original language of a human translation.

8   Conclusion and Future Work

In this work, we put translated text under the lens. We start by evaluating the impact of MTSs and source languages on the Machine Translation Detection task. We find that MTSs generate common patterns in the translated text that can be learned by a machine learning model trained using a single MTS. Furthermore, we show that the performance of the task is not significantly influenced by the MTS employed or the source language of the text. These results open the possibility to employ machine learning models trained solely on a subset of known MTSs or languages and identify text translated from any other MTSs or languages. Then, to the best of our knowledge, we introduce two new tasks: Machine Translator Identification and Source Language Identification. The goal of the Machine Translator Identification task (MTI) is to identify the MTS that has been used to translate a target text, while the Source Language Identification (SLI) task aims to identify the source language of a machine-translated text.

The models we propose for both tasks achieve an average F1-score of 88.5% and 78%, respectively, for the MTI and the SLI task. These last two tasks can help to characterize translated texts further and could be used as features for a classification task or give additional insights when studying potential threats. Our results, although they represent a first attempt to tackle the newly presented tasks, show that much more work can be done in this area.

While we achieve good performances, we believe there could be further improvement by using deep learning models that are particularly effective in NLP tasks, such as BERT, a pre-trained language representation model based on transformers [8]. However, the number of samples in the datasets should be increased to use deep learning techniques effectively. Furthermore, in our study, we perform all the experiments at the document level, using a mean of 1642.67 words. In the future, it would be interesting to propose the same tasks in a more challenging setting, using sentences rather than documents. This is particularly important since it makes it possible to evaluate very short texts. We consider only European languages (although with different origins) for the datasets: German, French, Italian, and Spanish. However, there are other languages, such as Arabic, Mandarin, or Hindi, that are widely used worldwide, and it could be interesting to expand the datasets and test the classifiers performances with the new data. Finally, with the recent popularity of Large Language Models (LLMs) such as ChatGPT, it could be interesting to verify if our model can still identify a text translated by ChatGPT and its original language, and also to introduce a new task for the detection of text generated by LLMs.

Acknowledgements. This work was supported in part by the MIUR under grant "Dipartimenti di eccellenza 2018-2022" of the Department of Computer Science of Sapienza University.

References

1. Aharoni, R., Koppel, M., Goldberg, Y.: Automatic detection of machine translated text and translation quality estimation. In: Proceedings of the 52nd Annual Meeting of the Association for Computational Linguistics (Vol. 2: Short Papers), pp. 289–295. Association for Computational Linguistics, Baltimore (2014). https://doi.org/10.3115/v1/P14-2048, https://aclanthology.org/P14-2048

2. Arase, Y., Zhou, M.: Machine translation detection from monolingual web-text. In: Proceedings of the 51st Annual Meeting of the Association for Computational Linguistics (Vol. 1: Long Papers), pp. 1597–1607. Association for Computational Linguistics, Sofia (2013). https://aclanthology.org/P13-1157

3. Battiti, R., Masulli, F.: Bfgs optimization for faster and automated supervised learning. In: International Neural Network Conference, pp. 757–760. Springer, Dordrecht (1990). https://doi.org/10.1007/978-94-009-0643-3_68

4. Bhardwaj, S., Alfonso Hermelo, D., Langlais, P., Bernier-Colborne, G., Goutte, C., Simard, M.: Human or neural translation? In: Proceedings of the 28th International Conference on Computational Linguistics, pp. 6553–6564. International Committee on Computational Linguistics, Barcelona (2020). https://doi.org/10.18653/v1/2020.coling-main.576, https://aclanthology.org/2020.coling-main.576

5. Bizzoni, Y., Juzek, T.S., España-Bonet, C., Dutta Chowdhury, K., van Genabith, J., Teich, E.: How human is machine translationese? comparing human and machine translations of text and speech. In: Proceedings of the 17th International

Conference on Spoken Language Translation, pp. 280–290. Association for Computational Linguistics (2020). https://doi.org/10.18653/v1/2020.iwslt-1.34, https://aclanthology.org/2020.iwslt-1.34

6. Cortes, C., Vapnik, V.: Support-vector networks. Mach. Learn. **20**(3), 273–297 (1995)

7. DeepL: Deepl translator (2021). https://www.deepl.com/pro-api

8. Devlin, J., Chang, M.W., Lee, K., Toutanova, K.: Bert: pre-training of deep bidirectional transformers for language understanding. arXiv preprint arXiv:1810.04805 (2018)

9. Forman, G., et al.: An extensive empirical study of feature selection metrics for text classification. J. Mach. Learn. Res. **3**, 1289–1305 (2003)

10. Freund, Y., Schapire, R.E.: A decision-theoretic generalization of on-line learning and an application to boosting. J. Comput. Syst. Sci. **55**(1), 119–139 (1997)

11. Gellerstam, M.: Translationese in swedish novels translated from English. In: Wollin, L., Lindquist, H. (eds.) Translation Studies in Scandinavia: Poceedings from the Scandinavian Symposium on Translation Theory (SSOTT) II, pp. 88–95. no. 75 in Lund Studies in English, CWK Gleerup, Lund (1986)

12. Google: Google translator (2021). https://cloud.google.com/translate

13. Google: Managing multi-regional and multilingual sites (2021). https://developers.google.com/search/docs/advanced/crawling/managing-multi-regional-sites

14. van Halteren, H.: Source language markers in EUROPARL translations. In: Proceedings of the 22nd International Conference on Computational Linguistics (2008), pp. 937–944. Coling 2008 Organizing Committee, Manchester, UK (2008). https://aclanthology.org/C08-1118

15. Hornik, K., Stinchcombe, M., White, H.: Multilayer feedforward networks are universal approximators. Neural Netw. **2**(5), 359–366 (1989)

16. Juuti, M., Sun, B., Mori, T., Asokan, N.: Stay on-topic: generating context-specific fake restaurant reviews. In: Lopez, J., Zhou, J., Soriano, M. (eds.) ESORICS 2018. LNCS, vol. 11098, pp. 132–151. Springer, Cham (2018). https://doi.org/10.1007/978-3-319-99073-6_7

17. Kacmarcik, G., Gamon, M.: Obfuscating document stylometry to preserve author anonymity. In: Proceedings of the COLING/ACL 2006 Main Conference Poster Sessions, pp. 444–451. Association for Computational Linguistics, Sydney (2006). https://aclanthology.org/P06-2058

18. Koehn, P.: Europarl: a parallel corpus for statistical machine translation. In: Proceedings of Machine Translation Summit X: Papers, pp. 79–86. Phuket, Thailand (2005). https://aclanthology.org/2005.mtsummit-papers.11

19. Koehn, P., et al.: Moses: open source toolkit for statistical machine translation. In: Proceedings of the 45th Annual Meeting of the Association for Computational Linguistics Companion Volume Proceedings of the Demo and Poster Sessions, pp. 177–180. Association for Computational Linguistics, Prague (2007). https://aclanthology.org/P07-2045

20. Koppel, M., Ordan, N.: Translationese and its dialects. In: Proceedings of the 49th Annual Meeting of the Association for Computational Linguistics: Human Language Technologies, pp. 1318–1326. Association for Computational Linguistics, Portland (2011). https://aclanthology.org/P11-1132

21. La Morgia, M., Mei, A., Nemmi, E., Raponi, S., Stefa, J.: Nationality and geolocation-based profiling in the dark (web). IEEE Trans. Serv. Comput. **15**(1), 429–441 (2019)

22. Labbé, C., Labbé, D.: Duplicate and fake publications in the scientific literature: how many SCIgen papers in computer science? Scientometrics **94**(1), 379–396 (2013). https://doi.org/10.1007/s11192-012-0781-y

23. Li, Y., Wang, R., Zhao, H.: A machine learning method to distinguish machine translation from human translation. In: Proceedings of the 29th Pacific Asia Conference on Language, Information and Computation: Posters, pp. 354–360, Shanghai, China (2015). https://aclanthology.org/Y15-2041

24. Lynch, G., Vogel, C.: Towards the automatic detection of the source language of a literary translation. In: Proceedings of COLING 2012: Posters, pp. 775–784. The COLING 2012 Organizing Committee, Mumbai, India (2012). https://aclanthology.org/C12-2076

25. Mahmood, A., Ahmad, F., Shafiq, Z., Srinivasan, P., Zaffar, F.: A girl has no name: Automated authorship obfuscation using mutant-x. Proc. Priv. Enhancing Technol. **2019**(4), 54–71 (2019)

26. Microsoft: Microsoft translator (2021). https://www.microsoft.com/translator/

27. Nguyen-Son, H.Q., Nguyen, H.H., Tieu, N.D.T., Yamagishi, J., Echizen, I.: Identifying computer-translated paragraphs using coherence features. In: Proceedings of the 32nd Pacific Asia Conference on Language, Information and Computation. Association for Computational Linguistics, Hong Kong (2018). https://aclanthology.org/Y18-1056

28. Nguyen-Son, H.Q., Thao, T., Hidano, S., Gupta, I., Kiyomoto, S.: Machine translated text detection through text similarity with round-trip translation. In: Proceedings of the 2021 Conference of the North American Chapter of the Association for Computational Linguistics: Human Language Technologies, pp. 5792–5797. Association for Computational Linguistics (2021). https://doi.org/10.18653/v1/2021.naacl-main.462, https://aclanthology.org/2021.naacl-main.462

29. Nguyen-Son, H.Q., Tieu, N.D.T., Nguyen, H.H., Yamagishi, J., Zen, I.E.: Identifying computer-generated text using statistical analysis. In: 2017 Asia-Pacific Signal and Information Processing Association Annual Summit and Conference (APSIPA ASC), pp. 1504–1511. IEEE (2017)

30. Padró, M., Padró, L.: Comparing methods for language identification. Procesamiento del lenguaje natural **33** (2004)

31. Papineni, K., Roukos, S., Ward, T., Zhu, W.J.: Bleu: a method for automatic evaluation of machine translation. In: Proceedings of the 40th Annual Meeting of the Association for Computational Linguistics, pp. 311–318. Association for Computational Linguistics, Philadelphia (2002). https://doi.org/10.3115/1073083.1073135, https://aclanthology.org/P02-1040/

32. Parmar, Y.S., Jahankhani, H.: Utilising machine learning against email phishing to detect malicious emails. In: Montasari, R., Jahankhani, H. (eds.) Artificial Intelligence in Cyber Security: Impact and Implications. ASTSA, pp. 73–102. Springer, Cham (2021). https://doi.org/10.1007/978-3-030-88040-8_3

33. Pedregosa, F., et al.: Scikit-learn: machine learning in Python. J. Mach. Learn. Res. **12**, 2825–2830 (2011)

34. Popescu, M.: Studying translationese at the character level. In: Proceedings of the International Conference Recent Advances in Natural Language Processing 2011, pp. 634–639. Association for Computational Linguistics, Hissar (2011), https://aclanthology.org/R11-1091

35. St, L., Wold, S., et al.: Analysis of variance (ANOVA). Chemom. Intell. Lab. Syst. **6**(4), 259–272 (1989)

36. Stamatatos, E.: Authorship attribution using text distortion. In: Proceedings of the 15th Conference of the European Chapter of the Association for Computational Linguistics, Vol. 1, Long Papers, pp. 1138–1149 (2017)

37. SystemsLab: Book dataset. https://github.com/SystemsLab-Sapienza/books-dataset

38. Tetreault, J., Blanchard, D., Cahill, A.: A report on the first native language identification shared task. In: Proceedings of the 8th Workshop on Innovative use of NLP for Building Educational Applications, pp. 48–57 (2013)

39. Wright, R.E.: Logistic regression (1995)

Geolet: An Interpretable Model for Trajectory Classification

Cristiano Landi[1(✉)], Francesco Spinnato[2,3], Riccardo Guidotti[1,3],
Anna Monreale[1,3], and Mirco Nanni[3]

[1] University of Pisa, Pisa, Italy
cristiano.landi@phd.unipi.it, {riccardo.guidotti,anna.monreale}@unipi.it
[2] Scuola Normale Superiore, Pisa, Italy
francesco.spinnato@sns.it
[3] ISTI-CNR, Pisa, Italy
{francesco.spinnato,riccardo.guidotti,anna.monreale,
mirco.nanni}@isti.cnr.it

Abstract. The large and diverse availability of mobility data enables the development of predictive models capable of recognizing various types of movements. Through a variety of GPS devices, any moving entity, animal, person, or vehicle can generate spatio-temporal trajectories. This data is used to infer migration patterns, manage traffic in large cities, and monitor the spread and impact of diseases, all critical situations that necessitate a thorough understanding of the underlying problem. Researchers, businesses, and governments use mobility data to make decisions that affect people's lives in many ways, employing accurate but opaque deep learning models that are difficult to interpret from a human standpoint. To address these limitations, we propose Geolet, a human-interpretable machine-learning model for trajectory classification. We use discriminative sub-trajectories extracted from mobility data to turn trajectories into a simplified representation that can be used as input by any machine learning classifier. We test our approach against state-of-the-art competitors on real-world datasets. Geolet outperforms black-box models in terms of accuracy while being orders of magnitude faster than its interpretable competitors.

Keywords: Trajectory Classification · Interpretable Machine Learning · Mobility Data Analysis · Explainable AI

1 Introduction

The increasing diffusion of GPS-capable electronic devices, such as mobile phones, vehicles, and trackers, contributes to generating massive amounts of mobility data [9]. In general, any moving entity can generate spatio-temporal trajectories, which companies, governments, and researchers use to address many crucial applications [1]. Thus, mobility data affect the livelihoods of millions of people.

One of the most common tasks in this field is trajectory classification, i.e., predicting the class label of an object based on its movement [5,9,14]. Trajectory

B. Crémilleux et al. (Eds.): IDA 2023, LNCS 13876, pp. 236–248, 2023.
https://doi.org/10.1007/978-3-031-30047-9_19

classifiers, for example, can differentiate between cars, taxis, buses, pedestrians, and bikes, recognize the movement of various animals, and infer people's jobs based on their routines. In [14] it is presented a survey comparing state-of-the-art trajectory classification approaches. The authors emphasize the main challenges in this field, namely the need for robust experimental evaluations across multiple datasets and the lack of advances in the state-of-the-art. Moreover, the majority of surveyed works are based on complex, black-box models such as Support Vector Machines (SVM), Multilayer Perceptrons (MLP), and deep Convolutional Neural Networks (CNN), which are inherently not interpretable from a human standpoint [7]. This can be a significant problem in high-stakes applications where the explanation aspect of machine learning models is critical for establishing trust in automated decision systems [11].

EXplainable Artificial Intelligence (XAI) for trajectories is an extremely under-explored topic in the literature. For this reason, we take inspiration from studies on XAI from time series [18], and specifically shapelets [22] to present the GEOgraphic ShapeLET classifier GEOLET, an interpretable classification approach for trajectory data. First, GEOLET uses geographic partitioning to segment the input data into subtrajectories. These subtrajectories are normalized and filtered in order to take only the most discriminative ones. They are then exploited to convert the input trajectories into a simplified, interpretable representation that can be used as input by any machine learning classifier. We evaluate GEOLET on five datasets and against state-of-the-art alternatives, considering multiple quantitative metrics. Furthermore, we qualitatively show that the proposed approach produces interpretable and easy-to-read explanations.

2   Related Works

The problem of trajectory classification consists in building a predictive model from labeled historical trajectories to classify new ones [6,9]. Trajectory classifiers can be divided into different families. Classical approaches usually extract *global*, or *local* features from the data, whereas modern approaches tend to directly process the raw trajectories with complex, deep learning-based models.

Global features-based approaches extract features like velocity change, duration, speed, etc. from the whole trajectory [14]; they can be highly effective for simple datasets, where similar properties are maintained throughout the entire path. However, these methods are insufficient for more articulated trajectories in which the target class is linked to an event occurring in a trajectory portion.

Local features-based approaches try to mitigate these problems by segmenting the trajectory into subtrajectories and extracting features from them. In [21], the authors extract statistical features from the segments of the trajectory, first globally and then locally. Finally, they compare Random Forest, Gradient Boosting Decision Tree and XGBoost as classification models. In [20], it is proposed a semi-supervised clustering approach coupled with a majority voting ensemble classifier to learn a metric that brings similar data closer and distances elements with different labels. The first two methods can be viewed as pseudo-interpretable procedures, depending on the classification model used after the dataset transformation. A Random Forest, for example, can be used to determine the average importance of each variable. However, the main issue is that interpretability varies depending on the complexity of the extracted features and the number of weak learners in the ensemble. To the best of our knowledge, the only fully interpretable trajectory classifier is MOVELETS [5]. Indeed, the idea behind MOVELETS is to extract discriminative segments from the trajectory and, similarly to the shapelet transform for time series [22], convert the dataset into a new representation that stores the shortest distances between each trajectory and subtrajectory. In line with shapelets, subtrajectories can be used to understand the logic of the classifier [18]. While promising, the proposed method is computationally complex as it generates all possible subtrajectories and is not suitable for large datasets. Furthermore, only the space dimension is used to compute the distance between trajectories and subtrajectories. Thus, the trajectories must be resampled to constant time intervals, and only equal-length subtrajectories can be compared.

Recently, neural networks have been used in trajectory classification approaches to achieve superior performance in a faster manner. Recurrent Neural Networks (RNN) and Convolutional Neural Networks (CNN), often used with time series data, can be easily extended to trajectories. In [8], the authors propose TraClets, a CNN-based method that represents a trajectory as an image and uses a CNN to solve the trajectory classification task. *MARC* [13] deals with trajectories augmented with semantic textual dimensions, exploiting the GPS data and information in the textual dimensions. Finally, Rocket [4], the state-of-the-art classifier for multivariate time series, can be easily applied to trajectories to achieve fast and extremely accurate performance. Unfortunately, Rocket, RNN, and CNN models lead to a non-interpretable prediction. For this reason, several XAI approaches have been proposed to address the issue. Still, they can only output explanations as saliency maps [3,15], highlighting the importance of each observation towards the classification.

Given the limitations of the literature, we propose a method for classifying trajectories based on local feature extraction. GEOLET attempts to overcome the interpretability limitations of black-box models, and optimize accuracy and runtime, which is often the main problem of feature extraction-based methods.

3   Background and Problem Setting

In this section, we define all the concepts necessary to understand our proposal. We define a trajectory as follows:

Definition 1 (Trajectory). A *trajectory* X is a sequence of spatio-temporal points $X = \{(\vec{x}_1, t_1), \ldots, (\vec{x}_m, t_m)\} \in \mathbb{R}^{m \times 3}$ where the spatial vectors $\vec{x}_j = (lat_j, long_j)$ are sorted by increasing time t_j, i.e., $\forall 1 \leq j < m$ we have $t_j < t_{j+1}$.

In a sense, trajectories can be viewed as multivariate time series containing two signals, i.e., the latitude and longitude, recorded at non-constant sampling rates [5,8,19]. A trajectory classification dataset is a set of trajectories with a vector of labels attached. Formally:

Definition 2 (Trajectory Classification Dataset). A *trajectory classification dataset* $\mathcal{D} = (\mathcal{X}, \mathbf{y}) \in \mathbb{R}^{n \times m \times 3}$ is a set of n trajectories, $\mathcal{X} = \{X_i \ldots, X_n\}$, with a vector of assigned labels (or classes), $\mathbf{y} = \{y_1, y_2, \ldots, y_n\} \in \mathbb{N}^n$.

For simplicity of notation, we use a single symbol m to denote the lengths of the trajectories, even if a trajectory dataset can contain instances having a different number of observations. We define the trajectory classification problem as follows:

Definition 3 (Trajectory Classification). Given a trajectory classification dataset $\mathcal{D}$, *trajectory classification* is the task of training a function f from the space of possible inputs to a probability distribution over the class values in $\mathbf{y}$.

The resulting trajectory classification function f takes as input a trajectory X and returns y according to what f learned, i.e., $y = f(X)$. In general, y can either be a discrete label or the probability of X belonging to a specific class.

Thus, given a trajectory classification dataset $\mathcal{D}$, our objective is to solve a trajectory classification problem by realizing an interpretable trajectory classification function f that allows to understand the reasons for a decision $y = f(X)$.

A fundamental aspect to introduce our proposal is the notion of *subtrajectory*:

Definition 4 (Subtrajectory). Given a trajectory X of length m, a subtrajectory $S = \{(\vec{s}_j, t_j), \ldots, (\vec{s}_{j+l}, t_{j+l})\}$, of length $l \leq m$, is an ordered sequence of consecutive values such that $1 \leq j \leq m - l + 1$.

Subtrajectories can be used for classification purposes, similarly to shapelets, by *selecting the most discriminative ones* w.r.t. the target label, depending on some statistical measure. *Mutual Information* [17] is commonly used for classification purposes, measuring the dependency between continuous and discrete variables. Once the most discriminative subtrajectories are found, the dataset can be transformed in a simpler representation, via the subtrajectory transform. Formally:

Definition 5 (Subtrajectory Transform). Given a trajectory dataset $\mathcal{X}$ and a set $\mathcal{S}$ containing h subtrajectories, the *Subtrajectory Transform* converts $\mathcal{X} \in \mathbb{R}^{n \times m \times 3}$ into a real-valued matrix $T \in \mathbb{R}^{n \times h}$, obtained by taking the *Best Fitting* of each trajectory $X \in \mathcal{X}$, and each subtrajectory $S \in \mathcal{S}$.

Usually, the best fitting of S in each X is computed by taking the minimum distance via a sliding window of length l. The most used distance functions to compare sequential data are the Euclidean distance and Dynamic Time Warping [2]. However, both have drawbacks when applied to trajectories. First, the Euclidean distance requires trajectories to have the same number of points, which is uncommon in real data. Secondly, both DTW and Euclidean distance implicitly need a constant sampling rate, which is not always guaranteed. For this reason, in our proposal, we adopt a distance specifically designed for trajectories, i.e., the *Interpolated Route Distance* (IRD) [19], which allows the comparison of trajectories having different lengths and sampling rates. IRD uses the *temporal*

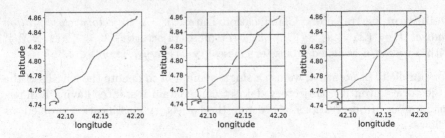

Fig. 1. Examples of partitioning. From left to right: original trajectory, Geohash, SAX.

dimension to align two trajectories and, if two observations do not occur at the same time-step, values are projected by interpolating the information. In other words, given two trajectories, IRD calculates the distance between them for each timestamp. If a timestamp is not present in the other time series, IRD uses the neighboring timestamps to interpolate the values and estimate a position.

4   Geolet

This section presents the GEOgraphic ShapeLET classifier (GEOLET), an interpretable classification approach for trajectory data. GEOLET is our answer to the problem of designing an interpretable trajectory classification function f for a trajectory classification task. GEOLET first *partitions* trajectories into multiple segments, yielding candidate subtrajectories. Secondly, it *normalizes* and *filters* them to produce a set of prototypical subtrajectories. Then, GEOLET *transforms* the dataset using the Subtrajectory Transform. Finally, any interpretable classification model can be used to classify the transformed data.

Partitioning. Several approaches can be used to partition a trajectory: binning approaches like *Symbolic Aggregate Approximation (SAX)* [10], or geographical ones like *Geohash* [16] (Fig. 1). SAX [10] is a discretization technique to convert time series into a sequence of symbols. It is usually applied by sliding window [12], creating a collection of SAX words that can be interpreted as time series subsequences. *We extend SAX to trajectories* by applying the approach independently to latitude and longitude signals, converting both into symbol sequences. In layman's terms, multiple coordinates in a trajectory are binned into a single symbol that represents an area. The converted signals are then used to generate a new symbol for each pair of observed symbols. The specific hyperparameters' configurations are detailed in Sect. 5. Another partitioning approach is *Geohash* [16], an indexing system encoding a rectangular geographic area into strings of letters and digits. Geohash divides the Earth into 32 regions via a bit array, associating each area with one of the symbols in [0-1a-z]. Then the process is repeated recursively until the algorithm reaches the desired accuracy.

Normalization. Following the partitioning phase, the segments must be normalized so that the domains of the various partitions overlap. This can be accomplished with a *Geohash normalization* or a *FirstPoint* normalization. In the Geohash normalization, the bottom-left vertex latitude and longitude of the Geohash cell are subtracted from the coordinates of each point of the subtrajectories. Intuitively, this is equivalent to overlapping each rectangle of the Geohash partitioning. On the other hand, in the FirstPoint normalization, the latitude and longitude of the first point of the subtrajectory are subtracted from the coordinates of each point of the subtrajectory. Intuitively, this is equivalent to overlapping the first point of each subtrajectory.

Filtering. After the partitioning and normalization phases, depending on the dimensionality of the data, we might end up with an enormous amount of subtrajectories. As a result, a filtering phase is carried out to reduce computational complexity and produce a smaller set of relevant subtrajectories. In this phase, subtrajectories are filtered by selecting a subset following some specific criterion. In the shapelet literature, these criteria can be unsupervised, such as random sampling and clustering, or supervised using statistical approaches, such as the Mutual Information or the Chi-squared test [12], that are used to find the subsequences that better discriminate between different classes. We experiment with both unsupervised and supervised approaches in Sect. 5.

Transform. Once a set of representative subtrajectories is found, the subtrajectory transform can be applied, transforming trajectories in a simpler representation, containing the Best Fitting (BF) between each trajectory in the original dataset and each extracted subtrajectory. For time series, the distance of choice is usually the Normalized Euclidean Distance (ED), however, as detailed in Sect. 3, this is not always the best choice for trajectories. Therefore, given a trajectory X of length m and a subtrajectory S of length l, the BF can be computed in different ways. For the Normalized Euclidean distance, a sliding window of size l is used to compare S with each subtrajectory of X. Formally,

$$BF_{ED}(X, S) = \min_{j=1}^{m-l+1} (ED(X_{j:j+l}, S))$$

where $X_{j:j+l}$ denotes a subtrajectory of X from j to $j + l$. On the other hand, defining the notion of best-fitting with DTW is not trivial. Indeed, using the same approach adopted for ED would limit the purpose for which DTW exists. Hence, we propose a similar approach, but where we use an expanded sliding window of length $l' > l$:

$$BF_{DTW}(X, S) = \min_{j=0}^{m-l'+1} (DTW(X_{j:j+l'}, S)).$$

Finally, since IRD exploits the time dimension to interpolate points when two time series do not have the same sampling rate, we calculate the sliding window size not w.r.t. the number of observations, but w.r.t. the time interval between the first observation of the subtrajectory S and its last timestamp t_l. Formally,

$$BF_{IRD}(X, S) = \min_{j=1}^{m-l+1} (IRD(X_{j:j+t_l}, S)).$$

Table 1. Datasets description.

	animals	vehicles	seabirds	geoLife	taxi
# trajectories	102	381	108	5,977	121,312
avg length (std)	173 (56)	601 (230)	2,904 (1162)	8,100 (15178)	55 (22)
avg Δtime (std)	3.75 (6)	83 (596)	100 (0)	3.88 (12)	15 (0)
Δ time min-75%-max	0-4-258	0-30-53,857	100-100-100	0-2-665	15-15-15
Target class (#classes)	species (3)	category (2)	species (3)	transport (2)	call type (3)

This formula describes the calculation of the best fitting with IRD, using a sliding window where the length is defined not as the number of features but as the time interval. The sliding window length must correspond to the minimum number of points necessary so that the trajectory time interval from j to $j + t_l$ is as close as possible to the subtrajectory's length.

Independently of the distance function adopted, the original dataset $\mathcal{X}$ is transformed in a simplified matrix representation T. We experiment both with continuous and discretized subtrajectories in Sect. 5. The transformed dataset T can be paired with any classification algorithm, having the advantage of a more interpretable data representation.

5   Experiments

We experiment with GEOLET quantitatively on five datasets and we report visual examples to show the benefits of an interpretable-by-design trajectory classifier.

Datasets. The trajectory classification datasets are described in Table 1. For animals the task consists in recognizing different species. For vehicles we want to distinguish between buses and trucks. For seabirds the task is recognizing flying trajectories of three species of seabirds. For geolife, due to the high number of classes and to the unbalancing, we simplify the problem to recognizing trajectories of public vs private means of transport. Finally, for taxi, the objective is to distinguish among different types of taxi calls within one month of observations. We highlight that, state-of-the-art interpretable classification methods are experimented only on very small datasets like `animal` and `vehicles`. Each dataset is divided in train/test with a ratio 70/30%.

Competitors. We compare GEOLET against two state-of-the-art methods, i.e., MOVELETS [5] and ROCKET [4]. MOVELETS, similarly to GEOLET, is an interpretable trajectory classifier that extracts discriminative subtrajectories and uses them to transform the dataset. The MOVELETS algorithm requires setting the minimum and maximum length of the generated subtrajectories. We use the default implementation values for `animals` and `vehicles`. Furthermore, we limit the maximum length to the logarithm of the number of maximum observations per trajectory for `seabirds`, `geolife`, and `taxi`. ROCKET is a not interpretable time series classifier that transforms the dataset by applying random convolutional kernels to generate multiple feature maps that capture different data trends. The only hyperparameter to choose for ROCKET is the number of convolutional kernels, which is set to 10,000 as recommended by the authors [4].

Geolet Parameters Setting.[1] For GEOLET, we use Geohash as a partitioning algorithm, FirstPoint normalization, Mutual Information (as implemented by `scikit-learn`) for filtering subtrajectories, and IRD as a distance measure. With this configuration, Geolet requires two hyperparameters: the Geohash precision $(prec)$, and the number of subtrajectories to extract (ns). We set the optimal parameters via grid-search on the training set[2].

Geolet Alternative Implementations. For a fair benchmarking of GEOLET, we devise some alternative versions that are still interpretable but extract explanations using different processes.

First, we compare the geographic-based segmentation of GEOLET against a purely SAX-based approach. For this purpose, we apply the SAX approximation, as detailed in Sect. 4. We name this baseline MRSQM-T because, as part of the filtering phase, we adopt MRSQM [12], a time series approach that extracts the top symbolic subsequences using the Chi-squared test. MRSQM-T randomly generates k configurations of the triples l, w, α where l is the size of the sliding window, w is the SAX word length, and α is the alphabet size. MRSQM-T generates these sets using the same seed to guarantee that the previous configurations remain the same as k increases. The optimal value of k is set to 25 for `animals` and 11 for `vehicles`. In the experiments, we observe that MRSQM-T achieves good accuracy but requires a great computational effort, resulting in high runtimes, even for these relatively simple datasets.

Secondly, we aim at comparing the supervised subtrajectory selection of GEOLET and MRSQM-T against an unsupervised one. For this purpose, we use a clustering approach to filter the extracted subtrajectories. Specifically, after Geohash segmentation and partitioning, prototypical subtrajectories are extracted through K-Medoid, using the Normalized Euclidean distance. Once the cluster centroids are extracted, they are compared using a sliding window to the original subsequences. Each trajectory is encoded with the identifier of the cluster centroids it contains. We name this baseline TRAC, Trajectory Approximation-based Classifier. TRAC requires four hyperparameters, i.e., the Geohash precision $prec$, the number of cluster k to use with K-Medoids (it also identifies the number of symbols in the alphabet), the sliding window length w, and the number of symbols subsequences top_{ss} to select based on the Mutual Information score. For each parameter, we performed a grid-search[3].

We highlight that, besides GEOLET, MRSQM-T and TRAC are original contributions and do not exist in the literature as interpretable trajectory classifiers.

---

[1] Code available at: github.com/cri98li/Geolet.

[2] `animals`: $prec = 2$ $ns = 21$; `vehicles`: $prec = 6$ $ns = 20$; `seabirds`: $prec = 5$ $ns = 50$; `geolife`: $prec = 6$ $ns = 50$; `taxi`: $prec = 5$ $ns = 50$.

[3] $prec \in [4, 5, 6, 7]$; $k \in [2, 5, 20, 100]$; $w \in [2, 3, 5]$; $top_{ss} \in [1, 2, 10, 50]$ on the training set. Hyperparameter choice does not significantly affect the method's performance. We found constant accuracy values for most of the hyperparameters tested. There were, however, peaks in the accuracy score for some values. Thus, for `animals` we set $prec = 4, w = 3$ and $top_{ss} = 2$. For the `vehicles` $prec = 6, w = 3$ and $top_{ss} = 10$.

Table 2. Performance scores, best values in bold, second best in italic.

		animals	vehicles	seabirds	geolife	taxi
accuracy	GEOLET	**0.935**	**0.965**	**0.967**	**0.861**	**0.578**
	ROCKET	0.871	0.928	0.667	0.733	0.566
	MOVELETS	0.563	0.921	0.718	–	–
	MRSQM-T	0.677	0.887	–	–	–
	TRAC	0.742	0.791	–	–	–
runtime	GEOLET	27.6 s	50.1 s	48 m	2.42 h	44 m
	ROCKET	**2.4 s**	**31.5 s**	**15.7 s**	**29.1 m**	**13.3 m**
	MOVELETS	25.7 s	141 m	126.9 s	–	–
	MRSQM-T	22.5 m	1.16 h	–	–	–
	TRAC	25.4 s	1.18 h	–	–	–

5.1   Classification Performance

Since GEOLET, MOVELETS, ROCKET, MRSQM-T and TRAC perform a transformation of the original data, any classification model can be applied to the transformed dataset. To compare the transformations fairly, we adopted the same effective model for all five approaches, i.e., a Random Forest classifier as implemented by the scikit-learn library. The best hyperparameters are found via grid-search with 10-fold cross-validation[4] on the training set.

Results in terms of accuracy and runtime are reported in Table 2[5]. We measure the execution time of each algorithm from the data preparation phase to the end of the dataset transformation. Hence, we exclude the time for training the final model. From a first glance, we can see that ROCKET is the method that takes the least time to execute. As for MOVELETS, we performed several attempts with the `geolife` and `taxi`, but all the tests ended with an "insufficient memory error". In addition, we recorded anomalous results with the `animals`, which we suspect was due to a bug in the original code. As for GEOLET, we can see that it manages to get the best results between these two methods, but it takes a longer execution time. The weakness of GEOLET compared to ROCKET and MOVELETS lies in the number of hyperparameters and configurations from which one can choose, which is discussed in Sect. 5.3. TRAC and MRSQM-T perform competitively w.r.t. MOVELETS in small datasets, but are both outperformed by GEOLET and ROCKET. Moreover, due to their high computational cost, they are hardly usable when dealing with real-world datasets.

5.2   Geolet Interpretability

This section provides an example of the kind of interpretable classification that GEOLET can provide. We apply GEOLET on `vehicles` with $prec = 4$ and Geohash as partitioning method, FirstPoint normalization, Normalized Euclidean

---

[4] n_estimators = range(300, 1500, 300), criterion = [gini, entropy], max_depth = range(2, 20, 3).

[5] Tests are performed on a machine with CPU: AMD Ryzen 9 3900X; RAM: 32 GB; OS: EndeavourOS Linux. Due to resource limitations, we used 20% of `geolife` and 70% of `taxi`.

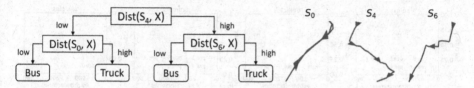

Fig. 2. GEOLET's Decision Tree for `vehicles` (left) and subtrajectories used (right).

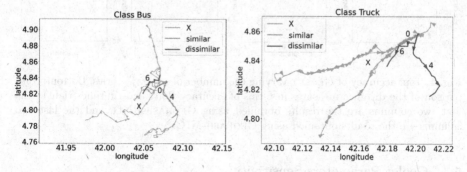

Fig. 3. Examples of the GEOLET explanation on two instances from `vehicles`. Left: instance of the class *Bus*. Right: instance of the class *Truck*.

distance and Mutual Information for the transform. Finally, we use a Decision Tree as a classification model as implemented by `scikit-learn`, which allows us to visualize the resulting model graphically and extract rules summarizing its decision boundaries. In particular, for `vehicles`, we identified the following rules:

$$r_1 = \{dist(X, S_4) \ is \ low \ \wedge \ dist(X, S_0) \ is \ low\} \rightarrow Bus$$
$$r_2 = \{dist(X, S_4) \ is \ low \ \wedge \ dist(X, S_0) \ is \ high\} \rightarrow Truck$$
$$r_3 = \{dist(X, S_4) \ is \ high \ \wedge \ dist(X, S_6) \ is \ low\} \rightarrow Bus$$
$$r_4 = \{dist(X, S_4) \ is \ high \ \wedge \ dist(X, S_6) \ is \ high\} \rightarrow Truck$$

We highlight that, to ease the understanding, we report "is high"/"is low" instead of the real distance because it is sufficient to understand the meaning of the rule without accounting for the specific threshold numbers. Specifically, "low" indicates that the distance measurement is below the split threshold value, and "high" indicates that the value exceeds it For instance, $dist(X, S_4) \leq 0.3$ is translated into $dist(X, S_4)$ *is low*. The decision tree and the subtrajectories are illustrated in Fig. 2. These rules show that the most representative subtrajectories are those with indices 0, 4, and 6. We can now understand the decisions of the classifier by visualizing where the subtrajectories fit within the trajectory. Figure 3 presents the classification of GEOLET for two instances. In particular, the instance belonging to the class Bus has segments very similar to subtrajectories 0 and 4, and are instead quite different from subtrajectory 6. On the other hand, the Truck instance contains almost perfectly the subtrajectory 0, but it is quite different from 4 and 6.

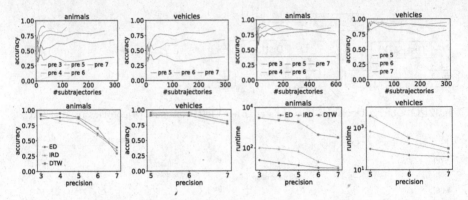

Fig. 4. Top: accuracy of GEOLET varying the number of subtrajectories. Bottom: comparison of the distance measures in terms of accuracy (left) and runtime (right). The first two columns are the results obtained using GEOHASH-RND, and the last two columns are the results obtained using GEOHASH-MIG.

5.3   Geolet Parameters Sensitivity

In general, it is extremely difficult to define a global heuristic for this approach. For this reason, we describe here our implementation choices and analyze how hyperparameters selection affects the GEOLET's results on animals and vehicles, providing some practical insights and guidelines.

Partitioning. In Fig. 4 we study how Geohash precision and number of subtrajectories affect accuracy. Also, we determine the importance of selecting trajectory using a well-founded criterion such as the Mutual Information Gain (GEOHASH-MIG), instead of simply selecting them randomly (GEOHASH-RND). IRD is used as distance, and FirstPoint is used as the normalization strategy. From the results, we can observe that, although random selection (GEOHASH-RND) leads to a worse result, it could be a great way to quickly determine the best precision for Geohash partitioning. The average runtime of GEOHASH-RND compared to GEOHASH-MIG turns out to be 13 times faster for animals and two times faster for vehicles. On the other hand, by selecting subtrajectories using Mutual Information (GEOHASH-MIG), we can achieve better results faster and with fewer subtrajectories. Regarding animals, we note that increasing the length of the subtrajectories improves the results.

Normalization. We study here the impact of using different normalization techniques, i.e., Geohash (GEOLET-GH) and FirstPoint (GEOLET-FP). Our experiments show that the accuracy of GEOLET-GH is 0.677 for animals and 0.791 for vehicles, while for GEOLET-FP is 0.935 for animals and 0.965 for vehicles. Therefore, we select FirstPoint as normalization for GEOLET.

Distance. Finally, we analyze the impact of different distance metrics on performance. Figure 4 (bottom) shows that the best distance for animals is DTW, while the best distance for vehicles is ED. However, when the computation time for the dataset transformation is considered, it is clear that larger datasets cannot use DTW. Thus, excluding DTW, IRD has the best accuracy score for `animals`, while it performs negligibly worse than ED for `vehicles`. As a result, our intuition is that: *(i)* for small datasets, the DTW is the best distance, *(ii)* for large datasets with consistent sample rates, ED is the best choice, while *(iii)* for large datasets with variegated sample rates, IRD is the best compromise between accuracy and runtime.

In summary, the most sensible hyperparameter of GEOLET is the precision.

6   Conclusion

We have presented GEOLET, an interpretable classifier for trajectory data. GEO-LET is able to transform trajectory data into a simplified representation that any classifier can use as an interpretable input source. We have shown that GEOLET outperforms state-of-the-art competitors in terms of accuracy while remaining competitive in terms of runtime. Besides, GEOLET is interpretable, returning subtrajectory-based explanations that are easily interpretable from a human standpoint. As future research directions, we intend to improve GEOLET's performance in terms of accuracy, runtime, and explainability. In this sense, many extensions are possible. Subtrajectories, can be improved by embedding properties such as scale and rotation invariance, resulting in a smaller set of prototypical and interpretable subsequences. Also, GEOLET can be extended to work with data that includes additional features like height and semantic textual dimensions, as well as data that uses different coordinate systems. To accomplish this, the modularity of the implementation can be used to introduce new distance measures, filtering approaches, normalization techniques, and partitioning methods. Finally, we want to investigate the regression and forecasting tasks, which are fundamental in this field but remain unexplored from an XAI standpoint.

Acknowledgment. This work is partially supported by the European Union NextGenerationEU programme under the fuding schemes PNRR-PE-AI scheme (M4C2, investment 1.3, line on Artificial Intelligence) FAIR (Future Artificial Intelligence Research), and "SoBigData.it - Strengthening the Italian RI for Social Mining and Big Data Analytics" - Prot. IR0000013. This work is partially supported by the European Community H2020 programme under the funding schemes: H2020-INFRAIA-2019-1: Res. Infr. G.A. 871042 *SoBigData++*, G.A. 761758 *Humane AI*, G.A. 952215 *TAILOR*, ERC-2018-ADG G.A. 834756 *XAI*, and CHIST-ERA-19-XAI-010 SAI, and by the Green.Dat.AI Horizon Europe research and innovation programme under the G.A. 101070416.

References

1. Andrienko, G.L., et al.: (So) big data and the transformation of the city. Int. J. Data Sci. Anal. **11**(4), 311–340 (2021)
2. Bellman, R., Kalaba, R.: On adaptive control processes. IRE Trans. Autom. Control. **4**(2), 1–9 (1959)

3. Bodria, F., Giannotti, F., Guidotti, R., Naretto, F., Pedreschi, D., Rinzivillo, S.: Benchmarking and survey of explanation methods for black box models. CoRR abs/2102.13076 (2021)

4. Dempster, A., Petitjean, F., Webb, G.I.: ROCKET: exceptionally fast and accurate time series classification using random convolutional kernels. Data Min. Knowl. Discov. **34**(5), 1454–1495 (2020)

5. Ferrero, C.A., Alvares, L.O., Zalewski, W., Bogorny, V.: MOVELETS: exploring relevant subtrajectories for robust trajectory classification. In: SAC, pp. 849–856. ACM (2018)

6. de Freitas, N.C.A., da Silva, T.L.C., de Macêdo, J.A.F., Junior, L.M.: Using deep learning for trajectory classification in imbalanced dataset. In: FLAIRS Conference (2021)

7. Guidotti, R., Monreale, A., Ruggieri, S., Turini, F., Giannotti, F., Pedreschi, D.: A survey of methods for explaining black box models. ACM Comput. Surv. (CSUR) **51**(5), 1–42 (2018)

8. Kontopoulos, I., Makris, A., Tserpes, K., Bogorny, V.: Traclets: harnessing the power of computer vision for trajectory classification (2022)

9. Lee, J., Han, J., Li, X., Gonzalez, H.: TraClass: trajectory classification using hierarchical region-based and trajectory-based clustering. Proc. VLDB Endow. **1**(1), 1081–1094 (2008)

10. Lin, J., Keogh, E.J., Lonardi, S., Chiu, B.Y.: A symbolic representation of time series, with implications for streaming algorithms. In: DMKD, pp. 2–11. ACM (2003)

11. Miller, T.: Explanation in artificial intelligence: insights from the social sciences. Artif. Intell. **267**, 1–38 (2019)

12. Nguyen, T.L., Ifrim, G.: Mrsqm: fast time series classification with symbolic representations. CoRR abs/2109.01036 (2021). arxiv.org/abs/2109.01036

13. Petry, L.M., da Silva, C.L., Esuli, A., Renso, C., Bogorny, V.: MARC: a robust method for multiple-aspect trajectory classification via space, time, and semantic embeddings. Int. J. Geogr. Inf. Sci. **34**(7), 1428–1450 (2020)

14. da Silva, C.L., Petry, L.M., Bogorny, V.: A survey and comparison of trajectory classification methods. In: BRACIS, pp. 788–793. IEEE (2019)

15. Simonyan, K., Vedaldi, A., Zisserman, A.: Deep inside convolutional networks: visualising image classification models and saliency maps (2014)

16. Suwardi, I.S., Dharma, D., Satya, D.P., Lestari, D.P.: Geohash index based spatial data model for corporate. In: 2015 International Conference on Electrical Engineering and Informatics (ICEEI), pp. 478–483. IEEE (2015)

17. Tan, P.N., Steinbach, M.S., Kumar, V.: Introduction to Data Mining. Pearson Education India, Noida (2016)

18. Theissler, A., Spinnato, F., Schlegel, U., Guidotti, R.: Explainable AI for time series classification: a review, taxonomy and research directions. IEEE Access **10**, 100700–100724 (2022)

19. Trasarti, R., Guidotti, R., Monreale, A., Giannotti, F.: Myway: location prediction via mobility profiling. Inf. Syst. **64**, 350–367 (2017)

20. Vouros, A., et al.: A generalised framework for detailed classification of swimming paths inside the morris water maze. Sci. Rep. **8**(1), 1–15 (2018)

21. Xiao, Z., Wang, Y., Fu, K., Wu, F.: Identifying different transportation modes from trajectory data using tree-based ensemble classifiers. ISPRS Int. J. Geo Inf. **6**(2), 57 (2017)

22. Ye, L., Keogh, E.J.: Time series shapelets: a new primitive for data mining. In: KDD, pp. 947–956. ACM (2009)

An Investigation of Structures Responsible for Gender Bias in BERT and DistilBERT

Thibaud Leteno[1(✉)], Antoine Gourru[1], Charlotte Laclau[2], and Christophe Gravier[1]

[1] Université Jean Monnet Saint-Étienne, CNRS, Institut d Optique Graduate School, Laboratoire Hubert Curien UMR 5516, 42023 Saint-Étienne, France
{thibaud.leteno,antoine.gourru,christophe.gravier}@univ-st-etienne.fr
[2] Télécom Paris, Institut Polytechnique de Paris, Paris, France
charlotte.laclau@telecom-paris.fr

Abstract. In recent years, large Transformer-based Pre-trained Language Models (PLM) have changed the Natural Language Processing (NLP) landscape, by pushing the performance boundaries of the state-of-the-art on a wide variety of tasks. However, this performance gain goes along with an increase in complexity, and as a result, the size of such models (up to billions of parameters) represents a constraint for their deployment on embedded devices or short-inference time tasks. To cope with this situation, compressed models emerged (e.g. Distil-BERT), democratizing their usage in a growing number of applications that impact our daily lives. A crucial issue is the fairness of the predictions made by both PLMs and their distilled counterparts. In this paper, we propose an empirical exploration of this problem by formalizing two questions: (1) Can we identify the neural mechanism(s) responsible for gender bias in BERT (and by extension DistilBERT)? (2) Does distillation tend to accentuate or mitigate gender bias (e.g. is DistilBERT more prone to gender bias than its *uncompressed version*, BERT)? Our findings are the following: (I) one cannot identify a specific layer that produces bias; (II) every attention head uniformly encodes bias; except in the context of underrepresented classes with a high imbalance of the sensitive attribute; (III) this subset of heads is different as we re-fine tune the network; (IV) bias is more homogeneously produced by the heads in the distilled model.

Keywords: Language Models · Fairness · Imbalance · Compression

1 Introduction

The introduction of large Pre-trained Language Models (PLM) has marked an important paradigm shift in Natural Language Processing (NLP). It leads to unprecedented progress in tasks such as machine translation, document classification [10], and multitasks text generation [29]. The strength of these approaches lies in their ability to produce contextual representations. They have been initially based on Recurrent Neural Networks (RNN) [5] and they have gradually integrated the Transformers model [34] as is the case for GPT3 [29] or BERT [10], for example. Compared to RNNs, Transformers can be parallelized, which opens the way, on one hand, to the use of ever-increasing training corpus (for example, GPT3 is trained on 45TB of data - almost the entire public web), and on the other hand, to the design of increasingly complex architectures (e.g., BERT large comprises 345 million parameters, BERT base 110 million). In a nutshell, Transformers [34] are founded on three key innovations: positional encoding, scaled dot product attention, and multi-head attention (we will come back to these elements in more detail in Sect. 2). As a result of a combination of all these elements, Transformers can learn an internal understanding of language automatically from the data. Despite their good performance on many different tasks, the use of these models in so-called sensitive applications or areas raised concerns over the past couple of years. Indeed, when decisions have an impact on individuals, for example in the medical and legal domains [9] or human resources [20], it becomes crucial to study the fairness of these models.

The core definition of fairness is still a hotly debated topic in the scientific community. In our work, we adopt the following commonly accepted definition [26]: *fairness refers to the absence of any prejudice or favoritism towards an individual or a group based on their intrinsic or acquired traits.* In machine learning, we assume that *unfairness* is the result of biased predictions (prejudice or favoritism), which are defined as elements that conduct a model to treat groups of individuals conditionally on some particular *protected* attributes, such as gender, race, or sexual orientation.

As an example, in human resources, the NLP-based recruitment task consists in analyzing and then selecting the relevant candidates. A lack of diversity inherent to the data, for instance, a corpus containing a large majority of male profiles (i.e. sample bias), will cause the model to maintain and accentuate a gender bias [33]. When handling simple linear models trained on reasonable size corpora, creating safeguards to avoid this type of bias is conceivable. With PLM, the characteristics that allow them to perform so well are numerous: the size of their training corpus, the number of parameters, and their ability to infer a fine-grained semantic from the data. However, they are also what make it difficult to prevent them from encoding societal biases [2].

Related Works. Several recent studies highlight fairness issues raised by models based on the Transformer architecture. These issues are observed in different levels of the NLP pipeline: text encoding [1,23], during the fine-tuning process [8], or simply as the potential harm caused on downstream tasks [23], with dedicated studies on language generation [32], document classification [3], toxicity detection, and sentiment analysis [19]. Besides measuring the fairness issue, *locating* the neural mechanism responsible for these issues is largely understudied and unsolved – locating such mechanisms would unlock the possibility for counter-measures in neural architectures. At the same time, a segment of research focused on compressing these large pre-trained models to attain similar performances with fewer parameters, so that running these models is more sustainable and more cost-effective. Several model compression techniques have been proposed, as discussed in [12] and namely the following compression families: pruning, quantization and distillation. The primer [24] increases the speed and generalization capacities by removing the less important model's weights with regard to the task, while quantization approximates the model's weights to reduce its complexity (e.g. reducing the numerical precision of the weights [22]). Finally, distillation [16] consists in training a smaller model (called *student model*) to mimic the predictions of the large PLM to distill (*teacher model*). In the present work, we focus on this latter, approach. One of the earliest model, DistilBERT [31] is able to reduce the number of parameters of BERT by 40% while maintaining 96% of accuracy in document classification. Looking at the impact of model distillation through fairness lenses has started to be investigated, mainly in the context of computer vision [17,18,25]. To summarize, their findings: i) compressed models impact underrepresented visual features directly related to bias, and ii) distilled models tend to accentuate discrimination already made by the teacher model. In NLP, fewer works have been conducted, and the conclusions are sometimes contradictory. While some works have shown that distilled versions of PLMs can exacerbate bias [7,28], other articles seem to reach an opposite conclusion [35]; in this latter, authors state that model distillation acts as a regularization technique allowing bias reduction.

Contribution. Based on existing results, we start from the postulate that PLMs, and more specifically BERT, encode undesirable bias. With a focus on the task of document classification on the Bias in Bio dataset, our objective is to identify the inner structure of the neural network architecture that produce bias, both for BERT and its distilled version DistilBERT. To this end, we design and conduct a series of experiments to verify the relation between models' fairness and their intermediate representation or the attention they carry to the embedding in different data balance setting.

Organisation. Section 2 provides background knowledge about BERT and DistilBERT. Section 3 presents the empirical protocols that we design. Section 4 details the technical setting and shows the obtained results of our experiments. Finally, we conclude in Sect. 5 and provide several perspectives unlocked by our experiments.

2   Preliminaries and Background

We study two PLM, BERT [10] and its distilled version, DistilBERT [31]. BERT is a general-purpose language model trained on masked language modeling task [1]. A small fraction of the words of each training document is masked, and the model is trained to reconstruct those masked words on a large amount of textual

---

[1] The model is also trained on a next sentence prediction task, but that is irrelevant in our work and therefore not presented here.

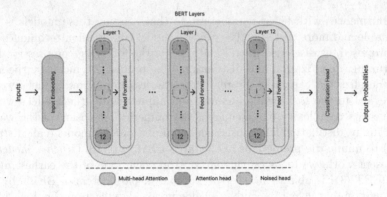

Fig. 1. Scheme of head's noising experiments

data. More precisely, the encoder part of a Transformer architecture takes a sentence, or short document, as input, and maps each token (word or subwords) to an initial representation space in $\mathbb{R}^{768}$.

The encoder then *contextualizes* these representations using a multi-head attention mechanism. The attention mechanism builds attention weights between each pair of words, based on their similarity in a latent space. These are then used to build new word representations by a simple weighted average operation. More precisely, this attention operation is computed for slices of dimensions in parallel before concatenation, this is why it is called multi-head. This is followed by a pass through a feed-forward neural network with residual connections. See Fig. 1 for an illustration of this operation. The output of the Transformer encoder is a matrix of size $L \times 768$, where L is the maximum document length of the model (512 for BERT) and 768 the number of dimensions. After pre-training on this task, the model can be *fine-tuned* on a downstream task, such as classification, by adding a linear layer on top of this model, that either inputs the final hidden state's "CLS" token (a special token corresponding to a representation of the sequence) or by pooling the representations of the words.

BERT, in its base version, has 110 M of parameters, 12 layers, and 12 attention heads. DistilBERT is a shallow version of BERT, trained with half the number of layers using distillation [16]. The principle of this compression method is to train a student model to replicate the behavior of the teacher. To do so, one feeds a dataset to the teacher to retrieve its predictions for each sample (the outputs of the teacher are soft targets, i.e. the probabilities over each class instead of the predicted label). On the other hand, the student receives the same input as training data and the predicted soft targets as training labels. The objective of the student is then to match the (soft) predictions of the teacher. The data used to train the student can either be unlabeled or labeled; in the second case, the true labels can also be fed to the student and a regularization term is added to the objective function to improve the student's performance. Using this approach, DistilBERT obtains up to 95% of the performance of BERT.

In our experiments, we use the pre-trained models from Hugging Face[2], more specifically, the Transformer models for sequence classification with a linear layer as classification head on top of the pooled output.

3   A Gender-Bias Neural Exploration Protocol for Language Models

3.1   Fine-Tuning Scenarios for Fairness Evaluation

In many real-life datasets, we observe gender ratio imbalance, making ML models outcomes prone to unfairness. For example, in the Bias in Bios dataset [6], more than 90% of nurses are women, while they are less than 2% to be a surgeon. In [11], authors show that training with imbalanced data allows the model to learn a correlation between the target label and sensitive attributes, therefore inducing bias. These observations are unsurprising: this kind of bias is considered to be extrinsic, and is caused by the data used during the fine-tuning. However, with pre-trained models, another type of bias emerges: intrinsic bias. They are encoded during the pre-training and are out of the control of the practitioner. The understanding of the neural behavior that leads to them remains unclear. In our work, we are primarily interested in understanding and exploring the inner operations of the Transformer architecture that are at stake in these findings.

To study in detail the effect of those biases, we fine-tune both BERT and DistilBERT on two sub-samples of the initial dataset: a balanced and an imbalanced one with regard to the sensitive groups (class imbalance remains identical for both datasets). These models will be referred to as Mi and Mb when fine-tuned with the imbalanced or the balanced dataset respectively. We believe that starting from the same models, but with different fine-tuning strategies will make it possible to make comparisons of the fairness of these models.

3.2   Attention and Hidden States Comparison (E1)

We first investigate the inner differences, induced by the fine-tuning process, between Mi and Mb models. (cf. Section 3.1). More precisely, we focus on the attention weights and the hidden states (the intermediate representations) between both models for similar input. For this first set of experiments, we propose to investigate fairness through the lens of the learning dynamic of the PLMs. Recent works [30] show that the first layers of deep architectures capture low-level information about the input data, and that the learned representations tend to become more abstract and finer when moving through the body of the network towards its heads. In [13], the authors specifically studied the dynamic of BERT fine-tuning and conclude that mainly the last layers are significantly changing, both their attention mode and the hidden representation that they produce. Our objective is then to verify the two following hypotheses.

---

[2] BERT: https://huggingface.co/docs/Transformers/model_doc/bert,
    DistilBERT: https://huggingface.co/docs/Transformers/model_doc/distilbert.

Hypothesis 1. *As layers specialize on the different granularity of the textual content, from grammatical to semantic aspects, we assume that monitoring the attention scores and hidden states of the successive layers of the models allows determining which one(s) is encoding bias.*

Hypothesis 2. *The distillation process implies that the student model will reproduce the behavior of the teacher, including biases in predictions. We assume that by reducing the depth, and hence the expressive power of the model, compression encourages amplified bias in the hidden representations.*

To proceed, we adapt the protocol of [13] to verify both Hypothesis 1 and 2. We first take a look at the modification of the similarity between tokens, where the attention is computed as a function of the similarity. In a second step, we look at how much hidden word representations are impacted by the model independently of the pairwise similarities.

Attention Values Comparison. The Jensen-Shannon divergence is a symmetrized version of the Kullback-Leibler divergence. It allows comparing two probability distributions P and Q. We propose to compare the attention of the two models layer-wise. Formally, we evaluate the divergence for each sample and for each head between the layers of two models (e.g., Mi and Mb). Let N be the number of examples in the evaluation set, H the number of attentions head, and W the number of tokens in a sequence. $A_i^h(token_t)$ and $A_b^h(token_t)$ are the attention scores for $token_t$ on head h respectively for models Mi and Mb. In this context, the JS divergence is defined as follows:

$$D_{JS}(\text{Mi}||\text{Mb}) = \frac{1}{N}\frac{1}{H}\sum_{n=1}^{N}\sum_{h=1}^{H}\frac{1}{W}\sum_{t=1}^{W}D_{JS}(A_i^h(token_t)||A_b^h(token_t)) \quad (1)$$

$D_{JS}(.||.) \in [0,1]$, where 0 indicates that the distributions are identical.

Hidden States Comparison. We compute the Singular Vector Canonical Correlation Analysis distance (SVCCA) [30] to observe the evolution of hidden states. SVCCA allows analyzing and comparing representations in deep learning models; in our case, the hidden representations produced by each Transformer layer. When computing SVCCA, we first perform a Singular Value Decomposition (SVD) of the representations produced by the two models for each input observation. Then, we compute the Canonical Correlation Analysis (CCA) [14] between the two subspaces created by the SVD to evaluate the correlation between the two representations and finally, condense the correlations obtained for each dimension into a distance.

Let c be the hidden size of the model and $\rho \in [0,1]$ the CCA. The SVCCA distance is defined as follows:

$$D_{SVCCA}(\text{Mi}||\text{Mb}) = 1 - \frac{1}{c}\sum_{j=1}^{c}\rho^{(j)} \quad (2)$$

$D_{SVCCA}(.||.) \in [0,1]$ with 0 meaning identical representations.

3.3 Head's Ablation (E2)

With multi-head attention, Transformers build for each head a different representation of the input embedding. We make the hypothesis that some representations might induce more biases than others. Thus, complementary to the previous experiments and in continuity with the goal of finding where are biases encoded in PLM, we successively ablate heads of the model to infer if some of them are responsible for biases in the model. The ablation is done by setting all its attention weights to 0 through all the layers, as shown in Fig. 1.

Hypothesis 3. *By ablating attention heads, we aim at removing the bias due to a given head and identify the ones' contributing to unfairness. In other words, we expect that when ablating a head responsible for bias, the model will obtain a better fairness score and reciprocally.*

In practice, we first fine-tune the model so that it learns the weights as in real-world applications. Then, we successively ablate heads and evaluate the performance and fairness of the model on new data to evaluate the bias encoded by the aforementioned heads. We are interested in the results of our models following two criteria: their predicting performance and their fairness.

Performance. The model performance is evaluated using a weighted version of the F-Score (since our target variable is multivalued).

More precisely, we compute the F-Score for each class, then compute the average weighted by the number of samples per class.

Gender Fairness. We are interested in group fairness, and several metrics have been proposed in the literature [4]. We choose the commonly used Equalized Odds (EO) [15], defined as follows $\mathbb{P}(\overline{y} = 1|y, S = 0) = \mathbb{P}(\overline{y} = 1|y, S = 1)$.
To ease the interpretation, we compute the difference version of EO given by

$$EO = |\mathbb{P}(\overline{y} = 1|y, S = 0) - \mathbb{P}(\overline{y} = 1|y, S = 1)|, \tag{3}$$

where $\overline{y}$ are the predictions, $y \in 0, 1$ are the true labels and S corresponds to the sensitive attributes (0 and 1 indicating the belonging to a sensitive group). $EO \in [0, 1]$ where a score closer to 0 indicates fairer predictions.

4 Experiments

4.1 Task and Dataset

In our experiments, we focus on a classification task and use a subset of the Bias in Bios dataset [6] called the Curriculum Vitae dataset[3]. It contains a set of short biographies associated with an occupation and a gender. The dataset is composed of 217,197 entries, and 28 professional occupations. The distribution of classes (occupations) and groups (genders) within each occupation is highly

---

[3] Dataset: https://www.kaggle.com/competitions/defi-ia-insa-toulouse/data.

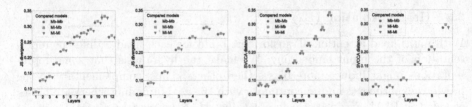

Fig. 2. JS divergence comparison for BERT (left) and DistilBERT (center-left) and SVCCA distance for BERT (center-right) and DistilBERT (right).

imbalanced. For example, class 19 corresponding to 'professor' represents 32.23% of the dataset and within the class, women represent 44.88% of the entries; class 23 corresponding to 'paralegal' represents 0.44% of the dataset, and women 84.17% of the class entries.

We create two versions of this dataset, a balanced version, and an imbalanced version, depending on the gender attribute. The former one is a subset, where for each class the largest sensitive group is truncated to equalize the proportion of individuals of each gender. The latter is a subset, where we reproduce the imbalance between gender observed in the initial dataset, but both groups are truncated to ensure that the number of samples in both subsets are equal.

Based on these two versions, we exploit the relationship between fairness and gender imbalance (cf. Section 3.1) to build two models Mi and Mb to further explore the mechanisms of bias. We evaluate the EO of BERT and DistilBERT fine-tuned on 70% of the samples, for both versions of the original dataset. In the sequel, we refer to these models as BERT_Mb and DistilBERT_Mb for the balanced versions, and BERT_Mi and DistilBERT_Mi for the imbalanced ones. To confirm our premise, we perform classification and observe an average EO over all classes three times higher for the imbalanced versions (0.13 vs. 0.42). Following this first experience, one might think that balancing the fine-tuning data is a sufficient and satisfactory solution to ensure fairness. However, before proceeding further, two remarks are in order. First, balancing the data is a first step in reducing bias, but it does not guarantee a fair model (EO above 0). Second, in many real-world scenarios, where multiple protected attributes can be observed simultaneously (e.g. women of color), this solution appears to be shortsighted as one cannot slice the data into more sub-population infinitely to rebalance classes.

4.2   Attention and Hidden States Comparison (E1)

We present the results of this experiment averaged over five random seeds in Fig. 2.

If we observe higher divergence between models Mi and Mb than between Mi and Mi or Mb and Mb on a given layer, we can assume this layer to be responsible for encoding bias. For the JS divergence, first we can note a similar pattern for both BERT and DistilBERT: the divergence increases as we move forward in the architecture, with a peak on the penultimate layer. For the SVCCA distance,

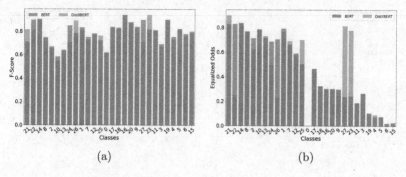

Fig. 3. (a) F-Score and (b) Equalized Odds for model `Mi` per class, without ablation and ordered by ascending ratio W/M.

the trend is similar, and we reach the highest value on the last layers. These findings are perfectly in line with the results of [13,27] claiming that fine-tuning mainly affects top layers. These first observations allow us to state that Distil-BERT follows the same learning dynamic than BERT during fine-tuning. Now, comparing [`Mi-Mi`, `Mb-Mb`] vs. `Mi-Mb`, we observe that the values for both metrics are slightly above on `Mi-Mb`, but not significantly. In addition, this difference is consistent over all layers. Finally, we observe the exact same behavior for Distil-BERT. These particular results are counterintuitive with our Hypotheses 1 and 2, and we cannot conclude that extrinsic bias makes some layers different with regard to internal representations and attention scores for both architectures.

4.3 Head's Ablation (E2)

Let us now look at the relationship between the heads and bias. Once the model is fine-tuned, we neutralize the heads in turn at inference time to estimate the bias they encode. Since the distribution of the classes is highly imbalanced, we evaluate the fairness of the model for each class individually.

Figure 3 shows the EO and F-Score for each class on the original model (`Mi`) (classes are sorted by ascending ratio women/men). Three observations can be drawn: firstly, for the highly imbalanced class (left-learning) EO is significantly higher than average; secondly, comparing BERT and DistilBERT, we see that for the majority of classes, we obtain an equivalent level of fairness, except for a few classes, where either one or the other is more biased. However, for the most imbalanced classes, DistilBERT is reaching a better level of fairness in comparison with BERT (lower or equal EO scores); thirdly, both architectures obtain comparable F-Score.

Now, we reproduce this evaluation twelve times (one time for each ablation), and observe different levels of variations for the EO depending on both the class and the head that is ablated. This implies that the representations produced by the attention head are different enough to impact the fairness of the models as assumed in Hypothesis 3. However, when the network is re-fine tuned we do

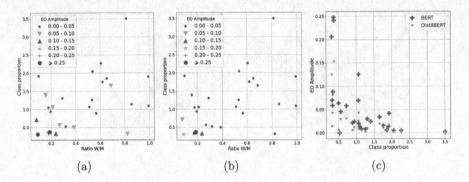

Fig. 4. EO amplitude for BERT's Mi (a), DistilBERT's Mi (b), BERT's and Distil-BERT's Mb (c).

not observe the same variations for a given head. For each class, we compute the difference of EO obtained when neutralizing a head gives the fairest score vs. the most unfair score and call it the amplitude. Let $EO_{class} = \{EO^1_{class}, ..., EO^{12}_{class}\}$ with EO^{head}_{class} the EO computed for a given class after noising a head. We defined the amplitude for a class as:

$$\text{amplitude}_{class} = \max_{head}(EO_{class}) - \min_{head}(EO_{class}) \tag{4}$$

Figure 4 shows the amplitude depending on the class imbalance and the ratio of women/men within the classes. The minority group is not the same for every class, thus, we compute the ratio as follows ratio W/M = $\min(\frac{\%women}{\%men}, \frac{\%men}{\%women}) \in [0,1]$. For better readability, we rescale the class proportion and amplitude vectors by taking $\log(\text{vector} + 1_n)$, n being the dimension of the vectors. We note, in Figs. 4a and 4b that the more a class suffers from double imbalance (underrepresented class and highly sensitive group imbalance) the more the heads will produce different representations, some being more biased than others. Also, when comparing the values of EO amplitude for BERT and DistilBERT, we observe that BERT is more sensitive to those scenarios than DistilBERT. According to Fig. 4c, where sensitive groups are balanced within the classes, the less a class is represented the higher the amplitude is, meaning that **BERT is generally more sensitive to class imbalance than Distil-BERT with regard to the homogeneity of head representations.** On the other hand, we evaluate the correlation between F-Score and EO, when ablating each head, and have not been able to establish a relation caused by the process.

5   Conclusion

This paper investigated the implication of inner elements of BERT-based models' architecture in bias encoding through empirical experiments on the Transformers' layers and attention heads. We also studied the attention carried by the

"CLS" token to the words of the sequence, specifically the pronouns 'he' and 'she', but also to the ones receiving the most attention from the aforesaid token, in an attempt to understand what the model was focusing on; this study did not lead us to any convincing results. Similarly, investigating the JS divergence and SVCCA distance between different layers (e.g. the 1 and 2) was not conclusive, we suspect it might since layers specialize on different aspects of the input text [21]. To summarize, we show that gender bias is not encoded in a specific layer or head. We also demonstrate that the distilled version of BERT, DistilBERT, is more robust to double imbalance of classes and sensitive groups than the original model. Even more specifically, we observe that the representations generated by the attention heads in such a context are more homogeneous for DistilBERT than for BERT in which some attention heads will be fair while others are very unfair. Thus, we advise giving special care to such patterns in the data but do not recommend ablating the heads producing more unfair representations since it could seriously harm the performance of the model. Finally, we recommend DistilBERT to the practitioner using datasets containing underrepresented classes with a high imbalance between sensitive groups, while cautiously evaluating class independently, using the protocol that we propose in this paper.

Acknowledgment. This work is funded by the french National Agency for Research (ANR) in the context of the Diké project (ANR-21-CE23-0026).

References

1. Basta, C., Costa-jussà, M.R., Casas, N.: Evaluating the underlying gender bias in contextualized word embeddings. In: The 1st Workshop on Gender Bias in NLP, pp. 33–39. ACL (2019)
2. Bender, E.M., Gebru, T., McMillan-Major, A., Shmitchell, S.: On the dangers of stochastic parrots: can language models be too big? In: FAccT, pp. 610–623 (2021)
3. Bhardwaj, R., Majumder, N., Poria, S.: Investigating gender bias in bert. Cognitive Computation (2020)
4. Caton, S., Haas, C.: Fairness in machine learning: a survey. arXiv preprint arXiv:2010.04053 (2020)
5. Dai, A.M., Le, Q.V.: Semi-supervised sequence learning. In: NeurIPS, p. 3079–3087 (2015)
6. De-Arteaga, M., et al.: Bias in bios: a case study of semantic representation bias in a high-stakes setting. In: FaccT, pp. 120–128 (2019)
7. Delobelle, P., Berendt, B.: Fairdistillation: mitigating stereotyping in language models. In: ECML-PKDD (2022)
8. Delobelle, P., Tokpo, E., Calders, T., Berendt, B.: Measuring fairness with biased rulers: a comparative study on bias metrics for pre-trained language models. In: Proceedings of NAACL, pp. 1693–1706 (2022)
9. Demner-Fushman, D., Chapman, W.W., McDonald, C.J.: What can natural language processing do for clinical decision support? J. Biomed. Inform. **42**(5), 760–772 (2009)
10. Devlin, J., Chang, M., Lee, K., Toutanova, K.: BERT: pre-training of deep bidirectional transformers for language understanding. In: Proceedings of NAACL-HLT, pp. 4171–4186 (2019)

11. Dixon, L., Li, J., Sorensen, J., Thain, N., Vasserman, L.: Measuring and mitigating unintended bias in text classification. In: Proceedings of the 2018 AAAI/ACM Conference on AI, Ethics, and Society, pp. 67–73 (2018)
12. Gupta, M., Agrawal, P.: Compression of deep learning models for text: a survey. ACM Trans. Knowl. Discov. Data 16(4), 1–55 (2022)
13. Hao, Y., Dong, L., Wei, F., Xu, K.: Investigating learning dynamics of BERT fine-tuning. In: Proceedings of the Conference of the Asia-Pacific Chapter of the ACL and the IJCNLP. Suzhou, China (2020)
14. Hardoon, D.R., Szedmak, S., Shawe-Taylor, J.: Canonical correlation analysis: an overview with application to learning methods. Neural Comput. 16(12), 2639–2664 (2004)
15. Hardt, M., Price, E., Srebro, N.: Equality of opportunity in supervised learning. In: NeurIPS, vol. 29 (2016)
16. Hinton, G., Vinyals, O., Dean, J.: Distilling the knowledge in a neural network. arxiv pre-print (2015)
17. Hooker, S., Courville, A., Clark, G., Dauphin, Y., Frome, A.: What do compressed deep neural networks forget? arxiv pre-print (2021)
18. Hooker, S., Moorosi, N., Clark, G., Bengio, S., Denton, E.: Characterising bias in compressed models. arXiv preprint arXiv:2010.03058 (2020)
19. Hutchinson, B., Prabhakaran, V., Denton, E., Webster, K., Zhong, Y., Denuyl, S.: Social biases in NLP models as barriers for persons with disabilities. In: ACL, pp. 5491–5501 (2020)
20. Jatobá, M., Santos, J., Gutierriz, I., Moscon, D., Fernandes, P.O., Teixeira, J.P.: Evolution of artificial intelligence research in human resources. Procedia Comput. Sci. 164, 137–142 (2019)
21. Jawahar, G., Sagot, B., Seddah, D.: What does bert learn about the structure of language? In: ACL (2019)
22. Kim, S., Gholami, A., Yao, Z., Mahoney, M.W., Keutzer, K.: I-bert: Integer-only bert quantization (2021)
23. Kurita, K., Vyas, N., Pareek, A., Black, A.W., Tsvetkov, Y.: Measuring bias in contextualized word representations. In: The 1st Workshop on Gender Bias in Natural Language Processing, pp. 166–172 (2019)
24. LeCun, Y., Denker, J., Solla, S.: Optimal brain damage. In: Advances in Neural Information Processing Systems, vol. 2 (1989)
25. Lukasik, M., Bhojanapalli, S., Menon, A.K., Kumar, S.: Teacher's pet: understanding and mitigating biases in distillation. arXiv preprint arXiv:2106.10494 (2021)
26. Mehrabi, N., Morstatter, F., Saxena, N., Lerman, K., Galstyan, A.: A survey on bias and fairness in machine learning. ACM Comput. Surv. 54(6), 1–35 (2021)
27. Merchant, A., Rahimtoroghi, E., Pavlick, E., Tenney, I.: What happens to BERT embeddings during fine-tuning? In: Proceedings of the 3rd BlackboxNLP Workshop, pp. 33–44. ACL (2020)
28. Radford, A., Narasimhan, K., Salimans, T., Sutskever, I., et al.: Improving language understanding by generative pre-training. OpenAI blog (2018)
29. Radford, A., Wu, J., Child, R., Luan, D., Amodei, D., Sutskever, I., et al.: Language models are unsupervised multitask learners. OpenAI blog 1(8), 9 (2019)
30. Raghu, M., Gilmer, J., Yosinski, J., Sohl-Dickstein, J.: Svcca: singular vector canonical correlation analysis for deep learning dynamics and interpretability. In: NeurIPS, vol. 30 (2017)
31. Sanh, V., Debut, L., Chaumond, J., Wolf, T.: Distilbert, a distilled version of bert: smaller, faster, cheaper and lighter. arxiv pre-print (2020)

32. Sheng, E., Chang, K., Natarajan, P., Peng, N.: The woman worked as a babysitter: on biases in language generation. In: EMNLP/IJCNLP, pp. 3405–3410 (2019)

33. Swinger, N., De-Arteaga, M., Heffernan IV, N.T., Leiserson, M.D., Kalai, A.T.: What are the biases in my word embedding? In: Proceedings of the AAAI/ACM Conference on AI, Ethics, and Society, pp. 305–311 (2019)

34. Vaswani, A., et al.: Attention is all you need. In: NeurIPS, vol. 30 (2017)

35. Xu, G., Hu, Q.: Can model compression improve NLP fairness. arXiv preprint arXiv:2201.08542 (2022)

Discovering Diverse Top-K Characteristic Lists

Antonio Lopez-Martinez-Carrasco[1(✉)], Hugo M. Proença[2], Jose M. Juarez[1], Matthijs van Leeuwen[2], and Manuel Campos[1,3]

[1] AIKE Research Group (INTICO), University of Murcia, Murcia, Spain
antoniolopezmc@um.es
[2] Leiden Institute of Advanced Computer Science, Leiden University, Leiden, The Netherlands
[3] Murcian Bio-Health Institute (IMIB-Arrixaca), Murcia, Spain

Abstract. In this work, we define the new problem of finding diverse top-k characteristic lists to provide different statistically robust explanations of the same dataset. This type of problem is often encountered in complex domains, such as medicine, in which a single model cannot consistently explain the already established ground truth, needing a diversity of models. We propose a solution for this new problem based on Subgroup Discovery (SD). Moreover, the diversity is described in terms of coverage and descriptions. The characteristic lists are obtained using an extension of SD, in which a subgroup identifies a set of relations between attributes (description) with respect to an attribute of interest (target). In particular, the generation of these characteristic lists is driven by the Minimum Description Length (MDL) principle, which is based on the idea that the best explanation of the data is the one that achieves the greatest compression. Finally, we also propose an algorithm called GMSL which is simple and easy to interpret and obtains a collection of diverse top-k characteristic lists.

Keywords: Subgroup Discovery · Subgroup List · the Minimum Description Length principle · Algorithm · Interpretable Machine Learning

1 Introduction

More and more often, Artificial Intelligence is required to generate readable, understandable and transparent models. In contrast to black-box machine learning (e.g., neural network models), interpretable machine learning is an increasing trend whose objective is to develop new methods and tools which allow humans to understand machine learning models and to interpret their results in many critical areas, such as medicine or economy. In this context, different research has been carried out [12,13].

The discovery of a collection of descriptions is helpful to better understand datasets. In this context, one type of collection is the characteristic list, whose

B. Crémilleux et al. (Eds.): IDA 2023, LNCS 13876, pp. 262–273, 2023.
https://doi.org/10.1007/978-3-031-30047-9_21

Fig. 1. Example of a description from a characteristic list.

purpose is to generalise an individual belonging to a specific category. These descriptions explain all typical features that characterize the individuals belonging to a specific category for a descriptive purpose [1]. An example of a description from a characteristic list is shown in Fig. 1.

The utilization of multiple characteristic lists is relevant because a single explanation of a target value is not always enough. A clear example of both needs is the clinical domain, in which a patient could have some diagnosis. A relevant task is to find all the risk factors that differentiate a diagnosis from others, not only from a predictive point of view, but from the descriptive point of view. However, a single characteristic list provides a limited explanation, having little value due to the possible lack of meaning from the clinical point of view. For example, a characteristic list automatically generated by some machine learning algorithm could not make sense for the clinicians and, therefore, be discarded.

Subgroup Discovery (SD) can be used as building block of characteristic lists. A subgroup identifies a relation between attributes (description) and an attribute of interest (target). Besides, subgroups can be used as local descriptive models that characterize subpopulations, in contrast to the whole population, in relation to the target attribute given a quality measure. However, only sets with few subgroups can be easily interpreted by an expert. To solve this problem, we can build a subgroup list model. We illustrate the advantages in Fig. 2, showing subgroups and subgroup lists extracted from the *zoo* dataset. On the one hand, a subgroup contains a set of selectors (i.e., a pattern or description) and it is generated when a quality measure given, e.g. Weighted Relative Accuracy (WRAcc), with respect to a target value is above a threshold. In the figure, the subgroup s1 "milk = yes" contains a single selector to define the class *type* = *'mammal'*. On the other hand, a subgroup list is an ordered collection of subgroups that explain the target value. In the figure, an example of four subgroup lists is depicted. Note that either a subgroup or a subgroup list provides an explanation of how to define a single class (*type* = *'mammal'* in this example), but not the others. Moreover, it is readable, understandable and has the potential to be interpretable.

A subgroup list can be interpreted as a decision list, since it is an ordered collection of subgroups of the form "else-if" (i.e., a subgroup description is only reached if all the above ones are not being true). Another model that can be used for the proposed problem is the decision set, which is formed by an unordered collection of subgroups of the form "if" (i.e., all subgroup descriptions apply independently). Although our objective is to describe data, both models can be used either for description or prediction tasks [7].

A great difficulty when creating a subgroup list is the large number of subgroups that are extracted. In the example shown in Fig. 2, there are 8,537,383 subgroups mined with an exhaustive SD algorithm that could be used to create subgroup lists. This can be solved using the Minimum Description Length (MDL) principle [6], a method of inductive inference whose fundamental idea is that the best explanation of the data is the one that achieves the greatest compression. In the context of SD and subgroup lists, it was shown that using the MDL principle is equivalent to performing a Bayesian statistical test and multiple hypothesis testing correction for every subgroup [10,11]. Thus, this leads to the discovery of statistically robust subgroups and subgroup lists.

Subgroup description	type= 'mammal' (Pos)	type= 'other' (Neg)
s1	milk='yes'	41	0
s2	venomous='no'	0	52
dr	-	0	8

Subgroup description	type= 'mammal' (Pos)	type= 'other' (Neg)
s1	backbone='yes', hair = 'yes'	39	0
s2	fins='no'	0	47
dr	-	2	13

Subgroup description	type= 'mammal' (Pos)	type= 'other' (Neg)
s1	eggs='no'	40	2
s2	backbone='yes'	1	41
s3	feathers='no'	0	17
dr	-	0	0

Subgroup description	type= 'mammal' (Pos)	type= 'other' (Neg)
s1	hair='yes'	39	4
s2	breathes='yes'	2	35
s3	toothed='yes'	0	14
dr	-	0	7

Fig. 2. An example of four subgroup lists generated from the *zoo* dataset (i.e., four different explanations of this dataset), being the target *type* = *'mammal'*. Notation: s1: subgroup1; dr: default rule; pos: positives; neg: negatives.

The main contributions of this research are: (1) the definition of the new problem of finding diverse top-k characteristic lists, and (2) a new algorithm called GMSL that solves this problem by using SD, the subgroup list model, and the MDL principle. Moreover, its results are simple, readable, understandable and statistically robust at the same time. This contribution improves the state of the art, since existing algorithms generate only one subgroup list and, therefore, different explanations for the same data are not possible. Note that, to the best of our knowledge, no algorithm in the literature combines SD and the MDL principle to generate diverse top-k subgroup lists.

The remainder of this paper is structured as follows: Sect. 2 defines the problem tackled in this research, while Sect. 3 shows and explains our proposal: the new algorithm called GMSL that generates diverse top-k subgroup lists. Moreover, Sect. 4 describes the configuration of the experiments carried out in this work and provides a discussion of the results obtained. Finally, Sect. 5 presents the conclusions reached after carrying out the research.

2   Problem Statement and Background

This section formalizes the problem tackled in this research, i.e., the generation of diverse top-k characteristic lists by using Subgroup Discovery (SD), the subgroup list model, and the Minimum Description Length (MDL) principle.

2.1   The Problem of Discovering Diverse Top-K Characteristic Lists

The fundamental concepts of this new problem are defined in this section.

First, an attribute a is a relation between an object property and its value. For example, $a = hair : no$. Moreover, the set of all unique values that an attribute can take is defined as the domain of the attribute and is denoted as $dom(a)$. Note that, depending on its domain, an attribute can be nominal or numeric. Second, an instance i is a tuple $i = (a_1, \ldots, a_m)$ of attributes, for example, $i = (milk : yes, hair : yes)$. Finally, a dataset d is a tuple $d = (i_1, \ldots, i_n)$ of instances. For example, $d = ((milk : yes, hair : yes), (milk : yes, hair : no))$. Note that we use the notation $v_{x,y}$ to indicate the value of the x-th instance i_x and of the y-th attribute a_y from a dataset d.

According to these basic definitions, the following ones can be given:

Definition 1 (Selector e). *Given an attribute a_y from a dataset d, a binary operator $\in \{=, \neq, <, >, \leq, \geq\}$ and a value w, being w in the domain of a_y, then a selector e is defined as a 3-tuple of the form $(a_y, operator, w)$.*

Informally, this means that a selector is a binary relation between an attribute from a dataset and one of its possible values, representing a property of a subset of instances from this dataset. Some examples of selectors are $e_1 = (age, >, 50)$ and $e_2 = (venomous, =, yes)$.

Definition 2 (Selector coverage). *Given an instance i_x, an attribute a_y and a selector $e = (a_y, operator, w \in dom(a_y))$, then i_x is covered by e if the binary expression "$v_{x,y}$ operator w" holds true. Otherwise, we say that it is not covered by e.*

Definition 3 (Pattern p). *A pattern p is a list of selectors $<e_1, \ldots, e_x>$ (i.e., a conjunction) in which all attributes of the selectors are different.*

Informally, this means that a pattern represents a list of properties of a subset of instances from a dataset.

Definition 4 (Pattern coverage). *Given an instance i and a pattern p, then i is covered by p if i is covered by e_x, $\forall e_x \in p$. Otherwise, we say that it is not covered by p.*

Definition 5 (Characteristic list l). *Given a dataset d and a selector e (denominated as category), then a characteristic list l is a collection of patterns $<p_1, \ldots, p_y>$ (each of them is denominated as "description") that allow to describe the instances from d belonging to e. Note that a characteristic list is used for a descriptive purpose.*

An example of description from a characteristic list is depicted in Fig. 1. This description is formed by a set of selectors that allow to describe the individuals belonging to the category $animal = 'turtle'$.

It is necessary to explain why "top-k" and "diversity" properties are essential in the new problem defined. Firstly, it is focused only on the generation of the top-k characteristic lists due to this generation is limited by the available computational capacity. This means that it is not feasible to carry out an exhaustive generation of all possible characteristic lists. Secondly, diversity is essential in this case, since multiple characteristic lists from l_1 to l_k will be generated and, therefore, it is necessary to ensure that they will be different and non-redundant. Diversity can be achieved both in terms of coverage and descriptions. The diversity in terms of coverage is considered when building a single characteristic list l_x to minimize the number of instances already covered by previous patterns. This means that, given two patterns $p_a \in l_x$ and $p_b \in l_x$, the instances covered by both patterns at the same time should be as few as possible. The diversity in terms of descriptions implies using different selectors and patterns in the different characteristic lists to ensure that the models provide multiple explanations of the same category or target value. This means that, given two characteristic lists l_x and l_y, then $\forall p_a$, $if\ p_a \in l_x$, $then\ p_a \notin l_y$.

Therefore, the new problem of discovering diverse top-k characteristic lists is defined as follows:

Definition 6 (Discovering diverse top-k characteristic lists problem). *Given a dataset d, a category e (in form of a selector) and the k maximum number of characteristic lists to generate, then the problem of discovering diverse top-k characteristic lists consists of generating a collection of characteristic lists $<l_1, \ldots, l_k>$ such that they are diverse and represent different explanations or perspectives of d in relation to e.*

Finally, the proposal carried out in this work (i.e., GMSL algorithm, which is explained in Sect. 3) solves this problem by using SD, the subgroup list model, and the MDL principle.

2.2 Subgroup Discovery

SD [2] is a supervised machine learning technique whose purpose is the identification of a set of relations between attributes (denominated as *description*) with respect to an attribute of interest (denominated as *target*). This technique is widely used for exploratory and descriptive data analysis and is also useful for obtaining general relations in a dataset and automatically generating hypotheses. In particular, SD helps to obtain groups of individuals that might overlap. However, as with many pattern mining techniques, SD experiences some problems such as pattern explosion or lack of statistical guarantees specially when using datasets with many attributes [9]. Therefore, configuring a list of the best subgroups that faithfully describes a dataset is not trivial.

Additionally, the fundamental concepts of SD are described as follows:

Given a pattern p and a selector e, a subgroup s is a pair (p, e) in which the pattern is denominated as 'description' and the selector is denominated as 'target'. Subgroups can be used for either a predictive purpose or a descriptive purpose (i.e., characteristic subgroups) [1]. Therefore, since our objective in this research is to describe and characterize individuals from a dataset, subgroups will be used as a fundamental part of characteristic lists (subgroup lists, in this case) with the objective of identifying all properties related to a specific category or target attribute. An example of subgroup is s = (<(shell, =, yes),(feathers, =, no),(backbone, =, yes)>, (turtle, =, yes)). Finally, given a subgroup s and a dataset d, a quality measure q is a function that computes one numeric value according to s and to certain characteristics from d [4]. Some examples of quality measures are Sensitivity, Piatetsky Shapiro or Weighted Relative Accuracy (WRAcc).

Following these definitions, given a dataset d, a quality measure q and a numeric value $threshold$, the subgroup discovery problem consists of exploring the search space of d in order to generate subgroups that have a value of q above $threshold$. Formally: $\mathcal{R} = \{(s, quality\_value)|quality\_value \geq threshold\}$.

Some examples of algorithms that generate individual subgroups are SD-Map [3], CN2-SD [8] or ID-Rsd [5], among others.

2.3 The Subgroup List Model

The subgroup list model was initially presented in [11] and, afterwards, expanded and detailed in [10]. A subgroup list is a collection of ordered subgroups followed by a default rule, whose objective is to partition the input data and to provide a description (i.e., an individual subgroup) of each of these partitions, except the last one (that corresponds to the default rule). While the default rule represents the dataset average and covers the instances that are well described by the dataset distribution, the subgroups cover the instances that are statistically different and interesting, compared to dataset distribution. Therefore, each instance of the input dataset can only be covered either by one individual subgroup or by the default rule. For example, if a subgroup list contains 10 subgroups, this means that the input dataset was partitioned into 11 subsets: the first 10 of them correspond to the 10 individual subgroups and the last one corresponds to the default rule. An example of subgroup list is shown in Fig. 3.

$subgroup_1$:	IF	$description_1$	THEN	$distribution_1\,(target)$
$subgroup_2$:	ELSE IF	$description_2$	THEN	$distribution_2\,(target)$
		$\vdots$		
$subgroup_w$:	ELSE IF	$description_w$	THEN	$distribution_w\,(target)$
$dataset$:	ELSE			$distribution\,(target)$

Fig. 3. Example of subgroup list with w subgroups.

2.4   The MDL Principle for Discovering a Single Subgroup List

According to the MDL principle, the best individual subgroup list is the one that compresses the data and the model the most, i.e., the simplest subgroup list that best fits the data. The authors of [10] defined the MDL encoding of the optimal subgroup list for a certain dataset.

However, as the problem of finding an optimal subgroup list is NP-hard, the authors of [10] also proposed a greedy approach that iteratively added one subgroup at the time to the subgroup list (after the last subgroup and before the default rule). According to this, given a dataset d, a subgroup list model M, and a subgroup candidate s, the best subgroup to add to a single subgroup list is the one that maximize the compression gain, which is defined as follows:

$$\Delta_\beta L(d, M \oplus s) = \frac{L(d, M) - L(d, M \oplus s)}{(n_s)^\beta} + \frac{L(M) - L(M \oplus s)}{(n_s)^\beta} \qquad (1)$$

Note that the $\oplus$ operator represents adding s at the end of M (before the default rule), and n_s is the number of instances covered by the description of s.

More details about $\Delta_\beta L$ and the β parameter can be found in [10], although the intuition is as follows: (1) a subgroup candidate that maximizes $\Delta_\beta L$ is maximizing a Bayesian proportions tests between the subgroup distribution and the dataset distribution while penalizing for larger descriptions; (2) $\Delta_\beta L > 0$ means there is more statistical evidence in favour of adding the subgroup candidate to the list than not adding it; and (3) β values closer to 0 prioritize subgroup candidates that cover more instances, while β values closer to 1 prioritize subgroup candidates that cover less instances.

Currently, state of the art only focuses on algorithms to discover a single subgroup list (e.g., SDD++ algorithm [10]). Therefore, they cannot return diverse top-k subgroup lists automatically.

3   GMSL Algorithm

In this work, we propose the Generation of Multiple Subgroup Lists algorithm (GMSL), whose purpose is to generate diverse top-k Subgroup Lists by combining SD and the MDL principle.

Our proposal is detailed in Algorithm 1, and it requires the following inputs: a dataset d, a collection of subgroup candidates C, the maximum number of subgroup lists to generate, and the normalization parameter β used by the compression gain $\Delta_\beta L$ (see Eq. 1). Besides, it is also necessary to state that the subgroup candidates from C could be generated with any algorithm and could be also filtered before executing GMSL algorithm.

The algorithm starts with the creation of the list $\mathcal{L}$, which has size max_sl and is initialized with empty subgroup lists (line 1). Next, we iterate through $\mathcal{L}$ (loop of the line 2), and for each subgroup list, continuous iterations through C are carried out in order to find the best subgroup candidate to add (lines 6–12). The compression gain for each current subgroup candidate is calculated with

---

Algorithm 1. GMSL algorithm.

Input: d { dataset } ; $\mathcal{C}$ { subgroup candidates } ; max_sl { maximum number of subgroup lists to generate ($\mathbb{N}$) } ; β { normalization parameter $\in [0, 1]$ }
Output: $\mathcal{L}$: collection of subgroup lists.
1: $\mathcal{L}$:= create a collection with max_sl empty subgroup lists.
2: **for each** $sl \in \mathcal{L}$ **do**
3:    **repeat**
4:       $best\_candidate := NULL$
5:       $bc\_comp\_gain := 0$
6:       **for each** $current\_candidate \in \mathcal{C}$ **do**
7:          $cc\_comp\_gain := \Delta_\beta L(d, sl \oplus current\_candidate)$
8:          **if** $cc\_comp\_gain > bc\_comp\_gain$ **then**
9:             $best\_candidate := current\_candidate$
10:            $bc\_comp\_gain := cc\_comp\_gain$
11:          **end if**
12:       **end for**
13:       **if** $best\_candidate \neq NULL$ **then**
14:          $sl := sl \oplus best\_candidate$
15:          $\mathcal{C}.delete(best\_candidate)$
16:          $\mathcal{C}.deleteRefinements(best\_candidate)$
17:       **end if**
18:    **until** $best\_candidate = NULL$
19: **end for**
20: **return** $\mathcal{L}$

---

the compression gain $\Delta_\beta L$ (line 7). The candidate with the highest compression gain will be selected (lines 8–11) and added to the current subgroup list (lines 13–17) until there are no subgroup candidates with positive compression gain. Finally, the algorithm returns the collection $\mathcal{L}$ containing max_sl subgroup lists. Note that computing the compression gain for each subgroup candidate using the MDL principle guarantees that all subgroups added to a subgroups list are statistically robust. Moreover, it is also relevant to remark that GMSL algorithm also encourages the generation of diverse subgroup lists to allow different explanations of the dataset.

Finally, we have to highlight how the algorithm generates diverse subgroup lists. In the first place, diversity in terms of coverage is guaranteed due to the utilization of the subgroup list model, since each instance of the input dataset can only be covered either by one individual subgroup or by the default rule. In the second place, diversity in terms of descriptions is achieved because each time that a subgroup candidate from $\mathcal{C}$ is added to a subgroup list, that subgroup and its refinements are deleted (lines 15 and 16). Therefore, each subgroup candidate appears at most once and the appearance of the same selectors in the different patterns is also minimized.

4   Experiments and Discussion

GMSL algorithm was implemented in `subgroups` python library[1]. The goal of the experiments carried out in this work was to validate our proposal in relation to the new defined problem (i.e., to verify whether GMSL algorithm can generate diverse top-k characteristic lists in form of subgroup lists). We used for this purpose the well-known *car-evaluation* dataset from UCI repository with *class = 'acc'* as target, meaning that the car is *acceptable* to be bought. The One Hot Encoding technique was applied to the dataset with the objective that attributes were binary. Therefore, this dataset had 1,728 instances and 18 attributes. After that, an exhaustive SD algorithm was executed using WRAcc quality measure and a threshold value of 0 (i.e., only subgroups whose WRAcc quality measure value is greater or equal than 0 were generated) and a maximum depth of 2. Note that any exhaustive SD algorithm could be applied in this point, since subgroups obtained by any exhaustive SD algorithm are always the same as long as the same quality measure and parameters are used. Finally, 302 subgroups were obtained. These subgroup candidates ($\mathcal{C}$) were the main input of GMSL algorithm to generate diverse top-k subgroup lists.

After carrying out the experiments described, diverse top-3 subgroup lists were generated, and they are represented in Fig. 4. For each one, the following elements are shown: (1) their individual subgroups and the default rule (denoted as `dr`), (2) the number of positive (i.e., such that the class is equal to 'acc') and negative (i.e., such that the class is not equal to 'acc') instances of the dataset, and (3) the cumulative sum of positive and negative instances covered by the subgroup list.

These three diverse subgroup lists shown in Fig. 4 represent different explanations of the same dataset. Different subgroups (i.e., different subgroup descriptions, which use different patterns) were used in the different subgroup lists. Therefore, different and diverse explanations were generated from the same data.

The figure also shows, for example, that the first and second subgroup lists include the original attribute *buying* (*buying_vhigh* and *buying_low* after applying One Hot Encoding), which is not used by the third subgroup list. Moreover, different attributes generate from the original *doors* attribute are used by all subgroup lists. Additionally, the first and second subgroup lists have 2 subgroups whose description has a single selector, while the third subgroup list has 3 subgroups whose description has a single selector. Besides, note that subgroups in a subgroup list need to be interpreted sequentially, since a subgroup list is ordered by definition.

According to the "cusum" value of the last subgroup (i.e., before the default rule), it can be observed that the first and third subgroup lists cover more positive examples than the second subgroup list. In the same way, the first subgroup list has fewer subgroups, being more general, while the second subgroup list has more subgroups, being more specific. It is relevant to note that, while subgroups are local model, subgroup lists are global model, since they cover the whole

---

[1] Source code available on: https://github.com/antoniolopezmc/subgroups.

	Subgroup description	Pos-Neg instances	Cusum
s1	doors_2='no', lug_boot_low='no'	384-384	384-384
s2	buying_vhigh='no'	0-720	384-1104
s3	buying_low='no'	0-240	384-1344
dr	-	0-0	384-1344

	Subgroup description	Pos-Neg instances	Cusum
s1	doors_2='no', lug_boot_high='yes'	204-180	204-180
s2	doors_2='no', lug_boot_med='no'	0-384	204-564
s3	doors_2='no', persons_small='no'	145-111	349-675
s4	persons_small='no'	0-384	349-1059
s5	lug_boot_high='yes'	0-64	349-1123
s6	buying_vhigh='yes', lug_boot_low='no'	0-48	349-1171
dr	-	35-173	384-1344

	Subgroup description	Pos-Neg instances	Cusum
s1	doors_4='yes', lug_boot_low='no'	198-186	198-186
s2	doors_more='no', lug_boot_low='no'	0-384	198-570
s3	lug_boot_low='no'	186-198	384-768
s4	maint_2='no'	0-432	384-1200
s5	doors_2='no'	0-96	384-1296
dr	-	0-48	384-1344

Fig. 4. Diverse top-3 subgroup lists generated from *car-evaluation* dataset (i.e., three different explanations of this dataset) with *class = 'acc'* as target, meaning that the car is *acceptable* to be bought. Notation: s1: subgroup1; dr: default rule; pos: positive instances; neg: negative instances; cusum: cumulative sum of pos/neg instances.

dataset. Moreover, subgroup list model is focused on a value of a target attribute. Additionally, each subgroup list has a different number of subgroups: the first subgroup list has three, the second subgroup list has six, and the third subgroup list has five.

It is necessary to remember that the collection of subgroup candidates is generated a-priori and, then, taken as an input by GMSL algorithm. Although this could penalize the algorithm in terms of memory consumption, it is also an advantage in term of flexibility, since it allows to prefilter this collection and to introduce domain knowledge. For example, some negative subgroups such as *doors_2 = 'no'* or *doors_more = 'no'* were generated from the *car-evaluation* dataset. However, they may not make sense for the user from the logical point of view, and consequently, they could be deleted before executing GMSL algorithm.

Note that subgroups from a subgroup list do not overlap by definition [10]. However, if we analyse each of these subgroups individually (i.e., without considering the subgroup list model), they could cover the same instances of the database.

In summary, we show for a particular case study that our proposal is suitable for solving the new problem defined initially, since it can discover diverse top-k characteristic lists in form of subgroup lists using SD and the MDL principle.

5 Conclusions

In this research, we defined the novel problem of discovering diverse top-k characteristic lists, which consists of providing users with the k best and diverse explanations of a dataset with a binary-target attribute.

To solve this problem, we proposed GMSL, an algorithm that takes a set of pre-computed subgroup candidates as input and returns a collection of diverse top-k subgroup lists. The goodness of fit is measured using the MDL principle and, moreover, diversity is defined in terms of coverage and descriptions. This way, we can provide different perspectives of the same data through the diverse top-k subgroup lists.

As shown in the examples, the results are simple and can be easily interpreted. To the best of our knowledge, this is the first proposal that uses SD and the MDL principle to solve the new defined problem.

Finally, future research could extend and improve the proposed algorithm in different ways, for example, by generating subgroup lists without a collection of subgroup candidates loaded a-priori. Moreover, the overlap between subgroups from a subgroup list could be also study in order to improve the model interpretability. Additionally, it would be interesting to extend the problem to a multiclass setting.

Acknowledgments. This work was partially funded by the CONFAINCE project (Ref: PID2021-122194OB-I00) by MCIN/AEI/10.13039/501100011033 and, as appropriate, by "ERDF A way of making Europe", by the "European Union" or by the "European Union NextGenerationEU/PRTR", and by the GRALENIA project (Ref: 2021/C005/00150055) supported by the Spanish Ministry of Economic Affairs and Digital Transformation, the Spanish Secretariat of State for Digitization and Artificial Intelligence, Red.es and by the NextGenerationEU funding. Moreover, this research was also partially funded by a national grant (Ref: FPU18/02220), financed by the Spanish Ministry of Science, Innovation and Universities (MCIU) and by a mobility grant (Ref: R-933/2021), financed by the University of Murcia.

References

1. Alkhatib, A., Boström, H., Vazirgiannis, M.: Explaining predictions by characteristic rules. In: Proceedings of European Conference on Machine Learning and Principles and Practice of Knowledge Discovery in Databases. Springer, Cham (2022)

2. Atzmueller, M.: Subgroup discovery - advanced review. WIREs: Data Min. Knowl. Discov. 5(1), 35–49 (2015)

3. Atzmueller, M., Puppe, F.: SD-Map – a fast algorithm for exhaustive subgroup discovery. In: Fürnkranz, J., Scheffer, T., Spiliopoulou, M. (eds.) PKDD 2006. LNCS (LNAI), vol. 4213, pp. 6–17. Springer, Heidelberg (2006). https://doi.org/10.1007/11871637_6

4. Duivesteijn, W., Knobbe, A.: Exploiting false discoveries - statistical validation of patterns and quality measures in subgroup discovery. In: IEEE 11th International Conference on Data Mining (ICDM 2011), pp. 151–160 (2011)

5. Grosskreutz, H., Paurat, D.: Fast and memory-efficient discovery of the top-k relevant subgroups in a reduced candidate space. In: Gunopulos, D., Hofmann, T., Malerba, D., Vazirgiannis, M. (eds.) ECML PKDD 2011. LNCS (LNAI), vol. 6911, pp. 533–548. Springer, Heidelberg (2011). https://doi.org/10.1007/978-3-642-23780-5_44

6. Grünwald, P.D.: The Minimum Description Length Principle, MIT Press Books, vol. 1. The MIT Press, Cambridge (2007)
7. Lakkaraju, H., Bach, S.H., Leskovec, J.: Interpretable decision sets: a joint framework for description and prediction. In: Proceedings of the 22nd ACM SIGKDD International Conference on Knowledge Discovery and Data Mining, pp. 1675–1684. KDD 2016, Association for Computing Machinery (2016)
8. Lavrač, N., Kavšek, B., Flach, P., Todorovski, L.: Subgroup discovery with CN2-SD. J. Mach. Learn. Res. **5**, 153–188 (2004)
9. van Leeuwen, M., Ukkonen, A.: Expect the unexpected – on the significance of subgroups. In: Calders, T., Ceci, M., Malerba, D. (eds.) DS 2016. LNCS (LNAI), vol. 9956, pp. 51–66. Springer, Cham (2016). https://doi.org/10.1007/978-3-319-46307-0_4
10. Proença, H.M., Grünwald, P., Bäck, T., van Leeuwen, M.: Robust subgroup discovery. Data Min. Knowl. Discovery **36**, 1885–1970 (2022)
11. Proença, H.M., Grünwald, P., Bäck, T., Leeuwen, M.V.: Discovering outstanding subgroup lists for numeric targets using MDL. In: Machine Learning and Knowledge Discovery in Databases (ECML PKDD 2020), pp. 19–35 (2021)
12. Semenova, L., Rudin, C., Parr, R.: On the existence of simpler machine learning models. In: ACM Conference on Fairness, Accountability, and Transparency, pp. 1827–1858. FAccT 2022, Association for Computing Machinery (2022)
13. Xin, R., Zhong, C., Chen, Z., Takagi, T., Seltzer, M.I., Rudin, C.: Exploring the whole rashomon set of sparse decision trees. ArXiv abs/2209.08040 (2022)

Online Influence Forest for Streaming Anomaly Detection

Inês Martins[1]([⊠]), João S. Resende[2], and João Gama[1]

[1] INESC TEC and Universidade do Porto, Porto, Portugal
inesmartins@fc.up.pt, jgama@fep.up.pt
[2] NOVA LINCS and Universidade Nova de Lisboa, Lisbon, Portugal
jresende@fct.unl.pt

Abstract. As the digital world grows, data is being collected at high speed on a continuous and real-time scale. Hence, the imposed imbalanced and evolving scenario that introduces learning from streaming data remains a challenge. As the research field is still open to consistent strategies that assess continuous and evolving data properties, this paper proposes an unsupervised, online, and incremental anomaly detection ensemble of influence trees that implement adaptive mechanisms to deal with inactive or saturated leaves. This proposal features the fourth standardized moment, also known as kurtosis, as the splitting criteria and the isolation score, Shannon's information content, and the influence function of an instance as the anomaly score. In addition to improving interpretability, this proposal is also evaluated on publicly available datasets, providing a detailed discussion of the results.

Keywords: Streaming data · Online · Incremental · Unsupervised · Anomaly detection · Ensemble · Kurtosis · Influence function

1 Introduction

The data revolution has branded the XXI century as the amount of data and heterogeneous platforms, responsible for mining information, constantly increase. Although this prospect provides meaningful patterns relevant in various fields such as healthcare and fraud detection, it also imposes privacy and security concerns, as well as efficient standardization to handle high speed and voluminous data, constantly expanding and evolving [1].

In anomaly detection, learning from data streams remains a challenge as it must consider an infinite and constantly changing nature that involves learning from imbalanced domains and forcing the evaluation process to encompass metrics that do not neglect the minority class [2]. Furthermore, the data flow depicted in most everyday scenarios matches the characteristics of a continuously evolving paradigm that introduces resource limitations and requirements for incremental and adaptive processing that delivers responses in a real-time fashion. As a result, concept drift, where the properties of the stream may change

B. Crémilleux et al. (Eds.): IDA 2023, LNCS 13876, pp. 274–286, 2023.
https://doi.org/10.1007/978-3-031-30047-9_22

over time, is a major point of discussion. An effective mechanism for alleviating concept drift and improving the representation of under-represented values is combining different base models in an ensemble approach. Ensemble methods for data stream mining have gained considerable popularity due to their high predictive capabilities, ability to confer robustness, and generalization [3].

Considering that anomalies are few and different compared to the rest of the data, the proposed method isolates anomalies, rather than profiling regular points, and attempts to determine the influence of each instance in the observed statistics in their groups. Unlike other methods that randomly select a splitting attribute, our approach favors the dimension that shows an increase in the fourth standardized moment (kurtosis), a measure of the heaviness of the tail of the distribution, as it is more likely to contain an outlier. This approach helps to tackle irrelevant dimensions that may lead to missing crucial anomalies [4].

In anomaly domains, it is relevant to identify abnormal or potentially defective events and localize the features that caused a distribution shift, which can be a critical step in the diagnosis. In this sense, our purpose is to design an online ensemble method that attempts to characterize the underlying distributions, isolating dynamics as they do not align with the expected behavior. Thus, the anomaly score is dictated by the complexity of the isolation process and the level of surprise given by the event's unpredictableness. Furthermore, as the influence of a sample that differs from the rest of the dataset tends to be larger than for normal points [5], the influence function of the proposed splitting heuristic will be used to score the deviation of an instance, measuring how deviating an example appears to be in a given distribution.

Therefore, the most significant contribution of this work is the design of a fully incremental and unsupervised anomaly detection strategy that focuses on identifying and curbing anomalous events by proposing online predictions where the algorithm responses are available sequentially over time. Moreover, to ensure a reliable representation of the evolving data characteristics that may lead to an obsolete model, this proposal also studies control mechanisms to examine the activity in the leaves and the consistency of the structure, that is, the ability to closely represent the observed behavior. Moreover, this procedure returns an anomaly score composed of three different metrics that could increase interpretability. Lastly, as it is imperative to attest to the effectiveness of the proposed methods in realistic scenarios, it also discusses and analyzes the results from testing this approach on publicly available real-time benchmark datasets with a distinctive number of points, dimensions, and anomalies.

Concisely, the paper is organized as follows: Sect. 2 provides a review of the current and most effective solutions that serve as motivation and inspiration to this work; Sect. 3 describes the implemented method from its principles to the basic unit of the ensemble and the anomaly score, completing with the pseudocode of an influence tree; Sect. 4 gathers the experimental trials and discussion of the results; finally, Sect. 5 closes this paper by stating final remarks and advancing future research directions and possible improvements.

2   Related Work

With the volume and speed of real-time data increasing, obtaining large amount of labeled data, specially in an imbalanced scenario, is a topic of interest. In recent years, the attention to methods such as autoencoders [6] or random forest [7] have changed toward unsupervised approaches such as isolation forests [8], an ensemble method, Local Outlier Factor (LOF) [9] as a density-based clustering solution, or One-class SVM [10], a kernel-based unsupervised learning technique.

Since the increasing search for real-time and adaptive streaming solutions, the community dedicated their scope to improving and adapting batch solutions to a continuous processing setting. As an example, *Pokrajak et al.* [11] proposed an incremental version of LOF, where the outlier factor is computed for each incoming data point, updating its statistics only with a few data points. Despite being an incremental method that can handle different densities and detect changes in data distributions, this solution demands high computational resources [12].

In real-time applications, predictions should be made online, where the algorithm identifies anomalies before incurring the actual event. Opposite to isolation trees, where both the split attribute and value are randomly selected to isolate abnormal instances at higher levels [8], *Putina et al.* [13] presents the Random Histogram Forest, an unsupervised and probabilistic approach, that builds a random forest based on the fourth central moment, also known as kurtosis, to guide the search for anomalous instances. In each leaf, the anomaly score, defined as the Shannon's information content, captures the likelihood of an example being an outlier [14]. Although it retains linear running time in the input size, this method does not implement an online streaming solution.

Although Isolation Forest [8] is an efficient method for anomaly detection with relatively low complexity, CPU, and time consumption, it requires all the data to build the forest, as well as pass over the dataset to assign an anomaly score. Thus, *Ding et al.* [15] adapted the isolation concept to streaming events using sliding windows. An important feature of this work is the ability to deal with concept drift by maintaining one input desired anomaly rate that determines if the detector is obsolete and if the latest data window should be used to build a new classifier.

Furthermore, *Tan et al.* [16] introduced a fast one-class anomaly detector for evolving data streams featuring an ensemble of random HS-Trees that does not require any data to build its structure. Unlikely Hoeffding Tree that induce decision trees and alter its structure dynamically by measuring the confidence of a splitting attribute heuristic as a new instance arrives [17], HS-Trees have a constant amortized time and memory complexity that records the mass profile of data operating with two consecutive windows where the learned profile is used to infer the anomaly scores of new data arriving in the latest window.

More recently, *Guha et al.* [4] proposed a non-parametric and unsupervised anomaly detection solution on streams based on the influence of an unseen point. This idea measures the externality imposed by that point by averaging the change in complexity. This ensemble of independent random-cut trees, named Robust Random-Cut Forest (RRCF), provides a dynamically maintained strategy that allows incremental updates with as few changes as possible. Comparatively to other proposals, where node split is uniformly chosen at random, RRCF determines the dimension to cut proportionally to the attributes' range, which makes this solution more resilient to irrelevant dimensions.

3   Online and Incremental Influence Forest

Similarly to the isolation-based method introduced by *Liu et al.* [8], this work recursively splits the data through a tree. In the original proposal, anomalies are expected to be quickly isolated, lying closer to the root, whereas normal instances are located deeper. This proposal attempts to identify the feature that influences the distribution's shape by measuring the concentration of values around the mean and the tails. The statistical measure that accounts for both peakedness, the concentration of probability mass around the mean, and heavy-tailedness, extreme values occurring with nonnegligible probability, is given by the standard fourth moment coefficient of kurtosis. The kurtosis of a random variable X ($K[X]$) is defined in Eq. 1, where μ and σ stand for the mean and standard deviation, respectively, and μ_4 represents the fourth central moment [18].

As it is perceived in Eq. 1, the standardized data is raised to the fourth power, which implies that instances within the region of the peak have a negligible contribution to the kurtosis score, while extreme observations outside the region of the peak (e.g., outliers) contribute the most. Moreover, since kurtosis is a standardized measure that describes the shape of the distribution, it is invariant to scale or location.

$$K[X] = E\left[\left(\frac{X-\mu}{\sigma}\right)^4\right] = \frac{E[(X-\mu)^4]}{E[(X-\mu)^2]^2} = \frac{\mu_4}{\sigma^4} \tag{1}$$

Furthermore, influence functions are a classic technique from robust statistics that assesses how the model parameters change as a training point significance is increased by an infinitesimal amount [19]. Hence, this technique promotes the knowledge of the impact of data contamination when a point mass or perturbation is added to a statistic value to deviate it from the expected distribution.

The kurtosis influence function, $IF(x; K(.))$, described in Eq. 2, which gives the name to this approach, provides a quantitative understanding of kurtosis ($K(.)$) when the contamination has occurred at point x. The expression, detailed by *Fiori et al.* [20], reveals that the contamination in both the tails and the center of the distribution increases this coefficient. Thus, as the influence function is unbounded, the kurtosis coefficient is sensitive to outlying values. Therefore, this formula estimates the contamination degree when an observation is added, helping to assess the impact of including a particular point and its degree of outlierness.

$$IF(x; K(.)) = \left(\left(\frac{x-\mu}{\sqrt{\mu_2}}\right)^2 - K(.)\right)^2 - K(.)(K(.)-1) - 4\frac{\mu_3}{\mu_2^{3/2}}\frac{x-\mu}{\sqrt{\mu_2}} \tag{2}$$

3.1   Influence Tree

Given a sample of data $X = x_1, ..., x_n$ of n instances from a d-variate distribution, to build a binary influence tree, the data space is recursively divided by selecting an attribute based on the heuristic measure, in this case, kurtosis, K. As this measure is expected to be affected by abnormal points, the highest value in this importance will indicate the presence of an outlier. However, it must be ensured that there is enough statistical evidence that the distribution has changed or that the number of instances processed is sufficient.

Similarly to the Hoeffding Trees [17], this approach wields Chebyshev's inequality, widely used in probability theory, to bound the tail probabilities of a random variable with finite variance. In particular, unlike other methods, this inequality can be applied to any distribution as long as it includes a defined variance and mean [21]. In other words, this will help to attest if, with a certain confidence, the heuristic measure has suffered an unexpected change.

Considering that X_a holds the highest observed K and, as depicted in Eq. 3, if the last observation added forced the $K(A)$ to differ from its mean in more than t units, the probability is, at most, the quotient of the variance and the squared value of its distance to the mean. In other words, if the difference from its mean is significantly higher than some value t, the attribute with the highest importance shows enough evidence that an extreme value has been added. Thus, it can also be used as a splitting attribute of the node.

$$Pr[|X - E[X]| \geq t] \leq \frac{V[X]}{t^2} \tag{3}$$

Concerning the splitting criteria, this approach is more similar to the Hoeffding trees proposal for mining high-speed data streams, which essays to guarantee, with high probability, that the attribute with the highest heuristic is the best choice [17]. In addition, this proposal is also inspired by the Random Histogram Forest that uses kurtosis as its splitting heuristic [13]. Furthermore, when it comes to the splitting value, similarly to most of the state-of-the-art approaches, this study randomly chooses a value in the range of values determined so far.

Finally, the leaves update the sufficient statistics for each attribute when an instance is added to the sample. These statistics include the variables to assess kurtosis, influence function, range, and the sample size of observed data points, filtering the incoming instances according to the observed dynamic, and only expanding when there is enough evidence that the distribution has shifted. In particular, these numbers are constant, which means the complexity does not depend on the number of instances, only on the number of attributes.

3.2   Dynamic Ensemble

One of the most common ensemble techniques is Bagging, which trains multiple base models with different points drawn by resampling the original dataset. In an online version, the forest trains several independent online influence trees to simulate the bootstrap process by sending a weight to each observation following a Poisson random variable [22]. This procedure adds another constant to define the number of trees in the forest that run in parallel, given their independent nature.

The online influence forest proposed in this work is structured incrementally for streaming data that is supposed to be continuous and infinite. As a single instance is not sufficient to make inferences about a population distribution, the nodes define a minimum number of instances. However, as the stream of events progresses, each tree structure cannot grow indefinitely, constraining the tree depth to a maximum depth bound, user-configurable, to limit the height of each tree. Consequently, as predictions are made online, the algorithm response becomes available as the event is being processed, leading to lower scores and not flagging anomalies until enough points have been examined.

Moreover, attesting if the tree structures are consistent with the dynamics present in the available sample, that is, whether they are considered obsolete, is also decisive to ensure that the ensemble is capturing the new data properties and maintaining its integrity. With the limited size of each tree, an indication that the structure is becoming saturated and unable to adapt to new instances is the number of leaves that reaches the maximum height and shows evidence that a split must occur. Therefore, the number of saturated leaves is supervised to determine when each structure must be redefined.

This proposal implements additional reframe strategies to control the forest's accordance with the data, as illustrated in Fig. 1. These strategies supervise the ability of the algorithm to reflect the current state of dynamics presented in the available data. Hence, the first strategy checks if, when a sample arrives at a leaf, the node is still active by checking the time between updates (Inactive Fig. 1). A leaf is considered inactive when, on average, it has been enough time to record twice the minimum number of instances in a node. In this case, it might suggest that the parent node has picked the wrong split, and the splitting value is reframed. Lastly, another approach has been studied to tackle the change in dynamic or when the tree is considered saturated (Saturation Fig. 1). This method maintains and reframes the tree structure from its root to the leaves by merging sibling nodes and replacing the parent node on higher levels. Thus, after the reevaluation, the number of leaves and levels reduce, and the original tree root, which holds the oldest distribution, is replaced.

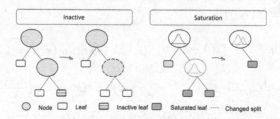

Fig. 1. Strategies to guarantee consonance between the tree structure and the observed data.

3.3 Anomaly Score

The anomaly score vital in unsupervised methods is another crucial component that attempts to quantify the degree of discrepancy from the expected behavior according to a set of principles. It is also the only way to comprehend how a particular decision has been made. In most cases, measuring each observation and assessing why it has been given a degree of unexpectedness can provide more insights about the problem at hand than the predictive performance. The demand for higher explainability levels arises as the incompleteness of the problem formalization increases [23]. In particular, cybersecurity and fault tolerance are some domains that often require high levels of interpretability.

As this system was inspired by isolation forest [8] and random histogram forest [13], integrating the influence function, each observation will be described with an isolation score, the Shannon information content, and the expected value of the difference between the influence function and its average, for each attribute.

In this regard, the output of our framework consists of a tuple specifying three metrics. Firstly, as defined in Eq. 4, the isolation score measures the average of the depth of each point from a collection of trees, $E[h(x)]$ and the average path length of an unsuccessful search in a binary search tree (BST). According to the expectation that anomalies will be filtered at higher levels, this formula returns a higher score for deviance values.

By defining a split based on the kurtosis statistic, when there is enough evidence that the distribution has changed, the leaves will become nodes, and the instances will not progress in the structure as the tree grows. The following scores will account for the density and the average poisoning when an observation is added to a leaf to survey the in-node distribution. Next, Eq. 5 calculates the Shannon information content, level of surprise, by measuring the probability of the cardinality of the leaf over the number of seen examples. Hence, anomalies will record higher levels of Shannon's information. Finally, the influence function is added to the equation. As this estimator deems the effect of adding one point to the distribution, this function returns the degree of contamination that a specific instance implies to the leaf. Thus, this statistic is related to the anomalousness degree of observation in a particular distribution. As this work is designed for multivariate analysis, the influence score, shown in Eq. 6, is given by the variability, over all attributes, of the kurtosis influence function when an example reaches a leaf $(IF(x; K(.)))$.

$$c(n) = 2H(n-1) - (2(n-1)/n) \qquad isolation(x; n) = 2^{-\frac{E[h(x)]}{c(n)}} \qquad (4)$$

$$P_{Leaf}[x] = \frac{|Leaf(x)|}{N} \qquad surprise(x) = \log\left(\frac{1}{P_{Leaf}[x]}\right) \qquad (5)$$

$$influence(x) = E[(IF(x; K(.)) - E[IF(x; K(.))])^2] \qquad (6)$$

Finally, the pseudo-code that illustrates the designed tree is represented in Algorithm 1.

Algorithm 1. kInfluence: Online and Incremental Influence Tree

---

Input : *node*: node of an influence tree;
        Ex: Example of a Stream;
        K(.): Splitting evaluation heuristic;
        δ: significance of choosing the correct splitting attribute;
        N_{min}: minimum sample size to test splitting significance;
        maxDepth: maximum depth a tree is allowed to grow.
Output: anomaly score metrics indexed by the row in stream
begin
    if *node is a leaf* **then**
        Update sufficient statistics (Subsection 3.1);
        Let $n \leftarrow$ sample size in leaf;
        if $n > N_{min}$ *and Ex is not empty* **then**
            if *node is inactive* **then**
                Reframe parent node (Figure 1);

            else
                Let X_a be the attribute with the highest $K(.)$;
                $p \leftarrow 1 - \frac{|K(A)-E[K(A)]|^2}{Var[K(A)]}$; (Equation 3)
                if $p < \delta$ **then**
                    Let $h \leftarrow$ depth of the tree and *saturation* $\leftarrow \frac{\#saturated\_leaves}{\#leaves}$;
                    if $h \geq maxDepth \wedge saturation > 0.5$ **then**
                        Check for consistency and reframe tree (Figure 1);

                    else
                        Split Attribute $\leftarrow X_a$;
                        Split Value $\leftarrow v \sim U(minX_a, maxX_a)$);
                        Let *node.left* $\leftarrow$ left child and *node.right* $\leftarrow$ right child;
        Let *score* $\leftarrow \{index : [isolation, shannon, influence]\}$ (Subsection 3.3)
    else
        score $\leftarrow kInfluence(node.left, Ex[node.X \leq node.v])$;
        score $\leftarrow kInfluence(node.right, Ex[node.X > node.v])$;
    return score;
end

---

4 Experimental Evaluation

Given the imbalanced nature of anomaly detection, the evaluation metric must be independent of the majority class. The precision, recall, and F1-score will be used in these experiments. While the recall is about completeness, concentrating on the percentage of correctly identified anomalies, precision calculates the rate of true positives over the detected anomalies, measuring the probability of correct detection of positive values and penalizing false alarms. Therefore, F_β is used to monitor several measures simultaneously. In this case, $\beta = 1$ assumes that precision and recall are equally important [2].

Table 1. Experimental trial metrics

Dataset				kInfluence		
Name	#points	#dim.	%outliers	Precision	Recall	F1
Ecoli[a]	336	7	2.6%	0.5	0.67	0.57
WBC[a]	278	30	5.6%	0.57	0.62	0.59
Ionosphere[a]	351	33	2.6%	0.30	0.55	0.39
Key Hold[b]	1883	1	0.006%	0.63	0.83	0.71
Key Updown[b]	5316	1	0.0008%	0.45	0.63	0.53
NYC[b]	10320	1	0.0005%	0.75	0.6	0.67

[a] http://odds.cs.stonybrook.edu/[NAME]-dataset/
[b] https://github.com/numenta/NAB/tree/master/data/realKnownCause/

Hence, an experimental evaluation was conducted on open-source datasets from different domains to attest to the performance of the proposed method. These examples, also considered in similar works, are available at Outlier Detection DataSets (ODDS) [24] and Numenta Anomaly Benchmark [25], a novel benchmark for evaluating online streaming anomaly detection applications. Table 1 summarizes the evaluation results where each row refers to a single dataset. The first four columns describe the data according to the number of instances, dimensionality, and the proportion of anomalies present. Besides the size difference in the first three datasets, these serve as multivariate analyses, and the last three as timeseries analyses. For these trials, as the algorithm parameters were kept constant, the procedures were conducted ten times to stabilize the outcomes, featuring a forest of 100 trees, 30 instances, and 95% confidence as the minimum number of points in the node and the probability to choose a split, respectively, and a maximum depth of 6. These values should be analyzed as a future direction, and each iteration will be plotted to understand this proposal's complexity and stabilizing times.

Although this proposal envisions an unsupervised learning method, an experimental evaluation, which should provide insights into how this approach behaves with distinct dimensions, sizes, or anomaly frequencies, compares the actual position of the outliers and the score information returned by our solution in a supervised manner. Given the online and incremental properties designed here, the algorithm requires a stabilizing time to accurately score points, as the first instances arriving will not be sufficient to make inferences. In this sense, to frame a realist scenario and not to compromise the performance, the outliers were randomly reorganized such that anomalies do not appear simultaneously or do not unfold in the first moments. As a result, only the timeseries datasets did not suffer any changes from the original form. Furthermore, as the similar approaches that inspired this work are designed with different characteristics or testing scenarios, their results will not be compared in this work. For the timeseries, despite working online and incrementally, our method missed one more anomaly than

RRCF [4] and added one more point as critical, rendering a higher false alarm rate or lower recall.

In a more detailed evaluation, as the last three datasets qualify as timeseries and facilitate the interpretation, they will be discussed closely, and the results and decision criteria will be analyzed.

The Key Hold dataset represents the timing of the key holds for several computer users, where the anomalies represent a change in the user. As critical anomalies are the ones that stand out amongst other points, it is possible to see different transitions reflecting an unexpected value for the key holds on that day. Figure 2a, from left to right, highlights the anomalies in red; the plot in the middle depicts the isolation, surprise, and influence score returned by our algorithm; and, finally, the last graph on the right investigates which observations classify with higher influence score as well as with higher surprise score. This figure shows that anomalies significantly differ from the rest of the values. Based on the definition of an anomaly, such points are more likely to appear on the upper side of the current trend. Furthermore, on the last graph, it is possible to see that the observations that score the highest ratings on the influence metric are usually points on the transition between values on similar timestamps, particularly after the first fortnight.

The Key Updown dataset describes keystrokes for several computer users, where the anomalies embody a change in the user. As assumed, abnormalities represent a significant transaction in their value. Figure 2b displays the anomalies in red; the metrics returned by our algorithm where, opposite from what is identified in the last trial, the influence score distinguishes points in the critical area. In particular, the last plot places outliers with influence, isolation, and surprise scores significantly more prominent than the surrounding observations. Moreover, as expected by the kurtosis and influence function, it is evident that orders that fall on the distribution's tails have their score increased, which is evident on the last graph where the tails of each timestamp are stressed as critical.

The NYC dataset corresponds to the number of NYC taxi passengers with five anomalies occurring during the NYC marathon, Thanksgiving, Christmas, New Year's day, and a snowstorm. The data file aggregates the total number of taxi passengers into 30-minute buckets. Therefore, to simplify, an anomaly often does not refer to a single observation but a time frame. Figure 2c illustrates the first timestamp with anomalous behavior issues. For instance, the first anomaly observed, the NYC marathon, has a lower value than the following numbers of passengers. From the metrics returned, depicted in the middle graph, there are not many anomalous points correctly predicted as critical. Since this dataset has many observations to be inspected with the naked eye, it is essential to examine the last plot to check which instances are flagged as dangerous. Thus, it is possible to see that the points with the most significant influence, isolation, and surprise scores have the highest number during the NYC marathon. Furthermore, despite their lower influence score, the isolation score spots the snowstorm as an outlier. The last anomaly correctly spotted was New Year's day, with a density and isolation score different from the peripheral. Another observation at the beginning of September registers similar scores as the New Year's day, which are identical by analyzing the recorded value. However, our approach was not able to detect Thanksgiving and Christmas days.

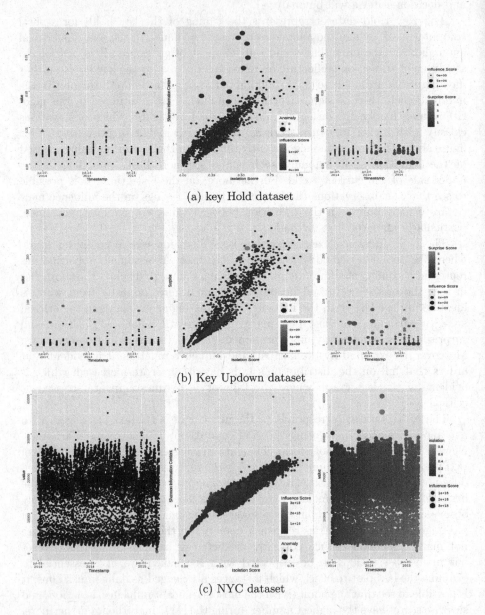

(a) key Hold dataset

(b) Key Updown dataset

(c) NYC dataset

Fig. 2. Exprimental plots

5   Conclusion and Future Directions

This paper proposes an online, incremental, and unsupervised forest for streaming anomaly detection that focuses on selecting the best attribute according to the kurtosis score. Praising the interpretability of the output, the model definition of an anomaly captures both the complexity of isolating an outlier, the surprise level when an instance reaches a node, and the contamination effect imposed by a discrepant observation.

Given this proposal's online and incremental nature, this approach is essential in studying anomaly detection in streaming data. Despite implementing methods to avoid inadequate splits or obsolete structures, the next step will be to study the ability to adapt to dynamic changes, as well as further evaluate the effects of the required parameters, from tuning the number of necessary instances in the node to the number of trees or maximum height of the structure. Furthermore, a future step will be to study the repeatability and comparison with the approaches that inspired this work on a similar testing evaluation scenario.

Therefore, the next future direction should include parameter tuning and the benefit of maintaining a window with the latest points to control the consistency of the forest while evaluating the impact on the false alarm rate and recall to maximize the performance, bearing in mind an extensive comparison with the identical studies in the literature.

Acknowledgements. This work has been supported by Fundação para a Ciência e Tecnologia (FCT), Portugal - 2021.04908.BD, NOVA LINCS - UIDB/04516/2020, CityCatalyst - POCI-01-0247-FEDER-046119, financed by FEDER, and by the CHIST-ERA grant CHIST-ERA-19-XAI-012, and project CHIST-ERA/0004/2019 and partially supported by the CHIST-ERA grant CHIST-ERA-19-XAI-012, funded by FCT. Also, this work is financed by the ERDF - European Regional Development Fund, through the Operational Programme for Competitiveness and Internationalisation - COMPETE 2020 Programme under the Portugal 2020 Partnership Agreement, within project City Analyser, with reference POCI-01-0247-FEDER-039924.

All the supports mentioned above are gratefully acknowledged.

References

1. Ramírez-Gallego, S., et al.: A survey on data preprocessing for data stream mining: current status and future directions. Neurocomputing **239**, 39–57 (2017)
2. Branco, P., Torgo, L., Ribeiro, R.P.: A survey of predictive modeling on imbalanced domains. ACM Comput. Surv. (CSUR) **49**(2), 1–50 (2016)
3. Gomes, H.M., Read, J., Bifet, A., Barddal, J.P., Gama, J.: Machine learning for streaming data: state of the art, challenges, and opportunities. ACM SIGKDD Explor. Newsl. **21**(2), 6–22 (2019)
4. Guha, S., Mishra, N., Roy, G., Schrijvers, O.: Robust random cut forest based anomaly detection on streams. In: International Conference on Machine Learning. PMLR, pp. 2712–2721 (2016)
5. Thimonier, H., Popineau, F., Rimmel, A., Doan, B.-L., Daniel, F.: Tracinad: measuring influence for anomaly detection. arXiv preprint arXiv:2205.01362 (2022)

6. Zhou, C., Paffenroth, R.C.: Anomaly detection with robust deep autoencoders. In: Proceedings of the 23rd ACM SIGKDD International Conference on Knowledge Discovery and Data Mining, pp. 665–674 (2017)

7. Breiman, L.: Random forests. Mach. Learn. **45**(1), 5–32 (2001)

8. Liu, F.T., Ting, K.M., Zhou, Z.-H.: Isolation forest. In: 8th IEEE International Conference on Data Mining. IEEE, vol. 2008, pp. 413–422 (2008)

9. Breunig, M.M., Kriegel, H.-P., Ng, R.T., Sander, J.: Lof: identifying density-based local outliers. In: Proceedings of the 2000 ACM SIGMOD International Conference on Management of Data, pp. 93–104 (2000)

10. Schölkopf, B.: Support vector method for novelty detection. In: Advances in Neural Information Processing Systems, vol. 12 (1999)

11. Pokrajac, D., Lazarevic, A., Latecki, L.J.: Incremental local outlier detection for data streams. In: IEEE Symposium on Computational Intelligence and Data Mining. IEEE, vol. 2007, pp. 504–515 (2007)

12. Salehi, M., Rashidi, L.: A survey on anomaly detection in evolving data: [with application to forest fire risk prediction]. ACM SIGKDD Explorations Newsl. **20**(1), 13–23 (2018)

13. Putina, A., Sozio, M., Rossi, D., Navarro, J.M.: Random histogram forest for unsupervised anomaly detection. In: 2020 IEEE International Conference on Data Mining (ICDM). IEEE, pp. 1226–1231 (2020)

14. Shannon, C.E.: A mathematical theory of communication. Bell Syst. Tech. J. **27**(3), 379–423 (1948)

15. Ding, Z., Fei, M.: An anomaly detection approach based on isolation forest algorithm for streaming data using sliding window. IFAC **46**(20), 12–17 (2013)

16. Tan, S., Ting, K., Liu, F.T.: Fast anomaly detection for streaming data. In: 22nd International Joint Conference on Artificial Intelligence, pp. 1511–1516 (2011). https://doi.org/10.5591/978-1-57735-516-8/IJCAI11-254

17. Domingos, P., Hulten, G.: Mining high-speed data streams. In: Proceeding of the 6th ACM SIGKDD International Conference on Knowledge Discovery and Data Mining (2002). https://doi.org/10.1145/347090.347107

18. Loperfido, N.: Kurtosis-based projection pursuit for outlier detection in financial time series. European J. Financ. **26**(2–3), 142–164 (2020)

19. Hampel, F.R.: The influence curve and its role in robust estimation. J. Am. Stat. Assoc. **69**(346), 383–393 (1974)

20. Fiori, A.M., Zenga, M.: The meaning of kurtosis, the influence function and an early intuition by l. faleschini, Statistica **65**(2), 135–144 (2005)

21. Lovric, M., et al.: International Encyclopedia of Statistical Science. Springer, Berlin (2011)

22. Oza, N.C., Russell, S.J.: Online bagging and boosting. In: International Workshop on Artificial Intelligence and Statistics. PMLR, pp. 229–236 (2001)

23. Doshi-Velez, F., Kim, B.: Towards a rigorous science of interpretable machine learning. arXiv preprint arXiv:1702.08608 (2017)

24. Rayana, S.: Odds library. http://odds.cs.stonybrook.edu/ (2016)

25. Lavin, A., Ahmad, S.: Evaluating real-time anomaly detection algorithms-the numenta anomaly benchmark. In: IEEE ICMLA, pp. 38–44 (2015)

APs: A Proxemic Framework for Social Media Interactions Modeling and Analysis

Maxime Masson[1]($\boxtimes$), Philippe Roose[1], Christian Sallaberry[1], Rodrigo Agerri[2], Marie-Noelle Bessagnet[1], and Annig Le Parc Lacayrelle[1]

[1] LIUPPA, E2S, University of Pau and Pays Adour (UPPA), Pau, France
{maxime.masson,philippe.roose,christian.sallaberry,
marie-noelle.bessagnet,annig.lacayrelle}@univ-pau.fr
[2] HiTZ Center - Ixa, University of the Basque Country UPV/EHU,
Donostia-San Sebastian, Spain
rodrigo.agerri@ehu.eus

Abstract. In this paper, we introduce a novel way to model and analyze social media interactions by leveraging the proxemics theory. Proxemics is the science that studies the effect of space and distance on interactions and behaviors. It is generally applied to the physical space but we hypothesize that adapting it to social media could provide a generic way to model and analyze the various kinds of interactions taking place in this virtual space. We designed a proxemic-based framework aiming to guide the analysis of data from a social media corpus that can be contextualized to a given application domain. We start by formally redefining proxemics in the context of social media and we leverage this redefinition to design a generic and extensible proxemic-based trajectory model dedicated to social media. We also propose novel proxemic distances applicable to this model. Finally, we experiment this proxemic framework on the field of tourism. The application to this use case demonstrates our framework's flexibility and effectiveness to model and analyze social media interactions.

Keywords: Social Media · Modeling · Proxemics · Social Web Analysis · Natural Language Processing

1 Introduction

In the last decade, we have witnessed significant growth and diversification of user-generated data sources. User-generated content (UGC) is a type of data that comes from regular people who voluntarily contribute to the community [10]. It can take many forms (e.g., *text, pictures, videos*) and originate from a wide variety of sources. Those are often found on the Web and range from traditional social media (generalists like *Twitter* or *Facebook*, or more specialized like *FourSquare*) to review sites (*TripAdvisor*, *Google Reviews*) as well as discussion forums or blogs. User-generated content is a significant opportunity

B. Crémilleux et al. (Eds.): IDA 2023, LNCS 13876, pp. 287–299, 2023.
https://doi.org/10.1007/978-3-031-30047-9_23

for researchers and companies, given that most of it, is free of charge and relatively easy to obtain [10]. Therefore, there is no longer a systematic need to buy commercial data or to conduct lengthy and costly data collection campaigns. Moreover, user-generated data sources are very diverse and cover a large number of application fields (such as *tourism, politics, housing*, etc.). Another benefit (which can also lead to multiple challenges) is the fact that those sources often contain massive amounts of data. This is the case when dealing with social media, one of the most prominent UGC source. In 2019, *Facebook* had more than 2 billion users and about 89% of young people in OECD countries were engaged in social networking online [19]. Social media have therefore become an essential resource for analyzing people's behavior around a wide variety of topics.

Due to the significant increase in online communications and the rise of user-generated content driven by social media, many research works are leveraging these new data sources to answer given requirements [20] using various methods and approaches. However, those are usually *ad hoc*, either intended for a given range of application domains or fit only for specific analysis requirements (e.g., designed to analyze particular behaviors, for example, *visitor flows* or *hate speech* only). Recently, we have seen more general processing methods being experimented [2] but they are usually low-level and complex to handle for non-specialist users (e.g., *raw Natural Language Processing modules*).

In this article, we propose a high-level framework aiming to guide the analysis of data from social media corpus for any application domain. It is driven by the theory of proxemics. Proxemics is the science that studies the organization of space and the effect of distances on interpersonal relationships [8]. It is based on the concepts of proxemic zones and dimensions and is generally applied to physical interactions [17]. We hypothesize that adapting proxemics to model and analyze social media interactions could be an effective way to provide a framework that is: (1) **generic** (e.g., *applicable to any social media*), (2) **flexible** (e.g., *adaptable to as much business requirements as possible*), (3) **domain-independent** (e.g., *compatible with any application domain*) and (4) **easy to manipulate** (e.g., *based on well known, tangible dimensions*).

The article is organized as follows. Firstly, we introduce our motivations and the reasons that led to the choice of proxemics as a basis for our framework. Secondly, we review the various applications of the proxemics theory and justify the choice of proxemics as the basis of our framework. Thirdly, our contributions are presented: (1) a formal redefinition of proxemics in the context of social media, (2) a generic model leveraging this redefinition and (3) 3 types of novel proxemic distances to characterize and evaluate interactions within social media. We finally experiment and evaluate our framework on social media *Twitter* with use cases coming from local tourism offices.

2   Background and Motivations

This work is carried out in the framework of a cross-border, regional project: the APs project (APs standing for *"Augmented Proxemics services"*). This project aims to collect, process, analyze and then value social media data related to the practice of tourism, visitor flows and the use of cultural heritage in the *Basque Country* area, a highly touristic territory spanning between France and Spain.

Currently, local tourism stakeholders (such as *tourism offices* or *destination marketing organization* - DMO) mostly use commercial datasets to study the practice of tourism in the region (*such as the ones from telephone companies or hotels*) but they are looking to diversify their data sources, in particular, by analyzing what is posted on social media. Several elements make social media attractive to study local tourism, they are: (1) massive (*large amounts of data are available*), (2) easy to access, (3) diverse (most aspects of tourism are covered, such as *activities, hotels, travel, etc.*), (4) inexpensive and (5) mostly up-to-date (*perpetual feed of new data*). To better illustrate the requirements of tourism stakeholders, here is an example of a scenario in which they want to make use of social media to help with decision making processes.

> *A beach-centered coastal city wants to diversify its touristic offer and is looking to invest in new types of activities. It has a limited budget and therefore wants to invest first in areas where the current tourists (who are coming primarily to enjoy the beaches) might be particularly interested in. To determine these, the tourism office uses social media to study tourists' behaviors.*

Although, as mentioned above, the main target of our project is tourism, we aim to move away from ad hoc processes and propose a framework that is generic and adaptable to any application domain. The APs framework life cycle (shown in Fig. 1) is divided in 4 major steps and leverage the proxemics theory to model and analyze social media data in order to answer various requirements, such as those described in the scenario above. A common model (*APs trajectory model*) is shared along all 4 steps, it is based on proxemics and will be described in Sect. 4.2. This model is central to our framework.

Collection (1) covers the entire process of finding and retrieving data, namely, it produces a corpus of social media posts from a specific dataset definition. In order to address this, we designed a generic and iterative methodology. It has been the subject of a separate paper, see [13] for more details. Users and posts collected in this step instantiate the raw layer of the model.

Transformation (2) refers to the various modifications and enrichments applied on previously collected data in order to increase their added value and prepare them for the next steps. Structured information extracted in this step enriches the model and helps to better characterize the posts.

Proxemic Analysis (3) leverages the previously instantiated and enriched model to compute proxemic metrics (called *distances*), those are raw indicators which are calculated depending on the requirements. This article will be mainly dedicated to this step, which is one of the originality of the APs framework.

Lastly, **Valuation** (4) allows the results of previous analysis (*raw indicators expressed as proxemic metrics*) to be viewed for end-users (*such as tourism stakeholders*). For tourism professionals, multidimensional maps could visualize trends and associations of themes and places in social media. We also envisage using the indicators produced as inputs for a tourism recommender system (*activities, places and itineraries*) or a system to connect tourists sharing interests.

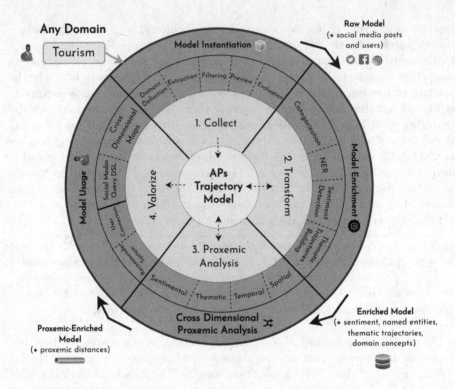

Fig. 1. Life cycle of the APs framework.

3 Related Work

A vast array of informative studies in the field of social media analysis exists, but these studies tend to be limited in two key areas. Firstly, numerous ones are tailored to a particular domain or specific requirements [1,5], making the methodology used challenging to apply to other use cases. Secondly, many of the studies carried out on social media analysis focus on specific stages of the analysis process and, therefore, do not propose a comprehensive approach. For example, NLP modules calibrated for short and informal messages [9]. Although some are generic, they do not provide a solution that covers the entire process, from data collection to usage by end-users. Thus, there is a need for a more holistic approach that encompasses the whole process of data analysis in social media. To build such approach, we leverage the proxemics theory.

Proxemics has its roots in the seminal work of the American anthropologist Edward T. Hall (*The hidden dimension*, 1966 [8]). He defines proxemics as *"the science that studies the organization of space and the effect of distance on interpersonal relations"* [8]. Hall is particularly interested in the notion of distance and the way it affects and regulates relations between individuals. He then goes further by proposing the notion of **proxemic zones** [8]. There are 4 core proxemic zones: (1) the **intimate zone** (*0 to 0.45m*) which is mainly used for physical contact, (2) the **personal zone** (*0.45 to 1.2m*) for interactions with very close people such as family or friends, (3) the **social zone**(*1.2 to 3.6m*) for discussions with strangers, and finally (4) the **public zone** (*more than 3.6m*) which is used when talking to a group. The perimeter of the zones given here is indicative, it can vary depending on many cultural, social and physical factors.

In 2006, Greenberg *et al.* extended Hall's definition of proxemics to introduce the notion of **proxemic dimensions** [6] (also called **DILMO** dimensions). They have identified 5 dimensions that can be used to express proxemics:

1. **Distance**: the measure of separation between several entities (*individuals, furnitures*). The distance allows determining the proxemic zones. It can be either continuous (e.g., *a distance in meters between two persons*) or discrete (e.g., *whether two persons are in the same room or not*) [6].
2. **Identity**: the set of features describing the individuality and the role of an entity in the space. For example *the name of a person, her age, gender, etc.*
3. **Location**: it describes qualitative aspects of the space [17]. The position of static entities (*like the layout of a room*) and dynamic ones (*individuals in the room*) in the space. It is usually expressed as x, y, z coordinates.
4. **Movement**: the change of location and orientation over time. It allows the proxemic space to become dynamic and evolves.
5. **Orientation**: the direction in which an entity is facing. Similarly to the distance, it can be continuous (e.g., *pitch, yaw and roll*) or discrete (e.g., *facing toward something, looking away from something, etc.*).

Proxemics can be studied at several levels: (1) the **individual level** (*how and why does an individual express specific traits and cognitive or affective states through her proxemic behavior* and (2) the **group level** (*how does the behavior of individuals has an effect on the group*) [14]).

Table 1. Overview of research works using proxemics for practical applications

Ref.	Space	Metrics	Level	Use of proxemics
[3] (2013)	Physical	Physical	Both	*Study teachers' behaviors.*
[23] (2012)	Physical	Physical	Individual	*Picture annotation.*
[18] (2014)	Physical	Physical	Individual	*Socially-aware robot navigation.*
[11] (2010)	Cyber	Physical	Individual	*Detect people reaction in a VR world.*
[21] (2021)	Cyber	Physical	Group	*Group behavior in a virtual workshop.*
[16] (2020)	Physical	Physical	Both	*Social distancing on human behavior.*
[15] (2021)	Physical	Physical	Individual	*Safety when using a VR headset.*
[7] (2021)	Cyber	Other	Individual	*Cybercrimes analysis.*
[12] (2019)	Cyber	Other	Individual	*Middleware configuration.*

Table 1 shows a handful of representative research works in which proxemics has been used. Most of these works use proxemics to analyze interactions in the **physical space** with **physical metrics**. For example, to help with the navigation of robots or drones [18] which is a case of individual-level proxemics. Proxemics has also been used in the field of education, to study the behavior of teachers and students in a classroom [3] both on an individual and collective level. Other use cases include picture annotation [23] or the evaluation of the impact of social distancing on people during the Covid-19 pandemic [16].

Other works have tried to apply proxemics to **cyberspaces** but still use **physical metrics**, for example in video games or in virtual reality worlds [11]. Recently, we also observed the use of proxemics to analyze **non-physical inter-actions** in cyberspace, this is called **digital proxemics** [7]. It is a novel research subject with few practical applications, it has been notably used for the analysis of cyber crimes and the enforcement of cyber law [7], as well as, to do reconfiguration in middleware [12]. We chose to use proxemics as the core of our framework due to several reasons:

- **Flexible**: proxemics is versatile and adaptable to various requirements. Its core concepts are broad and can be fitted to many use cases.
- **Fit for social media**: many aspects of proxemics can naturally be linked with social media components. The concept of *space* can be seen in many different ways (e.g., *physical space, VR space, social media space*).
- **Domain-independent**: it has no strong correlation to a specific domain, it is a very domain-unaware theory (*thematic genericity*).
- **Easy to manipulate**: Proxemic dimensions (such as *space, zones, movement, etc.*) are designed around the physical world. They are practical and tangible, therefore easier to understand, even by non-specialist users.

We will now explain how we redefined proxemics to provide a generic and flexible way to model and analyze interactions on social media.

4   A New Proxemic-Based Framework for Social Media

4.1   Formal Redefinition of Proxemics in the Context of Social Media

We start by formally redefining the theory of proxemics through its dimensions (DILMO), see Fig. 2. This is crucial because proxemics is primarily intended to be applied to physical interactions, whether in the real space (e.g., *a sensor detecting people moving in a room, etc.*) or in cyberspaces (e.g., *the proximity of characters in a video game or a virtual reality world, etc.*). Whereas, in the space of social media, interactions and zones are no longer physical. We define a social media **movement** m of size n as a tuple $m = (i, \{p_0, p_1, \ldots, p_n\})$

i is the **identity** associated with the movement, defined as:

$$i = (sm, u \lor (g, \{c_0, c_1, \ldots, c_e\}))$$

An identity can characterize (1) a single entity (with u the associated entity), for example, a particular social media user identified by their username. But also (2) a group of entities (with g the associated entity group defined by a number e of criteria c), such as users featuring common characteristics (e.g., *users considered as influencers because they have reached a certain number of followers*). Group identities have to be defined according to the studied requirements for the application domain (e.g., *what group of users are interesting to analyze for this domain*) and the limitations of the social media used (e.g., *which information can be extracted*). sm is the associated social media (e.g., *Twitter, etc..*).

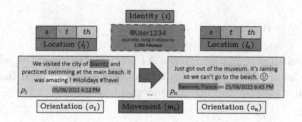

Fig. 2. Proxemics applied to social media posts

$\{p_0, p_1, \ldots, p_n\}$ is a list of n posts (*belonging to the associated identity i*) with:

$$p_i = (t, \{s_0, s_1, \ldots, s_j\}, \{th_0, th_1, \ldots, th_k\}, o)$$

t is the **temporal location** (*timestamp of when the post was issued*)

$\{s_0, s_1, \ldots, s_j\}$ is a set of j **spatial locations** s with: $s_i = (name, lat, lon)$ Spatial locations are instantiated from social media posts' geotags or by extracting named entities (*toponyms*) from their content. We define spatial locations by their *name* along with their GPS coordinates (*latitude* and *longitude*).

$\{th_0, th_1, \ldots, th_j\}$ is a set of k **thematic locations** th with:

$$th_i = (id, name, \{syn_0, syn_1, \ldots, syn_n\})$$

Thematic locations are domain-specific concepts extracted from the post content according to a hierarchy of concepts (e.g., *dictionary, thesaurus or ontology*) defining the domain of interest. Each thematic location th has a unique identifier id, a *name* along with a list of synonyms $\{syn_0, syn_1, \ldots, syn_n\}$.

Finally, o is the **orientation** defined by $o = (p, s)$ with p the polarity of the sentiment (*negative, positive, neutral*) expressed in the associated post and s the strength with $s \in [0, 100]$ (*0 being weak, 100 strong*).

Each movement m can be broken down into sub-movements called *trajectories*. The **spatial trajectory** (t_s) is the sequence of spatial locations s identified in a movement m. It provides a comprehensive overview of the places mentioned (e.g., *the places a person has visited*). It is a sequence of tuples because each post can be associated with any amount of spatial locations s.

$$t_s = (i, \{(s_0, s_1, \ldots, s_n), (s_0, s_1, \ldots, s_n), \ldots, (s_0, s_1, \ldots, s_n)\})$$

The **temporal trajectory** (t_t) is the sequence of temporal locations identified in a movement m. It is a sequence of temporal locations $temp$ (*we assume that a post can only be associated with a single timestamp*):

$$t_t = (i, \{temp_0, temp_1, \ldots, temp_n\})$$

The **thematic trajectory** (t_{th}) is the sequence of thematic locations th identified in a movement m. This movement highlights the domain-specific concepts associated with the identity (e.g., *the activities practiced by a certain category of tourists*). It is a sequence of tuples (*each post can have one or more thematic locations th*):

$$t_{th} = (i, \{(th_0, th_1, \ldots, th_n), (th_0, th_1, \ldots, th_n), \ldots, (th_0, th_1, \ldots, th_n)\})$$

This redefinition allows us to design the **APs trajectory model**, which leverage this formal redefinition of proxemics to model movements and interactions on social media in a generic and domain-independent manner.

4.2   The APs Trajectory Model

The APs trajectory model (*used in steps 2 and 3 of Fig.* 1) is designed in 5 parts, following the core dimensions of proxemics (DILMO). The UML class diagram of the model is shown in Fig. 3. This model is (1) **multidimensional** (*designed around the 5 dimensions of proxemic*), (2) **modular** (*use of all 5 dimensions is not mandatory, one can combine any number of them*) and (3) **extensible**.

The identity dimension allows modelling the studied population. Either a specific user (*IndividualIdentity*) or a group of users (*GroupIdentity*) featuring common characteristics or traits. The end-user has to define himself what groups of individuals he wants to study and define group identities accordingly.

The movement dimension provides the ordered sequence of posts belonging to a given identity. It gives a comprehensive view of an identity's activities on the chosen social media and allows linking posts together. It can, as mentioned earlier, be broken down into several sub-movements (*spatial, thematic, etc.*).

The location dimension models the posts along with their associated *Locations*. A given post can be in several locations at the same time, our model is therefore **ubiquitous**. Locations can be spatial, temporal or thematic. Thematic locations are defined by the end-user according to the studied domain and can be linked to external resources (*such as a domain-specific ontology, thesaurus or dictionary*). These resources provide additional information (e.g., *synonyms, definition, etc.*). Thematic locations can have relationships between them (e.g., *in the domain "tourism", "museum" is linked to "exhibition"*). When it comes to spatial locations, those are associated with a unique identifier linked to a geographic database. This allows for spatial locations to also feature relationships (e.g., *a city is within a region, itself within a country*). Locations are extensible, the end-user can define new types of locations if needed (e.g., a *political location* could be defined to model the political edge of various identities).

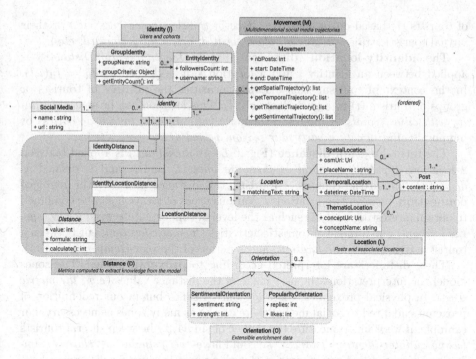

Fig. 3. The modular APs trajectory model using proxemics to model social media

The orientation dimension is also extensible. We have defined the *Sentimental Orientation* which models the overall sentiment of the associated post, but we could also use the *Popularity Orientation* which models the popularity of a given post on social media (*based on the number of replies and likes*).

The model comes with a set of constraints defined using OCL (*Object Constraint Language*) but we will not detail them here. Instead, we will now focus on the last proxemic dimension, the distance. This dimension differs from others because in the APs model, distances serve as indicators, computed metrics to extract knowledge and insights from the model.

4.3 Proxemic Distances for Social Media

We propose 3 proxemic distances applicable to our model (Fig. 3, *Distance*). These distances are intended to be raw indicators that allow the end-user to extract knowledge from the model to address specific analysis (*step 3 of Fig.* effig:wheel). The methods f used to calculate these distances are virtual and extensible, their implementations are **context-dependent** and differ depending on the use cases defined by the end-user.

The **inter-identity distance** (Fig. 3, *IdentityDistance*) is applied between 2 identities i_a and i_b and is defined as $d_i = f_i(i_a, i_b)$. In the context of tourism, this distance will measure the **similarity** between several tourists or groups of tourists (i) based on profiles (e.g., *type of traveler, home city, etc.*) or their virtual touristic trajectories (e.g., *places visited, activities performed, etc.*).

The **identity-location distance** (Fig. 3, *IdentityLocationDistance*) is applied between an identity i and a location l, it is defined as $d_{il} = f_{il}(i, l)$. In the context of tourism, this distance measures the **affinity** of tourists or groups of tourists (i) to a given location (l). This location can be spatial (*affinity with cities, points of interest*), thematic (*affinity with tourist activities*) or temporal (*affinity with a particular season, period of the week, etc.*).

The **inter-location distance** (Fig. 3, *LocationDistance*) is applied between 2 locations (*spatial, temporal or thematic ones*) l_a and l_b and is defined as $d_l = f_l(l_a, l_b)$. In the context of tourism, this distance measures the **level of connection** between two locations l. Various types of location can be combined to obtain interesting analysis such as the level of connection between a city (*spatial location*) and a range of touristic activities (*thematic locations*) or between a tourist activity (*thematic location*) and a period of the year (*temporal location*).

These distances are interpreted according to **proxemic zones**. The zones model the interpretations the user make of the distance values (*e.g., the metric used*). In physical proxemics, there are only 4 zones but in our redefinition of proxemics adapted to social media, there can be as many zones as necessary. For example, if we want to measure the degree of similarity between several tourists (*using an inter-identity distance*), we could have "*very similar*", "*relatively similar*" or "*not similar*" zones associated with a numeric range for distance values.

5   Experimentation

To experiment our APs framework, we chose the social media Twitter. We collected a corpus of posts originating from the *French Basque Coast*, which is broadly considered to be among the most touristic places in the region of interest of our project. The dataset we use contains 3,154 multilingual (*French, English and Spanish*) touristic tweets (*for details on the collection, see* [13]).

To instantiate our model, we rely on the *Thesaurus on Tourism & Leisure* of the *World Tourism Organization* [22] to define thematic locations. This resource covers roughly 1,300 touristic concepts, matched with tweets through entity linking. We use named entity recognition paired with *OpenStreetMap*, as well as geotags to capture spatial locations and the XLM-RoBERTa language model [4] to instantiate the orientation with the sentiment. Table 2 shows analysis requirements coming from a local tourism office and the versatility of proxemics to model them. By manipulating and crossing proxemic dimensions, we can easily express various domain requirements in a unified manner.

We will now try to model a more complex scenario (*the one from Sect. 2*) and evaluate it qualitatively. This scenario bends together 2 requirements from Table 2 (*row 1 and 2*). We are looking to establish which categories of touristic activities are mostly associated (*close* in terms of proxemic) with tourists going to the beach. The purpose is to give the beach-oriented city an idea of what types of beach-related touristic activities should be developed and invested in.

Table 2. Example of tourism analyses and how to model them using our framework

Analysis		D	I	L	M	O	Modeling
Sequence of touristic activities (e.g., what are activities often practiced sequentially)	DLM	•		•	•		**Distance** (*inter-locations*) between **touristic activities** (*thematic locations*) based on **sequences of activities** (*moving from one activity to another shorten the distance*). *Short distance = often chained activities*
Association of touristic activities (e.g., same day, same week)	DL	•		•			**Distance** (*inter-locations*) between **touristic activities** (*thematic locations*) based on the ubiquity between activities (*several activities in close time period*). *Short distance = activities often mentioned closely*
Attendance by city or activity (e.g., which tourist profile is mostly interested with which activity)	DIL	•	•	•			**Distance** (*identity-locations*) between **touristic activities** (*thematic locations*) or **cities** (*spatial locations*) and **types of tourist** (*group identities*) based on the frequency of mention. *The more a tourist group mention a given touristic activity or city, the shorter the distance between them is.*
Satisfaction of tourists about accommodations.	ILO		•	•		•	**Distance** (*identity-locations*) between **accommodation** (*spatial locations*) and **tourists' users** (*group identity*) based on the feelings expressed. *The more an accommodation (e.g.; hotel, camping, etc.) is mentioned positively, the shorter the distance is.*
Difference in behavior between old and young tourists.	DILMO	•	•	•	•	•	**Distance** (*inter-identity*) between **old tourists** (*group identity*) and **young tourists** (*group identity*) based on the similarity of their virtual **touristic trajectories**, as well as their common **touristic interests** (*sentimental orientation, what they both liked, disliked*).
Weather and time period impact on tourism practices.	DL	•		•			**Distance** (*inter-locations*) of **activities** (*thematic locations*) with **particular time periods** (*temporal locations*) or **weather** (*thematic locations*).

We set up an **inter-location distance** between **thematic locations**. To define those, we use relevant touristic concepts extracted from the WTO Thesaurus. The distance, normalized between 0 and 100, is defined as: $d_{location} = d_{association} + d_{sequence}$. $d_{association}$ represents the frequency two touristic concepts are found together in the same tweet (e.g., *a tourist tweets that he is* **surfing** *at the* **beach**). $d_{sequence}$ represents the frequency two touristic concepts have been found continuously in tourists' thematic movements (e.g., *a tourist tweets about being at the* **beach**, *then tweets again that he is* **surfing**).

Figure 4 shows the result of our modeling and analysis. We calculated distances between the thematic location *Beach* and each other thematic location, respectively (from the WTO Thesaurus). The size of each square displays the amount of time the thematic location was found in our dataset's posts and the black lines show the distance between locations (*thicker means shorter*).

At first sight, we can see several obvious short distances, such as the *Beach* location being very close to *Ocean*, *Summer* and *Sun*, but also to *Gliding Activities* (e.g., *Surfing*, etc.). It shows us that the distance we have implemented models phenomena and behaviors that are real and plausible.

Furthermore, other interesting analyses start to appear, like the relatively close distance of *Beach* with *Casino*. It seems that many beachgoers are also interested in the casino. The same goes for *Spectator Events* where street animations and spectacles also seem to be of great interest to beach tourists. All of these elements are valuable to help decision making processes of tourism stakeholders. Other representations around the tourism domain have been created regarding the use of soft mobilities, cross-border travel or business tourism in the region but we will not present them here due to space limitations. These representations provide domain experts, such as tourism offices, with deeper understanding and improved insight into the given domain of interest.

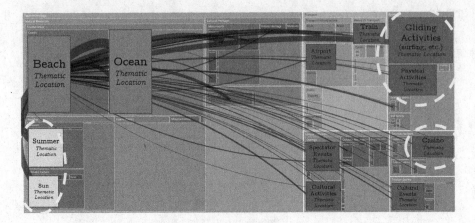

Fig. 4. Thematic map of the tourism space applied to tweets coming from a beach city

6   Conclusion

In this article, we introduced a novel proxemic framework dedicated to modeling and analyzing data from social media. This framework was designed as part of a local project aiming to analyze tourism practices in the *Basque Country*. To this purpose, we proposed a redefinition of the dimensions of proxemics fit for social media along with a multidimensional, modular and extensible proxemic-based trajectory model. We then defined 3 new proxemic distances that can be used to express the requirements of end-users. Our experiments with this framework in the domain of tourism have shown its flexibility to answer common requirements from tourism offices. We are now looking to experiment it on a larger social media corpus for several use cases: (1) to build a tourism recommender system able to suggest places, activities and itineraries fed by multiple proxemic distances, (2) for a tourist connection system (*connecting tourists with similar interests*) and (3) to use proxemics as a query language dedicated to social media corpus to abstract low-level IE (*Information Extraction*) concepts. Another step to take would be to experiment our framework on various domains (e.g., *politics, health*) to ensure its genericity.

References

1. Alalwan, A.A., Rana, N.P., Dwivedi, Y.K., Algharabat, R.: Social media in marketing: a review and analysis of the existing literature. Telemat. Inform. **34**(7), 1177–1190 (2017)
2. Albalawi, R., Yeap, T.H., Benyoucef, M.: Using topic modeling methods for short-text data: a comparative analysis. Front. Artif. Intell. **3**, 42 (2020)
3. Castañer, M., Camerino, O., Anguera, M.T., Jonsson, G.K.: Kinesics and proxemics communication of expert and novice PE teachers. Qual. Quant. **47**(4), 1813–1829 (2013). https://doi.org/10.1007/s11135-011-9628-5
4. Conneau, A., et al.: Unsupervised cross-lingual representation learning at scale. arXiv preprint arXiv:1911.02116 (2019)

5. González-Padilla, D.A., Tortolero-Blanco, L.: Social media influence in the COVID-19 pandemic. Int. Braz J Urol **46**, 120–124 (2020)
6. Greenberg, S., Marquardt, N., Ballendat, T., Diaz-Marino, R., Wang, M.: Proxemic interactions: the new ubicomp? Interactions **18**(1), 42–50 (2011)
7. Gunawan, A.B., Pratama, B., Sarwono, R.: Digital proxemics approach in cyber space analysis-a systematic literature review. ICIC Express Lett. **15**(2), 201–208 (2021)
8. Hall, E.T., Hall, E.T.: The hidden dimension, vol. 609. Anchor (1966)
9. Jiang, H., Hua, Y., Beeferman, D., Roy, D.: Annotating the Tweebank corpus on named entity recognition and building NLP models for social media analysis. arXiv preprint arXiv:2201.07281 (2022)
10. Krumm, J., Davies, N., Narayanaswami, C.: User-generated content. IEEE Pervasive Comput. **7**(4), 10–11 (2008)
11. Llobera, J., Spanlang, B., Ruffini, G., Slater, M.: Proxemics with multiple dynamic characters in an immersive virtual environment. ACM Trans. Appl. Percept. **8**(1), 1–12 (2010)
12. Luxey, A.: E-squads: a novel paradigm to build privacy-preserving ubiquitous applications. Ph.D. thesis, Université Rennes 1 (2019)
13. Masson, M., Sallaberry, C., Agerri, R., Bessagnet, M.N., Roose, P., Le Parc Lacayrelle, A.: A domain-independent method for thematic dataset building from social media: the case of tourism on Twitter. In: Chbeir, R., Huang, H., Silvestri, F., Manolopoulos, Y., Zhang, Y. (eds.) Web Information Systems Engineering (WISE 2022). LNCS, vol. 13724, pp. 11–20. Springer, Cham (2022). https://doi.org/10.1007/978-3-031-20891-1_2
14. McCall, C.: Mapping social interactions: the science of proxemics. In: Wöhr, M., Krach, S. (eds.) Social Behavior from Rodents to Humans. CTBN, vol. 30, pp. 295–308. Springer, Cham (2015). https://doi.org/10.1007/7854_2015_431
15. Medeiros, D., et al.: Promoting reality awareness in virtual reality through proxemics. In: 2021 IEEE Virtual Reality and 3D User Interfaces (VR), pp. 21–30 (2021)
16. Mehta, V.: The new proxemics: COVID-19, social distancing, and sociable space. J. Urban Des. **25**(6), 669–674 (2020)
17. Pérez, P., Roose, P., Cardinale, Y., Dalmau, M., Masson, D., Couture, N.: Mobile proxemic application development for smart environments. In: 18th International Conference on Advances in Mobile Computing and Multimedia, pp. 94–103 (2020)
18. Rios-Martinez, J., Spalanzani, A., Laugier, C.: From proxemics theory to socially-aware navigation: a survey. Int. J. Soc. Robot. **7**(2), 137–153 (2015)
19. Roser, M., Ritchie, H., Ortiz-Ospina, E.: Internet. Our World in Data (2015). https://ourworldindata.org/internet#the-rise-of-social-media
20. Shimada, K., Inoue, S., Maeda, H., Endo, T.: Analyzing tourism information on twitter for a local city. In: 2011 First ACIS International Symposium on Software and Network Engineering, pp. 61–66. IEEE (2011)
21. Williamson, J., Li, J., Vinayagamoorthy, V., Shamma, D.A., Cesar, P.: Proxemics and social interactions in an instrumented virtual reality workshop. In: Proceedings of the 2021 CHI Conference on Human Factors in Computing Systems (CHI 2021). Association for Computing Machinery, New York, NY, USA (2021)
22. World Tourism Organization: Thesaurus on tourism and leisure activities (2002)
23. Yang, Y., Baker, S., Kannan, A., Ramanan, D.: Recognizing proxemics in personal photos. In: 2012 IEEE Conference on Computer Vision and Pattern Recognition, pp. 3522–3529 (2012)

User Authentication via Multifaceted Mouse Movements and Outlier Exposure

Jennifer J. Matthiesen[1]([✉])(ID), Hanne Hastedt[2], and Ulf Brefeld[1]

[1] Leuphana University of Lüneburg, Lüneburg, Germany
{jennifer.matthiesen,brefeld}@leuphana.de
[2] Georg-August-University of Göttingen, Göttingen, Germany

Abstract. Gaining information about how users interact with systems is key to behavioural biometrics. Particularly mouse movements of users have been proven beneficial to authentication tasks for being inexpensive and non-intrusive. State-of-the-art approaches consider this problem an instance of supervised classification tasks. In this paper, we argue that the problem is actually closer to unsupervised one-class classification tasks. We thus propose to view behavioural user authentication as an unsupervised task and learn individual models using data from a single user only. We further show that, by being purely unsupervised, losses in performance can be counterbalanced by augmenting additional data into the training processes (outlier exposure). Empirical results show that our approach is very effective and outperforms the state-of-the-art in several performance metrics.

Keywords: User Authentication · Mouse Dynamics · Anomaly Detection

1 Introduction

User authentication is most commonly based on user-determinant keys, such as passwords or pin codes. In contrast to these traditional systems, biometric authentication [22,40] provides an additional layer of security. But these approaches come at the cost of storing sensitive data like fingerprints and require dedicated pieces of hardware.

An inexpensive and readily-available alternative is offered by *behavioural biometrics* [14]. In contrast to their biometric peers, these methods are non-intrusive and continuously analyze user behaviour for authentication during a session. Behavioural traits of users have been analysed in handwriting [9], voice [41], or keyboard and mouse dynamics [26]. Particularly the latter suggests itself for user authentication in computer-based systems since keyboard and mouse are considered standard equipment. While keystroke dynamics may contain sensitive personal information like passwords, mouse movements offer an implicit and non-sensitive measurement of idiosyncratic behaviour [17]. In fact, Rodden et al. [29] show that eye and mouse movement are significantly correlated and conclude that mouse movement serves as an appropriate proxy to address implicit user behaviour.

Mouse movement dynamics are typically handled in a fully-supervised multi-class or multi-label setup, where class labels are identified with user IDs. While being purely

B. Crémilleux et al. (Eds.): IDA 2023, LNCS 13876, pp. 300–313, 2023.
https://doi.org/10.1007/978-3-031-30047-9_24

supervised can add to predictive accuracy and detection performance, there are important limitations with this formalization. Firstly, the above approaches are often biased towards the data of other users present in training due to learning discriminating functions. Secondly, maintaining a multi-class approach in practice is close to being infeasible as every new user requires a full re-training of all models.

By contrast, we consider user authentication in an unsupervised approach, which learns a user's representations by using only the data of this very user. Learning individual models of normality for every user allows to create features which are independent from other users. This allows the model to generalise better with respect to the target user, especially in the presence of unknown users who have not been seen during training.

In this paper, we propose a novel methodology for user authentication using mouse dynamics and a deep one-class setup. Our contributions are as follows: (i) We phrase user authentication as a deep one-class machine learning problem that can enhance user authentication and (ii) show the effectiveness of using a multifaced input to extract appropriate features from mouse data. Finally, (iii) we visualise the individual and relevant parts of the user's mouse trajectory to gain an understanding of our approach, showing our model indeed focuses on characteristics previously known to be important from hand-crafted features but unattended in unsupervised approaches so far.

2 Related Work

Many studies have shown that mouse dynamics can deliver insights into a user's mood [42], satisfaction and frustration [6] or mental state [13]. Consequentially, mouse movement has received much attention in behavioural analyses [2,3,24,38].

The identification of users based on mouse strokes has been first investigated on the example of mouse-written signatures [9]. Many algorithmic approaches in dealing with mouse movement rely on hand-crafted features [11,14]. Such representations are often extended by peers to increase expressiveness and/or incorporate additional characteristics like the number of pauses or pause length [25]. Matthiesen et al. [25] showed that using a fixed feature set for every user does not cover all of them equally. Similar conclusions were made by [34]. Therefore, an individual feature set for every single user is required. Neural approaches try to overcome the dependency on hand-crafted features while mapping the input data to a feature space using corresponding objectives. Using mouse dynamics to tell users apart can have two main objectives: User identification and user authentication. Many of the previous work addresses the topic of user identification, which is to detect the right user in a set of all users [1,11,14,21,36]. The usual setup for this is a supervised multi-class classification. In contrast, user authentication underlies a binary decision, e.g. is the target user/ is not the target user [7,21,26,34,35,39]. Note that the common approach here is still supervised, i.e. using data of both classes. Chong et al. [7] are the first to investigate the potential of deep neural networks for mouse dynamics. They investigate multiple network architectures while casting the problem as a supervised multi-label problem. Applying a similar architecture, [1] propose a one-dimensional convolutional network (1D-CNN) for modelling temporal aspects of mouse movement. They train the models in a supervised one vs. rest manner using a binary-cross entropy loss.

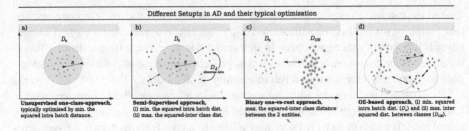

Fig. 1. Setups of AD, from only using the target class to incorporate an additional dataset.

Multi-class approaches imply retraining the whole model when adding/deleting users and are not feasible in dynamic environments. Binary one-vs-all strategies, on the other hand, are often biased towards the seen anomalies. Thus, we argue that an unsupervised anomaly detection (AD) method is more suitable for mouse-dynamics-based user authentication in real-world applications.

In the context of AD, unsupervised one-class approaches are appealing because they find minimum volume summarization of data at hand through hyperplanes [33] or hyperspheres [37]. Neural peers [20,30] introduce improvements by identifying anomalies through their alterity in feature space. This is often done by including additional data that are not part of the target concept, into the training process, for example by semi-supervised learning [16], pre-training [19,28], reference data [27,28] or outlier exposure (OE) [19]. We will make use of the latter in the remainder. An overview of common setups incorporating additional data into AD and their typical optimisation can be found in Fig. 1. For example, in contrast to a binary one-vs-rest strategy, the OE-approach extract rich descriptive features from the mouse trajectories instead of focussing on increasing the distance between the two entities (normal and anomalous data). Note that the concept of OE in AD can be found widely in the literature under various terms, such as reference dataset [27,28], auxiliary or OE dataset [19,31]. This auxiliary dataset enables the anomaly detector to generalize better for unseen data.

3   Representing Mouse Trajectories

Formally, a mouse trajectory is given by a sequence of spatial (x, y) coordinates ordered in time τ. In addition to the spatio-temporal information, mouse data contains events, for example $e \in \{\emptyset, c_L, c_R, c_M, s_{up}, s_{down}\}$, with left (c_L), right (c_R) or center (c_M) clicks, up $\{s_{up}\}$ and down $\{s_{down}\}$ scrolls, or $\emptyset$ in case there is no event. We represent a mouse trajectory as a sequence $x = \langle (\tau_1, x_1, y_1, e_1), \ldots, (\tau_T, x_T, y_T, e_T) \rangle$.

Mouse data records consist likewise of movements for interacting with the application as well as idiosyncratic movements. In the following, we aim at devising a representation that is as independent as possible from actual user interface (UI) and rather aim at capturing *how* a user moves the pointer to a certain location instead of where exactly an action has been performed. While such velocities are translation invariant, they also render certain patterns almost undetectable (e.g. loops).

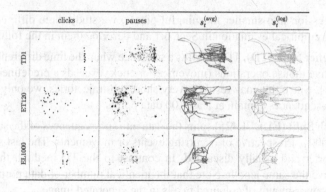

Fig. 2. Views of mouse movement with different splitting criterions.

3.1 Image-Based Tensor Representations

We propose to represent different *views* of mouse trajectories (e.g., trajectory, speed, click, pause) as an image, which allows to access shape of motion as well as characteristic patterns like loop or hesitation [4,7], see Fig. 2 for examples. A convolutional neural network (CNN) can detect edges very well and is therefore perfectly suited for such shapes. For the *trajectory view*, sub-sequences of the trajectory are re-scaled, plotted and saved as images. We adapt the size of the plot to the range of the respective trajectory to assure no bias from the positioning on the screen. Later, those images will be transformed into a multidimensional tensor to serve as an input for the network. To maintain the temporal information, we encode the *speed of the movement* with a colour interval, where the colour is determined by the actual speed of movement s_t at that position. To encode the speed value, we test two different normalisation approaches. We report on experiments with different normalisations in Sect. 6.1. Both ground on the speed $s_t = \frac{d_t}{\tau_t}$ of the movement, where d is the Euclidean distance, but are normalized (i) by the average speed: $s_t^{(avg)} = \frac{s_t}{\frac{1}{T}\sum_{t=1}^{T}\frac{d_t}{\tau_t}}$ and (ii) with a log-variant with

$\tilde{s}_t = \log(1 + s_t)$ $s_t^{(log)} = \frac{\tilde{s}_t - \tilde{s}_{\max}}{\tilde{s}_{\max} - \tilde{s}_{\min}}$, respectively, where $\tilde{s}_{max} = \max_t \log(1 + s_t)$ and $\tilde{s}_{min}$ analogously. A *click view* is simply containing indicators at click positions which are visualized by black crosses in the figure. The *pause view* contains the length of the pauses at the observed positions and is visualized by circles with radii corresponding to the length of the pause. We scale every pause so that the radius of the smallest pauses starts at a fixed radius. The different layers are aggregated to a multi-dimensional tensor, using one channel each, and serve as input to the model. We experiment with several combinations of input information and report the results in Table 1 (left).

3.2 Splitting Sessions

The overall sequences of our data cover a whole session of a user. Therefore we divide the total session into sub-sequences. The length of those sequences in the images is another aspect of representing mouse trajectory data. Since it is not obvious how to split a long user session into smaller meaningful pieces, we study three different splitting criteria in the empirical evaluation in Sect. 6.1 and describe them in the following.

Time Difference Split (TD). TD [7] splits a sequence when the time difference between two consecutive mouse operations (movement or click) exceeds a predefined threshold $\rho \in \{1s, 60s\}$. Since this may result in very short sub-trajectories, we only split if the resulting sub-sequences contain at least 100 data points.

Equal Length Split (EL). EL [25] splits the data into sub-sequences of the same length, $\omega \in \{200, 1000\}$, irrespective of occurring events or movements. The last sequence is naturally shorter and usually discarded. In contrast to the TD method, the resulting sequences have the same length. Note that the identical number of data points does not result in the same amount of coloured pixels in the generated image.

Equal Time Split (ET). There exist two main ways of recording trajectories: (i) record the position in equal time stamps, so that a static position will results in duplicates coordinates or (ii) recording on movements, meaning the distances between consecutive points will not be the same. Since the latter is the case in the *Balabit* data, we introduce an additional method: The Equal Time Split (ET). It is a temporal analogy to the previous splitting criterion and splits the trajectory after a fixed amount of time. We experiment with the thresholds $v \in \{10s, 120s\}$. Since the used mouse data is not recorded using equal time stamps but rather recorded on movements, this splitting method will not result in equally sized sub-sequences. Although this extension is straightforward, there does not seem to exist related work on this method.

4 Deep Anomaly Detection and Outlier Exposure

The main goal in an AD task is to obtain a feature map ϕ which separates the normal data from the outliers either linearly [33] or spherically [37]. Following the latter approach, a neural variant of the so-called Support Vector Data Description (SVDD), maps the data into a feature space $\mathcal{F}$ and finds the minimal enclosing sphere with center c and radius $R > 0$ that contains the majority of points. We derive two important characteristics for features for our one-class setup similar to [28].

(i) **Compactness.** Following the above-described cluster assumption [32], we want a similar feature representation extracted from the same class lie compactly in an enclosing hypersphere within feature space. Similar to the SVDD, the objective is to minimise the R of the hypersphere will result in reducing its volume and a more compact representation. However, if no further constraints are included this will directly result in the *hypersphere collapse* [8], when all data is mapped to the same point.

(ii) **Descriptiveness.** A feature map that gives rise to compact representations may not be rich enough to distinguish other users. We thus aim to devise a rich feature representation that is general enough to not only summarize individual user data well but, at the same time, allows us to identify other users because of their unique traits in moving the mouse. Thus, we need *descriptiveness* but cannot give up on compactness either (cf. [27]). Producing descriptive features is likewise a desired characteristic in multi-class classification, where those features would ensure a large inter-class distance.

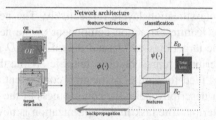

(a) The network ϕ extracts features from both Balabit and TWOS, while additional layers of ψ are used for the auxiliary data to calculate the descriptive loss E_D.

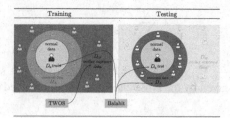

(b) For training we use the data of the target user (Balabit) and a sample of OE data (TWOS). For testing we use the legal users from Balabit.

Fig. 3. Depiction of the network (a) structure and of (b) the process for training and testing.

4.1 Outlier Exposure

The idea of OE originates in the observation that when learning a target concept, myriads of labelled examples exist that live in the same space but are known to *not* match the target concept [19]. While this insight borders on triviality, it is particularly powerful in unsupervised learning tasks like one-class and density estimation problems. Instead of only feeding observations of the desired target concept, additional data from possibly very different origins and sources is made available to the learner that now faces contrastive tasks: Ultimately, the goal is to provide a minimal description of the desired target concept but additionally, there is a classification problem that needs to be solved simultaneously using only the auxiliary data. The goal of the learning process is to identify a set of features that not only accounts for minimal description of the target concept but also induces high predictive accuracies on the auxiliary data. Note that OE-based approaches are generally classifies as unsupervised methods across literature [19,27,28,31] since the standard approach uses only data of one data (normal data) from the target dataset during training.

5 Authentication of Users

For the underlying architecture for our model, we consider the AlexNet CNN architecture [23]. We modify the network and separate it into the two parts ϕ und ψ for feature extraction and classification layers respectively. The network architecture is depicted in Fig. 3a. The feature extraction component of the network $\phi : \mathcal{X} \to \mathbb{R}^p$ acts on both sources, the target classes D_u and the OE data D_{OE}, to render learning compact as well as descriptive representations feasible. While the auxiliary data then branches into a standard feed-forward classification component $\psi : \mathbb{R}^p \to \mathbb{R}^{|\mathcal{Y}|}$ with a final softmax layer for descriptiveness, the compactness of the user data is evaluated by a variance-based criterion. The task in the optimization is now to find an appropriate feature extraction ϕ, such that the classification error and variance of user data is small, while simultaneously ensuring a compact representation of the features for D_u. This is achieved by deploying dedicated loss functions for controlling compactness and descriptiveness, respectively, and minimizing the two losses simultaneously (cf. [28]). The loss-controlling compactness measures the squared intra-batch distance

$$E_C = \frac{1}{N} \sum_{n=1}^{N} (\phi(\boldsymbol{x}_n; \theta) - \bar{\boldsymbol{x}}_{\neg n})^2, \tag{1}$$

with mean $\bar{\boldsymbol{x}}_{\neg n} = \frac{1}{N-1} \sum_{j \neq n} (\phi(\boldsymbol{x}_j; \theta))$ of the leave-one-out set $D_u \setminus \{\boldsymbol{x}_n\}$. The descriptiveness loss is given by the cross entropy over all involved classes $\mathcal{Y}$, given by

$$E_D = -\frac{1}{M} \sum_{m=1}^{M} \sum_{\bar{y} \in \mathcal{Y}} \delta_{\bar{y}, y_{N+m}} \log(\psi(\boldsymbol{x}_{N+m}; \theta)), \tag{2}$$

where δ is the Kronecker delta. The joint objective function for the entire architecture is given by aggregating Eqs. (1) and (2). We minimize $E_C(D_u) + \lambda E_D(D_{OE})$, where $\lambda > 0$ is a balancing term. In addition to user data D_u, we introduce an auxiliary and labelled M-sample $D_{OE} = \{(\boldsymbol{x}_{N+1}, y_{N+1}), \ldots, (\boldsymbol{x}_{N+M}, y_{N+M})\}$ with $\boldsymbol{x}_{N+m} \in \mathcal{X}$ and $y_{N+m} \in \mathcal{Y}$ for $1 \leq m \leq M$ and $\mathcal{Y}$ denotes the set of (arbitrary) class labels of the auxiliary data. Recall that both user observations $\boldsymbol{x}_n^{(u)}$ and auxiliary data $\boldsymbol{x}_{N+m}$ live in the same space $\mathcal{X}$ for all n, m. Here, we propose a neural architecture that combines learning a compact representation of target data D_u *and* a descriptive feature space on target and auxiliary data.

6   Empirical Results

We evaluate on the Balabit Mouse Dynamics Challenge [12] for sampling D_u and incorporate instances of the Wolf of SUTD (TWOS) [18] data as D_{OE}. *Balabit* contains mouse movements from 10 users from 65 sessions between 13640 and 83091 data points each, recorded during a set of unspecified but common administrative tasks. Since the screen resolution is not given, we normalize the trajectories based on the maximum coordinates. The TWOS data is the outcome of a gamified competition among competing companies over five days. It consists of 320 h of activity of 24 users and comprises a mouse, keyboard and other actions and logs, where we only use legal mouse movements in our experiments.

Setup. The two parts of the network ϕ and ψ are trained jointly with samples from both D_u and D_{OE}. For each user u, we train an individual model using only data

Table 1. Results of the experiments on different views (left) and splitting criteria (right). * Results are averaged over 9 users, since some images of user 07 resulted in numerical issues for those representations (see further details in Sect. 8).

	avg. AUC	avg. EER		Avg. No. Img.	avg. AUC	avg. EER
trajectory	0.670	0.420	TD1	1074	0.517	0.017
s_{avg}	0.656	0.268	TD60	77	0.576	0.476
s_{avg}, pause	0.698*	0.697*	ET10	2861	0.516	0.867
s_{log}, pause	0.710*	0.179*	ET120	344	0.606	0.600
s_{avg}, pause, click	0.755	0.272	EL200	950	0.723	0.409
s_{log}, pause, click	0.777	0.240	EL1000	188	0.777	0.240

from that user $D_{u_{train}}$ (Balabit) plus a sample from the 24 users as additional OE data D_{OE} (TWOS). More formally, let the size of a batch be n. Then, for every i^{th} sample $x_i^{D_u} \in \mathbb{R}^k$, where $1 \leq i \leq n$, we calculate the distance between the networks output and the rest of the batch. For every i^{th} sample $(x_i^{D_{OE}}, y_i) \in \mathbb{R}^k$, where $1 \leq i \leq n$, we calculate a loss for each class label y_i and sum the result. Hyperparameters are found via grid search and given by $\lambda = 1.0$, 300 epochs and a learning rate of $\eta = 0.0001$ on balanced batches containing 100 user and OE samples. We observe that a rather low learning rate results in better performance since it prevents overfitting on the OE data while still assuring convergence on the compactness loss. At test time, we use independent data of the target user $D_{u_{test}}$ and the nine remaining users from Balabit, similar to [15,20,28,30]. Note that the model has never seen the other users from *Balabit* during training. In this way, we ensure that the model can even distinguish from unseen users. This allows scalability and does not require retraining of the model, even if more new users are added.

6.1 Results

We first execute preliminary experiments, learning the optimal input and splitting strategy as described in Sect. 3.2. Using the resulting best-performing representation of mouse trajectories, we train the presented model in two different setups: (i) To show the influence of the utilisation of the OE data, we first train the model without the usage of the additional data. (ii) We build upon that and show the improvement in performance reached through the usage of OE data. We report average Areas under the ROC curve (AUCs) and equal error rates (EERs) over five repetitions.

Results for Optimal Representation. Table 1 (left) shows the results for different views (layers of input tensors), presented in Sect. 3. The results lay out that including likewise the pause and click *view* in the tensor leads to higher detection rates and the log-average performs slightly better than the global average. Table 1 (right) shows the results for different splitting criteria, also presented in Sect. 3. Firstly, the table nicely shows that the heuristics lead to considerably different numbers of training instances, however, recall that fewer instances contain longer parts of the respective user sessions.

Table 2. Detection performance per class. Results are retrieved over 5 random seeds.

	EER ($\downarrow$)				AUC ($\uparrow$)			
	[30] (features [25])	[30] (images)	ours D_u	ours $D_u \cup D_{OE}$	[30] (features [25])	[30] (images)	ours D_u	ours $D_u \cup D_{OE}$
user 07	0.302	0.399	0.414	0.181	**0.879**	0.542	0.714	0.857
user 09	0.207	0.608	0.451	0.237	**0.911**	0.474	0.661	0.80
user 12	0.629	0.263	0.325	0.090	0.250	0.607	0.594	**0.838**
user 15	0.532	0.437	0.152	0.219	0.426	0.550	0.555	**0.714**
user 16	0.552	0.492	0.418	0.418	0.424	0.515	0.682	**0.720**
user 20	0.402	0.463	0.716	0.290	**0.788**	0.496	0.490	0.728
user 21	0.476	0.332	0.281	0.138	0.548	0.554	0.625	**0.825**
user 23	0.619	0.403	0.216	0.305	0.267	0.577	0.662	**0.730**
user 29	0.619	0.500	0.533	0.320	0.346	0.472	0.7	**0.813**
user 35	0.609	0.420	0.265	0.200	0.283	0.554	0.662	**0.747**
Mean	**0.495**	**0.432**	**0.378**	**0.240**	**0.512**	**0.534**	**0.634**	**0.777**

While most splitting methods are just slightly better than random guessing, the EL split performs notably better. With $\omega = 1000$, decreases the EER by almost half.

Results for User Authentication. We now use the best-performing representation to compare to related work. As a baseline, we use the performance of the deepSVDD proposed in [30] as well as the user authentication approach using features proposed for Balabit from [25]. To show the influence of OE data, we also compare our approach to a variant that does not leverage OE data. To prevent hypersphere collapse, we incorporated an additional regularizer into Eq. (1) similar to [37]. To additionally show the benefit of our representation over state-of-the-art handcrafted features, we train a deepSVDD on features taken from [25] and another one on our tensor representation.

The results are shown in Table 2. Interestingly, the baseline on features and tensors leaves a mixed picture in terms of AUC (right part of the table). For users 07, 09, and 20, hand-crafted features outperform the tensor-based deepSVDD as well as the proposed approach. For the other users, the image-based representation is favourable, often by a large margin as seen for users 12, 23, or 35 which is also reflected by a slightly better average AUC over all users. However, even without including OE data, our proposed approach performs either on par or improves over the stronger baselines. This result impressively improved by including OE. Our proposed approach already constitutes an improvement in AUC by a factor of 1.2 over the baselines when no OE data is included in the training. Note that in this case, a regularizer has to be added to Eq. 1 to avoid a collapse of the hypersphere. The results for including OE are even better and raise the improvement in AUC by a factor of 1.5.

7   Visualisation of Important Information of the Mouse Dynamics

Since the performance using all three *views* and the splitting method EL1000 result in the best authentication performance, we can now investigate which parts of the input lead to creating compact and descriptive features for each user. To achieve this, we

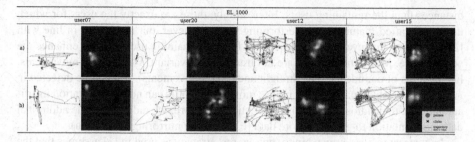

Fig. 4. Two examples per user of trajectories images and their LRP visualisation using the splitting criterion EL1000. We refer to the left example as (a) and the right as example (b).

utilise the layer-wise relevance propagation (LRP) [5]. It can help identify the parts of the input while highlighting input features that were decisive for the network's decision. The relevance R of every neuron is computed as follows: $R_i^{(l)} = \sum_j \frac{a_i w_{ij}}{\sum_{i'} a_{i'} w_{i'j}} R_j^{(l+1)}$, where $R_i^{(l)}$ and $R_i^{(l+1)}$ represent the relevance score of the neurons i and j in the layers l and $l + 1$ respectively. The activation of neuron i is represented as a_i and the weight connecting neuron i and j as w_{ij}. The LRP heatmap is then obtained by applying this principle to all layers. In addition, we implement the z^+-rule and a relevance filter as suggested in [10]: We adapt the threshold value for the filter to $k = 0.05$.

The results are shown in Fig. 4. Note that the trajectory is reconstructed for better legibility. The images used as input carry one channel per *view*, where different shadings are hard to detect for the human eye. The trajectory of user 07 (a) shows clearly the advantage of incorporating pauses into the input image. It can be seen that the pauses got a much higher relevance score than the clicks (black cross). Locally overlapping occurrences of pauses and clicks are likewise relevant. In contrast, the clicks are much more relevant to user 20 as can be seen in example a). When no clicks are made, the pauses are getting more relevant. In [7] only the plotted trajectory was used as input for the CNN. It was shown that the edges are the relevant element for the decision process of the network. Added pauses and clicks carry even more relevant information for user authentication and should not be left out in image-based deep learning approaches.

8 Discussion and Limitation

In this study, we cast user authentication base on mouse dynamics as a one-class problem. Multiple *views* of the trajectories are used as input to a CNN for extracting features using the objective of compactness and descriptiveness. Related work using deep neural networks for mouse trajectory data view the problem as a purely supervised task and often rely on pieces of information that is not always present, such as screen resolution [1,7]. We remove this implicit dependency on the screen to avoid identifying users based on their personal preferences or hardware but still report state-of-the-art results.

In our setup, we reached the best performance by equally weighted both losses, setting $\lambda = 1.0$. We did not detect a large difference in performance though. With the EL1000 split the trajectories of users 07, 09 and 20 cover much shorter (pixel-wise) distances than the remaining users. Interestingly, these are exactly the users for which the hand-crafted features were performing well. However, our results are in line with [25] and show that even shorter sequences for the remaining users did not enhance the performance. Mouse trajectories are not translation invariant. While some movements, like patterns of confidence (e.g. straight and direct movements), can still be detected in mirrored or rotated images, other mouse movement motifs can not be orientation invariant and lose their idiosyncratic characteristic. Therefore, we did not include additional data augmentation to generate more data (e.g. through mirroring or rotation).

In contrast to our setup, a binary one-vs-rest strategy as used in [1] assumes that the "rest" classes (e.g. anomaly samples) are representative for all other occurring anomalies. Often the same classes are used in training as well as testing, resulting in a high accuracy, but introducing a selection bias. Using an auxiliary dataset from the same field as the target dataset has been shown beneficial to increase authentication performance in mouse trajectories. Since the auxiliary dataset is just used for training but not for testing, the model even performs well when testing against trajectories of unseen users. In comparison to previous methods [25], the CNN-based model overcomes the dependency on hand-crafted features while learning to extract an individual feature set for every user.

For the presented approach, we compare different setups and *views*. To ensure a fair comparison we left the structure of the underlying model untouched. However, when taking the $[s_{avg},$ pause] *view* or the $[s_{log},$ pause] *view*, some images form user 07 caused numerical instabilities. There is no obvious visual difference in data between user 07 and other similar users. We excluded the models with these setups for further analysis and emphasize to utilise other combinations of trajectory representations when using this particular model.

9 Conclusion and Future Work

In this paper, we proposed an unsupervised learning approach for user authentication using only the data of one user for training. We showed that incorporating additional data can enhance the model's performance so that a distinction even to unknown users, which were never seen during training, becomes possible. This enables a deeper understanding of mouse cursor movements by visualising important key parts of the mouse trajectory for single users. Future research efforts should be directed to improve the discovery of mouse cursor motifs for individual users and their interplay with pauses. We thank the web-netz GmbH for funding this research and all former reviewers for the valuable feedback.

References

1. Antal, M., Fejér, N.: Mouse dynamics based user recognition using deep learning. Acta Universitatis Sapientiae, Informatica **12**, 39–50 (2020)
2. Arapakis, I., Lalmas, M., Valkanas, G.: Understanding within-content engagement through pattern analysis of mouse gestures. In: Proc. of the 23rd ACM CIKM, pp. 1439–1448 (2014)

3. Arapakis, I., Leiva, L.A.: Predicting user engagement with direct displays using mouse cursor information. In: Proc. of the 39th Int. ACM SIGIR Conf. on Research and Development in Inform. Retrieval, pp. 599–608. ACM (2016)
4. Atterer, R., Wnuk, M., Schmidt, A.: Knowing the user's every move: user activity tracking for website usability evaluation and implicit interaction. In: Proc. of the 15th Int. Conf. on WWW (2006)
5. Bach, S., Binder, A., Montavon, G., Klauschen, F., Müller, K.R., Samek, W.: On pixel-wise explanations for non-linear classifier decisions by layer-wise relevance propagation. PLoS ONE **10**(7), 1–46 (2015). https://doi.org/10.1371/journal.pone.0130140
6. Cepeda, C., et al.: Mouse tracking measures and movement patterns with application for online surveys. In: Holzinger, A., Kieseberg, P., Tjoa, A.M., Weippl, E. (eds.) CD-MAKE 2018. LNCS, vol. 11015, pp. 28–42. Springer, Cham (2018). https://doi.org/10.1007/978-3-319-99740-7_3
7. Chong, P., Elovici, Y., Binder, A.: User authentication based on mouse dynamics using deep neural networks: a comprehensive study. In: IEEE Transactions on Information Forensics and Security, vol. 15, pp. 1086–1101 (2020)
8. Chong, P., Ruff, L., Kloft, M., Binder, A.: Simple and effective prevention of mode collapse in deep one-class classification. CoRR abs/2001.08873 (2020)
9. Everitt, R., McOwan, P.: Java-based internet biometric authentication system. IEEE Trans. Pattern Anal. Mach. Intell. **25**(9), 1166–1172 (2003). https://doi.org/10.1109/TPAMI.2003.1227991
10. Fabi, K.: Layer-wise relevance propagation for pytorch (2021). https://github.com/KaiFabi/PyTorchRelevancePropagation
11. Feher, C., Elovici, Y., Moskovitch, R., Rokach, L., Schclar, A.: User identity verification via mouse dynamics. Inf. Sci. **201**, 19–36 (2012)
12. Fülöp, A., Kovács, L., Kurics, T., Windhager-Pokol, E.: Balabit mouse dynamics challenge data set (2016). https://github.com/balabit/Mouse-Dynamics-Challenge
13. Gajos, K., et al.: Computer mouse use captures ataxia and parkinsonism, enabling accurate measurement and detection. Mov. Disord. **35**(2), 354–358 (2019). https://doi.org/10.1002/mds.27915
14. Gamboa, H., Fred, A.: A behavioral biometric system based on human-computer interaction. In: Jain, A.K., Ratha, N.K. (eds.) Biometric Technology for Human Identification, vol. 5404, pp. 381–392. SPIE (2004)
15. Golan, I., El-Yaniv, R.: Deep anomaly detection using geometric transformations. In: Advances in Neural Information Processing Systems (NeurIPS) (2018)
16. Görnitz, N., Kloft, M., Rieck, K., Brefeld, U.: Toward supervised anomaly detection. J. Artif. Intell. Res. **46**, 235–262 (2013)
17. Haider, P., Chiarandini, L., Brefeld, U.: Discriminative clustering for market segmentation. In: Proc. of the ACM SIGKDD (2012)
18. Harilal, A., Toffalini, F., Castellanos, J., Guarnizo, J., Homoliak, I., Ochoa, M.: Twos: a dataset of malicious insider threat behavior based on a gamified competition. In: Proc. of MIST (MIST 2017), pp. 45–56. Association for Comp. Machinery, New York, NY, USA (2017)
19. Hendrycks, D., Mazeika, M., Dietterich, T.: Deep anomaly detection with outlier exposure. In: International Conference on Learning Representations (2019)
20. Hendrycks, D., Mazeika, M., Kadavath, S., Song, D.: Using self-supervised learning can improve model robustness and uncertainty. In: Wallach, H., Larochelle, H., Beygelzimer, A., d' Alché-Buc, F., Fox, E., Garnett, R. (eds.) Advances in Neural Information Processing Systems, vol. 32. Curran Associates, Inc. (2019)

21. Kaixin, W., Hongri, L., Bailing, W., Shujie, H., Jia, S.: A user authentication and identification model based on mouse dynamics. In: Proceedings of the 6th Int. Conf. on Information Engineering (ICIE 2017), Association for Computing Machinery, New York, NY, USA (2017)

22. Komarinski, P.: Automated fingerprint identification systems (AFIS). Elsevier (2005)

23. Krizhevsky, A., Sutskever, I., Hinton, G.E.: Imagenet classification with deep convolutional neural networks. In: Pereira, F., Burges, C., Bottou, L., Weinberger, K. (eds.) Advances in Neural Information Processing Systems, vol. 25. Curran Associates, Inc. (2012)

24. Lagun, D., Ageev, M., Guo, Q., Agichtein, E.: Discovering common motifs in cursor movement data for improving web search. In: Proc. of the 7th ACM Int. Conf. on Web Search and Data Mining, pp. 183–192. ACM (2014)

25. Matthiesen, J.J., Brefeld, U.: Assessing user behavior by mouse movements. In: Stephanidis, C., Antona, M. (eds.) HCII 2020. CCIS, vol. 1224, pp. 68–75. Springer, Cham (2020). https://doi.org/10.1007/978-3-030-50726-8_9

26. Mondal, S., Bours, P.: A study on continuous authentication using a combination of keystroke and mouse biometrics. Neurocomputing 230, 1–22 (2017)

27. Perera, P., Patel, V.M.: Deep transfer learning for multiple class novelty detection. CoRR abs/1903.02196 (2019)

28. Perera, P., Patel, V.M.: Learning deep features for one-class classification. IEEE Trans. Image Process. 28(11), 5450–5463 (2019)

29. Rodden, K., Fu, X., Aula, A., Spiro, I.: Eye-mouse coordination patterns on web search results pages. In: Extended Abstracts on Human Factors in Computing Systems (CHI EA 2008), pp. 2997–3002. Assoc. for Computing Machinery, New York, NY, USA (2008)

30. Ruff, L., et al.: Deep one-class classification. In: ICML, pp. 4393–4402 (2018)

31. Ruff, L., Vandermeulen, R.A., Franks, B.J., Müller, K.R., Kloft, M.: Rethinking assumptions in deep anomaly detection (2020)

32. Ruff, L., et al.: Deep semi-supervised anomaly detection. In: Int. Conf. on Learning Representations (2020)

33. Schölkopf, B., Platt, J.C., Shawe-Taylor, J., Smola, A.J., Williamson, R.C.: Estimating the support of a high-dimensional distribution. Neural Comput. 13(7), 1443–1471 (2001)

34. Shen, C., Cai, Z., Guan, X., Du, Y., Maxion, R.A.: User authentication through mouse dynamics. In: IEEE Trans. Inf. Forensics Security, vol. 8, pp. 16–30 (2013)

35. Shen, C., Cai, Z., Liu, X., Guan, X., Maxion, R.A.: MouseIdentity: modeling mouse-interaction behavior for a user verification system. IEEE Trans. Hum.-Mach. Syst. 46(5), 734–748 (2016)

36. Shen, C., Cai, Z., Maxion, R.A., Xiang, G., Guan, X.: Comparing classification algorithm for mouse dynamics based user identification. In: 2012 IEEE Fifth International Conference on Biometrics: Theory, Applications and Systems (BTAS), pp. 61–66 (2012)

37. Tax, D.M., Duin, R.P.: Support vector data description. ML 54(1), 45–66 (2004)

38. Tzafilkou, K., Protogeros, N.: Mouse behavioral patterns and keystroke dynamics in end-user development: what can they tell us about users' behavioral attributes? In: Computers in Human Behavior, vol. 83, pp. 288–305 (2018)

39. Wei, A., Zhao, Y., Cai, Z.: A deep learning approach to web bot detection using mouse behavioral biometrics. In: Sun, Z., He, R., Feng, J., Shan, S., Guo, Z. (eds.) CCBR 2019. LNCS, vol. 11818, pp. 388–395. Springer, Cham (2019). https://doi.org/10.1007/978-3-030-31456-9_43

40. Yao, B., Ai, H., Lao, S.: Person-specific face recognition in unconstrained environments: a combination of offline and online learning. In: 2008 8th IEEE International Conference on Automatic Face & Gesture Recognition, pp. 1–8 (2008)

41. Zhang, L., Tan, S., Yang, J., Chen, Y.: VoiceLive: a phoneme localization based liveness detection for voice authentication on smartphones. In: Proceedings of the 2016 ACM SIGSAC Conference on Computer and Communications Security (CCS 2016), pp. 1080–1091. Association for Computing Machinery, New York, NY, USA (2016)
42. Zimmermann, P., Guttormsen, S., Danuser, B., Gomez, P.: Affective computing - measuring mood with mouse and keyboard. Int. J. Occup. Saf. Ergon. **9**, 539–551 (2003)

Explaining Black Box Reinforcement Learning Agents Through Counterfactual Policies

Maria Movin[1,2](✉) (iD), Guilherme Dinis Junior[1,2](iD), Jaakko Hollmén[2](iD),
and Panagiotis Papapetrou[2](iD)

[1] Spotify, Stockholm, Sweden
[2] Stockholm University, Stockholm, Sweden
maria.movin@dsv.su.se

Abstract. Despite the increased attention to explainable AI, explainability methods for understanding reinforcement learning (RL) agents have not been extensively studied. Failing to understand the agent's behavior may cause reduced productivity in human-agent collaborations, or mistrust in automated RL systems. RL agents are trained to optimize a long term cumulative reward, and in this work we formulate a novel problem on how to generate explanations on when an agent could have taken another action to optimize an alternative reward. More concretely, we aim at answering the question: *What does an RL agent need to do differently to achieve an alternative target outcome?* We introduce the concept of a counterfactual policy, as a policy trained to explain in which states a black box agent could have taken an alternative action to achieve another desired outcome. The usefulness of counterfactual policies is demonstrated in two experiments with different use-cases, and the results suggest that our solution can provide interpretable explanations.

Keywords: Explainable AI (XAI) · Reinforcement Learning · Counterfactual Explanations

1 Introduction

Reinforcement learning (RL) is an area of machine learning in which learning is based on rewarding desired behaviors and penalizing undesired ones. Unlike classical machine learning, RL is an active learning method in which an RL agent learns through trial and error, with the goal of taking actions in an environment maximizing a cumulative future reward [11]. An agent's decisions are defined by a policy, which dictates the actions to be taken in a given state. For example, in a healthcare setting, an agent acting on behalf of a medical practitioner can decide what treatment (action) to recommend to an observed patient situation (state), given the treatment policy it has learned by observing other state-action pairs. RL has successfully been applied in several areas, such as games [10], recommendation systems [1], and in healthcare decision support systems [8].

B. Crémilleux et al. (Eds.): IDA 2023, LNCS 13876, pp. 314–326, 2023.
https://doi.org/10.1007/978-3-031-30047-9_25

Despite the recent advancements in the field of RL, the decision-making processes of RL agents are challenging for humans to understand. The policies of RL agents can be more or less interpretable depending on the problem complexity and the policy function used; The increased complexity of the problems for which RL is applied, and the use of neural networks as approximate policy functions, have led to the less interpretable policies. In these cases, the policies can be seen as black boxes for which only the input and the output are known, but little is known about the inner workings of the model. The field of explainability for RL has emerged to provide methods for explaining RL agents with the goal of improving transparency and establishing trust in these systems [4]. However, few studies have investigated or studied the use of *counterfactual explanations* for RL [9]. Counterfactual explanations are a specific form of explanations that provide answers to what-if questions of the form: *what-if the world would have looked different, how would that impact the outcome?* [12]. The area of explainable machine learning has adopted and successfully applied counterfactual explanations, especially in classical machine learning tasks, such as classification and regression [6]. At the same time, in the field of RL, earlier work has focused on generating counterfactual *state* explanations, i.e., how a state needs to be minimally changed (e.g., an image in an Atari game) so that the agent (i.e., a player) takes another action [9], resulting in a set of counterfactual states. Hence, the focus in [9] is mainly on how changes to a state affect the actions of the agent, rather than explaining how the chosen policy affects the long-term goal of the agent.

In this paper, we focus on the following question: *What does an RL agent need to do differently to achieve an alternative target outcome?*, which, to the best of our knowledge, has not been explored and tackled in the current RL literature. More concretely, we introduce the concept of *counterfactual policy* as a policy that can explain in which states the agent could have taken a different action to get to another desired outcome. Given an agent that follows a black box policy, hereafter referred to as a black box agent, the goal is to create a counterfactual policy that learns when an intervention needs to be done on the black box agent's actions, and learn what action should be taken instead. Importantly, the goal is *not* to find a more optimal policy but rather a policy that is optimized for *explaining* in which key states, i.e. the most important states, the black box agent could do something differently to reach a different outcome. We do not limit our method to only one state, but instead look at the full trajectory of the black box agent. Explanations can be given in our method in two ways: (1) online, through direct feedback on the black box agent's actions, and (2) offline, through statistics on disagreement states, which we define as the states for which the counterfactual policy needed to intervene on the black box policy's actions.

Motivational Real-World Example. To give an example on when counterfactual policies could be useful in practice we consider the case of a black box RL agent providing treatment recommendations to patients in a health care decision-support setting. The black box agent can be optimized for some different rewards, both long and short term rewards: e.g., few side effects of the medication and cure of the patient. We now focus on the question: *How would*

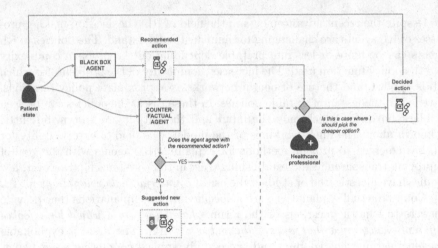

Fig. 1. Using a counterfactual policy online in a decision support system. In this example a healthcare provider uses a black box agent to make recommendations on treatments to patients. A counterfactual policy could, in this case, be used to explain in which key cases the black box policy could select less expensive treatments, while still curing the patient.

the agent need to act differently to recommend less expensive treatments with the same beneficial effects on the patient's health? In this case, our goal is to create a counterfactual policy that can explain for which key cases (i.e., key states) other medications (e.g., actions) could result in cheaper but still curing treatments for the patients. Acknowledge here that the goal of the counterfactual policy is to explain the black box policy, rather than being a better performing policy on the black box agent's main task (i.e., having few side effects and curing the patient). We search for a counterfactual policy that disagrees with the black box policy as few times as possible to only disturb the decision support system in a few key cases, while still leading to cheaper treatments. Furthermore, there is value of providing this information both when the decision-support system is in use online (as illustrated in Fig. 1), and as an evaluation summary of one or more agents before deciding to launch one of them online at the health-care provider.

Contributions. Our contributions are threefold: (1) We introduce and formalize the problem of generating counterfactual explanations describing the alternative actions an RL agent could have taken to reach another outcome, (2) we provide an algorithmic solution to the problem based on counterfactual policies, and (3) we instantiate the problem with two use-cases from the following application areas: medical treatment recommendation and cliff walking; we additionally demonstrate the utility of our formulation and solution by presenting the generated explanations for these use-cases.[1]

---

[1] Our code is available at https://github.com/dsv-data-science/rl-counterfactual-policy-explanations.git.

2   Related Work

The importance of understanding RL agents has been highlighted in several papers and some different approaches have been explored [4]. The concept of critical states was introduced by Huang et al. [5] as a way to increase trust in RL agents. The critical states are states in which an agent greatly prefers a few actions, and performance would be greatly affected if the agent were to take a random action. The idea is that the action the agent takes in critical states can help the users to get a better mental model of the policy and work as an explanation on the agent's behavior. Instead of focusing on critical states, a recent work has shown the benefit of using disagreement states as a summary when comparing two agents [2]. Amitai et al. [2] showed that using states in which two policies disagree provide summaries that are easier to understand for users compared to summaries based on critical states. The use of disagreement states, and contrastive explanations are similar to our work. However, our work does not focus on generating summaries to compare two agents, but rather explanations on how a black box agent can reach another outcome.

The idea of counterfactual explanations for RL agents has been explored in earlier works [3,9]. Olson et al. [9] aimed at providing explanations in the form of what needs to change in a state for the RL agent to take another action. They used a deep generative model to generate new counterfactual states in which the agent takes another action, and tested the method on Atari games. Instead of creating synthesized states from deep generative models, Frost el al. [3] used states that can be reached by valid actions, to make sure the states were reachable, and provided the full trajectory from those states as explanations. Different from our study, none of these studies aimed at explaining what the agent itself needs to do differently to achieve a target goal.

3   Explaining RL Agents Through Counterfactual Policies

3.1   Reinforcement Learning Preliminaries

RL is different from other machine learning techniques in that an agent is trying to learn the optimal way of interacting with an environment by trial and error, rather than learning from labels (supervised learning), or patterns in the data (unsupervised learning). The RL problem to be solved and its environment are often described as a Markov Decision Process (MDP), defined by $\{\mathcal{S}, \mathcal{A}, \mathbf{P}_{a,s}, R_a, \gamma\}$, where $\mathcal{S}$ is the state space, $\mathcal{A}$ is the action space, $\mathbf{P}_{a,s}$ is the transition probability matrix to a new state given the previous state s and an action a, R_a is the reward in the current state given action a, and $\gamma \in [0,1]$ is the discount factor which weighs the importance of future rewards. In this work, we focus on episodic reinforcement learning for which the agent interacts with the environment in episodes with a given start and end state for each episode. For each step in the episode, the agent chooses an action based on the current state, the environment provides the next state, based on $\mathbf{P}_{a,s}$, alongside a reward based on R_a. The actions of the agent are provided by a policy $\pi(a|s)$

that decides which action a to take at a given state s. The general goal in RL is to find a policy that optimizes the cumulative reward given the environment of the MDP. A Q-value function for a policy π, $Q_\pi(s,a)$, represents the expected discounted cumulative reward if the agent performs action a at state s, and then follows policy π. Q-learning is an RL algorithm that iteratively updates a Q-table with the Q-values to improve the behavior of the agent after exploring and exploiting the possible actions in the state space [11].

3.2   Problem Formulation

Consider a fixed black box policy π_{bb} trained to solve an episodic problem in a known environment described by a MDP, with $\mathrm{MDP}_{bb} = \{\mathcal{S}_{bb}, \mathcal{A}_{bb}, \mathbf{P}_{bb}, R_{bb}, \gamma_{bb}\}$. Our goal is to generate a counterfactual policy π_c based on a MDP that is related to MDP_{bb}, which we denote as $MDP_c = \{\mathcal{S}_c, \mathcal{A}_c, \mathbf{P}_c, R_c, \gamma_c\}$. The goal of the counterfactual policy is to explain in which states $s \in \mathcal{S}_{bb}$ the black box policy π_{bb} takes a non-optimal action $a \in \mathcal{A}_{bb}$. The action can either be non-optimal in regards to the reward function R_{bb} of the black box policy, or with respect to another reward function, denoted as $R_{outcome}$, designed for an alternative outcome. The explanation is done through counterfactual *disagreement states*, defined as the states for which the black box policy π_{bb} and the counterfactual policy π_c choose different actions, i.e., $\pi_c(a_c|s) \neq \pi_{bb}(a_{bb}|s)$.

Since it is important that the counterfactual policy π_c is not just another *better* policy, but rather a policy that can *explain* the black box policy π_{bb}, we also introduce the concept of *degree of disagreement*, defined as the number of disagreement states between two policies. Hence, we search for a counterfactual policy π_c for which the degree of disagreement between π_c and the black box policy π_{bb} is minimized, while still reaching the target goal. This is similar to the sparsity metric commonly used for counterfactual explanations in other machine learning areas [6].

Problem 1. (**Explanations through Counterfactual Policies**) Given an environment described by a MDP, an agent A_{bb} following a black box policy π_{bb}, and an alternative reward function $R_{outcome}$, our goal is to find a counterfactual policy π_c with a minimum degree of disagreement with π_{bb} so that $R_{outcome}$ is maximized.

In other words, our objective is to generate *sparse* explanations on when A_{bb} should take a different action than what is suggested by π_{bb}, in order to receive a higher alternative reward based on $R_{outcome}$.

3.3   Counterfactual Policy Generation

Defining the Environment for the Counterfactual Policy. To train the counterfactual policy we first define the MDP of the policy, MDP_c, which relates to the MDP of the black box, MDP_{bb}:

Algorithm 1. Learning the Counterfactual Policy with Q-Learning

---

Require: π_{bb}, step size $\alpha \in (0,1]$, small $\epsilon > 0$
  Initialize $Q((s, a_{bb}), a)$ for all $(s, a_{bb}) \in S_c, a \in A_c$ arbitrarily
  $Q((goal\ state, .).) \leftarrow 0$
  for all episodes **do**
    $s \leftarrow$ start state
    for all steps in episode **do**
      $a_{bb} \leftarrow \pi_{bb}(a|s)$
      Chose action a using an ϵ-greedy policy derived from $Q((s, a_{bb}), a)$
      Take action a, observe R, next state s' and next black box action a'_{bb}
      $Q((s, a_{bb}), a) \leftarrow Q((s, a_{bb}), a) + \alpha[R + \gamma \max_a Q((s', a'_{bb}), a) - Q((s, a_{bb}), a)]$
      $s \leftarrow s'$

- S_c: a state space which consists of the combination of all states S_{bb} and actions A_{bb} that the black box policy can take. The size of the state space is, hence, $|S_{bb} \times A_{bb}|$.
- A_c: equal to A_{bb}; the both policies use the same discrete action space.
- $\mathbf{P}_c$: equal to $\mathbf{P}_{bb}$; the same state transition probability as for the black box agent.
- R_c: a reward function that entails two parts, one that controls the main goal of the task and one that controls the number of disagreements. We present more details on how it is defined in the next section.
- γ: the discount factor which weighs the importance of future rewards.

Defining the Counterfactual Reward Function. One of the most important parts in generating the counterfactual policy is the design of the reward function R_c, as it needs to represent the target goal of the explanation. To aid this, we define R_c to be a sum of two reward functions, i.e., $R_{outcome}$ and $R_{disagree}$ (Eq. 1). $R_{outcome}$ needs to be designed to help the agent succeed with the alternative outcome, while $R_{disagree}$ needs to penalize the counterfactual policy when disagreeing with the black box policy.

$$R_c = R_{outcome} + R_{disagree} \tag{1}$$

For example, if we want to understand how a certain black box policy could take safer actions, we need to define a reward function for the counterfactual policy that penalizes unsafe actions ($R_{outcome}$). Additionally, we prefer sparse explanations, and thus we seek a counterfactual policy that disagrees with the black box policy as few times as possible. This is achieved by defining a reward function that additionally penalizes the counterfactual policy each time it disagrees with the black box policy ($R_{disagree}$). Our objective, hence, boils down to maximizing the reward function expressed by Eq. 1.

Learning and Using the Counterfactual Policy. When the problem is formulated into the MDP we optimize the policy using Q-learning as described in Algorithm 1. The trained counterfactual policy can thereafter be used to give online explanations to the black box model by providing online feedback on

---

Algorithm 2. Algorithmic Overview of our Method

Require: $\pi_{bb}, \pi_c, \mathbf{P}$, online

    $s \leftarrow$ *start state*

    disagreement_states $\leftarrow []$

    while $s \neq$ *goal_state* **do**

        $a_{bb} \leftarrow \pi_{bb}(a|s)$

        $a_c \leftarrow \pi_c(a|s, a_{bb})$ ▷ CF policy evaluate action from BB policy

        if $a_{bb} = a_c$ **then**

            $a_{chosen} \leftarrow a_{bb}$

        else if online $=$ True **then**   ▷ Online with human-in-the-loop

            $a_{chosen} \leftarrow a_{bb}$ or a_c based on human decision

        else if online $=$ False **then**   ▷ Offline to generate statistics

            disagreement_states append $((s, a_{bb}, a_c))$

            $a_{chosen} \leftarrow a_{bb}$ or a_c randomly   ▷ To search both paths

        $s \leftarrow s_{t+1}$ based on $\mathbf{P}[s_{t+1}|a_{chosen}, s]$

    return *disagreement_states*

---

when it disagrees with the black box policy, or to provide offline explanations in the form of statistics of the disagreement states. An overview of our method is presented in Algorithm 2. The offline statistics are gathered by running an evaluation over a set of episodes. Since we want to make sure to explore a diverse set of trajectories to understand both where the disagreement states lie, and how the black box policy behaves after an intervention, we randomly choose between following the black box or counterfactual policy when there is disagreement.

4 Experiments

We provide two instantiations of our problem and demonstrate the proposed solution. The two use-cases we explore are taken from healthcare decision support and grid world cliff walking, respectively.

4.1 Synthetic Medication

We introduce a synthetic use-case for which we have full control over the environment and have predefined the optimal actions in each state. The use-case is inspired by decision support systems in healthcare. More specifically, we assume we have a patient and a doctor (either a human or a decision support system) that decides the medication for the patient. The goal is to cure the patient as quickly and with as few medications as possible. The doctor can choose between two different medications, i.e., a green or a red pill. The patients are cured when they first are provided a red pill, followed by a green pill.

Instantiation of π_{bb}. The problem and environment from the black box policy's π_{bb} view can be formulated as a MDP in the following way:

– S_{bb}: a discrete state space of three possible states {*ill, under medication, cured*} describing the patient's state.

- $\mathcal{A}_{bb}$: a discrete action space {green pill, red pill, wait} describing the actions the doctor/policy can take in the different states.
- $\mathbf{P}_{bb}$: the transition probability matrix. This problem is deterministic and is described by moving to the next state only if the right pill in the sequence is taken, otherwise the patient goes back to the state of being ill (and starts over on medication).
- R_{bb}: the reward function in this problem is set to minimize the amount of medication given to the patient, and thus the agent gets '-1' for each step taken, plus one extra '-1' if the wrong action is taken (i.e., wrong medication).
- γ: the discount factor is set to 1 for simplicity.

We test three different fixed black box policies $\pi_{bb}(a|s)$: (1) always providing a green pill π_{bb}^g, (2) randomly choosing an action π_{bb}^r, and (3) training the policy using Q-learning and solving the problem optimally π_{bb}^o. To better understand how these three policies can cure patients quicker and with fewer pills we apply our method to generate explanations through counterfactual policies.

Instantiation of π_c. The problem for the counterfactual policy's π_c view can be described by a MDP that is related to the MDP of the black box policy. Recall from our formulation that the action space, $\mathcal{A}$, and the transition probabilities, $\mathbf{P}$, are the same for both MDPs. The rest of the MDP is instantiated as follows:

- $\mathcal{S}_c$: a discrete state space describing the patient's state, i.e., {ill, under medication, cured}, combined with the black box policy's action space, i.e., {green pill, red pill, wait}.
- R_c: $R_{outcome}$ is set to the reward for the black box R_{bb}, and $R_{disagree}$ is set to -1 for each time the counterfactual policy disagrees with the black box policy.
- γ: the discount factor, which we set to 1 in this experiment.

Learning the Counterfactual Policy. We train three counterfactual policies $(\pi_c^g, \pi_c^r, \pi_c^o)$, one for each black box policy, using 1000 randomly sampled episodes using Q-learning ($\epsilon = 0.5$, $\alpha = 0.1$) as described in Algorithm 1. The disagreement states were thereafter evaluated offline during another 1000 episodes as described in Algorithm 2.

Explanations. In Table 1 we present the disagreement rates for the three different black box policies. As expected, overall the counterfactual policy π_c^g disagreed with π_{bb}^g half of the times (when π_{bb}^g should have chosen the red pill), π_c^r had overall disagreements with π_{bb}^r two thirds of the time (π_{bb}^r randomly took the right action a third of the time), while the optimal policy, π_{bb}^o did not have any disagreements with its corresponding counterfactual policy π_c^o. The overall disagreement rate provides some understanding of the models, but to really get an explanation on where the policies should have taken a different action we also present the disagreement states (Table 1). This result shows that π_c^g disagreed with π_{bb}^g 100% of the times in the Ill state where π_{bb}^g needs to take a red pill to move to the Under medication state, while π_c^r disagreed with the random policy π_{bb}^r equally often in both states.

Table 1. Disagreement rate between the black box policies (π_{bb}) and corresponding counterfactual policies

Disagreement state	Always green π_{bb}^g	Random π_{bb}^r	Optimal π_{bb}^o
Overall	0.5	0.67	0.0
Ill	1.0	0.67	0.0
Under medication	0.0	0.67	0.0

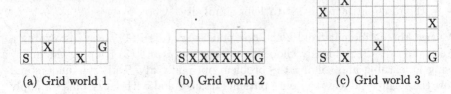

(a) Grid world 1   (b) Grid world 2   (c) Grid world 3

Fig. 2. Grid world cliff walking maps. The goal for the RL agent is to walk from the start state (S) to the goal state (G), without falling into one of the cliffs (X).

4.2 Grid World Cliff Walking

In the synthetic medication example we showed that our method works as expected on a trivial problem with few actions and states. To increase the complexity we include the cliff walking grid world example presented in [11]. It is an episodic problem where an agent's goal is to walk in a grid from the start state (S) to the goal state (G). If the agent falls into one of the cliffs (X) that are present on the grid, it needs to start from S again. There are different approaches on how to train an agent to perform this walk. One way is to find the *shortest path*, i.e., it can be trained to take as few steps as possible, while another way is to find a *safe path*, i.e., it can prefer to take a route that is as far from the cliffs as possible. This is an interesting example for this paper, since we can use it to evaluate cases when the counterfactual policy has another outcome goal compared to the black box policy ($R_{outcome} \neq R_{bb}$). We will run our experiments over three different grids, presented in Fig. 2. The idea is to get counterfactual explanations on what a black box agent trained to walk the short path needs to do differently to become safer. This will be achieved by providing disagreement states generated by a counterfactual policy trained to explain how the black box model can act so that it takes a safer path, but with as few disagreements between the policies as possible.

Instantiation of π_{bb}. The problem and environment for the black box agent can be formulated as follows:

- $\mathcal{S}_{bb}$: a discrete state space describing the agent's position on the grid.
- $\mathcal{A}_{bb}$: a discrete action space {*up, down, left, right*} describing the actions the agent can take in the different states.

- $\mathbf{P}_{bb}$: the state transition probability, which includes constraints, such that the agent cannot move outside of the grid and if it falls into the cliff it transitions back to the start state S.
- R_{bb}: The agent receives a reward of '-1' for each step it takes, and in addition '-100' if it falls into a cliff.
- γ: the discount factor is set to 1 in this experiment.

Instantiation of π_c. The problem for the counterfactual policy can be formulated as follows (remember that $\mathcal{A}_c = \mathcal{A}_{bb}$ and $\mathbf{P}_c = \mathbf{P}_{bb}$):

- $\mathcal{S}_c$: a discrete state space describing the black box agent's position on the grid combined with the 4 actions the black box agent can take.
- R_c: The instantiation of reward functions is described in detail below.
- γ: the discount factor is set to 1 in this experiment.

Instantiation of R_c. The goal of the first part of the reward function, $R_{outcome}$, is to reward *safe* actions while still taking as few steps as possible. The change we introduce compared to R_{bb} is to add a penalty that is based on the distance of the agent to the nearest cliff point (cp) on every move (Eq. 2). This way, we incentive moving towards the goal while remaining as far as possible from any cliffs. The second part of the reward function, $R_{disagree}$, is set to -1 for each time the counterfactual policy disagrees with the black box policy.

$$R_{outcome} = R_{bb} - \frac{1}{\arg\min_{cp} D(s, cp)} \tag{2}$$

Learning the Counterfactual Policy. We run counterfactual policy learning for the grid maps depicted in Fig. 2. For each grid, we learn a black box policy using Q-learning for 5000 episodes ($\epsilon = 0.1, \alpha = 0.1$), followed by learning a counterfactual policy using Q-learning for 1000 episodes ($\epsilon = 0.5, \alpha = 0.1$) (Algorithm 1).

Explanations. The counterfactual trajectories are presented in Fig. 3, where we visualize the path of the black box policy and that suggested by the counterfactual policy in disagreement states. These were collected by sampling 500 episodes, where in disagreement states we choose to either follow the black box or the counterfactual policy at random. Across all three maps, the counterfactual policy suggests changes that make the agent steer clear of cliffs. Yet, there are other patterns uncovered from the disagreement states, such as the one in Fig. 3e, where the black box policy can get locked in loop forever, and the counterfactual policy suggests actions to help it achieve its goal.

4.3   Lessons Learned

In this paper, we made a first attempt to explain RL agents by generating counterfactual policies. A counterfactual policy can provide informative knowledge of a black box policy by learning when to intervene on the black box policy's chosen actions to achieve a target goal. Nonetheless, the field of explainable RL

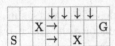

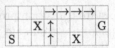

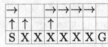

(a) GW 1 - π_{bb} on disagreement states   (b) GW 1 - π_c on disagreement states

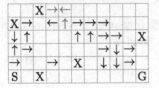

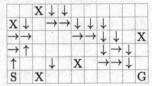

(c) GW 2 - π_{bb} on disagreement states   (d) GW 2 - π_c on disagreement states

(e) GW 3 - π_{bb} on disagreement states   (f) GW 3 - π_c on disagreement states

Fig. 3. Grid world (GW) maps: arrows represent disagreement states between a black box policy, π_{bb}, and a counterfactual policy, π_c, trained to explain where π_{bb} can be safer.

is still largely unexplored [4] and this work can be used as a starting point for further investigation into the area of explaining RL agents. The idea of using a counterfactual policy to explain a black box policy is novel and provides several benefits over earlier RL explainability methods. First, it provides explanations through the full trajectory of an episodic problem. This differs from the work by Olson et al. [9] since they simplified the problem into a classification task by only considering one state at the time. Our work considers the full trajectory and the outcome of the black box agent. Second, we focus on the agent rather than the environment which makes it easier to understand what the agent is doing wrong rather than how the environment impacts the agent.

One of the challenging parts of our method is to design the right reward function. It is non-trivial to balance between the number of disagreements and the main optimization goal. In some cases it might be extremely important to reach the optimal target reward with the number of disagreements being less important. However, in other cases, the number of disagreements might be of higher importance, and thus it needs to be tuned based on the problem at hand. This is equivalent to how counterfactual explainability methods are tuned for other machine learning tasks, such as classification and regression, where an explanation that changes few features (sparsity) is preferable, however, optimizing too heavily for sparsity might lead to less robust and faithful explanations [7]. The other challenge lies in the cardinality of the state space of our counterfactual policy, $|\mathcal{S}_{bb} \times \mathcal{A}_{bb}|$. For moderately sized state or action spaces, it can significantly affect the computational efficiency of learning a counterfactual policy.

In this work we have limited our experiments to tabular episodic RL problems with finite state and action spaces. A natural extension would be to explore continuous state spaces and explain more complex and deep RL agents. Expressing disagreement states and counterfactual trajectories is more challenging in that setting, and one alternative is to discretize continuous problems.

5 Conclusions

In this paper we formulated the problem of generating counterfactual policies for reinforcement learning (RL) and provided an algorithmic solution for solving the problem. We instantiated our formulation and solution using two use-cases and demonstrated how our solution can explain black box RL agents through counterfactual policies. Since the ultimate goal of any explainable RL method is to be useful for humans to understand RL agents, an interesting next step would be to evaluate the usefulness of our method in a user study. This would help understand our method's advantages but also its limitations which can provide ideas on how it can be improved.

Acknowledgements. Special thanks to docent Jussi Karlgren working at Spotify, who provided us with valuable early feedback on the project and thorough feedback on the final paper.

References

1. Afsar, M.M., Crump, T., Far, B.: Reinforcement learning based recommender systems: a survey. ACM Comput. Surv. **55**(7), 1–38 (2021)
2. Amitai, Y., Amir, O.: Summarizing policy disagreements for agent comparison. In: Proceedings of the 36th AAAI Conference on Artificial Intelligence (2022)
3. Frost, J., Watkins, O., Weiner, E., et al.: Explaining reinforcement learning policies through counterfactual trajectories. arXiv preprint arXiv:2201.12462v1 (2021)
4. Heuillet, A., Couthouis, F., Díaz-Rodríguez, N.: Explainability in deep reinforcement learning. Knowl.-Based Syst. **214**, 106685 (2021)
5. Huang, S.H., Bhatia, K., Abbeel, P., Dragan, A.D.: Establishing appropriate trust via critical states. In: 2018 IEEE/RSJ International Conference on Intelligent Robots and Systems (IROS), pp. 3929–3936. IEEE (2018)
6. Karimi, A.H., Barthe, G., Schölkopf, B., Valera, I.: A survey of algorithmic recourse: contrastive explanations and consequential recommendations. ACM Comput. Surv. **55**(5), 1–29 (2022)
7. Laugel, T., Lesot, M.J., Marsala, C., Detyniecki, M.: Issues with post-hoc counterfactual explanations: a discussion. arXiv preprint arXiv:1906.04774 (2019)
8. Liu, S., See, K.C., Ngiam, K.Y., Celi, L.A., Sun, X., Feng, M.: Reinforcement learning for clinical decision support in critical care: comprehensive review. J. Med. Internet Res. **22**(7), e18477 (2020)
9. Olson, M.L., Khanna, R., Neal, L., Li, F., Wong, W.K.: Counterfactual state explanations for reinforcement learning agents via generative deep learning. Artif. Intell. **295**, 103455 (2021)

10. Silver, D., Huang, A., Maddison, C.J., et al.: Mastering the game of go with deep neural networks and tree search. Nature **529**, 484–489 (2016)
11. Sutton, R.S., Barto, A.G.: Reinforcement learning: an introduction, 2nd edn. Adaptive Computation and Machine Learning Series, The MIT Press (2018)
12. Wachter, S., Mittelstadt, B., Russell, C.: Counterfactual explanations without opening the black box: automated decisions and the GDPR. Harv. J. L. Tech. **31**, 841 (2017)

A GNN-Based Architecture for Group Detection from Spatio-Temporal Trajectory Data

Maedeh Nasri[1]([⊠]) [ID], Zhizhou Fang[2] [ID], Mitra Baratchi[2] [ID],
Gwenn Englebienne[3] [ID], Shenghui Wang[3] [ID], Alexander Koutamanis[4] [ID],
and Carolien Rieffe[1,3,5] [ID]

[1] Department of Developmental Psychology, Leiden University,
Leiden, The Netherlands
m.nasri@fsw.leidenuniv.nl
[2] Leiden Institute of Advanced Computing, Leiden University,
Leiden, The Netherlands
[3] Faculty of Electrical Engineering, Mathematics and Computer Science,
University of Twente, Enschede, The Netherlands
[4] Faculty of Architecture and the Built Environment, Delft University of Technology,
Delft, The Netherlands
[5] Department of Psychology and Human Development, University College London,
London, UK

Abstract. Detecting and analyzing group behavior from spatio-temporal trajectories is an interesting topic in various domains, such as autonomous driving, urban computing, and social sciences. This paper revisits the group detection problem from spatio-temporal trajectories and proposes "WavenetNRI", a graph neural network (GNN) based method. The proposed WavenetNRI extends the previously proposed neural relational inference (NRI) method (an unsupervised learning approach for inferring interactions from observational data) in two directions: (1) symmetric edge features and edge updating processes are applied to generate symmetric edge representations corresponding to the symmetric binary group relationships; (2) a gated dilated residual causal convolutional (GD-RCC) block is adopted to capture both short and long dependency of the edge feature sequences. We evaluated the performance of the proposed model on three simulation datasets and three real-world pedestrian datasets, using the Group Mitre metric to measure the quality of the predicted groups. We compared WavenetNRI with four baseline methods, including two clustering-based and two classification-based methods. In these experiments, NRI and WavenetNRI outperformed all other baselines on the group-interaction simulation datasets, while NRI performed slightly better than WavenetNRI. On the pedestrian datasets, the WavenetNRI outperformed other classification-based baselines. However, it did not compete against the clustering-based methods. Our ablation study showed that while both proposed changes cannot be effective at the same time, either of them can improve the performance of the original NRI on one dataset type.

Keywords: Group detection · Spatio-temporal data · Deep learning

© The Author(s), under exclusive license to Springer Nature Switzerland AG 2023
B. Crémilleux et al. (Eds.): IDA 2023, LNCS 13876, pp. 327–339, 2023.
https://doi.org/10.1007/978-3-031-30047-9_26

1   Introduction

Detecting group behaviors based on users' spatio-temporal trajectories has numerous social and urban applications [1, 4, 10]. For example, detecting groups of pupils playing in a schoolyard facilitates psychologists in understanding pupils' social behavior [10]. Most previous studies in group detection tasks relied on heavy feature engineering [13, 16]. These approaches extract selected features from raw trajectory data based on domain knowledge specific to an application area. This restricts generalization to other similar problems. This approach may also ignore informative underlying spatio-temporal patterns that are present in the raw data.

Recently, graph neural networks (GNNs) showed strong potential for relational reasoning [2]. GNNs could be used in group detection by modeling agents (or members of a community) as nodes and their relationships as edges. For example, Thompson et al. [14] proposed a graph convolutional network (GCN) to detect conversational groups among static agents involved in the same conversation.

In contrast with static groups, moving groups might dynamically change their distance from other groups in the same environment. This adds extra challenges to the group detection task. Kipf et al. [7] proposed a GNN-based method, Neural Relational Inference (NRI), which applied a GNN to infer the interactions between moving particles given their spatio-temporal sequences in a physical system. In this work, the interactions in a physical system are assumed constant among certain pairs of particles over the given time window. In a realistic social group setting, however, individuals often change their interaction partners. This renders the group detection problem a more challenging task compared to the interaction detection tasks considered by Kipf et al. [7]. For example, while the atoms in a molecule constantly interact with particular atoms over time, children playing in a playground might switch their playmates.

The strong performance of the NRI model in recovering the ground-truth interaction graphs makes it a suitable candidate to be further investigated in group detection tasks. The current study extends the original NRI method in two directions to extend its use from interaction detection to the more complex and realistic social group detection task: (1) We propose a GNN architecture for capturing both short and long dependence in the group detection task where the interactions between agents may change over time. For this purpose, the 1D convolutional layer in NRI is replaced with a gated dilated residual causal convolutional (GD-RCC) block, as proposed by Wavenet [11]. (2) The original NRI builds and updates edge features by simply concatenating the node features, which does not satisfy the symmetric property of group relationships. We propose using symmetric temporal edge features and symmetric edge updating to tackle this problem.

Overall, this paper makes the following contributions:

- We propose a framework for group detection building upon the NRI interaction detection method. Our framework can capture short and long depen-

dencies in the spatio-temporal data and can satisfy the symmetric property
of group behavior.
- We extend NRI by applying the Louvain community detection algorithm to
  transform the predicted interactions into predicted groups.
- We evaluate our group detection framework using three group-interaction
  simulation datasets and three pedestrian datasets and further compare our
  method against four state-of-the-art methods.
- We investigate the effectiveness of our two proposed changes, namely, the
  GD-RCC block and symmetric temporal edge feature with symmetric edge
  updating processes, on the original NRI in an ablation study.

The rest of the paper is organized as follows. In Sect. 2, we formulate the group
detection problem. Section 3 discusses the related works. We present our proposed methodology in Sect. 4. In Sect. 5, the experiments are discussed. Finally,
Sect. 6 presents conclusions and future research directions.

2   Problem Formulation

Assume given the spatio-temporal trajectories of N agents in a time window
with a duration of T time steps, where the spatio-temporal measurements (e.g.,
position, speed, acceleration, etc.) of each agent $i \in 1, ..., N$ at a time step
$t \in 1, ..., T$ is denoted by X_i^t and the spatio-temporal sequences of all agents are
denoted by $X_{1:N}^t$. The goal is to detect groups $C = \{c_j | j = 1, ..., K\}$ of agents,
where $K \leq N$ is the number of groups, assuming that the group relationships
are constant in a time window, while agents could interact with other agents
from a different group. We aim to learn the probability of pairwise interactions
$\hat{I}$ between agents within the time window given $X_{1:N}^{1:T}$, i.e., $P(\hat{I}|X)$, such that
the predicted pairwise interactions reflect the group memberships of agents in
community detection algorithms.

Our proposed method to solve this problem employs a GNN encoder to predict pairwise interactions $\hat{I}$. The Louvain community detection algorithm [3]
transforms the predicted pairwise interactions $\hat{I}$ into predicted groups $\hat{C}$. We
train the GNN encoder in a supervised way using the ground-truth pairwise
group relationships G where $G_{(i,j)} = 1$ denotes that agent i and agent j are
in the same group and otherwise $G_{(i,j)} = 0$. In the training phase, the goal is
to minimize the difference between G and $\hat{I}$ by minimizing the weighted cross-
entropy loss function.

3   Related Work

This section discusses the related work in group detection algorithms and further
explores studies that proposed GNN models for spatio-temporal data.

Group Detection: Many previous studies in group detection tasks are based on
classic machine learning methods with hand-crafted features [13,16]. Yamaguchi

et al. [16] proposed an SVM-based framework applying normalized histograms of distances, velocity, and direction features to classify the binary group detection. Using supervised clustering, Solera et al. [13] proposed a structural SVM [15] framework to find groups of pedestrians based on hand-crafted features, e.g., distance, motion causality, trajectory shape, and paths convergence. Despite acceptable results, generating hand-crafted features needs domain knowledge. Besides, these features usually depend on particular data types and applications, e.g., the features created for detecting pedestrians walking on streets may not apply to other complex social settings (e.g., children playing).

To address this problem, many recent studies proposed deep learning-based methods. In GD-GAN [5], an LSTM-based generator predicts future trajectories. In this work, groups are detected by clustering the hidden states of this LSTM-based generator. Contrary to GD-GAN, which predicts future trajectories, our work predicts the pairwise interactions using a GNN encoder. This is beneficial because it can be directly trained with the ground-truth group relationships without special optimization algorithms, such as the Block-coordinate Frank-Wolfe (BCFW) algorithm, in a computationally efficient way [9].

GNN for Spatio-Temporal Data: Most GNN-based works for spatio-temporal data, such as TrafficGraphNet [8], focus on improving the performance of forecasting tasks. This approach learns the node representations by aggregating the nodes' neighborhoods and does not directly model the pairwise interactions or group relationships needed for group detection tasks. Methods such as NRI [7] that focus on predicting the edges between nodes can denote the interaction or relation types between nodes. For instance, the encoder part of NRI [7] applies a GNN-encoder to predict the interaction types between particles in a physical system. In our study, we extended the encoder part of NRI, which predicts the interactions between agents for a group detection task.

4   Methodology

In this section, we first present the interaction model implemented using a GNN encoder. Next, the two main proposed features of this model, (i) symmetric edge features and (ii) GD-RCC, are each discussed separately. We employ a GNN encoder, based on NRI [7], and a GD-RDCC block, based on Wavenet [11], to create our proposed model "WavenetNRI".

4.1   GNN Encoder: Interactions Modelling

The core part of the proposed method is a GNN encoder proposed in NRI [7], which predicts the distribution of the interaction and non-interaction edges. In NRI, the initial edge features and edge updating are implemented by concatenating the features of the end nodes as follows:

$$e^t_{(i,j)} = [X^t_i, X^t_j](t \in 1, ..., T), \quad h^1_j = f_v(\sum_{i \neq j} h^1_{(i,j)}), \quad h^2_{(i,j)} = f_e([h^1_i, h^1_j]) \quad (1)$$

where the spatio-temporal sequence of agent i at period of time $t \in 1, ..., T$ is denoted by X^t_i. The initial edge feature of the agents i and j at time step t is denoted by $e^t_{(i,j)}$. $[\cdot, \cdot]$ denotes concatenation. h^1_j and f_e denotes node representation of the agent j and edge updating function, respectively. The edge and node updating functions f_e and f_v are multilayer perceptrons (MLPs). NRI further applies a 1D convolutional layer with attentive pooling to transform the edge sequence $e^t_{(i,j)}$ into the vector representations of edges $h^1_{(i,j)} = \sum_t a^t s^t$, where a and s are attention score and edge representation, respectively (details are shown in Fig. 1).

There are several limitations in the GNN encoder of the original NRI method: (1) Building and updating edge features, and representations by simply concatenating the node features (shown in Eq. 1) cannot explicitly model the spatial differences of agents. Furthermore, the results of this concatenation are not symmetric, which may not satisfy the symmetric nature of group relationships. (2) Using only one convolutional layer may not capture the long-term interactions of the sequences of edge features. To tackle these limitations, we made the following changes to the original NRI:

- We included the spatial differences between agents and temporal increments in the initial temporal edge features $e^t_{(i,j)}$ and updated the edge features by element-wise product of the end nodes' representations. Consequently, the final edge vector representations $h^2_{(i,j)}$ are symmetric and capture both spatial differences between the agents and their movements (explained in Sect. 4.2).
- We replaced the single 1D convolutional layer in NRI with a GD-RCC block based on Wavenet model [11] to learn the temporary edge features and capture both short and long-term interactions of the edge feature sequences (explained in Sect. 4.3).

4.2 Symmetric Edge Features and Updating

In our proposed method, the edge features are constructed by concatenating the spatial differences of the node measurements and the temporal increments, which is formulated as follows:

$$e^t_{(i,j)} = [\|X^t_i - X^t_j\|, \Delta X^t_i \odot \Delta X^t_j], \quad t \in 1, ..., T-1, \quad \Delta X^t_i = X^{t+1}_i - X^t_i \quad (2)$$

where the Euclidean distance between agent i and agent j is denoted by $\|X^t_i - X^t_j\|$ and is used to model the spatial difference between agents and their movements (temporal increments). The element-wise production of the increments of the two agents is denoted by $\Delta X^t_i \odot \Delta X^t_j$. We achieve two benefits with this formulation: (i) the temporal edge $e^t_{(i,j)}$ captures both the spatial difference between agent i and agent j as well as the temporal increments of the agents; (ii) the edge features are symmetric, i.e., $e^t_{(i,j)} = e^t_{(j,i)}$, corresponding to the symmetric properties of the pairwise group relationships.

The edge sequences $e^t_{(i,j)}$ are passed to a GD-RCC block to get the vector representations of edges, denoted by $h^1_{(i,j)}$. For a node j, the vector representation

$h^1_{(i,j)}$ of incoming edges are aggregated and fed to a node updating function f_v to get a higher level node representation h^1_j of the node j, which is the same as the node updating process in NRI as shown in Eq. 1.

These node representations are combined by element-wise production and fed to another neural network f_e to get final edge representations $h^2_{(i,j)}$, which represents the logits of categorical distributions of edges, shown in Eq. 3. Through this process, the final edge representation $h^2_{(i,j)}$ captures not only the interaction between node i and node j but also the interactions of node i and node j with other nodes [7].

$$h^1_j = f_v(\sum_{i \neq j} h^1_{(i,j)}), \quad h^2_{(i,j)} = f_e([h^1_{(i,j)}, h^1_i \odot h^1_j]) \tag{3}$$

After supervised training, a community detection algorithm is applied to the interaction graphs to find clusters denoting groups.

4.3   GD-RCC Block

A GD-RCC block [11] is used to transform the edge sequences $e^t_{(i,j)}$ into the vector representation $h^1_{(i,j)}$. The causal convolution preserves the order of the edge sequences by using features from past time steps. With dilated convolutional kernels, the receptive fields are expanded exponentially by staking convolutional layers [11]. The skip connection, a 1D CNN, solves the gradient vanishing problem when increasing the number of layers [6]. The gating activation function, as formulated in Eq. 4, regulates the information flow and performs significantly better than rectified linear activation (ReLU) [11]:

$$e_{l+1} = tanh(W^1_l * e_l) \odot \sigma(W^2_l * e_l) \tag{4}$$

where l is the layer index. W^1_l and W^2_l are two different learnable 1D-convolution parameters of the layer l; e_l denotes the hidden states of edge features of the layer l. $*$ denotes the convolutional operation. σ and $\odot$ denote the Sigmoid function and element-wise multiplication, respectively.

A 1D convolutional layer with attentive pooling over all time steps is applied afterward to get the vector representations of the edges $h^1_{(i,j)}$. This process is visualized in Fig. 1.

During the supervised training phase, the ground-truth pairwise group relationships $G_{(i,j)}$ are used as labels; i.e., $G_{(i,j)} = 1$ denotes agent i and agent j are in the same group while $G_{(i,j)} = 0$ denotes otherwise. Due to the imbalanced distribution of the labels, the weighted cross-entropy $H(\hat{I}, G)$, as described in Eq. 5, is used as a loss function in which the rare labels are assigned higher weights:

$$H(\hat{I}, G) = -\sum_{(i,j)} [w_G G_{(i,j)} log(I_{(i,j),2}) + w_{NG}(1 - G_{(i,j)}) log(I_{(i,j),1})], \tag{5}$$

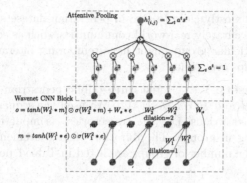

Fig. 1. 1D GD-RCC CNN block (green dashed line block) with Attentive Pooling (red dashed line block) calculated over the sequence of edges e^t. The edge feature sequences $e^{1:T}$ will be fed into a 1D GD-RCC CNN block with skip connections to get hidden states $o^{1:T}$. Here each m^i denotes a node in the first hidden layer. W_l^1 and W_l^2 denote two different learnable convolutional parameters of the layer l (the blue arrows). W_s denotes the 1D CNN skip connection (the green arrow). The hidden states $o^{1:T}$ will be fed into two 1D CNNs f_{pred} (predicts the edge representation s^t (the red arrows)) and f_{score} (predicts the attention score a^t (the yellow arrows)). The vector representations of edges is $h_{(i,j)}^1$. (Color figure online)

where $w_G = \frac{n_G + n_{NG}}{2n_G}$ and $w_{NG} = \frac{n_G + n_{NG}}{2n_{NG}}$ denote the weight of the group label and the weight of the non-group label, respectively. While n_G and n_{NG} are the number of group labels and non-group labels in the training dataset, respectively. By minimizing the weighted cross-entropy, the encoder is optimized to identify the "interaction" versus "no interaction" relation between agents.

5 Experiments

We studied the performance of our method on two types of datasets, i.e., real-world and simulated datasets. Before presenting the results, we first discuss these dataset types, the evaluation metrics, baseline measures, and the experimental setup.

5.1 Dataset

We trained and validated our model on three simulation datasets and three real-world pedestrian datasets. In pedestrian datasets, people walk in different group settings without interacting with other group members. In contrast, in the simulation datasets, cross-group interactions between particles are possible. The real-world datasets have been widely used by other researchers. Due to the lack of interaction between different groups in the pedestrian datasets, the developed methods can be tailored only to improve performance on these datasets and often are not applicable in real-world scenarios. Therefore, we chose to use simulation data, in addition to the pedestrian datasets, to increase the diversity of the datasets by considering the probability of cross-group interactions.

This probability is mostly close to zero in pedestrian datasets. This enabled us to simulate more accurately real-world communities, such as schoolyards, where pupils from a particular group might have temporary interactions with peers from different groups over time.

Pedestrian Datasets: We selected three public pedestrian datasets, namely *zara01*, *BIWI ETH* and *BIWI Hotel* [12]. We used the sequences of annotated locations of the pedestrians, i.e., the trajectories, as input features to detect pedestrians walking in groups. The duration of measurement, the number of pedestrians, and the number of groups are listed in Table 1 per dataset.

Table 1. The specification of pedestrian datasets

Dataset name	Duration(s)	Number of pedestrians	Number of groups
zara01	360.4	148	45
BIWI ETH	713.4	360	65
BIWI Hotel	722.4	389	41

Group-Interaction Simulation Datasets: To simulate group interactions, we extended the spring simulator introduced by Kipf et al. [7], which simulates the movement of particles randomly connected by a spring in a 2D box. We extended this simulation by defining groups of particles such that particles within a group have a higher probability of having interaction. In our proposed group-interaction simulation, the probability that particle v_i and particle v_j interact with each other given their group relation $G_{(i,j)}$ is formulated as follows:

$$P(I_{(i,j)} = 1|G_{(i,j)}) = 1 - exp(-a(G_{(i,j)} + b)), (a > 0, b > 0) \qquad (6)$$

where interaction and group relationship between particles v_i and v_j is denoted by $I_{(i,j)}$ and $G_{(i,j)}$, respectively. $G_{(i,j)} = 1$ if v_i and v_j are in the same group otherwise $G_{(i,j)} = 0$. The values of a and b control group interaction and non-group interaction probabilities. Specifically, the value of a controls the overall magnitude of the probabilities, and the value of b impacts the non-group interaction probability. The specification of the three simulation datasets is described in Table 2. Each dataset has 2500 simulations, which include the locations and velocities of the particles over time. The duration of each simulation is 20 s, corresponding to 50 time steps.

5.2   Evaluation Metrics

We applied Group Mitre $\Delta_{GM}(C, \hat{C})$ [13] to measure the quality of the predicted groups, where C and $\hat{C}$ are disjoint sets denoting the true groups and predicted groups, respectively. The exact procedure for calculating the Group Mitre (precision and recall) is presented in the work of Solera et al. [13], and we omitted the details due to the limit in space.

Table 2. The specification of the group-interaction simulation datasets.

Dataset	Number of particles	a	b	Probability of group interaction	Probability of non-group interaction
Simulation I	5	3	0.02	95.3%	5.8%
Simulation II	10	3	0.02	95.3%	5.8%
Simulation III	10	3	0.05	95.7%	13.9%

5.3 Baselines

We compared the results of our method with the following four baselines:

- **ATTR** [16] is a classification-based method that adopts a linear SVM to classify the binary group relationships based on hand-crafted histograms of distance, direction, and velocity. The regularisation parameter is set to 10.
- **S-SVM** [13] is a clustering-based method that uses a structured SVM to predict the pairwise similarities of the agents and further applies a correlation clustering component to predict the clusters. S-SVM is trained with the BCFW [9] algorithm. The regularisation parameter is set to 10.
- **GD-GAN** [5] is a clustering-based method that adopts an LSTM-based GAN to predict the future trajectory of agents. The DBSCAN algorithm is applied to the hidden states of the LSTM to find the groups. The dimensions of hidden states are set to 256.
- **NRI** [7] is a classification-based method extended by applying the Louvain community detection algorithm to transform the predicted pairwise interactions to the clusters denoting groups. The kernel size of the 1D convolutional layer is set to 5. The node updating and edge updating processes are MLPs with a hidden dimension of 256.

5.4 Experiment Settings

In our experiments, we set the kernel size of the GD-RCC block to five in Eq. 4. The hidden dimension size of the node and edge functions in Eq. 3 were set to 256. The stochastic gradient descent with a momentum equal to 0.9 was applied for optimization. The code to generate the group-interaction simulation datasets and to implement WavenetNRI is available in the Github repository[1].

5.5 Results

In this section, the results of our experiments are discussed. In each dataset, 60% of the samples were randomly chosen for training; 20% were randomly chosen as validation, and the remaining 20% were testing set. The results of both group-interaction simulation datasets and pedestrian datasets are listed in Table 3.

---

[1] https://github.com/fatcatZF/WavenetNRI.

According to Table 3, NRI and WavenetNRI outperformed all other baselines, and NRI performed slightly better than WavenetNRI on simulation datasets in both recall and precision of group mitre Δ_{GW}. While on pedestrian datasets, GD-GAN [5] outperformed all other methods in both measures. The proposed WavenetNRI could outperform the original NRI [7] and ATTR [16] as the two classification-based baselines.

Concerning the impact of the population size (comparing *Simulation I* and *Simulation II*), we observed that by increasing the number of particles in simulation datasets, both precision and recall were decreased for all methods, except for NRI [7]. The same behavior was observed regarding the probability of non-group interactions (comparing *Simulation I* and *Simulation III*).

Furthermore, we calculated the average pairwise Euclidean distance between the group and non-group members of the two datasets. Our investigation of the differences between these two types of datasets showed that in the pedestrian datasets, the pairwise average Euclidean distances between group members (0.950 m) were much lower than those from different groups (4.698 m), i.e., the pedestrians were closer to their group members than other groups. While in the group-interaction simulation datasets, the differences between the Euclidean distances of the same groups (1.039 m) and that of different groups (1.725 m) were not significant.

Thus, distinguishing between group members and non-group members is more challenging in the simulation datasets compared with pedestrian datasets. Moreover, the fact that baselines do not generalize to simulation datasets suggests that available research might not be applicable to real-world scenarios where there is a chance for cross-group interactions.

Table 3. Experimental results of recall (R) and precision (P) based on Group Mitre Δ_{GW}. The best average values of recall and precision are highlighted with bold text.

	Simulation I		Simulation II		Simulation III		zara01		ETH		Hotel	
	R	P	R	P	R	P	R	P	R	P	R	P
ATTR [16]	0.579 ±0.017	0.481 ±0.020	0.512 ±0.009	0.388 ±0.015	0.511 ±0.006	0.386 ±0.005	0.889 ±0.076	0.879 ±0.077	0.745 ±0.067	0.746 ±0.087	0.833 ±0.072	0.841 ±0.068
S-SVM [13]	0.664 ±0.075	0.600 ±0.067	0.529 ±0.039	0.413 ±0.017	0.459 ±0.037	0.382 ±0.030	0.893 ±0.026	0.906 ±0.033	0.887 ±0.027	0.911 ±0.021	0.925 ±0.024	0.927 ±0.030
GD-GAN [5]	0.531 ±0.003	0.430 ±0.004	0.514 ±0.003	0.383 ±0.004	0.512 ±0.003	0.383 ±0.004	**0.949** ±0.046	**0.934** ±0.051	**0.931** ±0.037	**0.950** ±0.028	**0.925** ±0.084	**0.944** ±0.058
NRI [7]	**0.995** ±0.002	**0.994** ±0.003	**0.997** ±0.002	**0.994** ±0.002	**0.998** ±0.001	**0.996** ±0.001	0.801 ±0.096	0.737 ±0.108	0.663 ±0.083	0.669 ±0.080	0.577 ±0.122	0.565 ±0.122
Wavenet-NRI	0.990 ±0.010	0.988 ±0.013	0.985 ±0.005	0.970 ±0.010	0.986 ±0.004	0.972 ±0.007	0.893 ±0.090	0.900 ±0.107	0.793 ±0.078	0.815 ±0.079	0.748 ±0.106	0.790 ±0.086

5.6   Ablation Study

Our proposed approach applied two changes to the original NRI (i.e., adding symmetric edge features and symmetric edge updating process and the GD-RCC block). In this section, we explored the effects of these changes by performing an ablation study. To test the impact of the symmetric edge features and symmetric edge updating process, the same 1D convolutional as the original NRI with the symmetric edge features and the symmetric edge updating process was applied. This model is called "NRI-Symmetric". To test the effects of the GD-RCC block, "Wavenet-GD-RCC" was designed, which used the GD-RCC block with the same edge features and edge updating process as the original NRI. We compared the performance of these two methods with the proposed WavenetNRI and the original NRI on the simulation and pedestrian datasets. The results of both experiments are listed in Table 4. According to the results listed in Table 4, the Wavenet-GD-RCC performed slightly better than NRI, while the performance of NRI-Symmetric was lower than NRI. Therefore, the GD-RCC block could slightly improve the performance of NRI on the group-interaction datasets, and the symmetric edges and symmetric edge updating process negatively affected the original NRI. Additionally, the NRI-Symmetric performed better than the NRI, and Wavenet-GD-RCC performed similarly to NRI on the pedestrian data sets. Therefore, the symmetric edge features with the symmetric edge updating process could improve the performance of NRI on the pedestrian data sets, and the GD-RCC block did not significantly affect NRI's performance. Thus, the results were consistent per dataset type but not overall. We also noticed that either change could add value to one dataset category. As discussed earlier, the complexity of the simulation datasets in the behavior and interactions of the group members and non-group members might explain the inconsistent performance in these two types of datasets.

Table 4. Ablation study results of recall (R) and precision (P) based on Group Mitre Δ_{GW}. The best average values of recall and precision are highlighted with bold text.

	Simulation I		Simulation II		Simulation III		zara01		ETH		Hotel	
	R	P	R	P	R	P	R	P	R	P	R	P
NRI [7]	0.995 ±0.002	0.994 ±0.003	0.997 ±0.002	0.994 ±0.002	**0.998** ±0.001	0.996 ±0.001	0.801 ±0.096	0.737 ±0.108	0.663 ±0.083	0.669 ±0.080	0.577 ±0.122	0.565 ±0.122
NRI-Symmetric	0.990 ±0.004	0.987 ±0.006	0.981 ±0.007	0.964 ±0.013	0.981 ±0.007	0.961 ±0.009	0.851 ±0.093	0.813 ±0.091	0.679 ±0.094	0.686 ±0.096	0.708 ±0.121	0.739 ±0.115
Wavenet-GD-RCC	**0.998** ±0.002	**0.997** ±0.001	**0.999** ±0.001	**0.997** ±0.002	**0.998** ±0.001	**0.997** ±0.001	0.719 ±0.138	0.625 ±0.165	0.542 ±0.146	0.530 ±0.147	0.566 ±0.169	0.554 ±0.163
Wavenet NRI	0.990 ±0.010	0.988 ±0.013	0.985 ±0.005	0.970 ±0.010	0.986 ±0.004	0.972 ±0.007	**0.893** ±0.090	**0.900** ±0.107	**0.793** ±0.078	**0.815** ±0.079	**0.748** ±0.106	**0.790** ±0.086

6   Discussion and Conclusions

The present study explored the application of GNN by extending the NRI model [7] for group detection in two directions: (1) by applying symmetric edge features with symmetric edge updating processes and (2) by replacing the 1D convolution layer with a GD-RCC block, as proposed by Wavenet [11]. We compared the performance of WavenetNRI with other baselines on the three group-interaction simulation datasets and three pedestrian datasets. NRI and WavenetNRI outperformed all other baselines on the group-interaction simulation datasets. Although the pedestrian datasets were captured in real-world setups, the simulation datasets were better reflecting complex group interactions with larger groups, which stresses the importance of the obtained results. On the pedestrian datasets, although our proposed method did not compete against the clustering-based baselines, i.e., GD-GAN [5] and S-SVM [13], it outperformed classification-based methods, i.e., ATTR [16] and the original NRI [7]. Yet, baseline methods did not generalize very well to the simulation datasets. We further evaluated the effects of our changes to the original NRI in the ablation study. We found that on the group-interaction data sets, the GD-RCC block slightly improved the performance of NRI. Simultaneously, the symmetric edge features with symmetric edge updating processes negatively affected the performance of NRI. On the pedestrian data sets, the symmetric edge features with symmetric edge updating processes improved the performance of NRI, while the GD-RCC block had no significant effect on NRI.

Our analysis demonstrates that WavenetNRI is highly effective at predicting pairwise interactions, which ultimately reflect the group memberships of agents in an interacting environment. One drawback of the proposed method is its dependency on ground truth data. Unsupervised methods such as GD-GAN are preferable if ground truth is not available for a particular study. Many real-world communities, such as sports clubs and schoolyards, can be understood as a dynamic interacting system, where applying a trained WavenetNRI model can be helpful in predicting group memberships within the system.

The current study can be improved by investigating how to adapt the proposed neural network design more efficiently to different datasets using meta-learning. Additionally, it is worth studying how to extend the proposed classification-based method to a supervised clustering task. And finally, designing a fully supervised model by adding a final layer to classify nodes into the group they belong to could be investigated in the future.

References

1. Baratchi, M., Meratnia, N., Havinga, P.J.: On the use of mobility data for discovery and description of social ties. In: Proceedings of the IEEE/ACM International Conference on Advances in Social Networks Analysis and Mining, pp. 1229–1236 (2013)
2. Battaglia, P.W., et al.: Relational inductive biases, deep learning, and graph networks. arXiv preprint arXiv:1806.01261 (2018)
3. Blondel, V.D., Guillaume, J.L., Lambiotte, R., Lefebvre, E.: Fast unfolding of communities in large networks. J. Stat. Mech: Theory Exp. **2008**(10), P10008 (2008)
4. Chon, Y., Kim, S., Lee, S., Kim, D., Kim, Y., Cha, H.: Sensing WiFi packets in the air: practicality and implications in urban mobility monitoring. In: Proceedings of the ACM International Joint Conference on Pervasive and Ubiquitous Computing, pp. 189–200 (2014)
5. Fernando, T., Denman, S., Sridharan, S., Fookes, C.: GD-GAN: generative adversarial networks for trajectory prediction and group detection in crowds. In: Jawahar, C.V., Li, H., Mori, G., Schindler, K. (eds.) ACCV 2018. LNCS, vol. 11361, pp. 314–330. Springer, Cham (2019). https://doi.org/10.1007/978-3-030-20887-5_20

6. He, K., Zhang, X., Ren, S., Sun, J.: Deep residual learning for image recognition. In: Proceedings of the IEEE Conference on Computer Vision and Pattern Recognition, pp. 770–778 (2016)
7. Kipf, T., Fetaya, E., Wang, K.C., Welling, M., Zemel, R.: Neural relational inference for interacting systems. In: Proceedings of the International Conference on Machine Learning, pp. 2688–2697. PMLR (2018)
8. Kumar, S., Gu, Y., Hoang, J., Haynes, G.C., Marchetti-Bowick, M.: Interaction-based trajectory prediction over a hybrid traffic graph. In: Proceedings of the IEEE/RSJ International Conference on Intelligent Robots and Systems, pp. 5530–5535. IEEE (2021)
9. Lacoste-Julien, S., Jaggi, M., Schmidt, M., Pletscher, P.: Block-coordinate Frank-Wolfe optimization for structural SVMs. In: Proceedings of the International Conference on Machine Learning, pp. 53–61. PMLR (2013)
10. Nasri, M., et al.: A novel data-driven approach to examine children's movements and social behaviour in schoolyard environments. Children 9(8), 1177 (2022)
11. Oord, A.V.D., et al.: WaveNet: a generative model for raw audio. arXiv preprint arXiv:1609.03499 (2016)
12. Pellegrini, S., Ess, A., Schindler, K., Van Gool, L.: You'll never walk alone: modeling social behavior for multi-target tracking. In: Proceedings of the IEEE International Conference on Computer Vision, pp. 261–268. IEEE (2009)
13. Solera, F., Calderara, S., Cucchiara, R.: Socially constrained structural learning for groups detection in crowd. IEEE Trans. Pattern Anal. Mach. Intell. 38(5), 995–1008 (2015)
14. Thompson, S., Gupta, A., Gupta, A.W., Chen, A., Vázquez, M.: Conversational group detection with graph neural networks. In: Proceedings of the International Conference on Multimodal Interaction, pp. 248–252 (2021)
15. Tsochantaridis, I., Hofmann, T., Joachims, T., Altun, Y.: Support vector machine learning for interdependent and structured output spaces. In: Proceedings of the International Conference on Machine Learning, p. 104 (2004)
16. Yamaguchi, K., Berg, A.C., Ortiz, L.E., Berg, T.L.: Who are you with and where are you going? In: Proceedings of the IEEE Conference on Computer Vision and Pattern Recognition, pp. 1345–1352. IEEE (2011)

Discovering Rule Lists with Preferred Variables

Ioanna Papagianni[✉][iD] and Matthijs van Leeuwen[iD]

LIACS, Leiden University, Leiden, The Netherlands
{i.papagianni,m.van.leeuwen}@liacs.leidenuniv.nl

Abstract. Interpretable machine learning focuses on learning models that are inherently understandable by humans. Even such interpretable models, however, must be trustworthy for domain experts to adopt them. This requires not only accurate predictions, but also reliable explanations that do not contradict a domain expert's knowledge. When considering rule-based models, for example, rules may include certain variables either due to artefacts in the data, or due to the search heuristics used. When such rules are provided as explanations, this may lead to distrust.

We investigate whether human guidance could benefit interpretable machine learning when it comes to learning models that provide both accurate predictions and reliable explanations. The form of knowledge that we consider is that of *preferred variables*, i.e., variables that the domain expert deems important enough to be given higher priority than the other variables. We study this question for the task of multiclass classification, use probabilistic rule lists as interpretable models, and use the minimum description length (MDL) principle for model selection.

We propose S-CLASSY, an algorithm based on beam search that learns rule lists and takes preferred variables into account. We compare S-CLASSY to its baseline method, i.e., without using preferred variables, and empirically demonstrate that adding preferred variables does not harm predictive performance, while it does result in the preferred variables being used in rules higher up in the learned rule lists.

Keywords: Classification · Probabilistic rule lists · Minimum description length (MDL) principle · Human-guided machine learning

1 Introduction

Explainable Artificial Intelligence (XAI) and interpretable machine learning [10] are important topics that currently attract a lot of attention within and outside the academic community. Although the two fields are clearly related in that both aim to provide explanations for predictions (or other outcomes) given by AI systems, they usually refer to slightly different approaches. XAI approaches typically attempt to provide post-hoc explanations for predictions [16], which can be done for any type of predictive model—whether it's a complex, 'black box' model such as a neural network, or a simpler model such as a linear regression

B. Crémilleux et al. (Eds.): IDA 2023, LNCS 13876, pp. 340–352, 2023.
https://doi.org/10.1007/978-3-031-30047-9_27

model. Interpretable machine learning, on the other hand, focuses on learning *interpretable models*, models that are inherently understandable by humans—such as linear regression models and rule- and tree-based models.

In application domains where high-stake decisions are made, such as law and health care, predictive models are used for *decision support*, i.e., assisting the domain experts making decisions rather than autonomously making decisions. This calls for *human-centred AI*, where machine learning models augment human experts rather than replace them. This leads to extra requirements on models and algorithms: for domain experts to adopt an AI system, it must be trustworthy, i.e., it must not only provide accurate predictions, but also reliable explanations.

Providing reliable explanations is by no means a simple feat. Models are often learned from relatively small datasets—especially in high-stake settings where data is typically expensive—and certain associations may be perceived as more reliable than others. In health care, for example, a medical doctor will only trust explanations using patient properties of which they think a relationship with the target variable is plausible. Explanations that contradict a domain expert's knowledge, in contrast, are likely to be detrimental to their trust.

We argue that interpretable machine learning has an advantage over XAI in such settings, because interpretable models make it easier to explain how and why predictions are made. Nevertheless, this does not imply that the predictions made by interpretable models are "right for the right reasons" [15]. When considering rule-based models, for example, rules may be based on certain variables either due to associations in parts of the data, or due to the search heuristics used. When such rules are provided as explanations, this will lead to distrust.

Approach and Contributions. In this paper we investigate whether human guidance could benefit interpretable machine learning when it comes to learning models that provide both accurate predictions and reliable explanations. More specifically, we study whether prior knowledge provided by a domain expert may lead to models consistent with that knowledge. This can be seen as an instance of *informed machine learning* [17], in which prior knowledge (given as, e.g., knowledge graphs or human feedback) informs the learning process.

The form of knowledge that we consider is that of *preferred variables*, i.e., variables that the domain expert deems important enough to be given higher priority than the other variables while learning a predictive model. The idea is that specifying detailed knowledge is often hard, but experts will usually have a good idea of which variables they expect to be the most informative with regard to the variable of interest, for which predictions are to be made.

We consider the task of multiclass classification, because it is one of the most commonly studied and practically used machine learning tasks. As models we use probabilistic rule lists, because they are interpretable and we have recently introduced algorithms for finding good rule lists using the minimum description length (MDL) principle as model selection criterion [11,12]. That is, we use compression as optimisation criterion, which has as most notable advantages that it makes hyper-parameter tuning unnecessary and avoids overfitting.

After discussing related work in Sect. 2, we motivate and formalise the problem of discovering rule lists with preferred variables in Sect. 3. Following this, in

Sect. 4 we propose S-CLASSY, an algorithm based on beam search that learns rule lists and takes preferred variables into account by first exploring and considering rules that include at least one preferred variable. Section 5 empirically investigates the effect of having preferred variables on compression, runtime, predictive accuracy, overfitting, and rule list size. For this we simulate external knowledge by ranking variables by feature importances that we obtained with random forests. We compare S-CLASSY to its baseline method BCLASSY, which does not use the preferred variables, on commonly used benchmark datasets. The results demonstrate that augmenting the rule learning process with background knowledge—in the form of preferred variables—does not harm any of the major evaluation criteria, while it does result in the preferred variables being used in rules higher up in the rule list.

2   Related Work

Based on the type of model used, rule learning can be roughly divided in *rule list* learning, *rule set* learning, and mixtures of both. Well-known classification algorithms such as RIPPER [3], C4.5 [13], FURIA [6], and unordered-CN2 [2] use an ordered *one-vs-all* approach to learn rules for the multiclass classification problem, as a result of which they essentially return ordered lists of rule sets. Such lists of sets are harder to interpret than 'plain' rule lists or rule sets. CBA [8] uses large numbers of association rules, which also hampers interpretability.

Direct rule set learning methods include IDS [7] and DRS [19], but unlike our approach they are not probabilistic. TURS [18] is a recent method for learning 'truly unordered' probabilistic rule sets, using a surrogate score to tackle incomplete rule sets. Another recent approach is CLASSY [12], a state-of-the-art algorithm for learning ordered rule lists. It discovers probabilistic rules for multinomial targets with both categorical and quantitative predictive variables. Both TURS and CLASSY use the minimum description length (MDL) principle [5] as model selection criterion to select rules that compress the data well but have a relatively low model complexity. While CLASSY uses a pre-mined set of candidate patterns, Proença et al. [11] later proposed SSD++, an improved algorithm that directly finds good rule lists using beam search for candidate generation. Although SSD++ was introduced for subgroup list discovery, it can just as well be be used for classification; we will call this beam search version BCLASSY.

As far as we are aware, how to influence search in rule learning using background knowledge has hardly been studied. In subgroup discovery, IDSD [4] is an interactive search where the user can influence the beam of a beam search by providing feedback (like/dislike). This results in erratic search behaviour though.

3   Rule Learning with Preferred Variables

We start with important definitions and notation in Subsect. 3.1, after which we introduce the problem statement in Subsect. 3.2 and briefly summarise the model and data encoding that we will use in Subsect. 3.3.

3.1 Data, Rules, and Rule Lists

Let $D = (X, Y)$ be a supervised dataset, consisting of a dataset X and a (multi)class label vector Y. Let $\mathcal{X}$ and $\mathcal{Y}$ be the instance space and the set of all $|\mathcal{Y}|$ classes, respectively. Let $V = \{v_1, v_2, ..., v_m\}$ be the set of all $m = |V|$ variables in X, with each v_i representing a one-dimensional variable with domain $\text{dom}(v_i)$. Each $(x, y) \in D$ is a record, where instance $x = (x_1, x_2, ..., x_m) \in \mathcal{X}$ is a vector of values with $x_i \in \text{dom}(v_i)$ for each v_i, and $y \in \mathcal{Y}$ is the class label belonging to the instance. Dataset D has $n = |D|$ records.

We are interested in learning rules from data. Here, a *rule* r is a conditional statement that links occurrences of patterns to class probabilities. More precisely, a rule is a pair $r = (p, \pi(p))$, where antecedent p is a pattern and its consequent is a probability distribution $\pi(p)$. A *pattern* is a conjunction of conditions over variables, e.g., $p = [v_1 = \text{'A'} \wedge v_3 = 0]$. Further, $\pi(p)$ is a categorical probability distribution $\pi = (\pi^{y_1}, \pi^{y_2}, ..., \pi^{y_{|\mathcal{Y}|}})$ over all class labels $\mathcal{Y}$. An example rule could be *if* $[v_1 = \text{'A'} \wedge v_3 = 0]$ *then* $\pi^{y_1} = 0.85, \pi^{y_2} = 0.05, \pi^{y_3} = 0.10$.

A *probabilistic rule list* (PRL) R is an ordered list of $l + 1$ rules $(r_1, r_2, ..., r_l, r_\emptyset)$, where the last rule in the list, $r_\emptyset$, is called the default rule. It has the empty set as antecedent and is assigned a probability distribution $\pi_\emptyset$.

The *usage* of a pattern $p \in R$ is the number of its occurrences in a dataset D, disregarding all instances that were covered by previous patterns in R, i.e.,

$$usg(p_i \mid R, D) = |\{x \subset D \mid p_i \sqsubseteq x \wedge (\bigwedge_{\forall j < i} p_j \not\sqsubseteq x)\}|, \tag{1}$$

where $p \sqsubseteq x$ denotes that pattern p *occurs* in instance x, i.e., x satisfies all conditions in p, and $\not\sqsubseteq$ is the reverse. The *label-oriented usage* of a pattern $p_i \in R$ is the number of pattern occurrences in dataset D that correspond to class label l, where $D^{y=l} = \{(x, y) \subset D \mid y = l\}$ is the subset of D where class label l occurs:

$$usg(p_i \mid R, D^{y=l}) = |\{x \subset D^{y=l} \mid p_i \sqsubseteq x \wedge (\bigwedge_{\forall j < i} p_j \not\sqsubseteq x)\}|. \tag{2}$$

We consider the problem of rule learning for *multiclass (or multinomial) classification*, meaning that it is our aim to learn a rule list from a given supervised dataset such that it can accurately predict the class labels for unseen instances.

3.2 Problem Statement

As mentioned in the previous section, the minimum description length (MDL) principle [5] has previously been successfully used for rule learning [12,18]. Informally, the principle states that the best model is the one that best compresses the data together with the model. Formally, given a (training) supervised dataset D and a corresponding model class $\mathcal{R}$, consisting of all possible rule lists for D, the optimal rule list R^* is given by

$$R^* = \underset{R \in \mathcal{R}}{\arg\min}\, L(D, R) = \underset{R \in \mathcal{R}}{\arg\min}\, [L(R) + L(Y|X, R)], \tag{3}$$

where $L(R)$ is the encoded length, in bits, of the rule list and $L(Y|X, R)$ is the encoded length, in bits, of class labels Y given data X and rule list R.

This is the same problem formalisation as was previously used for CLASSY [12], and was shown to result in compact rule lists that performed well in terms of predictive performance. One advantage of using the MDL principle is that it automatically protects from overfitting by balancing model complexity with goodness of fit, hence cross-validation for hyperparameter tuning is not necessary.

Since finding the optimal rule list R^* is a hard problem, heuristic algorithms— such as CLASSY and BCLASSY—are used in practice. Although predictive performance of the resulting rule lists may be excellent, the patterns used may be less than ideal to a domain expert due to two reasons: 1) the optimal rule list may not be found due to the use of heuristic search; and 2) under certain circumstances multiple variables may lead to equally 'good' rules, in which case one of those is arbitrarily chosen and used. The latter may happen, for example, when two variables are strongly associated. For high-stake decisions it is crucial that a model uses the 'right' variables for a prediction though, so that the patterns can be served to domain experts as explanations and gain their trust.

Because of the second reason, improving the learning algorithms is unlikely to ever completely address this issue: in practice only a limited sample of data is available, and that may contain insufficient information to be able to choose the 'right' variables. We therefore argue that it may be needed to integrate *external knowledge* in the learning process in order to obtain more reliable explanations.

As an initial step in this direction, we investigate whether injecting limited expert knowledge may be helpful in guiding the heuristic search to rule lists that are at least equally predictive but use patterns that are potentially more informative to domain experts than if no such knowledge is provided.

More specifically, we assume that we have access to a *domain expert* who is knowledgeable on the domain of the classification problem under consideration. The domain expert specifies a (small) set of *preferred variables* $U \subset V$ of which they are convinced they could and should be used for predicting the target variable Y. The preferred variables should be given higher priority during the search for a rule list than the remaining variables, i.e., $V \setminus U$, meaning that they should be considered for pattern growth first. Note that this does not mean that the preferred variables should be used regardless of the data; if the domain expert is wrong, this should not result in models with poor predictive performance.

3.3   Encoding

For the code length of the model and the code length of the data given the model, i.e., $L(R)$ and $(Y|X, R)$, respectively, we use the same encoding as used by CLASSY [12]. We here only provide a brief overview.

Model Encoding. We use the *universal code for integers*[1] $L_{\mathbb{N}}(i)$ to penalise for rule length, while the *uniform code* provides a means to assign equal-length

---

[1] $L_{\mathbb{N}}(i) = \log^* i + \log \lambda$, where $\log^* i = \log i + \log \log i + \dots$ and constant $\lambda \approx 2.865064..$

codes to variables and values of variables. The length of a pattern p_i is given by the number of conditions in that pattern encoded by the universal code for integers, and then each condition c_j in p_i is encoded using uniform codes for the variables and values, i.e., $L(p_i) = L_\mathbb{N}(|p_i|) + \sum_{c_j \in p_i} (\log |V| + \log \mathrm{dom}(v_j))$. Now, the total length of a probabilistic rule list R is computed as the sum of the number of rules and the lengths of the individual patterns, given by

$$L(R) = L_\mathbb{N}(|R|) + \sum_{p_i \in R} L(p_i). \tag{4}$$

Data Encoding. For the encoding of the class label vector the *prequential plug-in code* is used, which at each stage is the optimal code given the data so far (i.e., it *sequentially* predicts the next symbol). It is given by

$$\pi_{plug-in}(y_i = l | Y_{i-1}) := \frac{|\{y \in Y_{i-1} | y = l\}| + \epsilon}{\sum_{k \in y} |\{y \in Y_{i-1} | y = k\}| + \epsilon}, \tag{5}$$

where y_i is the ith class label, $Y_{i-1} = \{y_1, ..., y_{i-1}\}$ is the sequence of the $i - 1$ first class labels, and $\epsilon = 1$ (for a uniform prior). The above can be expressed by the maximum likelihood estimator (MLE) π_i^l for any probability $\pi(y = l | p_i)$, any rule p_i and any class label l. The Laplace smoothing (pseudocount ϵ) is added to the equation of the maximum likelihood estimator to all label-oriented usages to avoid probabilities of zero. Then, the *smoothed* MLE is formalised as

$$\pi_i^l = \frac{usg(p_i | R, D^{y=l}) + \epsilon}{usg(p_i | R, D) + |\mathcal{Y}|\epsilon}. \tag{6}$$

4 Beam Search with Preferred Variables

Rule learning is generally a hard problem, and finding the MDL-optimal rule list is certainly hard—heuristic algorithms are therefore common practice. The original CLASSY algorithm [12] iteratively selected patterns from a pre-mined candidate set. The SSD++ algorithm [11] improved on this by means of a beam search; although SSD++ is aimed at finding subgroup sets, Proença's dissertation [9] has shown that rule learning and subgroup discovery are closely related. We here employ the SSD++ beam search algorithm for learning rule lists as in CLASSY, and dub this beam search variant of the algorithm BCLASSY. Using the beam search has several advantages: 1) there is no need to pre-mine candidates, allowing to prune the search space as the search progresses; 2) on-the-fly discretisation can be used, giving better results for quantitative variables [11]. An additional advantage that is of particular importance to us is that the beam search allows to guide the search using preferred variables.

We propose S-CLASSY, a greedy algorithm based on BCLASSY that starts with a rule list consisting of only the default rule and iteratively adds rules until compression cannot be improved, taking into account the preferred variables as given by the domain expert. Before we describe the algorithm in detail we briefly summarise how compression gain is computed.

Algorithm 1. S-CLASSY algorithm

Input: Dataset D, set of preferred variables S, beam width ω_b, maximum pattern length $|r|_{max}$, minimum support threshold m_s
Output: Probabilistic rule list R

1: $R \leftarrow [r_\emptyset]$ ▷ Start with default rule
2: **while** True **do**   ▷ Repeat while compression can be improved
3:      $Cands \leftarrow \emptyset$
4:      **for** $s \in S$ **do**   ▷ Perform beam search for each preferred variable
5:          $Cands \leftarrow Cands \cup BeamSearch(s, R, D, \omega_b, |r|_{max}, m_s)$
6:      **end for**
7:      **if** $Cands = \emptyset$ **then**   ▷ No candidates? Perform full beam search
8:          $Cands \leftarrow BeamSearch(\emptyset, R, D, \omega_b, |r|_{max}, m_s)$
9:      **end if**
10:     $r \leftarrow \text{argmax}_{r' \in Cands}\, \delta L(D, R \oplus r')$ ▷ Pick rule that maximises normalised gain
11:     **if** $\delta L(D, R \oplus r) > 0$ **then**
12:        $R \leftarrow R \oplus r$ ▷ Add rule to rule list
13:        $S \leftarrow S \setminus VariablesInPattern(r)$ ▷ Update preferred variable set
14:     **else**
15:        **return** R ▷ Return final rule list
16:     **end if**
17: **end while**

Compression Gain. To find a good rule list according to Eq 3, in each iteration we aim to find that rule that improves compression the most. To this end, absolute compression gain ΔL is defined as $\Delta L(D, R \oplus r) = L(D, R) - L(D, R \oplus r)$, i.e., the number of bits gained by adding a rule r to rule list R. Since greedy search combined with absolute compression gain has been shown to favour fewer rules that are less accurate but cover more instances [12], we also use the *normalised compression gain*. The normalised gain $\delta L(D, R \oplus r)$ is the absolute gain normalised by the usage of the corresponding pattern p, $\delta L(D, R \oplus r) = \frac{\Delta L(D, R \oplus r)}{usg(p \mid R, D)}$.

Algorithm. S-CLASSY is given by Algorithm 1. It starts by initialising a rule list to the default rule (Ln 1). Then, one rule is added in each iteration of the main loop (Ln 2–17) until no rule that improves compression can be found and the resulting rule list is returned (Ln 15). In each iteration, first a beam search is done for each preferred variable (Ln 4–6), starting the search from the given preferred variable s (i.e., it only considers patterns that include a condition on s). If S is empty or the previous beam searches did not result in any candidate rules, then a beam search considering all possible rules is conducted (Ln 7–9). In all calls to the beam search, the beam width, maximum pattern length, and minimum support threshold hyper-parameters are given to constrain the search.

After all beam search procedures have been completed, the candidate rule with the largest normalised compression gain is selected (Ln 10). If its compression gain is larger than 0 (Ln 11), it improves overall compression (Eq. 3) and is thus added to the rule list (Ln 12). Note that all rules are added at the end of the rule list, but just before the default rule (which is always updated to reflect

Table 1. *Dataset characteristics*: number of records ($|D|$); total number of values for all v_i, denoted $|\mathbf{x}|$; number of variables m; number of class labels ($|\mathcal{Y}|$); size of set S, denoted K. Further, $*$ indicates dataset characteristics after the removal of rows with *NaN* values. *Relative compression* ($L\%$) and *runtime* (*sec*) are averaged over all 10 folds using the top-K set of preferred variables S.

Dataset	Characteristics					$\mu(L\%)$		$\mu(sec)$	
	$|D|$	$|\mathbf{x}|$	m	$|\mathcal{Y}|$	K	S-Classy	bClassy	S-Classy	bClassy
Breast*	683	16	9	2	1	**0.62**	**0.62**	**0.12**	0.13
Cong. voting*	231	34	16	2	2	**0.75**	**0.75**	**0.37**	0.42
Dermatology*	358	49	12	6	2	**0.47**	**0.47**	**3.09**	3.11
Heart*	297	50	13	5	2	0.12	0.13	**1.14**	1.21
Ionosphere	351	157	34	2	4	**0.39**	**0.39**	**3.35**	3.48
Iris	150	19	4	3	1	**0.75**	**0.75**	**0.31**	0.33
Led7	3200	24	7	10	1	**0.51**	**0.51**	6.21	**6.11**
Letter	20000	102	16	26	2	**0.50**	**0.50**	**822.08**	823.55
Mushroom*	5644	80	21	2	3	**0.97**	**0.97**	**1.94**	1.99
Pen digits	10992	86	16	10	2	**0.84**	**0.84**	294.91	**294.09**
Pima Indians	768	38	8	2	1	**0.10**	**0.10**	**0.01**	**0.01**
Tic-tac-toe	958	29	9	2	1	**0.46**	**0.46**	**1.22**	1.26
Waveform	5000	101	21	3	3	**0.44**	**0.44**	**42.75**	42.76
Wine	178	68	13	3	3	0.62	**0.64**	**2.78**	2.94

the class distribution of the uncovered instances). Finally, the set of preferred variables is updated: any of the preferred variables used in the pattern of rule r are removed from S, and they will not be given priority in further iterations.

Note that the algorithm only adds rules that improve overall compression. In that sense the preferred variables can help guide the search, but if no viable rules using those variables are found the algorithm falls back to using other variables—if the knowledge provided by a domain expert is not in agreement with the evidence in the data, then this cannot have a negative impact.

5 Experiments

All experiments[2] use a minimum support threshold of $m_s = 5\%$ and maximum pattern length of $|r|_{max} = 4$, following the baseline comparisons of Proença [12].

Data. We evaluate our algorithm using 14 discrete-valued datasets publicly available from LUCS/KDD[3], see Table 1 for their characteristics. We randomise the order of the instances before splitting into folds for 10-fold cross-validation.

---

[2] The source code is available at: https://github.com/ioannapap/S-CLASSY.

[3] https://cgi.csc.liv.ac.uk/~frans/KDD/Software/LUCS-KDD-DN/DataSets/dataSets.html#datasets.

Simulating Knowledge. As no expert knowledge is available for these datasets and we aim for reproducible results, we choose to simulate expert knowledge. For fairness we do not wish to use BCLASSY or other interpretable models for this. Instead we use Random Forest (RF), which is widely implemented and often used in real world applications [1,14]. We train a random forest on the entire dataset and rank all variables based on the entropy-based feature importance scores. Next, we select a set S of K variables as preferred variables, where K depends on the number of variables in a dataset. We use $K = m \cdot 10\%$, rounded upwards, which is a small but substantial percentage of the total number of variables m. For our empirical evaluation, we consider three sets of variables as preferred variables to be given as input: 1) top-K, the K highest ranked variables; bottom-K, the K lowest ranked variables; and 3) random-K, K variables that are selected uniformly at random from V (once, before all experiments are done).

Evaluation Criteria. We evaluate the algorithm on 1) compression, 2) runtime, 3) classification performance, 4) overfitting, and 5) interpretability. For the compression criterion, we calculate *relative compression* gain as

$$L\% = 1 - \frac{L(D, R)}{L(D, \{\emptyset\})}. \tag{7}$$

That is, it is the compressed size of the data given the final rule list $L(D, R)$ over the compressed size of the data given the rule list with only the default rule $L(D, \{\emptyset\})$. We subtract the fraction from 1, so that the closer to 1 the relative compression gets, the better. We use a timer to measure runtime in seconds for every fold and then average it over all 10 folds ($\mu(sec)$). Similar to [12], we check the *Area Under the ROC Curve* (AUC) to quantify the *classification performance*. We weigh per class binary $AUCs$ with their marginal frequencies since the majority of the datasets are multinomial.

Overfitting is here evaluated as the mean absolute difference in AUC between the train and test set, i.e., $|\mu(AUC)_{train} - \mu(AUC)_{test}|$. How to evaluate the interpretability of a rule-based model is a complex topic on itself. A minimum requirement for a rule list to be interpretable is that it needs to be small, i.e., it must contain relatively few rules that are not too long. We therefore quantify *interpretability* using average rule length ($\mu|r|$) and the average number of rules in a rule list ($\mu|R|$). Lastly, we are interested in investigating whether the preferred variables influenced the rules learned. For this we use the frequencies f and positions of the preferred variables in the learned rules. Specifically, we care mostly about the frequency of preferred variables in the first rule of each rule list, which we annotate by $f@1$ and average over all folds.

5.1   Results

Regarding relative compression and runtime, we recognise that when the top-K preferred variable set S is used, S-CLASSY performs similarly to BCLASSY, see Table 1. S-CLASSY's runtime is slightly lower overall. Our algorithm also ranks first regarding accuracy when using the top-K, as seen in Table 2. This is a

Table 2. Mean (μ) results per dataset using 10-fold cross-validation, with fixed $m_s = 5\%$, $\omega_b = 100$ and $|r|_{max} = 4$. Bottom, random and top are the three different sets of simulated 'preferred variables' used for S-CLASSY. AUC is the *Area Under the ROC Curve* in test set, $|\mu(AUC)_{train} - \mu(AUC)_{test}|$ is the overfitting. Then, $f@1$ and $\max(f@1)$ is the frequency of *all* preferred variables and the maximum (highest) frequency of a preferred variable respectively, both at 1st position of the rule list, for S-CLASSY (abbreviated S-CL) and BCLASSY (abbreviated BCL).

Dataset	$\mu(AUC)_{test}$					$\|\mu(AUC)_{train} - \mu(AUC)_{test}\|$					$\mu(f@1)$		$\max(f@1)$	
	S-CLASSY			BCLASSY	RF	S-CLASSY			BCLASSY	RF	S-CL	BCL	S-CL	BCL
	bottom	random	top			bottom	random	top			top		top	
Breast	0.50	0.94	0.94	0.94	**0.95**	**0**	0.007	0.007	0.007	0.007	**0**	1	1	1	1
Cong. voting	**0.97**	**0.97**	**0.97**	0.97	0.96	**0.002**	**0.002**	**0.002**	0.002	0.007	**0.50**	0.50	1	1
Dermatology	0.50	**0.86**	**0.86**	0.86	0.73	0	0.029	0.034	0.029	0.006	**0.50**	0	**0.60**	0
Heart	0.50	0.66	**0.67**	0.66	**0.67**	0	0.018	0.018	0.019	0.012	0.95	1	1	1
Ionosphere	0.54	0.84	0.85	0.85	**0.90**	**0.004**	0.081	0.072	0.075	0.009	0.25	**0.47**	1	1
Iris	**0.95**	**0.95**	**0.95**	0.95	0.95	0.019	0.019	0.019	0.019	0.022	1	0	1	0
Led7	**0.84**	**0.84**	**0.84**	0.84	0.50	0.005	0.005	0.005	0.005	**0.001**	1	1	1	1
Letter	**0.79**	**0.79**	**0.79**	0.79	0.50	0.007	0.008	0.007	0.011	**0**	1	0.45	1	0.90
Mushroom	0.50	1	**1**	1	0.99	0	0	**0**	0	0	0.33	0.33	1	1
Pen digits	0.97	0.97	**0.98**	0.97	0.50	0.010	0.010	0.011	0.011	**0**	0.33	0.33	1	1
Pima Indians	0.50	0.50	0.66	0.66	**0.67**	0	0	0.001	0.001	0.012	1	1	1	1
Tic-tac-toe	**0.87**	**0.87**	**0.87**	0.87	0.64	**0.009**	**0.009**	**0.009**	0.009	0.025	1	1	1	1
Waveform	0.82	0.82	**0.83**	0.82	0.77	0.026	0.026	0.026	0.026	**0.006**	0.57	0.50	1	1
Wine	0.92	0.92	0.92	0.92	**0.94**	0.058	0.058	0.058	0.058	**0.034**	0.37	0.33	1	0.90
$rank_{all}$	2.86	1.86	**1.29**	1.57	3.14	**1.36**	2.21	2.36	2.57	2.36	**1.14**	1.36	1	1.29

good result, as it shows that adding preferred variables does not harm predictive performance and may even benefit it. When using the bottom-K variables as background knowledge, performance is worse than for the other variants and also worse than BCLASSY. This indicates that providing preferred variables is not entirely without risk: despite our goals, poorly chosen variables may result in worse predictive performance. All variants of S-CLASSY perform better overall than RF $(rank_{all})$[4], which is interesting because the RF-based feature importance does benefit S-CLASSY. With regard to overfitting, bottom and random sets do well but that is to be expected; poor predictions on training data are still poor on test data. More interestingly, the results suggest that providing informative preferred variables potentially leads to less overfitting, as S-CLASSY with 'top' evaluation sets ranks higher than BCLASSY and equal to RF with regard to overfitting.

Clearly, the average number[5] of conditions and rules that were discovered in all different sets of S-CLASSY and BCLASSY, presented in Table 3, show little to no difference with BCLASSY scoring overall first. However, when we calculate the *Jaccard distance*[6] between the conditions of BCLASSY rules and S-CLASSY rules per rule position, we discover that the rules are in fact different. The non-

---

[4] $Rank_{all}$ (smaller is better) is the average rank over all datasets.

[5] The lowest (> 0) $\mu|r|$, $\mu|R|$ the better, 0 is treated as the worst.

[6] For *Jaccard distance*, the closer to 0 the more similar and the closer to 1 the more different (preferred state).

Table 3. Mean (μ) number of 1) conditions in a rule r ($|r|$); and 2) rules in a rule list R ($|R|$) per experiment, using 10-fold cross validation with fixed $m_s = 5\%$, $\omega_b = 100$ and $|r|_{max} = 4$. Bottom, random and top are the simulated sets of 'preferred variables' used for S-CLASSY. The mean (μ) *Jaccard distance* is calculated between each simulated variable set with BCLASSY.

| Dataset | $\mu|r|$ | | | | $\mu|R|$ | | | | $\mu(Jaccard\ distance)$ | | |
| | S-CLASSY | | | BCLASSY | S-CLASSY | | | BCLASSY | S-CLASSY | | |
	bottom	random	top		bottom	random	top		bottom	random	top
Breast	0	4	4	4	0	2	2	2	1	0	0
Cong. voting	2	1	1	1	1	1	1	1	0.47	0	0
Dermatology	1	2	2	2	1	7	7	7	1	0.07	0.34
Heart	0	3	3	2	0	2	2	2	1	0.37	0.37
Ionosphere	1	2	2	2	1	6	5	5	1	0.66	0.04
Iris	2	2	2	2	2	2	2	2	0.23	0.23	0.47
Led7	3	3	3	3	21	21	21	21	0	0	0
Letter	4	4	4	4	153	150	151	151	0.91	0.67	0.68
Mushroom	1	2	2	2	1	2	5	5	1	0.85	0
Pen digits	4	4	4	4	77	73	76	76	0.94	0.94	0
Pima Indians	0	0	1	1	0	0	1	1	1	1	0
Tic-tac-toe	4	3	3	3	5	5	5	5	0.13	0.24	0
Waveform	4	4	3	4	39	39	40	39	0.76	0.81	0.27
Wine	3	3	3	2	4	4	4	4	0.08	0.08	0.08
$rank_{all}$	2.14	1.57	1.36	**1.29**	2	1.5	1.64	**1.43**	**1.21**	1.71	2.21

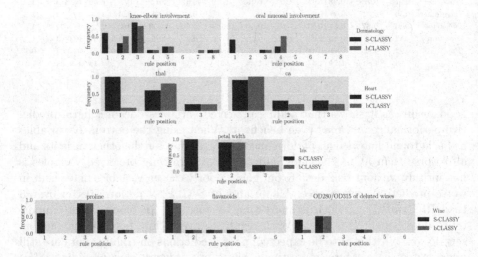

Fig. 1. Mean (μ) frequency of top preferred variables per rule position in the rule list using 10-fold cross validation in Dermatology, Heart, Iris and Wine dataset.

zero *Jaccard distances* shown in Table 3 are further explained by Table 2, where we present the mean frequency f of all variables $s \in S$ in the top-K set and the *maximum* frequency f of the most used preferred variable at the first rule position in the rule list. Moreover, Fig. 1 shows—for four datasets—in more detail the differences in the variables used in the rules learned by top-K S-CLASSY and BCLASSY, making it visible that our algorithm manages to include and combine the preferred variables more often than BCLASSY as earlier as possible. These results not only demonstrate that our method manages to learn different rules in the rule list from its predecessor algorithm, but at the same time 1) ensure that the preferred variables are incorporated at the very beginning of our rule list, and 2) keep competitive, and even in some experiments higher, classification performance.

6   Conclusions and Future Work

We argued that human guidance might be beneficial to interpretable machine learning, especially in settings where predictions need to be accurate as well as reliable and trustworthy explanations are needed. We investigated whether this is the case for the problem of multiclass classification and used rule lists as models. The form of knowledge that we considered is that of *preferred variables*, i.e., variables that the domain expert deems important enough to be given higher priority in the learning process than the other variables.

We proposed S-CLASSY, an algorithm based on beam search that learns rule lists and takes preferred variables into account by first only exploring rules that include one of the preferred variables. An empirical comparison of S-CLASSY to its baseline method, i.e., without using preferred variables, demonstrated that adding preferred variables does not harm predictive performance, while it does result in the preferred variables being used in rules higher up in the learned rule lists. From this we conclude that human guidance might indeed be beneficial to rule learning, for predictive accuracy but also for learning the 'right' rules.

We consider this only to be a first step towards human-guided rule learning. In the future, interesting directions would be to examine other model classes, such as (unordered) rule sets, and expand the background knowledge language, e.g., by allowing constraints based on conditions or patterns, or based on other properties of a rule-based model. A more extensive study on the consequences of using specified/preferred variables in terms of classification performance and interpretability is also worth pursuing. In addition, we aim to evaluate our approach with real-world case studies involving actual domain knowledge provided by domain experts. Finally, we think that interactive rule learning is a promising avenue for future research.

Acknowledgements. This work is supported by Project 4 of the Digital Twin research programme, a TTW Perspectief programme with project number P18-03 that is primarily financed by the Dutch Research Council (NWO).

References

1. Chaudhary, A., Kolhe, S., Kamal, R.: An improved random forest classifier for multi-class classification. Inf. Proc. Agric. **3**(4), 215–222 (2016)
2. Clark, P., Boswell, R.: Rule induction with CN2: some recent improvements. In: Kodratoff, Y. (ed.) EWSL 1991. LNCS, vol. 482, pp. 151–163. Springer, Heidelberg (1991). https://doi.org/10.1007/BFb0017011
3. Cohen, W.W.: Fast effective rule induction. In: Machine Learning (1995)

4. Dzyuba, V., van Leeuwen, M.: Interactive discovery of interesting subgroup sets. In: Tucker, A., Höppner, F., Siebes, A., Swift, S. (eds.) IDA 2013. LNCS, vol. 8207, pp. 150–161. Springer, Heidelberg (2013). https://doi.org/10.1007/978-3-642-41398-8_14
5. Grünwald, P.D.: The Minimum Description Length Principle (Adaptive Computation and Machine Learning) (2007)
6. Hühn, J., Hüllermeier, E.: FURIA: an algorithm for unordered fuzzy rule induction. Data Min. Knowl. Disc. 19(3), 293–319 (2009). https://doi.org/10.1007/s10618-009-0131-8
7. Lakkaraju, H., Bach, S.H., Leskovec, J.: Interpretable decision sets: a joint framework for description and prediction. In: ACM SIGKDD, pp. 1675–1684 (2016)
8. Liu, B., Hsu, W., Ma, Y.: Integrating classification and association rule mining. In: ACM SIGKDD (1998)
9. Proenca, H.M.: Robust rules for prediction and description. Ph.D. thesis, Leiden University (2021)
10. Molnar, C.: Interpretable Machine Learning. Lulu.com, Morrisville (2020)
11. Proença, H.M., Grünwald, P., Bäck, T., van Leeuwen, M.: Robust subgroup discovery. Data Min. Knowl. Disc. 36(5), 1885–1970 (2022). https://doi.org/10.1007/s10618-022-00856-x
12. Proença, H.M., van Leeuwen, M.: Interpretable multiclass classification by MDL-based rule lists. Inf. Sci. 512, 1372–1393 (2020)
13. Quinlan, J.: C4.5: Programs for Machine Learning (2014)
14. Sarica, A., Cerasa, A., Quattrone, A.: Random forest algorithm for the classification of neuroimaging data in Alzheimer's disease: a systematic review. Front. Aging Neurosci. 9, 329 (2017)
15. Schramowski, P., et al.: Making deep neural networks right for the right scientific reasons by interacting with their explanations. Nat. Mach. Intell. 2(8), 476–486 (2020)
16. Sokol, K., Flach, P.: Explainability is in the mind of the beholder: establishing the foundations of explainable artificial intelligence. arXiv preprint arXiv:2112.14466 (2021)
17. Von Rueden, L., et al.: Informed machine learning-a taxonomy and survey of integrating knowledge into learning systems. arXiv:1903.12394 (2019)
18. Yang, L., van Leeuwen, M.: Truly unordered probabilistic rule sets for multi-class classification. In: ECMLPKDD (2022)
19. Zhang, G., Gionis, A.: Diverse rule sets. In: ACM SIGKDD, pp. 1532–1541 (2020)

Don't Start Your Data Labeling from Scratch: OpSaLa - Optimized Data Sampling Before Labeling

Andraž Pelicon[1,2(✉)], Syrielle Montariol[1], and Petra Kralj Novak[3]

[1] Department of Knowledge Technologies, Jožef Stefan Institute,
Ljubljana, Slovenia
[2] Jožef Stefan International Postgraduate School, Ljubljana, Slovenia
andraz.pelicon@ijs.si
[3] Central European University, Vienna, Austria
novakpe@ceu.edu

Abstract. Many text classification tasks face a severe class imbalance problem that limits the ability to train high-performance models. This is partly due to the small number of instances in the minority class, so that the minority class patterns are not well-represented. A common approach in such cases is to resort to data augmentation techniques; however, these have shown mixed results on text data. Our proposed solution is to Optimize the data Sampling prior to Labeling (OpSaLa) to obtain overrepresented minority class(es) in the training dataset. We evaluate our approach on three real-world hate speech datasets and compare it to four commonly used approaches: training on the "natural" class distribution, a class weighting approach, and two oversampling approaches: minority oversampling and backtranslation. Our results confirm that the OpSaLa approach yields better models while the labeling budget stays the same.

1 Introduction

A general assumption in machine learning is that the training data should match the natural distribution of the data [21]. However in many cases the natural distribution of the data is severely skewed, leading to models that are unable to generalize well on the minority classes due to the low amount of training instances.

One example of a text classification task where data skewness is a common phenomenon is hate speech detection, a task which has recently attracted interest both in industrial context as well as for conducting social science experiments on large corpora [2,6]. Several works have however noted the relatively low amount of hate speech content compared to respectful content on social media platforms. Consequently, studies report low performance on the minority hateful classes; in the extreme cases, as much as 40% of instances containing hate speech are misclassified [4].

B. Crémilleux et al. (Eds.): IDA 2023, LNCS 13876, pp. 353–365, 2023.
https://doi.org/10.1007/978-3-031-30047-9_28

A common modeling approach in the presence of unbalanced class distribution is to resort to one of the popular data oversampling and augmentation techniques. However, several recent works have shown that for a large range of text classification tasks and datasets, common data augmentation methods have limited effect on performance when state-of-the-art modelling techniques based on large pre-trained language models are used [10,17]. Furthermore, the performance gain of training on the augmented examples is considerably lower than training on the same amount of real-world examples [23]. In order to improve the performance, especially for the underrepresented classes, real examples rich with patterns representative of each class are needed in the training dataset.

In this work, we propose the Optimized Data Sampling before Labelling (OpSaLa) approach for tackling text classification problems with severely unbalanced natural class distributions. The idea of this approach is to use external information sources that aid us in sampling more real examples of the minority class(es) at the data selection for labeling phase. This way, we achieve a more balanced class distribution for training the model as well as obtain a training dataset richer with real patterns for the minority class. The proposed approach is to be used early in the data preparation stage before any labeling takes place.

While the OpSaLa approach is very simple and has proven to be applicable in practice, we thoroughly evaluate its benefits and implications. In this paper, we address the following methodological research questions:

- Is OpSaLa better compared to the default train-test split?
- How does OpSaLa compare to the commonly used cost-sensitive reweighting schemes as well as state-of-the-art data oversampling and augmentation approaches?

To address the above research questions, we perform experiments on three relatively large (cca. 50.000 training data instances each) real-world hate speech datasets, in three languages and from two social media platforms. Our results show that with our OpSaLa approach, we are able to train better performing models than we would if we were training directly on an unbiased (more skewed) training set. We further show that our approach also yields better models when compared to other data oversampling and augmentation approaches. Due to the variety and sizes of the datasets as well as the use of the state-of-the-art methods, we are confident our results are general and applicable to similar text classification tasks.

2 Related Work

The assumption that training and test data should follow the same distribution for machine learning was recently reevaluated. A study performed on tabular data with classical machine learning methods has shown that changing the minority class distribution is beneficial for the final model performance on the minority class [21]. A comprehensive survey identifying the challenges of handling imbalanced class problems during classification process using machine learning algorithms is presented in [1]. In the specific case of text classification, there are three main approaches to alleviate the impact of unbalanced class distribution: data resampling, cost-sensitive re-weighting and data augmentation.

Cost-sensitive re-weighting refers to tuning the loss function of neural networks to increase the misclassification cost of minority class instances compared to the other classes. A widely used strategy is to assign weights which are inversely proportional to the class frequency [7]. A more elaborate reweighting scheme [3] suggests to count instances in a small neighbouring region as one instance for the purposes of weight calculation. However, this scheme mainly leads to improvements for problems with large (>10) number of classes.

Data resampling uses sampling techniques on the input labeled training set to output a transformed data set with a (more) balanced class distribution. Two common approaches to data resampling are oversampling the minority class and undersampling the majority class. A study comparing these approaches using classical machine learning techniques in the hate speech domain [14] shows that undersampling the majority class generally leads to better results. However, in the presence of severe class imbalance, undersampling can remove a large number of samples and severely reduce the training set size. This can lead to high loss of information which negatively impacts the overall performance of the final model. A recent study has introduced a dynamic sampling approach, where the distribution of the classes varies cyclically at each batch during training [12], improving over classical "static" resampling.

Data augmentation approaches increase the number of instances of the minority class by generating different synthetic examples using real ones. These are slightly modified versions of the original instances or newly created synthetic instances. Several data augmentation approaches were developed specifically for textual data: among other, using WordNet as a dictionary to randomly replace words/phrases with their synonyms in an instance [25], replacing words using the nearest neighbour of the word from a given word embedding [20], or replacing random words in a sentence based on the predictions from a BERT model conditioned on the label for a particular instance [24]. In [16], backtranslation between two languages was used as an augmentation method for a sentiment analysis task. Compared to resampling and cost-sensitive approaches, data augmentation approaches introduce knowledge from external resources in the training dataset, thus including more variance by introducing syntactic changes to the original instance.

Our work focuses on text classification settings where large corpora of unlabeled data are typically available (for example, social media data). We argue that when dealing with class-unbalanced domains (like hate speech), the decision on which data to select for labeling is important and influences the downstream task. Similar to resampling approaches, we aim at a more even class distribution in our training set (compared to the natural class unbalance). A benefit of our approach is that our training dataset is richer in *real* minority class examples, and we show in our experiments that this improves the performance of state-of-the-art transformer-based text classification models.

3   Methodology

We propose the Optimized Data Sampling before Labelling (OpSaLa) approach. It addresses text classification with unbalanced class distribution when vast amount of unlabeled data is available. The goal is to obtain a minority-class-rich training dataset. The OpSaLa approach is depicted in Fig. 1. The main differences to the usual data annotation procedures is as follows:

1. Collect more unlabeled data than you plan to label.
2. Use available external sources relevant to your task to guide the sampling towards an overrepresented minority class(es) dataset for the training set.
3. Separately, sample randomly the examples for the test set.

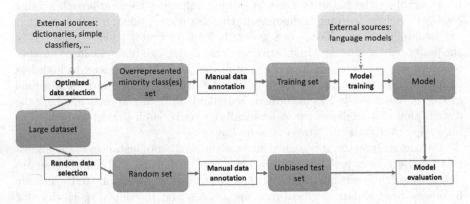

Fig. 1. OpSaLa schema. Schema describing the optimized data selection before labeling approach (OpSaLa): We create a training set with over-represented minority class and a randomly sampled test set.

With the first (1) step, we collect as much relevant unlabeled data as possible, ideally orders of magnitude more than our labeling budget permits to label and thus construct an unlabeled corpus. In step (2) we use external information sources to select training instances to be labeled from this unlabeled corpus. The purpose of step (2) is to guide our sampling in choosing instances which are more likely to belong to the underrepresented classes before any human-labelling step is performed. The OpSaLa method does not strictly propose which external sources of information to use during sampling. These can be as simple as a list of keywords with which we can filter instances that are more likely to belong to the underrepresented class. Alternatively, we can use more sophisticated methods; we may for example use dictionary methods or even existing computational approaches to produce soft labels for our unlabeled instances. We detail the external resources used for the experiments in this paper in Sect. 4.1.

Step (3) deals with sampling and labeling our test set, which we also sample from the gathered unlabeled collection. It is crucial that the test set is sampled

Table 1. Datasets description in terms of number of annotated instances in the training and test set as well as distributions of the training and test set. Per-class distributions are reported in the following order: Acceptable, Inappropriate, Offensive, Violent.

Language	Train set size	Test set size	Train set distribution (%)	Test set distribution (%)
English	51 655	10 759	51.3 / 1.7 / 44.5 / 2.5	71.6 / 1 /26.9 / 0.5
Italian	59 870	10 536	64.9 / 4.6 / 27.3 / 3.2	75.7 / 3.7 / 19.3 / 1.3
Slovenian	50 000	10 000	61.1 / 3.8 / 34.3 / 0.8	66.4 / 1.4 / 31.8 / 0.4

randomly so it retains the natural class distribution of the data. This way, we can objectively evaluate the inference capabilities of our models on unseen data which come from the *natural, real* class distribution. In this work, all methods are evaluated on this set with the natural distribution of classes, that we call the *Natural test set*.

4   Experimental Framework

We perform extensive evaluation of our proposed OpSaLa data sampling approach by comparing it to a train-test split baseline (Sect. 4.3) and to other commonly-used approaches for handling class imbalance in text classification (Sect. 4.4): class weighting, minority class oversampling and backtranslation oversampling. Our experiments are performed on three datasets (Sect. 4.1) using state-of-the-art transformer language models (Sect. 4.2).

4.1   Datasets and External Resources

We used the proposed OpSaLa approach to construct three labeled hate speech datasets. We collected social media posts in three languages, English, Italian and Slovenian, from two social media platforms, YouTube and Twitter. In our current setting, we trained three probabilistic binary offensive speech classifiers on publicly available data as external resources for the OpSaLa approach. We used the FRENK data [9] for Slovenian and English, and a dataset of hate speech against immigrants for Italian [15]. Although the class distribution for these data sets is also skewed, we did not resample the data for training the classifiers. The labels from these classifiers were only used to guide data selection for training set labeling, so weak (better than random) classifiers were sufficient for this phase. For the training set selection, we apply the classifiers to our collection of unlabeled data and select instances with an increased probability of being offensive. Instances for our test set are randomly selected from a later time period than the training set. All datasets were then labeled by human observers. The statistics for the datasets can be found in Table 1.

In this work, we model the hate speech task as a 4-class problem distinguishing between the following four speech types:

- **Acceptable**: does not present inappropriate, offensive or violent elements.
- **Inappropriate**: contains terms that are obscene or vulgar; but the text is not directed at any specific target.
- **Offensive**: includes offensive generalizations, contempt, dehumanization, or indirect offensive remarks.
- **Violent**: threatens, indulges, desires or calls for physical violence against a target; it also includes calling for, denying or glorifying war crimes and crimes against humanity.

Each instance in the datasets is labeled by two annotators. The impact of having diamond standard labels (i.e., the labels are not consolidated) instead of gold standard labels has been thoroughly explored in our previous work [8].

4.2   Classification Models

We perform the experiments with Transformer language models. For each language, we fine-tune a monolingual language model for hate speech detection on our dataset. For the English YouTube dataset, we use the English base version of the BERT language model [5], for Italian YouTube dataset, we use the Alberto model [13], while for the Slovenian Twitter data, we use the Sloberta model [19]. All the models have similar architectures based on the original BERT model.

Fine-tuning of the language models is performed end-to-end in a standard way for classification tasks as presented in [5]. We perform the training of the models using the HuggingFace Transformers library [22]. In this work, we do not perform any hyperparameter optimization; instead, all the models are fine-tuned using a common set of hyperparameters. The set of hyperparameters is the default one used in the Transformers library and was optimized for a large selection of classical NLP tasks. We train the models for 3 epochs and select the best model based on the validation set score.

We tokenize the textual input for the neural models with the respective language models' tokenizers. After tokenizing all inputs, their maximum length is set to 256 tokens. Longer sequences are truncated, while shorter sequences are zero-padded.

We evaluate the performance using three standard machine learning measures, namely Accuracy, macro-averaged F1 score and F1 score for the minority class. Additionally, we use the Krippendorff's Alpha reliability measure. This metric was originally developed to measure the agreement between human annotators, but can also be used to measure the agreement between a classification model prediction and a gold standard. The main advantage of Krippendorff's Alpha compared to macro F1 score is that it takes ordering of classes into account and has the agreement by chance as the baseline. The measure ranges form -1 to 1 where -1 denotes systematic disagreement, 1 denotes perfect agreement and 0 denotes agreement by chance.

4.3   OpSaLa vs. Training on the Natural Distribution

For each dataset, we create an *OpSaLa-optimized train set* and a *Natural test set* on which we evaluate all the trained models, as described in Sect. 3. As a baseline, we first compare model performance of models trained on OpSaLa-optimized train datasets to the performance of models trained on a train set with the natural distribution. For each dataset, we obtain this *Natural train set* by resampling our OpSaLa-optimized training set so that it matches the distribution of the *natural* test set, which reflects the natural distribution of the data. To control for the double annotations present in the collected datasets, we include both labels into the training set during resampling if it does not violate the distribution requirements. In order to control for the differences in sizes between the Natural and OpSaLa-optimized training set, we reduce the size of our OpSaLa training set by randomly removing instances but keeping the same optimized class distribution. We then test both models on the Natural test set.

   We repeat the experiment 10 times to control for variability in data instances which are chosen to be included in the resampled training sets. We test the two settings for statistically significant differences using Mann-Whitney U nonparametric test [11]. We preselect the significance value of the statistical test and set it to p = 0.01. Results are presented in Table 2.

4.4   OpSaLa vs. Approaches for Class Imbalance Handling

We compare our OpSaLa approach to other commonly used approaches for alleviating high class imbalance in text classification: class weighting, minority class oversampling and backtranslation oversampling. To control for the effect of dataset size on the results, the class oversampling and backtranslation oversampling approaches are compared to training on the original OpSaLa-optimized training set without any undersampling. For this reason, the results are obtained only on one training sample and are reported separately.

Sample weighting is an approach where the goal is to change the impact of training instances on the optimization based on the class they belong to. These weights are applied to the loss function during training. In order to compensate for the impact of class imbalance, instances of the minority classes are weighted more than the instances of the majority classes.

   The experimental setting mirrors the one described in Sect. 4.4. Using the Natural train set, we associate instances of each class to their appropriate weight. More precisely, we weight each instance belonging to a given class with the inverse of the number of samples in this class: $W_C = \frac{1}{N_C}$, where W_C represents the weight associated with the instances of the given class and N_C represents the number of samples in the training set for this class. Using this reweighting schema we train the model on the Natural train set and compare the performance with the performance on the resized OpSaLa-optimized training set.

Minority class oversampling randomly selects instances from the minority class and reintroduces them in the training set. Starting from the *Natural train set*, we iteratively sample instances from the minority class and reintroduce them into the training set thus getting a more balanced class distribution. To control for the size, the resampling is performed until we reach the same dataset size as the original OpSaLa-optimized train set. We then train the model on this oversampled training set and compare its performance to the model trained on the original OpSaLa-optimized train set. Results are presented in Table 3.

Backtranslation oversampling is a data augmentation approach which randomly selects instances from the minority class, translates them into another language and then translates them back into the original language. The backtranslated examples are then reintroduced into the training set. Similarly to the minority class oversampling, we start from the *Natural train set*, iteratively sampling instances from the minority class, backtranslating them before adding them into this training set. To control for the size, this resampling is performed until the new train set reaches the same size as the original OpSaLa-optimized train set.

To perform backtranslation of instances, we use the OPUS-MT neural machine translation models [18]. For each source language in our datasets, we select the best target language for translation based on the reported ROGUE scores of the neural machine translation models for the source-target language pair. For the Slovenian dataset, we select Russian as target language; for Italian, we select French; and for English, we select Italian to perform the translation. Results are presented in Table 3.

5   Results

The results (see Table 2) show that training the models on the training set optimized using the proposed OpSaLa approach yields consistently better results when compared with training on training set with the natural class distribution. Training on the OpSaLa-optimized training set yields models with better overall performance as measured with the Krippendorf's Alpha and macro F1 scores. We observe statistically significant differences in terms of macro F1 score on all three datasets and statistically significant differences in terms of Krippendorf's Alpha on two out of three datasets. Furthermore, we observe that models trained on the OpSaLa-optimized training set are better at classifying the underrepresented minority class. We observe statistically significant differences in terms of F1 score for the minority class on two out of three datasets. Higher accuracy scores are observed for models trained on training set with natural class distribution, however these must be considered with caution as in highly skewed data settings accuracy tends to favor models that perform well on majority class and may be misleading for overall model performance.

We generally observe that training on the OpSaLa-optimized training set leads to better performing models when comparing training on the optimized

Table 2. OpSaLa vs. natural class distribution and class reweighting. Results comparing models trained on OpSaLa-optimized training set with models trained on Natural training set, and with models trained using class reweighting. * denotes statistically significant results between OpSaLa and natural distribution while † denotes statistically significant results between OpSaLa and class reweighting. Both settings were tested with Mann-Whitney U test.

	OpSaLa	Natural distribution	Class reweighting
YouTube English			
Alpha	**0.5903** ± 0.006*†	0.5439 ± 0.008	0.5636 ± 0.006
F1 minority	**0.2833** ± 0.031*†	0.0612 ± 0.048	0.2215 ± 0.043
F1 macro	**0.5325** ± 0.015*	0.4498 ± 0.011	0.5140 ± 0.012
Accuracy	0.8232 ± 0.004*†	**0.8236** ± 0.001	0.8017± 0.005
YouTube Italian			
Alpha	**0.5716** ± 0.010*†	0.5428 ± 0.009	0.5423 ± 0.005
F1 minority	**0.4369** ± 0.016*†	0.3512 ± 0.034	0.3898 ± 0.021
F1 macro	**0.6296** ± 0.006*†	0.5994 ± 0.008	0.5991 ± 0.006
Accuracy	0.8380 ± 0.001†	**0.8391** ± 0.002	0.7990 ± 0.004
Twitter Slovenian			
Alpha	**0.5720** ± 0.023†	0.5676 ± 0.007	0.4665 ± 0.223
F1 minority	**0.1916** ± 0.130	0.0239 ± 0.037	0.1103 ± 0.112
F1 macro	**0.5546** ± 0.045*†	0.4904 ± 0.013	0.4652 ± 0.106
Accuracy	0.8054 ± 0.008†	**0.8075** ± 0.002	0.7544 ± 0.035

training set against other post-hoc oversampling approaches (see Table 3). On English and Italian YouTube datasets, the models trained on the OpSaLa-optimized training set achieve better overall performance as well as better performance for the underrepresented minority class when compared with both backtranslation an minority oversampling approaches. On the Slovenian Twitter dataset, we observe less clear benefits form training on the OpSaLa-optimized training dataset.

We also observe that training on the OpSaLa-optimized training set leads to better performing models when comparing to training with class reweighting. The results are again consistent across all three datasets, showing that we train models which perform better overall and are also more accurate when detecting instances from the underrepresented class. The results are presented in Table 2.

6   Discussion

Our analysis shows that optimizing the training dataset selection using the proposed OpSaLa approach is beneficial for text classification problems where we observe high class imbalance. The results show that we are able to train models

Table 3. OpSaLa vs. oversampling and backtranslation. Results comparing models trained on OpSaLa-optimized training set with models trained on training set, augmented using minority and backtranslation oversampling. Th two oversampling approaches are compared to training on the original OpSaLa-optimized training set without any undersampling. For this reason, results for only one sample are reported.

	OpSaLa	Minority oversampling	Backtranslation oversampling
YouTube English			
Alpha	**0.5869**	0.5416	0.5421
F1 minority	**0.3300**	0.2121	0.2158
F1 macro	**0.5581**	0.5047	0.4899
Accuracy	0.8195	**0.8227**	0.8196
YouTube Italian			
Alpha	**0.5854**	0.5357	0.5347
F1 minority	**0.4697**	0.4406	0.3681
F1 macro	**0.6397**	0.6215	0.5913
Accuracy	**0.8386**	0.8349	0.8314
Twitter Slovenian			
Alpha	**0.5873**	0.5646	0.5589
F1 minority	0.2080	**0.3130**	0.2857
F1 macro	0.5661	**0.5735**	0.5528
Accuracy	**0.8077**	0.8056	0.8055

which perform better for underrepresented classes without sacrificing the overall performance of the model. Our experiments also show that the OpSala approach is better compared to state-of-the-art approaches for handling class imbalance in text classification. Training on the OpSaLa-optimized training set yields better performing models than training on oversampled training sets with artificially augmented instances. This observation further implies that for textual data, examples sampled from the real world cannot be effectively replaced with synthetic ones. Furthermore, by using the OpSaLa approach, we make better use of time and resources used for labeling as we are able to train better models for the same labeling budget compared to training on a randomly sampled training set.

Our results confirm the assumption of the OpSaLa approach: intentional manipulation of training data sampling leads to better performing models even though we are effectively biasing the distribution of our training set. We show empirically that even small changes in the class distribution can have significant impact on the performance of our final model. The class distributions in our OpSaLa-optimized training sets do not drastically differ from the natural class distributions and are far from balanced. Yet they yield a significant beneficial effect. We assume this phenomenon is due to the optimized training set being richer in patterns which are representative of the minority class, leading to better performance of the final model.

In our application, we used simple probabilistic binary offensive classification models to obtain toxicity labels for each instance. A potential downside of this approach is the bias that the external information sources introduce in the training set. We argue, however, that this bias is properly controlled for by the requirement that the test set is randomly sampled. Excessive bias from external resources would result in reduced performance on the test set compared to training on a natural train set.

A potential limitation of our study is the fact that the OpSaLa approach was not tested against a completely unbiased training set. While our initial experiment compares models trained on the OpSaLa-optimized training set and a training set with natural class distribution, it should be noted that the training set with natural class distribution is artificially created from the OpSaLa-optimized sample which could suggest that some bias is introduced in terms of the diversity of patterns for each class. Given that our results are tested on a completely unbiased test set, we argue that the impact of such bias is minimal. Even so, a follow-up experiment with a randomly sampled and manually labeled training set could further solidify our conclusions.

7 Conclusion

This work presents OpSaLa, an approach to optimizing the data selection step before labeling so that minority classes are better represented in the training set. It is applicable in classification scenarios with severely unbalanced natural class distributions. We have shown that, given the same annotation budget, training on an OpSaLa-optimized training set improves the classification performance on minority classes without reducing the overall performance of the model. It also outperforms state-of-the-art methods that aim to compensate for class imbalance: class weighting, minority oversampling and backtranslation. Our experiments in hate speech text classification have shown that OpSaLa is effective and efficient in producing information-rich training datasets that improve the overall model performance without increasing the data annotation budget.

Acknowledgements. The authors acknowledge the financial support from the Slovenian Research Agency for research core funding for the programme Knowledge Technologies (No. P2-0103) and for the project Hate speech in contemporary conceptualizations of nationalism, racism, gender and migration (J5-3102). This research was performed in the scope of the RobaCOFI project (Robust and adaptable comment filtering) funded through the EU H2020 project AI4Media.

References

1. Ali, H., Salleh, M.N.M., Saedudin, R., Hussain, K., Mushtaq, M.F.: Imbalance class problems in data mining: a review. Indones. J. Electr. Eng. Comput. Sci. **14**(3), 1560–1571 (2019)

2. Cinelli, M., Pelicon, A., Mozetič, I., Quattrociocchi, W., Novak, P.K., Zollo, F.: Dynamics of online hate and misinformation. Sci. Rep. **11**(1), 1–12 (2021)
3. Cui, Y., Jia, M., Lin, T.Y., Song, Y., Belongie, S.: Class-balanced loss based on effective number of samples. In: Proceedings of the IEEE/CVF Conference on Computer Vision and Pattern Recognition, pp. 9268–9277 (2019)
4. Davidson, T., Warmsley, D., Macy, M., Weber, I.: Automated hate speech detection and the problem of offensive language. In: Proceedings of the International AAAI Conference on Web and Social Media, vol. 11 (2017)
5. Devlin, J., Chang, M.W., Lee, K., Toutanova, K.: Bert: pre-training of deep bidirectional transformers for language understanding. arXiv:1810.04805 (2018)
6. Evkoski, B., Pelicon, A., Mozetič, I., Ljubešić, N., Kralj Novak, P.: Retweet communities reveal the main sources of hate speech. PLoS ONE **17**(3), e0265602 (2022)
7. Huang, C., Li, Y., Loy, C.C., Tang, X.: Learning deep representation for imbalanced classification. In: Proceedings of the IEEE Conference on Computer Vision and Pattern Recognition, pp. 5375–5384 (2016)
8. Kralj Novak, P., Scantamburlo, T., Pelicon, A., Cinelli, M., Mozetic, I., Zollo, F.: Handling disagreement in hate speech modelling. In: Information Processing and Management of Uncertainty in Knowledge-Based Systems. IPMU 2022. CCIS, vol. 1602. Springer, Cham (2022). https://doi.org/10.1007/978-3-031-08974-9_54
9. Ljubešić, N., Fišer, D., Erjavec, T.: The FRENK datasets of socially unacceptable discourse in Slovene and English (2019). arXiv:1906.02045
10. Longpre, S., Wang, Y., DuBois, C.: How effective is task-agnostic data augmentation for pretrained transformers? In: Findings of the Association for Computational Linguistics: EMNLP 2020, pp. 4401–4411. ACL, Online, November 2020
11. Mann, H.B., Whitney, D.R.: On a test of whether one of two random variables is stochastically larger than the other. Ann. Math. Stat. **18**, 50–60 (1947)
12. Montariol, S., Simon, É., Riabi, A., Seddah, D.: Fine-tuning and sampling strategies for multimodal role labeling of entities under class imbalance. In: Proceedings of the CONSTRAINT Workshop, pp. 55–65 (2022)
13. Polignano, M., Basile, P., De Gemmis, M., Semeraro, G., Basile, V.: Alberto: Italian BERT language understanding model for NLP challenging tasks based on tweets. In: 6th Italian Conference on Computational Linguistics, vol. 2481, pp. 1–6 (2019)
14. Rathpisey, H., Adji, T.B.: Handling imbalance issue in hate speech classification using sampling-based methods. In: ICSITech, pp. 193–198. IEEE (2019)
15. Sanguinetti, M., Poletto, F., Bosco, C., Patti, V., Stranisci, M.: An Italian Twitter corpus of hate speech against immigrants. In: LREC (2018)
16. Shleifer, S.: Low resource text classification with ulmfit and backtranslation. arXiv:1903.09244 (2019)
17. Stepišnik-Perdih, T., Pelicon, A., Škrlj, B., Žnidaršič, M., Lončarski, I., Pollak, S.: Sentiment classification by incorporating background knowledge from financial ontologies. In: Proceedings of the 4th FNP Workshop (2022, to appear)
18. Tiedemann, J., Thottingal, S., et al.: OPUS-MT-Building open translation services for the world. In: Proceedings of the 22nd Annual Conference of the European Association for Machine Translation (2020)
19. Ulčar, M., Robnik-Šikonja, M.: SloBERTa: slovene monolingual large pretrained masked language model (2021)
20. Wang, W.Y., Yang, D.: That's so annoying!!!: a lexical and frame-semantic embedding based data augmentation approach to automatic categorization of annoying behaviors using# petpeeve tweets. In: EMNLP, pp. 2557–2563 (2015)
21. Weiss, G.M., Provost, F.: Learning when training data are costly: the effect of class distribution on tree induction. J. Artif. Intell. Res. **19**, 315–354 (2003)

22. Wolf, T., et al.: HuggingFace's Transformers: State-of-the-art Natural Language Processing. ArXiv abs/1910.03771 (2019)
23. Wong, S.C., Gatt, A., Stamatescu, V., McDonnell, M.D.: Understanding data augmentation for classification: when to warp? In: International Conference on DICTA, pp. 1–6. IEEE (2016)
24. Wu, X., Lv, S., Zang, L., Han, J,, Hu, S.: Conditional BERT contextual augmentation. In: Computational Science – ICCS 2019. ICCS 2019. LNCS, vol. 11539, pp. 84–95. Springer, Cham (2019). https://doi.org/10.1007/978-3-030-22747-0_7
25. Zhang, X., Zhao, J., LeCun, Y.: Character-level convolutional networks for text classification. Adv. Neural Inf. Process. Syst. **28**, 649–657 (2015)

The Other Side of Compression: Measuring Bias in Pruned Transformers

Irina Proskurina[✉], Guillaume Metzler, and Julien Velcin

Université de Lyon, Lyon 2, ERIC UR3083, Bron, France
irina.proskurina@univ-lyon2.fr

Abstract. Social media platforms have become popular worldwide. Online discussion forums attract users because of their easy access, speech freedom, and ease of communication. Yet there are also possible negative aspects of such communication, including hostile and hate language. While fast and effective solutions for detecting inappropriate language online are constantly being developed, there is little research focusing on the bias of compressed language models that are commonly used nowadays. In this work, we evaluate bias in compressed models trained on Gab and Twitter speech data and estimate to which extent these pruned models capture the relevant context when classifying the input text as hateful, offensive or neutral. Results of our experiments show that transformer-based encoders with 70% or fewer preserved weights are prone to gender, racial, and religious identity-based bias, even if the performance loss is insignificant. We suggest a supervised attention mechanism to counter bias amplification using ground truth per-token hate speech annotation. The proposed method allows pruning BERT, RoBERTa and their distilled versions up to 50% while preserving 90% of their initial performance according to bias and plausibility scores.

Keywords: Hate speech recognition · Model fairness · Structured pruning · Compressing transformers

1 Introduction

The spread of offensive speech in social media is considered a precursor of numerous existing social issues, such as the distortion of victims' portrayal in society, social tension, dissemination of entrenched stereotypes, provoking hostility and hate crime, not to mention the mental toll. Rational content moderation and filtering in social networks is the primary tool for preventing these consequences of offensive speech. Given the number of everyday social media posts, the need for automated content monitoring looks inevitable. Automated solutions also help to prevent moral damage and the negative impact of disturbing texts on annotators [20]. Recently, algorithmic moderation has become a ubiquitous tool for the vast majority of social networks, including Facebook, YouTube and Twitter. Nevertheless, existing challenges of the hate speech detection task form a stumbling block to guaranteeing accurate and unbiased models' predictions. Context

B. Crémilleux et al. (Eds.): IDA 2023, LNCS 13876, pp. 366–378, 2023.
https://doi.org/10.1007/978-3-031-30047-9_29

sensitivity and an unclear author's intention are the main challenges at the data annotation stage. These factors are the primary sources of the annotators' disagreement during the dataset creation. And the annotation bias in data influences learning bias accumulated when training a classifier, so the risk of the annotators' bias inheritance increases. In the case of hate speech classification, there is a risk of unintended identity-based bias. For example, non-hateful texts containing mentions of gender, nationality or other protected attributes can be classified as a harmful utterances. The cases of biased decision-making are governed by law. For example, the social media platforms that signed the EU hate speech code [1] have to delete posts using offensive and inappropriate language within 24 h. Given the number of everyday posts to check, automated moderation system feedback delay is highly restricted. For that reason, accelerated and compressed models receive more attention for the task.

Our paper presents one of the first attempts to analyze biased outcomes of compression in the context of hate and offensive language detection. In particular, we analyze the impact of encoder layer pruning in pre-trained Transformer Language Models (LMs, in short). Removing layers does not require additional fine-tuning and allows for explaining the contribution of the encoder blocks to model decision-making. We analyse the layers' contribution to rational model decision-making in terms of performance and fairness.[1]

The main contributions of this work are the following: *(i)* We measure identity-based bias in pruned Transformer LMs. *(ii)* We study which group of encoder layers (bottom, middle or upper) can be efficiently pruned without biased outcomes. *(iii)* We propose word-level supervision in pruned Transformer LMs as a debiasing method.

The paper is organized as follows. First, we report an analysis of related literature in Sect. 2. Section 3 provides the definition of pruning strategies, supervised token-wise attention learning methodology, and a list of evaluation criteria[2]. Section 4 provides the results and analysis of bias evaluation in compressed models.

2   Related Work

Inappropriate language with identity-targeted insults posted online provokes the dissemination of stereotypes about minority members [9]. To prevent hate and offensive language from being posted, automated hate speech detectors and filters are used [3]. The detectors vary depending on the task solved, such as profanity, individual cyberbullying, sexism, harassment, and othering language recognition.

Early research works approach the tasks using statistical and machine learning models trained on a suite of linguistic features extracted from text [7,21,23]. Recently, pre-trained Transformer LMs predominated over conventional machine learning methods [13]. Despite being efficient in a range of tasks associated

---

[1] The implementation of the experiments can be found at https://github.com/upun aprosk/fair-pruning.

[2] In our work, we use token-wise and word-level supervision interchangeably.

with hate speech classification, Transformer LMs can lack generalisation ability, increasing the risks of unintended bias [19,24]. There is little research studying whether compression could amplify bias, though novel model compression techniques in NLP are constantly being developed. Compression can be achieved, for instance, through pruning some parts of the Transformer LMs: neurons, heads, layers [4,14].

At the same time, in other fields, recent research shows that even when compressed models perform on par with the baselines, the predictions of pruned models can become considerably disproportionate and skewed. For example, the image features underrepresented in the training data could be misclassified by the compressed models [5].

To the best of our knowledge, our work is one of the first attempts to analyse bias amplification in compressed models in the context of a hate speech classification task. We transfer the hypothesis from the related work [5] to a compression impact study in Transformer LMs: if the impact of compression is uniform, then the shift in scores achieved on the texts mentioning a target community t should also be uniform compared to the overall scores shift β. That forms our null hypothesis H_0:

$$H_0 : \beta_0^t - \beta_0 = \beta_c^t - \beta_c$$
$$H_1 : \beta_0^t - \beta_0 \neq \beta_c^t - \beta_c, \tag{1}$$

where β is an overall score, the superscript t is used to denote the score on texts mentioning community t, the subscript 0 is used for the scores of non-pruned full models, and the subscript c is used to denote the compressed models. We use fairness-related measures as β. We use the Wilcoxon Mann Whitney test to decide whether to accept the null hypothesis or an alternative one H_1, that the compression is not uniform and there is a relative difference in fairness for particular target subgroup t across 10 experiment runs.

3   Methodology

We approach the hate speech detection problem as a supervised multi-class classification with three classes: hate, offensive, and neutral. In this section, we first elaborate on Transformer LMs background and our pruning techniques and explain the motivation behind these compression strategies. Afterwards, we describe the experimental setup, including data, baselines, and evaluation criteria.

3.1   Transformer Background and Models

BERT is a Transformer LM known for achieving state-of-the-art results in various tasks, including hate language detection [12]. The BERT model configuration is defined by the number of encoder layers L and attention heads H. Each

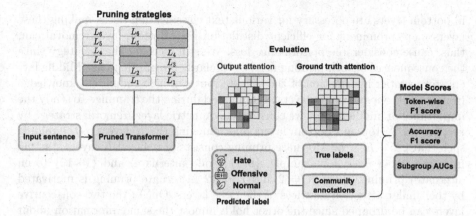

Fig. 1. End-to-end experimental pipeline. We prune the model by removing the layers, then use output attentions and predicted labels to evaluate Token F1 score, Accuracy/F1 scores, and Subgroup/BPSN/BNSP AUCs.

attention head receives a matrix $X_{n \times d}$ as an input with row-wise token representation, where n is the number of tokens in the input sequence, and d is the representation dimension. The output of the head is an updated matrix X_{out}:

$$X_{out} = W^A(XW^V),$$

where $W^A = \text{softmax}(\frac{(XW^Q)(XW^K)^T}{\sqrt{d}}) \in \mathbb{R}^{n \times n}$ is matrix with attention weights, and W^Q, W^V, W^K are projection matrices, the weights updated during the training. We consider a Transformer LM configuration defined by L encoder layers (blocks): $\{l_1, l_2, \ldots l_L\}$ and H attention heads.

3.2 Pruning Techniques

Following recently proposed pruning approaches, allowing for probing the importance of the layers [17], we explore six layer removal strategies: top, bottom, symmetric, alternate (odd and even), and contribution-based. Finally, we prune K of the layers selected via the pruning strategy, where $K = 2, 4, 6$ for architectures with $L = 12$ layers and $K = 1, 2, 3$ for $L = 6$ layers models. The end-to-end pipeline of experiments is illustrated in Fig. 1.

Each pruning strategy is motivated by the redundancy of the layer that shows the relevancy of linguistic signals that the layer brings up. The syntactic and semantic information from the text is captured between the middle and upper layers [15]. However, the latter are more affected by the fine-tuning [11] and can be indifferent to decision-making. Therefore, the *top* pruning strategy for removing K upper layers (i.e., close to the model output) from pre-trained models could prevent overfitting issues. Surface features of the text being captured in bottom layers are necessary for various text classification tasks, making these layers more prominent for efficient distillation [22]. Bottom layer removal can, thus, cause considerable performance loss. We still consider that strategy since the consequences of such pruning need to be clarified regarding bias. Middle layers store syntax information of the text [6], but can hold redundant knowledge from the bottom and upper layers accumulated during the training. To study the importance of middle layers, we consider a symmetric layers removal strategy by keeping the X top and bottom layers and removing the K layers in the middle such that $2X + K = L$. Alternate pruning consists of removal K layers starting from the upper ones; for example, $\{9, 11\}$ (odd alternate) and $\{10, 12\}$ (even alternate) pruning, when $K = 2$ and $L = 12$. Alternate pruning is motivated by the similar attention matrices of the close layers. One of the two consecutive layers can be dropped since the other holds almost the same information about the input text.

Lastly, we also estimate each layer's contribution to the decision-making. Given an input text sequence s_i, we measure the contribution of the layer l with cosine similarity between an input and output representations of the [CLS]-token, corresponding to the input sequence representation:

$$\phi_{s_i}(l) = \cos(Z_{l-1}, Z_l),$$

where Z_l is a vector of hidden states of the layer l, corresponding to the [CLS]-token[3]. We average the values over the validation texts. We prune K layers for each model with the highest contribution scores. Based on the obtained contribution scores, we consider the following layers removal lists for the models: BERT $\{5, 10, 9, 7, 2, 4\}$, RoBERTa $\{1, 2, 6, 8, 9, 4\}$, DistilBERT $\{2, 3, 4\}$, DistilRoBERTa $\{6, 2, 3\}$.

The efficiency of pruning models following the observed strategies depends on the number of pruned layers. The ratio of removed layers decreases the number of model parameters, resulting in fine-tuning speed-up [17].

3.3   Debiasing Approach

For these experiments, we use attention weights W^A to interpret model-decision. We suggest a debiasing approach that prompts the model to assign the larger weights to truly important tokens for the prediction, i.e. word-level supervision. For that, we change the loss computed during the training:

$$Loss_{\sum} = Loss_{pred} + \lambda Loss_{attn}, \tag{2}$$

where $Loss_{pred}$ is conventional cross-entropy classification loss, $Loss_{attn}$ is attention loss, computed based on the rationales provided along with data annotations, and λ is a hyper-parameter regulating the contribution of attention loss to the overall loss. Here, we use ground truth attention for calculating attention

---

[3] That token is used for classification in Transformer LMs.

loss, which we introduced above (2) that is also depicted in Fig. 1. We calculate the difference between the final hidden state corresponding to [CLS]-token and ground truth attentions (rationales). At the same time, using ground truth attention makes the model focus on truly important tokens (ground truth rationales) for classification, reducing bias in models [10]. So, we treat fine-tuning with supervised attention on training set to compensate for the knowledge lost during compression and simultaneously prevent biased outcomes in compressed models.

3.4 Experimental Setup

We use state-of-the-art Transformer LMs for our experiments: base uncased configurations ($L = 12$, $H = 12$) of BERT [22] and RoBERTa [8], and their distilled [18] versions ($L = 6$, $H = 12$): DistilBERT, DistilRoBERTa. As the baselines, we use LMs fine-tuned for ten epochs with the batch size 16 and learning rate $2 \cdot 10^{-5}$ on training data. We use the benchmark dataset for explainable offensive and hate language detection HATEXPLAIN [10]. That dataset contains 20,148 posts collected from Twitter and Gab, each labelled as hateful, offensive, or normal. The dataset was annotated through crowdsourcing and contains extra annotations: hate and offence target communities and textual highlights, marked by annotators as reasoning for decision labelling, i.e. rationales. Rationales are represented as binary arrays, with one corresponding to the words marked by annotators as the ones influencing their labelling decision (offensive, hate or normal) and 0 for the rest of the words. To our knowledge, no other datasets have a similar range of annotations. For the experiments devoted to debiasing, we consider the following ranges of hyper-parameter, regulating the contribution of attention loss to the overall loss: $\lambda \in \{10^{\{-2,-1,0\}}\}$.

3.5 Evaluation

We use the train, development and test stratified split provided along with the dataset for the three following steps: models fine-tuning (train), hyper-parameters search (development), and evaluation (test).

We use a suite of evaluation metrics when establishing the baselines [10]. We report accuracy and macro F1-score reflecting the ability of the model to distinguish between hate, offensive and normal classes.

We measure identity-based bias in pruned models with the threshold-agnostic fairness metrics first introduced in [2]. These measures are AUC scores on the selected subset of the data. In particular, the data is divided into four domains: D^+, D_t^+, D^-, and D_t^-, where D^+ are posts labelled as hateful or offensive, D^- are normal posts, and D_t are the posts mentioning target community t. We use the following metrics: (1) Subgroup AUC = $\mathrm{AUC}(D_t^+ \cup D_t^-)$, (2) Background Positive Subgroup Negative BPSN = $\mathrm{AUC}(D_{\backslash t}^+ \cup D_t^-)$, and (3) Background Negative Subgroup Positive BNSP = $\mathrm{AUC}(D_{\backslash t}^- \cup D_t^+)$. Here, Background refers to the texts not mentioning the community t. BPSN (BNSP) measures the false-positive (false-negative) rates for the texts mentioning target community t. We report aggregated scores for communities computed with Generalized Mean of Bias (GMB):

$$GMB\,(m_t) = \left(\frac{1}{N} \sum_{s=1}^{N} m_t^p\right)^{\frac{1}{p}},$$

where m_t is a bias metric calculated for community t, N is a number of communities, and p is a constant exponent. We set $p = -5$ to emphasize the contribution of the lowest values m_t to the generalized score. The value p that is used is the same as the one used by the authors of the dataset [10].

Lastly, we estimate whether the models focus on relevant context when making the predictions. For that, we compare the context marked by annotators as influencing their class labelling decision, i.e. aforementioned (ground truth) rationales, and the model output rationales (Fig. 1). As for the model output rationales, we select top-5 tokens with the largest attention weights. Given ground truth rationales, we compute the token F1-scores calculation based on precision and recall for model output rationales. The token F1 score refers to the plausibility suite of metrics [10].

4   Results

4.1   Pruning Impact

We find a typical pattern across the layer removal strategies: pruning leads to unintended identity-based bias, and the risks of unethical predictions increase with the ratio of pruned weights. Furthermore, layer removal provokes statistically significant differences in community-level fairness between a range of compressed and non-compressed models. Table 1 reports the results obtained for different models when pruning upper layers. The token F1 scores are low; the rationales annotation procedure can explain that. Most tokens can be labelled as 0, including articles, prepositions and other probably related tokens to the hate span. Low per-token alignment between predicted and ground truth rationales also provokes high variance in the Token F1 scores.

We find similar trends according to fairness loss between different pruning strategies and present results only for the upper layers of pruning. We observe that the disparate effect of pruning on a target-level basis is less common for BERT than for RoBERTa. For BERT, the maximum number of target communities with statistically significant difference scores shift is 4 (out of 10 most frequent communities in the data). In contrast, for RoBERTa, that number is maximum and equal to 6. DistilBERT is more robust to pruning in terms of both fairness and performance. DistilRoBERTa is also less sensitive to pruning, but only in terms of performance. We also find that the disproportionate effect of pruning takes place even when maintaining up to 90% of the original performance (for instance, that is the case of DistilBERT with 3/6 layers removed). That shows that there is also another side of pruning: performance loss does not necessarily go along with fairness loss.

Table 1. Performance of original and pruned models on HATEXPLAIN test set. Layers correspond to the number of **upper** layers left. For the pruned models, we report the number of target communities for which the assumption H_0, formulated in (1), of compression uniform impact, is rejected, which means the compression has increased the biases.

Model	Layers	F1 score	Token F1 score	Count Signif Target Classes		
				Subgroup	BNSP	BPSN
BERT	12/12	$67.28_{\pm 0.13}$	$48.58_{3.28}$	-	-	-
	10/12	$65.31_{\pm 0.17}$	$38.35_{\pm 4.11}$	2	0	1
	8/12	$64.82_{\pm 0.15}$	$32.57_{\pm 4.06}$	2	0	2
	6/12	$63.46_{\pm 0.21}$	$34.4_{\pm 3.87}$	4	0	2
DistilBERT	6/6	$66.19_{\pm 0.44}$	$43.31_{\pm 3.42}$	-	-	-
	5/6	$66.08_{\pm 0.62}$	$42.77_{\pm 4.13}$	0	0	0
	4/6	$65.66_{\pm 0.51}$	$42.1_{\pm 3.98}$	3	0	1
	3/6	$64.31_{\pm 0.83}$	$39.81_{\pm 4.22}$	3	1	2
RoBERTa	12/12	$83.42_{\pm 0.4}$	$46.64_{\pm 3.51}$	-	-	-
	10/12	$81.46_{\pm 0.41}$	$39.37_{\pm 4.61}$	4	2	2
	8/12	$78.67_{\pm 0.58}$	$38.49_{\pm 4.23}$	6	3	4
	6/12	$77.08_{\pm 0.33}$	$24.47_{\pm 4.08}$	6	5	5
DistilRoBERTa	6/6	$82.02_{\pm 0.36}$	$42.08_{\pm 5.24}$	-	-	-
	5/6	$81.08_{\pm 0.4}$	$33.2_{\pm 4.75}$	3	0	2
	4/6	$77.06_{\pm 0.48}$	$32.76_{\pm 5.21}$	3	2	4
	3/6	$74.05_{\pm 0.43}$	$32.6_{\pm 4.61}$	6	5	6

In Fig. 2a, we plot BERT and RoBERTa Subgroup AUC scores for the ten most frequent communities in data. We find that pruning disproportionately affects some subgroups. For example, for RoBERTa with two last layers removed, there is a subgroup AUC score gain for some subgroups compared to the original model (Asian, Hispanic); for other cases, the score decreases considerably (Jewish, Refugee). At the same time, for BERT, the results are mostly stable, except for Women, Arab, and a few other subgroups. We also observe that there is sometimes an improvement between compressed and non-compressed BERT and RoBERTa models. We suggest that this is due to the dynamics of fine-tuning: some layers could learn wrong features from text and add bias. The results for distilled models are displayed in Fig. 2b. The general trend is the same for distilled models: fairness steadily decreases with an increase in the number of removed layers. Figure 3 shows Subgroup AUC scores when removing bottom layers. We do not report results for other pruning techniques for the lack of space. The general pattern of fairness loss is the same for the bottom layer pruning strategy.

4.2 Debiasing with Word-Level Supervision

The reported token F1 scores (Table 1, column 3) drop with an increasing number of pruned layers across all the models. That means that pruned models

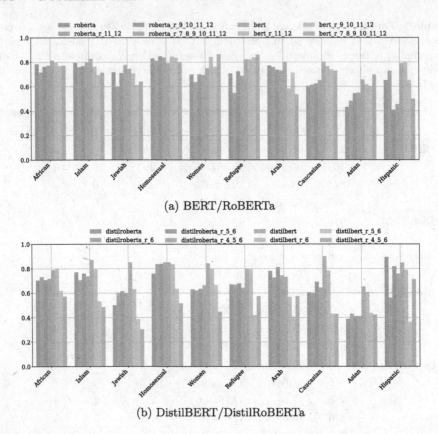

(a) BERT/RoBERTa

(b) DistilBERT/DistilRoBERTa

Fig. 2. Community-wise Subgroup AUC scores on HATEXPLAIN test set. $r*$ = set of **upper** removed layers.

pay less attention to important contexts when making predictions. Recall that critical context is defined by ground truth rationales provided along with data annotations. We suppose that supervised attention learning can compensate for that loss during fine-tuning of the pruned model. We conduct the experiments on the models with the maximum of layers removed: pruned BERT with $L = 6$ and RoBERTa and distilled models with $L = 3$. Table 2 and Table 3 report fairness scores obtained for the models when pruning the upper and bottom layers. We present the scores for two strategies for the lack of space; the scores obtained when pruning other layers fall under the conclusion we draw.

We find that supervised attention reduces bias for all the models; the fairness improvement is substantial for non-distilled models: +0.172 for RoBERTa and +0.213 for pruned BERT when using $\lambda = 1$ (in comparison to models trained without attention learning). However, the performance loss is substantial for values greater than 1, so we do not report that result. For distilled models, the maximum improvements are +0.028 for DistilBERT and +0.03 for Distil-RoBERTa.

Table 2. Performance and fairness scores (Subgroup AUC) of models trained with word-level supervision. The numbers in parentheses represent the ratio of the layers preserved when pruning **upper** layers. $\lambda = 0$ stands for non-supervised attention learning.

Model	λ	F1 score	Token F1 score	Subgroup AUC
BERT (6/12)	0	$63.46_{\pm 0.21}$	$34.4_{\pm 3.87}$	$0.59_{\pm 0.01}$
	0.01	$65.12_{\pm 0.38}$	$36.3_{\pm 4.01}$	$0.707_{\pm 0.11}$
	0.1	$65.92_{\pm 0.24}$	$39.26_{\pm 3.91}$	$0.784_{\pm 0.07}$
	1	$66.61_{\pm 0.17}$	$45.54_{\pm 3.29}$	$0.803_{\pm 0.12}$
DistilBERT (3/6)	0	$64.31_{\pm 0.83}$	$39.81_{\pm 4.22}$	$0.768_{\pm 0.24}$
	0.01	$64.35_{\pm 0.51}$	$40.4_{\pm 3.04}$	$0.748_{\pm 0.16}$
	0.1	$65.11_{\pm 0.7}$	$41.03_{\pm 3.28}$	$0.794_{\pm 0.31}$
	1	$66.71_{\pm 0.22}$	$42.67_{\pm 3.14}$	$0.796_{\pm 0.28}$
RoBERTa (6/12)	0	$77.08_{\pm 0.33}$	$24.47_{\pm 4.08}$	$0.519_{\pm 0.21}$
	0.01	$80.86_{\pm 0.22}$	$33.19_{\pm 3.28}$	$0.612_{\pm 0.29}$
	0.1	$78.58_{\pm 0.23}$	$36.49_{\pm 4.11}$	$0.681_{\pm 0.17}$
	1	$82.38_{\pm 0.26}$	$40.52_{\pm 3.81}$	$0.691_{\pm 0.14}$
DistilRoBERTa (3/6)	0	$71.05_{\pm 0.43}$	$32.6_{\pm 4.61}$	$0.62_{\pm 0.08}$
	0.01	$79.14_{\pm 0.47}$	$34.41_{\pm 4.11}$	$0.634_{\pm 0.04}$
	0.1	$81.25_{\pm 0.33}$	$36.51_{\pm 3.5}$	$0.635_{\pm 0.08}$
	1	$81.96_{\pm 0.51}$	$43.02_{\pm 4.14}$	$0.65_{\pm 0.09}$

We also report F1 scores showing how supervised attention learning improves performance, similar to fairness increase. The scores are on par with the baselines when using $\lambda = 1$. We show that the debiasing conducted via supervised attention learning improves all models' fairness scores.

5   Conclusion

In this work, we conducted two chains of experiments to analyse the effect of Transformer LMs pruning in the context of hate speech classification tasks. We performed the experiments on a dataset containing Twitter and Gab data. First, we analysed the effect of pruning in terms of both fairness and performance loss for BERT, RoBERTa, and their distilled versions. We also estimated to which extent the pruned models rely on relevant context when making predictions. Our results show that removing any layer from Transformer LMs results in fairness loss even when the performance loss could be negligible. We statistically prove that there is a deviation in target community-level predictions when removing the layers from the models. Second, we conduct supervised attention-learning experiments to reduce bias in pruned models. Our results show that fairness score improvement depends on the hyper-parameter regulating the addition of attention loss to the overall loss. The pruned models achieve the best scores when $\lambda = 1$.

From the theoretical perspective, our work suggests a new research direction, focusing on fairness loss that should not be ignored when designing and evaluating compressed models, including the classification task. We also suggest using supervised attention learning to compensate for the knowledge lost for pruned models. That correspondingly highlights the usefulness of relevant context annotations when designing the dataset.

The main limitations of our work are caused by the scope of data we use to study compression impact. Due to the demand for other datasets with similar fine-grained supervision, we are working on building new datasets in other languages. Future work may focus on compression impacts study (tensor decomposition, quantization, parameters sharing) and other debiasing techniques. The latter can be applied to the original model before the compression to estimate the consequences of initial bias in compressed versions. When compared to other debiasing approaches, the results of current research may serve as the baselines. The results of our work can also be used for further linguistic analysis, focusing on functional attributes of text [16].

Acknowledgements. This work was funded by the ANR project Dikè (grant number ANR-21-CE23-0026-02).

Appendix

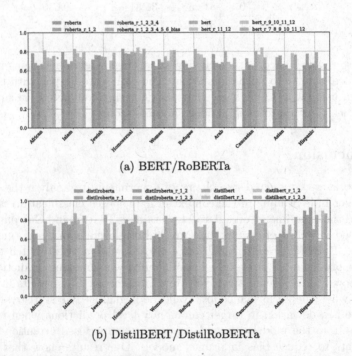

(a) BERT/RoBERTa

(b) DistilBERT/DistilRoBERTa

Fig. 3. Community-wise Subgroup AUC scores on HATEXPLAIN test set. $r* =$ set of **bottom** removed layers.

Table 3. Performance and fairness scores (Subgroup AUC) of models trained with word-level supervision. The numbers in parentheses represent the ratio of the layers preserved when pruning **bottom** layers. $\lambda = 0$ stands for non-supervised attention learning.

Model	λ	F1 score	Token F1 score	Subgroup AUC
BERT (6/12)	0	$62.97_{\pm 0.11}$	$30.5_{\pm 5.02}$	$0.52_{\pm 0.09}$
	0.01	$62.5_{\pm 0.18}$	$33.2_{\pm 4.67}$	$0.54_{\pm 0.07}$
	0.1	$63.25_{\pm 0.24}$	$34.05_{\pm 4.47}$	$0.591_{\pm 0.12}$
	1	$65.93_{\pm 0.26}$	$35.77_{\pm 3.88}$	$0.692_{\pm 0.54}$
DistilBERT (3/6)	0	$64.22_{\pm 0.36}$	$37.18_{\pm 4.04}$	$0.738_{\pm 0.17}$
	0.01	$63.08_{\pm 0.27}$	$38.07_{\pm 4.71}$	$0.736_{\pm 0.23}$
	0.1	$63.32_{\pm 0.4}$	$40.11_{\pm 3.96}$	$0.75_{\pm 0.09}$
	1	$64.1_{\pm 0.28}$	$40.05_{\pm 2.88}$	$0.791_{\pm 0.22}$
RoBERTa (6/12)	0	$78.18_{\pm 0.32}$	$25.32_{\pm 4.51}$	$0.683_{\pm 0.31}$
	0.01	$78.77_{\pm 0.29}$	$29.9_{\pm 4.42}$	$0.669_{\pm 0.34}$
	0.1	$78.92_{\pm 0.35}$	$31.54_{\pm 4.06}$	$0.684_{\pm 0.28}$
	1	$79.98_{\pm 0.32}$	$39.06_{2.88}$	$0.693_{\pm 0.31}$
DistilRoBERTa (3/6)	0	$78.13_{\pm 0.48}$	$34.18_{\pm 3.85}$	$0.625_{\pm 0.27}$
	0.01	$77.05_{\pm 0.53}$	$36.05_{\pm 4.06}$	$0.618_{\pm 0.14}$
	0.1	$78.21_{\pm 0.41}$	$43.61_{\pm 3.92}$	$0.626_{\pm 0.11}$
	1	$78.83_{\pm 0.36}$	$44.5_{\pm 2.92}$	$0.643_{\pm 0.15}$

References

1. Bisht, A., Singh, A., Bhadauria, H., Virmani, J., et al.: Detection of hate speech and offensive language in twitter data using LSTM model. In: Jain, S., Paul, S. (eds.) Recent Trends in Image and Signal Processing in Computer Vision. AISC, vol. 1124, pp. 243–264. Springer, Singapore (2020). https://doi.org/10.1007/978-981-15-2740-1_17
2. Borkan, D., Dixon, L., Sorensen, J.S., Thain, N., Vasserman, L.: Nuanced metrics for measuring unintended bias with real data for text classification. In: Companion Proceedings of The 2019 World Wide Web Conference (2019)
3. Fortuna, P., Nunes, S.: A survey on automatic detection of hate speech in text. ACM Comput. Surv. (CSUR) **51**(4), 1–30 (2018)
4. Gupta, M., Varma, V., Damani, S., Narahari, K.N.: Compression of deep learning models for NLP. In: Proceedings of the 29th ACM International Conference on Information & Knowledge Management, pp. 3507–3508. CIKM 2020, Association for Computing Machinery, New York, NY, USA (2020). https://doi.org/10.1145/3340531.3412171
5. Hooker, S., Courville, A., Clark, G., Dauphin, Y., Frome, A.: What do compressed deep neural networks forget? arXiv preprint arXiv:1911.05248 (2019)
6. Jawahar, G., Sagot, B., Seddah, D.: What does BERT learn about the structure of language? In: Proceedings of the 57th Annual Meeting of the Association for Com-

putational Linguistics, pp. 3651–3657. Association for Computational Linguistics, Florence, Italy, July 2019. https://doi.org/10.18653/v1/P19-1356

7. Lima, L., Reis, J.C., Melo, P., Murai, F., Benevenuto, F.: Characterizing (un) moderated textual data in social systems. In: 2020 IEEE/ACM International Conference on Advances in Social Networks Analysis and Mining (ASONAM), pp. 430–434. IEEE (2020)

8. Liu, Y., et al.: Roberta: a robustly optimized BERT pretraining approach. CoRR abs/1907.11692 (2019). http://arxiv.org/abs/1907.11692

9. Maass, A., Cadinu, M.: Stereotype threat: when minority members underperform. Eur. Rev. Soc. Psychol. **14**(1), 243–275 (2003). https://doi.org/10.1080/10463280340000072

10. Mathew, B., Saha, P., Yimam, S.M., Biemann, C., Goyal, P., Mukherjee, A.: Hatexplain: a benchmark dataset for explainable hate speech detection. In: Proceedings of the AAAI Conference on Artificial Intelligence, vol. 35, pp. 14867–14875 (2021)

11. Merchant, A., Rahimtoroghi, E., Pavlick, E., Tenney, I.: What happens to BERT embeddings during fine-tuning? arXiv preprint arXiv:2004.14448 (2020)

12. Mozafari, M., Farahbakhsh, R., Crespi, N.: Hate speech detection and racial bias mitigation in social media based on BERT model. PLoS ONE **15**(8), e0237861 (2020)

13. Mutanga, R.T., Naicker, N., Olugbara, O.O.: Hate speech detection in twitter using transformer methods. Int. J. Adv. Comput. Sci. Appl. **11**(9) (2020)

14. Neill, J.O.: An overview of neural network compression. arXiv preprint arXiv:2006.03669 (2020)

15. Niu, J., Lu, W., Penn, G.: Does BERT rediscover a classical NLP pipeline? In: Proceedings of the 29th International Conference on Computational Linguistics, pp. 3143–3153 (2022)

16. Röttger, P., Vidgen, B., Nguyen, D., Waseem, Z., Margetts, H., Pierrehumbert, J.B.: Hatecheck: functional tests for hate speech detection models. arXiv preprint arXiv:2012.15606 (2020)

17. Sajjad, H., Dalvi, F., Durrani, N., Nakov, P.: Poor man's BERT: smaller and faster transformer models. CoRR abs/2004.03844 (2020)

18. Sanh, V., Debut, L., Chaumond, J., Wolf, T.: DistilBERT, a distilled version of BERT: smaller, faster, cheaper and lighter. ArXiv abs/1910.01108 (2019)

19. Soares, I.B., Wei, D., Ramamurthy, K.N., Singh, M., Yurochkin, M.: Your fairness may vary: pretrained language model fairness in toxic text classification. In: Annual Meeting of the Association for Computational Linguistics (2022)

20. Steiger, M., Bharucha, T.J., Venkatagiri, S., Riedl, M.J., Lease, M.: The psychological well-being of content moderators. In: Proceedings of the 2021 CHI Conference on Human Factors in Computing Systems, ACM, May 2021. https://doi.org/10.1145/3411764.3445092

21. Waseem, Z., Hovy, D.: Hateful symbols or hateful people? Predictive features for hate speech detection on twitter. In: Proceedings of the NAACL Student Research Workshop, pp. 88–93 (2016)

22. Xu, C., Zhou, W., Ge, T., Wei, F., Zhou, M.: Bert-of-theseus: compressing BERT by progressive module replacing. arXiv preprint arXiv:2002.02925 (2020)

23. Xu, J.M., Jun, K.S., Zhu, X., Bellmore, A.: Learning from bullying traces in social media. In: Proceedings of the 2012 Conference of the North American Chapter of the Association for Computational Linguistics: Human Language Technologies, pp. 656–666 (2012)

24. Yin, W., Zubiaga, A.: Towards generalisable hate speech detection: a review on obstacles and solutions. PeerJ Comput. Sci. **7**, e598 (2021)

Dropping Incomplete Records is (not so) Straightforward

Rianne M. Schouten, Victoria Taşcău[✉], Gabriel G. Ziegler, Davide Casano,
Marco Ardizzone, and Michael-Angelos Erotokritou

Eindhoven University of Technology, Eindhoven, The Netherlands
r.m.schouten@tue.nl, {v.tascau,g.gomes.ziegler,d.casano,m.ardizzone,
m.a.erotokritou}@student.tue.nl

Abstract. A straightforward approach to handling missing values is
dropping incomplete records from the dataset. However, for many forms
of missingness, this method is known to affect the center and spread of
the data distribution. In this paper, we perform an extensive empirical
evaluation of the effect of the drop method on the data distribution.
In particular, we analyze two scenarios that are likely to occur in practice but are not often considered in simulation studies: 1) when features
are skewed rather than symmetrically distributed and 2) when multiple
forms of missingness occur simultaneously in one feature. Furthermore,
we investigate implications of the drop method for classification accuracy and demonstrate that dropping incomplete records is doubtful, even
when test cases are dropped as well.

Keywords: Missing data · dropping incomplete records · skewness

1 Introduction

A straightforward approach to handling missing values is dropping incomplete
records from the dataset [3]. For some forms of missingness, this method is known
to affect the data distribution by creating a shift in the mean of the distribution
and by influencing its standard deviation. In other situations, dropping incomplete records merely reduces the dataset size, although this could bring about
new problems such as imprecise statistical estimates or lack of training data
[11,16].

For the student, scientist or engineer, dropping incomplete records allows
to quickly move forward with developing the desired machine learning model.
However, such a model may not live up to its expectations, albeit because after
deployment incomplete cases that are dropped cannot be predicted nor classified.
At the same time, incomplete training data could make the development of an
AI system conceptually or practically impossible [14].

In this paper, we perform an extensive empirical evaluation of the effects of
missing data. The goal of our investigation is twofold. First, we add to existing knowledge by studying how measures of the center and spread of the data

B. Crémilleux et al. (Eds.): IDA 2023, LNCS 13876, pp. 379–391, 2023.
https://doi.org/10.1007/978-3-031-30047-9_30

distribution are affected when 1) features are skewed rather than symmetrically distributed and 2) several forms of missingness occur simultaneously rather than separately. Both situations are likely to occur in practice but are often not considered in analyses of missing data problems.

Second, we investigate implications of the drop method for classification accuracy. In the study of missing data, incomplete datasets are generally randomly split into training and test data [7–9]. Although such an approach does justice to a development process that should align with the situation after deployment, it prohibits the direct investigation of effects of missing data and only allows the study of imputation methods. On the other hand, when test data is a specific selection of complete records in an incomplete dataset, the distribution of the training data may differ from test or application data, creating issues such as concept drift [22]. In this paper, we demonstrate how missing data shifts the observed data distribution, that some forms of missingness behave unexpectedly when the distribution is skewed and that classification accuracy is affected whether or not you drop incomplete test records.

2   Background

2.1   Preliminaries

We consider a d-dimensional space $\mathcal{X} = \{\mathcal{X}_1, \mathcal{X}_2, ..., \mathcal{X}_{d-1}, \mathcal{Y}\}$ and let $\mathbf{X} = (X_1, X_2, ..., X_{d-1}, Y)$ be a random variable taking values in $\mathcal{X}$. Note that we write X_j and Y to distinguish predictor variables from the assigned outcome variable, but we assume that all variables have a joint distribution $P(\mathbf{X})$. Next, we define a *complete* dataset $D \in \mathbb{R}^{n \times d} = \{\mathbf{x}^1, \mathbf{x}^2, ..., \mathbf{x}^n\}$ to be a collection of n independent and identically distributed realizations of $\mathbf{X}$.

Furthermore, we define a missing data indicator $R \in \{0, 1\}^{n \times d}$ that reveals whether values in D are missing or not. Here, $r_{ij} = 1$ when d_{ij} is observed and $r_{ij} = 0$ when d_{ij} is missing for all $i \in \{1, 2, ..., n\}$ and $j \in \{1, 2, ..., d\}$. We distinguish the hypothetically complete dataset D from the masked, or *incomplete*, dataset by denoting the latter with by $\tilde{D} = \{\tilde{\mathbf{x}}^1, \tilde{\mathbf{x}}^2, ..., \tilde{\mathbf{x}}^n\}$.

In this paper, we investigate the effect of dropping incomplete records for various missing data scenarios. Essentially, the procedure discards all observations that have at least one missing value. We denote the resulting dataset by $\bar{D} = \{\tilde{\mathbf{x}}^i | \mathbf{r}^i = \mathbf{1}\}$; the vector of indications for case i should be an all-ones vector. Denoting the sample size of the dropped dataset by $\bar{n}$, the overall missingness percentage is defined as $\rho = 100\frac{\bar{n}}{n}$.

2.2   Missing Data Mechanisms

In the study of missing data, the process that governs the probability that certain values are missing is called the *missing data model* or *missing data mechanism* [3,11,15]. It is helpful to understand which missing data mechanisms are present in order to choose appropriate missing value treatments. Rubin [15] distinguishes the following three missing data mechanisms.

First, data is said to be Missing Completely At Random (MCAR) if the probability of being missing is unrelated to observed and missing data distribution: $P(R|D_{\mathrm{obs}}, D_{\mathrm{mis}}, \psi) = P(R|\psi)$. With MCAR every data value has the same, fixed probability of being missing, denoted by ψ. Consequently, observed and missing data distributions will be similar, and a method such as dropping incomplete records will allow for the estimation of unbiased statistical parameters [3].

Second, data is Missing At Random (MAR) if observed data governs the missingness probabilities: $P(R|D_{\mathrm{obs}}, D_{\mathrm{mis}}, \psi) = P(R|D_{\mathrm{obs}}, \psi)$. Here, observed and missing data distribution may be different and statistical inferences based on observed data alone may be severely biased. However, under the MAR assumption, observed data contains all information necessary to model the missing data; $P(D_{\mathrm{mis}}|D_{\mathrm{obs}}, R)$. This concept of *ignorability* is an important starting point for many imputation methods [5].

When data is neither MCAR nor MAR, information about the missing values is missing from the dataset. Then, data is Missing Not At Random (MNAR). In other words, the probability to be missing depends on the missing value itself: $P(R|D_{\mathrm{obs}}, D_{\mathrm{mis}}, \psi)$.

2.3   Missing Data Types

The missing data model can be any function that maps a numerical value to a probability: $f : x \mapsto p$, with $p \in [0, 1]$. In practice, when performing experiments with missing data, the logistic or sigmoid function $f_{\mathrm{logistic}}(x) = \frac{1}{1+\exp^{-x}}$ is a convenient choice, partially because for any normally distributed input vector $\mathbf{x} = \{x^1, x^2, ..., x^n\}$, the sum of n Bernoulli trials with success probabilities $\mathbf{p} = \{p^1, p^2, ..., p^n\}$ equals $n\bar{p}$ with $\bar{p} = \frac{1}{n}\sum_{i=1}^{n} p^i = 0.5$ [13]. In practice, this means that 50% of the records will be incomplete.

Recently, [17] proposed a multivariate amputation procedure that allows for easy control of missing data characteristics such as missing data mechanisms, percentage and patterns (every unique row in R is considered a pattern). In addition, they distinguish four versions of the logistic function that allows the researcher to control what part of the data distribution will be masked. These versions are called missing data *types* and can be seen in Fig. 1.

The *right* missingness type is the normal logistic function and assigns high probabilities to large values: $f_{\mathrm{right}}(x) = f_{\mathrm{logistic}}(x)$. The opposite is the *left* missingness type: $f_{\mathrm{left}}(x) = f_{\mathrm{logistic}}(-x)$. Furthermore, *tail* and *mid* missingness assign high probabilities to values in the tails and center of the distribution respectively: $f_{\mathrm{tail}}(x) = f_{\mathrm{logistic}}(|x| - 0.75)$ and $f_{\mathrm{mid}}(x) = f_{\mathrm{logistic}}(-|x| + 0.75)$. Here, 0.75 is a fixed value that ensures $\rho = 50\%$ missingness (other percentages are easily obtained by shifting the logistic functions horizontally). All these missing data types reflect real-world scenarios such as survey questions not being answered for extreme values or medical tests not being executed for 'average' patients. In Sect. 5, we show that the effect of dropping incomplete records varies for these missingness types [17,18].

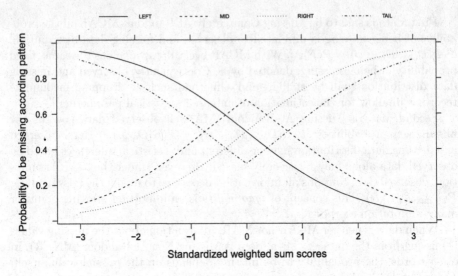

Fig. 1. Four missing data types according to [17]. Standardized weighted sum scores are linear combinations of observed data. In our notation, we use the general term **x**.

3   Related Work

Handling missing values by dropping incomplete records is also known as complete-case analysis or listwise deletion. Especially in the domain of statistics, the effect of complete-case analysis on the validity of statistical estimates has been studied substantially [2,4,10,11,21]. In the machine learning domain, dropping incomplete records is generally accepted if missingness percentages are small or missing values are evenly divided over the data distribution [1,8]. However, it is not part of empirical studies simply because dropped test cases cannot be evaluated [7–9].

We consider our paper to build on work by [18]. Schouten et al. [18] investigate whether the correlation between data features influences the effect of missing data on estimates of the mean, standard deviation and correlation. They find that for an estimate of the mean, when data correlations are small, the effects of MAR missingness converge towards those of MCAR missingness; in contrast, for large data correlations MAR behaves like MNAR.

Furthermore, [18] compare the effects of right, left, tail and mid missingness types. They empirically show that right and left missingness affect the center of the distribution, whereas tail and mid missingness affect the spread of the distribution. The larger the data correlation, the more these effects appear. These results are interesting because they show that for some forms of MAR and MNAR missingness, depending on the statistical quantity of interest, dropping incomplete records may not be as harmful as we may think.

In this paper, we evaluate the behavior of missing data for two scenarios that are likely to occur in practice but have not been studied empirically: 1) the effect of skewness and 2) the simultaneous presence of multiple mechanisms. We furthermore investigate the drop method from a machine learning point of view by analyzing its effect on classification accuracy.

4   Experimental Design

We design two experiments. First, we perform a synthetic data experiment to investigate the effect of skewness. Thereafter, Sect. 4.1 outlines our experiments with a real-world, public dataset. All our experimental code and results are available at https://github.com/Research-Topics-in-Data-Mining/missingness-effect-complete-dataset.

Synthetic dataset generation is done by drawing a dataset H with $n = 10\,000$ observations from a multivariate normal distribution $\mathbf{X} \sim \mathcal{N}(\mu, \Sigma)$ with mean vector $\mu_H = [10, 10]$ and covariance matrix $\Sigma_H = [1, 0.5; 0.5, 1]$. In a copy $H' = H$ we then create right-directed skewness in the first feature by squaring all values larger than 3 standard deviations from the center. Formally, for $i \in \{1, 2, ..., n\}, A^i = (H_1^i)^2$ if $H_1^i > f_{\text{mean}}(H_1) + 3f_{\text{std}}(H_1)$ and $A^i = H_1^i$ otherwise. Then, $H_1' = A^1 \cap A^2 \cap ... \cap A^n$. The amount of skewness can be calculated by $f_{\text{skew}}(D_j) = 3(f_{\text{mean}}(D_j) - f_{\text{median}}(D_j))/f_{\text{std}}(D_j)$ [13]. With our approach, the average skewness is 0.11.

To ensure fair comparison between datasets H and H', we standardize both datasets into D and D' respectively such that $\mu_D = \mu_{D'} = [0, 0]$ and $\Sigma_D = \Sigma_{D'} = I_2^{-1}$. Subsequently, for both D and D' separately, we generate $\rho = 50\%$ missingness in the first feature for all combinations of missing data mechanisms and the four missingness types; resulting in 9 scenarios: MCAR, $4 \times$ MAR, $4 \times$ MNAR. MAR missingness is created by using the observed values in the second feature. For the exact procedure, we apply the multivariate amputation procedure implemented in function ampute [17] in R.

We evaluate effects of missing data on the center and spread of the distribution by calculating the difference between the dropped and the complete dataset for two measures of the center, the mean and median, and two measures of the spread, the standard deviation and interquartile range. We do this for the skewed and non-skewed data separately. For instance, the mean shift for the non-skewed data is $\varphi_{\text{mean shift non-skewed}} = f_{\text{mean}}(D_1) - f_{\text{mean}}(\bar{D}_1)$. We repeat the experiment $T = 1\,000$ times.

4.1   Real-world Data Experiment

The real-world public Breast Cancer dataset[1] [12,20] contains 10 predictor features and 1 binary outcome variable for $n = 569$ cases. We generate a simple missing data pattern where missing values occur in one feature. Specifically, we

---

[1] https://archive.ics.uci.edu/ml/datasets/breast+cancer+wisconsin+(diagnostic).

decide to ampute *smoothness* based on observed data in feature *symmetry*; the two features have a correlation coefficient of 0.6 and smoothness has a medium importance in a Random Forests (RF) classification model.

We investigate the effect of the simultaneous occurrence of missing data mechanisms. Note that such a concurrent existence can happen in several ways. For instance, multiple patterns exist where each pattern follows a different mechanism, or the missingness probabilities come from both observed and unobserved data. We choose an option where missing values occur in one feature; yet for each mechanism some other subsection of the data is used to determine which cases should be amputed. We use the implementation of the multivariate amputation procedure in Python: `pyampute` [19]).

Specifically, MCAR, MAR and MNAR mixtures are created by varying the missingness percentages for each mechanism $\rho_{mcar}, \rho_{mar}, \rho_{mnar} \in \{0, 10, 20\}$. Consequently, we obtain 26 configurations (the scenario 0-0-0 will not generate any missing values), where some configurations contain a single mechanism and others mixtures of 2 or 3 mechanisms. We sequentially perform the experiment for right, left, mid and tail missingness types (thus, a MAR-MNAR mixture follows the same type), and repeat every simulation scenario $T = 1\,000$ times. Our evaluation metrics are the same as the ones described for the synthetic data experiment in Sect. 4. The true mean, median, standard deviation and interquartile range of the complete smoothness feature are 0.096, 0.096, 0.014 and 0.018 respectively.

Next, we investigate implications of the drop method for classification accuracy. In this study, we apply random forests using the Scikit-learn library in Python with a maximum tree depth of 3 and no tuned hyperparameters. This random forest has an accuracy of 0.936 on the complete dataset. The incomplete dataset is analyzed using two scenarios as shown in Fig. 2:

a) An incomplete dataset is randomly split into training and test data, and incomplete records are dropped from both sets.[2]

b) The test data is a selection of complete records in an incomplete dataset. During development, incomplete cases will be dropped from the training data.

Although the first scenario is not applicable during deployment, it may still reveal interesting patterns since the smoothness feature is skewed with a factor 0.10. Consequently, different missingness types may affect the observed data distribution differently. The second scenario is generally not applied in practice (at least, we hope so), but provides an excellent way of studying the extent to which distribution shift affects classification accuracy.

---

[2] N.B.: in the general case, this may affect training and test distribution, but it is unclear how. Homogeneity might increase, but the data might also become more scattered and hence variance might increase. Since the distribution can be affected in a wide variety of possible ways, we will simply ignore this effect; note that technically this might affect the definition of accuracy.

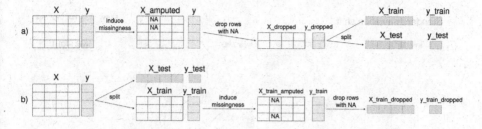

Fig. 2. Two scenarios for splitting incomplete data into train and test set.

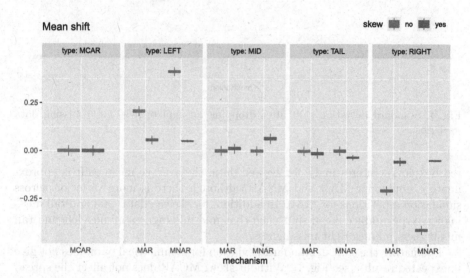

Fig. 3. Mean shift after dropping incomplete rows for 9 missing data scenarios.

5   Results

All results can also be found in our Github repository.

5.1   Results for Skewness

Figure 3 shows the mean shift for all 9 simulation settings; symmetrical data in red and skewed data in green. Without skew, results confirm existing knowledge that MCAR, and mid and tail types of MAR and MNAR missingness do not create mean shift. In contrast, left and right types of MAR and MNAR missingness shift the mean to the right (positive shift) and to the left (negative shift) respectively. MNAR generates more shift than MAR missingness.

When data has right-directed skewness, left and right missingness types create less mean shift than in the case of symmetrical data (compare the green and

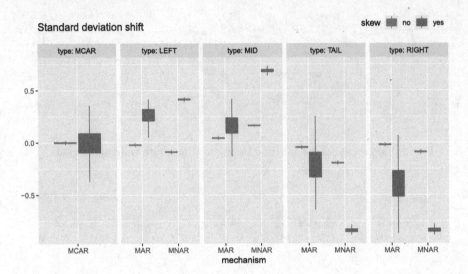

Fig. 4. Standard deviation shift after dropping incomplete rows for 9 missing data scenarios.

red boxplots). Interestingly, for skewed data, the average mean shift is approximately similar for MAR and MNAR, although there is more variation across simulation repetitions for MAR. In addition, for skewed data, mid and tail missingness types induce mean shift such that mid missingness mimics left and tail missingness mimics right missingness.

Evaluating the shift in standard deviation for symmetrical data does not give unexpected results (see Fig. 4). Without skew, MCAR does not affect the spread of the distribution, while mid and tail missingness types respectively increase and decrease the standard deviation. Furthermore, for both left and right types of missingness the standard deviation is reduced.

Interestingly, for left missingness, right-directed skewness *increases* the standard deviation rather than decreasing it (compare green with red boxplots in second panel). Furthermore, when data is skewed, effects of missing data on the standard deviation vary more between simulation repetitions, especially for MCAR and MAR mechanisms. For measures of the median and interquartile range, results are similar as in Figs. 3 and 4 but less extreme.

5.2   Results for Mixtures of Mechanisms

We present the average mean shift over $T = 1\,000$ repetitions for mixtures of MCAR, MAR and MNAR mechanisms in Fig. 5. Naturally, for right missingness, we see that the higher the missingness percentage, the more the mean shifts. This increase is larger for MNAR than for MAR missingness. For instance, for 10% MNAR missingness (center of the figure), the increase per 10% of MAR missingness is around 0.0005 (from light orange to dark orange). In contrast,

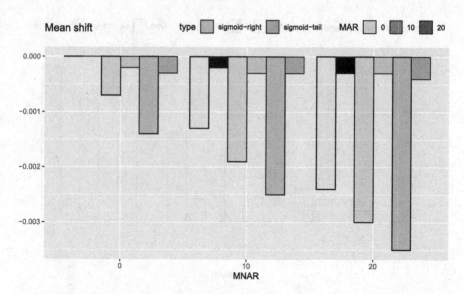

Fig. 5. Average mean shift for mixtures of MCAR, MAR and MNAR mechanisms. MCAR missingness is fixed to 10%.

for 10% MAR missingness (the very light orange bars), the increase per 10% of MNAR missingness is around 0.001. On average, the mean shift for MNAR missingness is twice the amount of the shift for MAR missingness.

Figure 5 demonstrates that the effect of combining multiple mechanisms is additive (rather than, for example, multiplicative). For instance, compare the mixture of 10% MNAR and 10% MNAR (medium orange, center of the figure) with a single MNAR mechanism of 20% (light orange, right side). It turns out that the former creates an approximate mean shift of 0.001 + 0.0005 = 0.0015; the latter shifts 2 · 0.001 = 0.002. In addition, a higher missingness percentage may not necessarily result in more mean shift. For instance, the combination of 10% MNAR and 20% MAR (dark orange, center of the figure) shifts the mean with 0.001 + 2 · 0.0005 = 0.002, but a pure 20% MNAR mechanism has a similar effect. Our findings confirm that not only the missingness percentage determines the extent to which the mean shifts, but especially the occurrence of certain (combinations of) missingness mechanisms play a role.

Figure 6 displays the effects of left and mid missingness types on the standard deviation. Similarly as when we evaluated mean shift, combining multiple mechanisms has an additive effect. For instance, for mid missingness, 10% MNAR combined with 20% MAR increases the standard deviation with 0.0008. A pure 20% MNAR mechanism induces the same amount of shift.

Interestingly, a pure left-type of MNAR missingness *decreases* the standard deviation (light orange bars show negative shift). In contrast, combinations of MNAR and MAR missingness *increase* the standard deviation. Here, it seems that the MNAR component behaves as if the smoothness feature is symmetrically

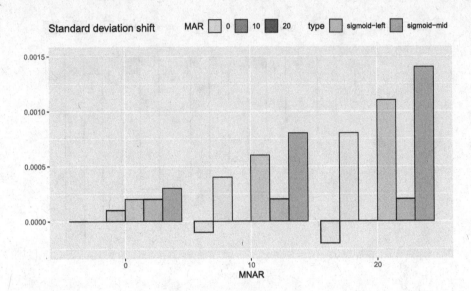

Fig. 6. Average standard deviation shift for mixtures of MCAR, MAR and MNAR mechanisms. MCAR missingness is fixed to 10%.

distributed, whereas the MAR component seems to be affected by the right-directed skewness in the feature (see Sect. 5.1). This may be explained by the fact that for MAR, missingness probabilities depend on observed data in the *symmetry* feature, which is skewed as well.

5.3   Results for Classification Accuracy

We present results for our investigation of the effect of the drop method on classification accuracy in Table 1. We present the correlation between the absolute shift and classification accuracy. Interestingly, when incomplete data is randomly split in training and test data and incomplete records are dropped in both datasets (split scenario a in Fig. 2), all significant correlations are positive (orange values), which means that a larger shift will increase the classification accuracy. This finding is rather counterintuitive, but there may possibly be an effect on the symmetry of the data such that records that are difficult to predict are dropped from both training and test data.

When the test data is a selection of complete records from an incomplete dataset (split scenario b in Fig. 2), all significant correlations are negative (teal values). Here, the larger the shift, the lower the classification accuracy. These findings confirm intuitions; if training data has a different data distribution than test data, classification accuracy will decrease.

Table 1. Correlation between the absolute mean and standard deviation shift and classification accuracy for two train-test split scenarios. MCAR = 0%, SEs are 0.011 in all scenarios, non-significant correlations are italicized. Orange and teal values present an increase and decrease in accuracy respectively.

sigmoid	mean shift accuracy		std shift accuracy	
	a	b	a	b
right	*0.009*	-0.165	*0.009*	-0.081
left	0.254	*0.006*	*0.017*	*0.014*
tail	*0.015*	-0.046	0.107	*-0.009*
mid	0.123	-0.034	0.191	-0.088

6   Discussion and Conclusion

Dropping incomplete records is straightforward; at least when there is no doubt about the effects on the center and spread of the data distribution. However, we demonstrated that when data contains right-directed skewness, tail and mid missingness types induce mean shift, left missingness increases rather than decreases the standard deviation, and the effects of MAR missingness fluctuate substantially. We furthermore showed that when multiple missing data mechanisms occur simultaneously, their effects on the data distribution are additive.

We evaluated the relation between dropping incomplete records and classification accuracy using a Random Forests (RFs) classification model. In reality, RFs are able to handle missing data by making surrogate splits or by treating missing values as a separate category. Moreover, [6] connect RFs to Probabilistic Circuits and propose Generative Forests (GeFs); a family of models that could handle incomplete features internally. Nevertheless, in this paper, our interest is not in creating the best classification model, but rather to show relations between data distribution shift and accuracy.

We found that classification accuracy decreases when training and test data are not identically distributed. Alternatively, when incomplete records are dropped before making a train-test split, accuracy increased. A possible explanation is that in such a situation, records that were difficult to predict were dropped. Note that classification accuracy changes when a record crosses the decision threshold; subtle differences in accuracy may be better detectable by evaluating a prediction model.

In sum, we showed that dropping incomplete records alters the data distribution considerably; some changes are straightforward, others are not. In general, our findings have implications for popular imputation methods such as mean imputation since imputations based on shifted data may transform the data structure even more.

6.1 Limitations and Future Work

This paper expands upon a long tradition of missing data research, historically driven by statisticians more than data miners. We explore the three canonical missing data mechanisms MCAR, MAR and MNAR as proposed by Rubin [15], and consider the four missing data types from [17] as also illustrated in Fig. 1. We believe that the conclusions we draw here are valid and well-supported by a rigorous set of experiments, but these experiments do come with some limitations. The experiments are run on variations of only a single dataset, apart from the class label all attributes are real-valued, the experiments employ only a single classifier, only a single feature has missing values, only unimodal distributions are investigated, further parameter sensitivity analyses could be imagined (for instance, do the conclusions change when the missingness rate is varied, or when missingness depends on the class label?). It is apparent that more work on this topic is to be done in the (near) future.

Acknowledgments. Many thanks to dr. Wouter Duivesteijn and prof. Mykola Pechenizkiy for their continuous support in all possible ways. Thank you Hilde Weerts for being a sparring partner.

References

1. Acuna, E., Rodriguez, C.: The treatment of missing values and its effect on classifier accuracy. In: Banks, D., McMorris, F.R., Arabie, P., Gaul, W. (eds.) Classification, Clustering, and Data Mining Applications. Studies in Classification, Data Analysis, and Knowledge Organisation, pp. 639–647. Springer, Berlin, Heidelberg (2004). https://doi.org/10.1007/978-3-642-17103-1_60
2. Brand, J.P., van Buuren, S., Groothuis-Oudshoorn, K., Gelsema, E.S.: A toolkit in SAS for the evaluation of multiple imputation methods. Stat. Neerl. **57**(1), 36–45 (2003)
3. van Buuren, S.: Flexible Imputation of Missing Data, 2nd edn. Chapman and Hall/CRC, Boca Raton (2018)
4. van Buuren, S., Brand, J.P., Groothuis-Oudshoorn, C.G., Rubin, D.B.: Fully conditional specification in multivariate imputation. J. Stat. Comput. Simul. **76**(12), 1049–1064 (2006)
5. van Buuren, S., Groothuis-Oudshoorn, K.: MICE: multivariate imputation by chained equations in R. J. Stat. Softw. **45**, 1–67 (2011)
6. Correia, A., Peharz, R., de Campos, C.P.: Joints in random forests. Adv. Neural Inf. Process. Syst. **33**, 11404–11415 (2020)
7. García-Laencina, P.J., Sancho-Gómez, J.L., Figueiras-Vidal, A.R.: Pattern classification with missing data: a review. Neural Comput. Appl. **19**(2), 263–282 (2010)
8. Garciarena, U., Santana, R.: An extensive analysis of the interaction between missing data types, imputation methods, and supervised classifiers. Expert Syst. Appl. **89**, 52–65 (2017)
9. Hoogland, J., et al.: Handling missing predictor values when validating and applying a prediction model to new patients. Stat. Med. **39**(25), 3591–3607 (2020)
10. Little, R.J.: Regression with missing X's: a review. J. Am. Stat. Assoc. **87**(420), 1227–1237 (1992)

11. Little, R.J., Rubin, D.B.: Statistical Analysis with Missing Data, Wiley Series in Probability and Statistics, vol. 793. Wiley, Hoboken (2019)
12. Mangasarian, O.L., Street, W.N., Wolberg, W.H.: Breast cancer diagnosis and prognosis via linear programming. Oper. Res. **43**(4), 570–577 (1995)
13. Miller, I., Miller, M., Freund, J.E.: John E. Freund's Mathematical Statistics, 6th edn. Prentice Hall, Upper Saddle River, N.J. (1999)
14. Raji, I.D., Kumar, I.E., Horowitz, A., Selbst, A.: The fallacy of AI functionality. In: ACM Conference on Fairness, Accountability, and Transparency, pp. 959–972 (2022)
15. Rubin, D.B.: Inference and missing data. Biometrika **63**(3), 581–592 (1976)
16. Schafer, J.L., Graham, J.W.: Missing data: our view of the state of the art. Psychol. Methods **7**(2), 147 (2002)
17. Schouten, R.M., Lugtig, P., Vink, G.: Generating missing values for simulation purposes: a multivariate amputation procedure. J. Stat. Comput. Simul. **88**(15), 2909–2930 (2018)
18. Schouten, R.M., Vink, G.: The dance of the mechanisms: how observed information influences the validity of missingness assumptions. Sociol. Methods Res. **50**(3), 1243–1258 (2021)
19. Schouten, R.M., Zamanzadeh, D., Singh, P.: pyampute: a python library for data amputation, August 2022. https://doi.org/10.25080/majora-212e5952-03e
20. Street, W.N., Wolberg, W.H., Mangasarian, O.L.: Nuclear feature extraction for breast tumor diagnosis. In: Acharya, R.S., Goldgof, D.B. (eds.) Biomedical Image Processing and Biomedical Visualization. Society of Photo-Optical Instrumentation Engineers (SPIE) Conference Series, vol. 1905, pp. 861–870, July 1993
21. Toutenburg, H., Srivastava, V.K.: Shalabh: amputation versus imputation of missing values through ratio method in sample surveys. Stat. Pap. **49**(2), 237–247 (2008)
22. Žliobaitė, I., Pechenizkiy, M., Gama, J.: An overview of concept drift applications. In: Japkowicz, N., Stefanowski, J. (eds.) Big Data Analysis: New Algorithms for a New Society. SBD, vol. 16, pp. 91–114. Springer, Cham (2016). https://doi.org/10.1007/978-3-319-26989-4_4

Meta-learning for Automated Selection of Anomaly Detectors for Semi-supervised Datasets

David Schubert[1(✉)], Pritha Gupta[2], and Marcel Wever[3]

[1] Software Engineering and IT Security, Fraunhofer IEM, Paderborn, Germany
`david.schubert@iem.fraunhofer.de`
[2] Department of Computer Science, Paderborn University, Paderborn, Germany
[3] MCML, LMU Munich, Munich, Germany

Abstract. In anomaly detection, a prominent task is to induce a model to identify anomalies learned solely based on normal data. Generally, one is interested in finding an anomaly detector that correctly identifies anomalies, i.e., data points that do not belong to the normal class, without raising too many false alarms. Which anomaly detector is best suited depends on the dataset at hand and thus needs to be tailored. The quality of an anomaly detector may be assessed via confusion-based metrics such as the Matthews correlation coefficient (MCC). However, since during training only normal data is available in a semi-supervised setting, such metrics are not accessible. To facilitate automated machine learning for anomaly detectors, we propose to employ meta-learning to predict MCC scores using the metrics that can be computed with normal data only and order anomaly detectors using the predicted scores for selection. First promising results can be obtained considering the hypervolume and the false positive rate as meta-features.

Keywords: Anomaly Detection · Meta-learning · AutoML

1 Introduction

Automated machine learning (AutoML) [1] refers to the vision of automating the data science process emerging from the unmatched demand for expertise in data science and machine learning in particular. A substantial part of AutoML literature is concerned with the selection, configuration, and composition of machine learning algorithms tailored for a certain task, consisting of a dataset and a performance measure [2]. After first promising results for standard (binary or multiclass) classification and regression tasks on tabular data could be achieved [3–5], AutoML quickly spread to further data types, learning problems, and tasks, such as image classification [6], multi-label classification [7], remaining useful lifetime estimation in predictive maintenance [8], online learning [9], natural language processing [10], and multi-modal data [11]. While the aforementioned extensions of AutoML require supervision in terms of labels being provided for fitting the

© The Author(s), under exclusive license to Springer Nature Switzerland AG 2023
B. Crémilleux et al. (Eds.): IDA 2023, LNCS 13876, pp. 392–405, 2023.
https://doi.org/10.1007/978-3-031-30047-9_31

models, more recently, AutoML methods for dealing with unsupervised or semi-supervised machine learning tasks are proposed as well.

One such problem class deals with the detection of outliers, sometimes also referred to as anomalies. While performance complementarity can also be observed for anomaly detectors, i.e., for different datasets, different anomaly detection methods perform best, this setting imposes severe challenges for adopting AutoML methods to determine the most suitable anomaly detection method. First, if the provided dataset contains both classes at all, i.e., normal and anomalous, their frequency is typically unbalanced. More specifically, it is usually assumed that anomalies occur only with very low frequency. Second, the dataset at hand may comprise no anomalies at all, and the anomaly detection method is meant to detect any new data points that deviate from the training data and thus represent anomalies [12,13], e.g., in intrusion detection. In the following, we refer to this setting as being semi-supervised. For commonly available sampling-based AutoML approaches, the latter data situation is difficult to handle since they rely on probing sampled algorithms and hyperparameter values for the given data. For example, a solution candidate is evaluated via cross-validation and the average performance for a performance measure of interest is observed, e.g., error rate for classification or mean squared error for regression. Obviously, in the domain of anomaly detection, we are interested in finding a model that can detect anomalies reliably but also classifies normal data points as such. Performance measures such as Matthews correlation coefficient (MCC) or the area under the ROC curve are considered to account for the imbalance between normal and anomalous data points. However, these performance measures require the presence of both types of data points, normal *and* anomalous data points, such that they cannot be computed in the semi-supervised setting mentioned above.

In this paper, we investigate the use of meta-learning to overcome the issue of evaluating algorithms and their hyperparameter values for a performance measure of interest, which may require data points of both classes. To this end, we assess the predictive power of two metrics that can be computed with normal data points only: hypervolume of anomaly detectors and the false positive rate (FPR). In our experimental section, we find promising results when utilizing these metrics in terms of meta-features for landmarking and feature descriptions of anomaly detectors. In the following, we formally introduce the problem that we target in this publication.

2   CASH for Semi-supervised Anomaly Detection

Let $\mathcal{X}$ and $\mathcal{Y} = \{0, 1\}$ be an instance space and a target space respectively. Furthermore, we assume instances $x \in \mathcal{X}$ to be (non-deterministically) associated with a label $y \in \mathcal{Y}$. In the setting of anomaly detection, we associate instances x with $y = 1$ in case it is an anomaly and $y = 0$ if it is normal. Moreover, anomalies are assumed to occur only seldomly, i.e., $y = 1$ is a minority class, and during training only normal data points are provided. This setting is also occasionally referred to as semi-supervised anomaly detection [12]. For convenience, let $\{0\} =: \mathcal{Y}_0 \subset \mathcal{Y}$. We seek to learn a hypothesis $h : \mathcal{X} \to \mathcal{Y}$ from training data given in the form of $\mathcal{D} = \{(x_i, y_i) \in \mathcal{X} \times \mathcal{Y}_0 \mid 1 \leq i \leq n\}$ that is able to discriminate well between normal and anomalous data points.

Let be $\mathcal{A} := \{A^{(1)}, A^{(2)}, \dots A^{(m)}\}$ a set of anomaly detectors with corresponding hyperparameter spaces $\Lambda^{(1)}, \Lambda^{(2)}, \dots \Lambda^{(m)}$. Additionally, a dataset of training instances $\mathcal{D} := \{(\mathcal{X}, \mathcal{Y}_0)\}_{i=1}^n$, and a performance measure $m : \mathcal{Y} \times \mathcal{Y} \to \mathbb{R}$, we aim to find the most suitable anomaly detector A^* together with its hyperparameter setting λ^* with respect to m:

$$A^*_{\lambda^*} \in \underset{A^{(i)} \in \mathcal{A}, \lambda \in \Lambda^{(i)}}{\arg\max} \; \mathbb{E}_{(x,y) \sim P} \left[m(A^{(i)}_\lambda(\mathcal{D}, x), y) \right], \tag{1}$$

where $A^{(i)}_\lambda$ is trained on training data $\mathcal{D}$ and makes a prediction on x which is then compared to the ground truth y and without loss of generality m is to be maximized. This problem is better known as the combined algorithm selection and hyperparameter optimization (CASH) problem as initially formalized in [3].

To find such an $A^*_{\lambda^*}$, in sampling-based AutoML systems, one would split the dataset $\mathcal{D}$ into subsets of training and validation data $\mathcal{D}_{train}$ and $\mathcal{D}_{val}$, respectively, and use the performance on the validation data as an estimate of the true generalization performance:

$$\widehat{A}^*_{\lambda^*} \in \underset{A^{(i)} \in \mathcal{A}, \lambda \in \Lambda^{(i)}}{\arg\max} \; \mathbb{E}_{\mathcal{D}_{train}, \mathcal{D}_{val} = (x_j, y_j)_{j=1}^s} \left[\frac{1}{s} \sum_{j=1}^s m(A^{(i)}_\lambda(\mathcal{D}_{train}, x_j), y_j) \right]. \tag{2}$$

3   Related Work

To the best of our knowledge, there are no other publications that focus on the CASH problem in a semi-supervised setting. Therefore, we refer to approaches that target related problems, i.e., unsupervised approaches and semi-supervised hyperparameter optimization approaches in this section.

In [14], Zhao et al. present MetaOD, an approach to unsupervised outlier model selection. It is based on the construction of meta-features for a corpus of training datasets with outlier labels and the performance of over 300 models for each dataset. Similarly to our evaluation, Zhao et al. use the meta-features and performance values to train a performance predictor, which is used to choose high-performing models for new datasets. However, we focus on the semi-supervised setting of outlier detection rendering landmark features meant to capture information regarding potential outliers not applicable.

Putina et al. introduce AutoAD [15], a framework with the same purpose as MetaOD but without utilizing meta-learning. They measure the performance of an anomaly detector via metrics applied to the data before and after removing the top-ranked anomalies predicted by the detector in question. Again, this approach is not applicable to our semi-supervised setting as the employed metrics imply the existence of anomalies in the tails of the data distribution. We do not assume any anomalies being present in the datasets for which an anomaly detector is to be optimized. Other approaches like PyODDS [16], TODS [17], AutoOD [18], and LSCP [19] utilize supervised performance metrics rendering them not applicable to our semi-supervised setting too.

Tax et al. have a body of works evolving from the idea of the *hypervolume* of one-class classifiers as being the hypervolume of the subspace that a one-class classifier predicts to belong to the target class [20–22]. In [21], they aim to optimize the hyperparameters of the support vector data description. For this purpose, they estimate the hypervolume of the support vector data description and use a linear combination of this estimation and the FPR (error on the target class) as performance metric for optimization. The linear combination encompasses a trade-off parameter that has to be set manually.

Our approach is motivated by this idea to a large extent. However, we focus on the more general CASH problem. To this end, we evaluate the utilization of the hypervolume and the FPR as features description of anomaly detectors and for landmarking purposes. Furthermore, we train a meta-model for the prediction of the quality of anomaly detectors on unseen datasets.

4 Meta-Learning for Selecting Anomaly Detectors

According to the assumption that $\mathcal{D}$ only contains normal data, i.e., data points of the form (x_j, y_j) with $y_j \in \mathcal{Y}_0$, performance measures that quantify how accurately an anomaly detector may identify anomalies, e.g., via MCC, precision, recall, or AUC, cannot be assessed for evaluation. Hence, such measures can neither be employed by AutoML systems to search for the most appropriate anomaly detector for a given data set.

To overcome this issue, we aim to substitute the performance measure m in Eq. 2 by a surrogate model $\widehat{m}$, which provided a feature description $f_A \in \mathbb{R}^k$ of algorithm $A^{(i)}$ together with its hyperparameter setting λ and a feature description $f_{\mathcal{D}} \in \mathbb{R}^l$ of the dataset $\mathcal{D}$ in question predicts the measure of interest. To this end, the surrogate is built on datasets for which anomalies are actually known, and hence a performance value can be computed. Then, we can substitute m by $\widehat{m}$ in Eq. 2 to obtain

$$\widehat{A}^*_{\widehat{\lambda}^*} \in \underset{A^{(i)} \in \mathcal{A}, \lambda \in \Lambda^{(i)}}{\arg\max} \ \widehat{m}(f_A(A^{(i)}, \lambda), f_{\mathcal{D}}(\mathcal{D})).$$

In the following, we consider mainly two types of meta-features to describe algorithms as well as datasets (in terms of landmarking features): hypervolume and FPR.

Hypervolume The hypervolume can be considered as a means to describe how tightly an anomaly detector fits the normal data points. With a smaller hypervolume, chances are low that anomalies are missed, whereas a larger hypervolume may lead to anomalies not being identified as such.

False Positive Rate The FPR assesses how many data points in the training data are falsely classified as anomalies, i.e., the anomaly detector would raise a false alarm in this case.

It is easy to see that minimizing both the hypervolume and the FPR conflicts with each other. Therefore, we seek to find a method yielding a suitable trade-off between the two measures.

5   Experimental Evaluation

In the following, we assess the predictive power of the hypervolume and the FPR for the task of selecting a suitable anomaly detector for an unseen dataset. Particularly, we try to answer the following research questions:

RQ1 Does the combination of the hypervolume and FPR meta-features yield an advantage over utilizing only one of the features?

RQ2 Is the combination of the hypervolume and FPR meta-features reasonably informative to select configurations of anomaly detectors?

To answer these research questions, we consider a linear combination of the features to order anomaly detectors and evaluate the general usefulness of these metrics in terms of a feature description of the detectors. Furthermore, we train a meta-model to predict the MCC score of anomaly detectors. This model gets as input landmarking features resulting from the hypervolume and FPR of a fixed portfolio of anomaly detectors and the hypervolume and FPR of the anomaly detector in question.

Figure 1 shows the process that we follow for our experiments. We continuously refer to this figure in the course of this section. All experiments of the evaluation were executed on virtual machines with Ubuntu 20.04.4, 16 cores (Intel Xeon E5-2695 v3), and 128 GB RAM. Sections 5.1 and 5.2 elaborate on the assimilation of the meta-learning dataset and a correlation analysis, respectively. Thereafter, Sect. 5.3 discusses the results of the evaluation.

5.1   Meta-learning Dataset Assimilation

We base our meta-learning datasets on a number of datasets for outlier detection. The source of these datasets is the collection of Tax et al. [24]. The estimation of the hypervolume of anomaly detectors, which we explain in the context of the corresponding activity, scales poorly with an increasing number of dimensions. Thus, we only select those datasets that have eight features or less, resulting in 15 Base Datasets, which can be inferred from Table 1. We execute the Meta-learning Dataset Assimilation activity iteratively for all these datasets.

One of our basic assumptions is that outliers are the minority class (cf. Sect. 2). Not all of the datasets reflect this assumption. Thus, we Sub-sample Outliers to ensure that anomalies constitute between 5% and 10% of the total

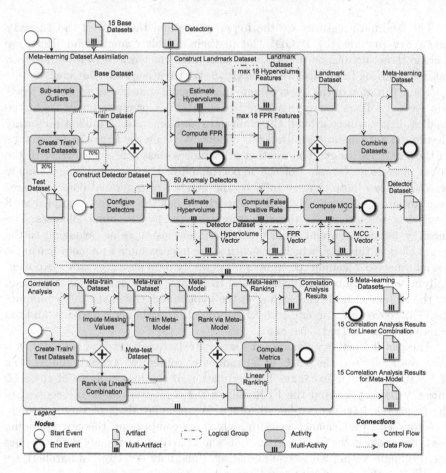

Fig. 1. BPMN Diagram [23] of the Experiments

dataset. We do this only for those datasets, where outliers originally amount to more than 10% of the dataset. To Create the Train and Test Datasets, we shuffle the Base Dataset and perform a stratified split into Train (70%) and Test Datasets (20%). The remaining 10% are kept for future experiments that are not in the focus of this publication. Furthermore, we remove all outliers from the Train dataset to match our semi-supervised setting. Lastly, we scale the Train and Test Datasets by subtracting the mean and scaling the features independently to the interquartile range with respect to the Train Dataset.

Figure 1 shows that the control flow forks at this point into the Construct Landmark Dataset and Construct Detector Dataset activities. These activities extract our meta-features, which are meant to characterize datasets and anomaly detectors, respectively.

The landmark features are the hypervolume and the FPR of the `Anomaly Detectors` provided by PyOD[1] [25] in their default configuration. Here, we exclude those algorithms that do not finish after one week of runtime for each dataset resulting in at most 18 `Anomaly Detectors` per `Base Dataset`.

The estimation of the hypervolume that a model spans with respect to a certain dataset follows the approach of Tax et al. [22]. Correspondingly, we fit the smallest hyper-sphere around the train data and generate 35 million uniformly distributed data instances within this sphere. We train the `Anomaly Detectors` on the `Train Dataset`. The fraction of instances classified as belonging to the target class is an estimate of the hypervolume of the corresponding anomaly detector. Please note that we inferred the number of generated data instances from initial experiments. Here, it is important that the sphere is sampled with a sufficient resolution as, otherwise, the estimated hypervolume collapses. The `Anomaly Detector` would predict only very few instances as belonging to the target class in this case, which means that the hypervolume as a meta-feature is not informative. Meaning the number of necessary data instances grows exponentially with the number of dimensions of the `Train Dataset`. This is the reason for the aforementioned bad scaling of the hypervolume estimation. To compute the FPR, we execute a monte-carlo cross validation with a 30% test-size and ten repetitions on the `Train Dataset`.

The parallel activity `Construct Detector Dataset` extracts the hypervolume, the FPR, and the MCC of 50 randomly configured `Anomaly Detectors` of PyOD per `Train Dataset`. The estimation of the hypervolume of these 50 `Anomaly Detectors` and the FPR follows the computations in the context of the `Landmark Dataset` explained above. We use the MCC of these anomaly detectors as a performance measure, which we compute on the corresponding `Test Dataset`. If the feature extraction via a particular anomaly detector does not terminate within 5 h (while executing 4 anomaly detectors in parallel), we replace it with a newly configured anomaly detector.

The last activity is `Combine Datasets`, which uses the `Landmark Dataset` and the `Detector Dataset` to create the `Meta-learning Dataset`. We combine the features of the former datasets such that each instance of the resulting `Meta-learning Dataset` encompasses all landmarking features of the `Train Dataset` and the detector features of one particular anomaly detector.

5.2   Correlation Analysis

We execute the `Correlation Analysis` iteratively for all 15 `Meta-learning Datasets`. The `Meta-train` and `Meta-test Datasets` are created in a leave-one-out fashion. Meaning, in each iteration of the `Correlation Analysis` one `Meta-learning Dataset` is used as `Meta-test Dataset` and the remainder is merged and used as `Meta-train Dataset`. Since not all of the 18 `Anomaly Detectors` terminate for all datasets in the context of the `Compute Landmark Dataset` activity, we remove all landmark features that have no values in the `Meta-test Dataset` and scale the remaining features individually to be in [0, 1].

---

[1] https://pyod.readthedocs.io/ in version 0.9.7.

After the creation of these datasets, the control flow forks again. The lower control flow ranks the `Anomaly Detectors` covered by the `Meta-test Dataset` according to the simple linear combination of their detector features as a score:

$$lc(A_\lambda, \mathcal{D}) = 1 - \frac{hypervolume(A_\lambda, \mathcal{D}) + FPR(A_\lambda, \mathcal{D})}{2}.$$

The upper control flow ranks the `Anomaly Detectors` via a meta-model that predicts the MCC of the `Anomaly Detectors` covered by the `Meta-test Dataset` based on the landmarking and detector hypervolume and FPR features. We opt for a mean strategy to impute missing values. Furthermore, we use the `Meta-train Dataset` to train a random forest regressor (`Meta-Model`) in default parameterization[2]. Thereafter, we rank the `Anomaly Detectors` of the `Meta-test Dataset` using the predicted MCC of the `Meta-Model` as a score.

We execute the `Compute Metrics` activity for the `Linear Ranking` and the `Meta-learn Ranking` separately. At the beginning of the activity, we scale the MCC to the $[0, 1]$ interval to avoid coping with negative scores. As metrics, we use the regret@k, Kendalls rank correlation coefficient (τ) [26], and the normalized discounted cumulative gain (NDCG) [27], which are metrics that refer to rankings. We base our evaluation on ranking metrics because a correct order of the anomaly detectors' scores is more important to guide an AutoML system than precise prediction of an anomaly detectors performance, e.g., in terms of the mean squared error. While a high precision in the predicted quality is certainly desirable, it is not necessary for choosing the most promising out of a set of anomaly detectors. Here, a precise ranking is sufficient.

The regret@k compares the performance of the best model within a top-k ranking with the actual best model known for a dataset and reports the absolute difference. In our case, it refers to the scaled MCC and gives an intuition about the performance of the top-ranked anomaly detectors.

τ measures the correspondence between two rankings, which are the ranking given by the method in question and the optimal ranking given by the true MCC values. τ ranges in the interval $[-1, 1]$ where negative values indicate a negative correlation of the rankings and positive values a positive correlation. We utilize the b-version of τ, which accounts for ties.

Similarly, the NDCG is a measure to compare a predicted ranking to an optimal one. In comparison to τ, the actual scores (in our case the scaled MCC values) influence the NDCG and not only the ranking inferred from those scores. The NDCG ranges in the $[0, 1]$ interval where values close to 1 denote a high quality of the predicted ranking.

---

[2] https://scikit-learn.org/1.1/modules/generated/sklearn.ensemble.RandomForestRegressor.html.

5.3   Results

In this section, we discuss the performance of the Linear Ranking and
Meta-learn Ranking approaches introduced in Sect. 5.2. Due to the lack of
directly comparable approaches (cf. Sect. 3), we introduce three different baselines. Table 1 reports on the performance of the approaches with respect to the
metrics introduced in Sect. 5.2. Additionally, it reports the maximum, mean,
and minimum of the MCC of the anomaly detectors considered for prediction
scaled to [0, 1] to give an impression of the corresponding distribution. Please
note that the NDCG is generally not particularly insightful and is reported for
completeness.

The first baseline approach is based on randomization. Here, we report the
mean values of 50 randomized rankings. The second approach utilizes the FPR
detector feature introduced in Sect. 5.1 and ranks the 50 anomaly detectors per
dataset with the inversed FPR $(1 - FPR(A_\lambda))$ as score. The last baseline approach works analogously to the second one but uses the inverse of the hypervolume detector feature.

Table 1 shows the conflicting nature of the FPR and hypervolume features
already mentioned in Sect. 4, as good τ performances for one approach are typically accompanied by rather poor ones for the other. Additionally, the regret
and τ indicate that the FPR is more informative than the hypervolume. One
possible reason for this is that the FPR directly influences the MCC but the
hypervolume only if it correlates with the false negative rate. This is not necessarily the case, e.g., if an anomaly detector exactly covers the hypersphere,
which in turn does not cover any anomalies. Furthermore, the hypervolume is
very sensitive to noisy features. An anomaly detector can expand in the direction
of such features and drastically increase its volume without affecting its MCC.
Please note that τ of the randomized approach is not meaningful on its own,
with the mean converging against zero as expected.

Referring to **RQ1** (cf. Sect. 5), Table 1 shows that the linear combination
of hypervolume and FPR detector features outperforms the separate baseline
approaches introduced before. For 5 datasets the linear rankings have the top
detector in the first place and for two more datasets it is within the top-5 ranked
detectors. Furthermore, we see that the mean and median of τ indicate a positive
association in relation to optimal rankings, which is better than the corresponding performances of the baseline approaches.

Please note that neither the FPR nor the hypervolume-based approach would
be able to guide an AutoML system that can freely choose hyperparameter values from the configuration space definition. The AutoML system would favor
detectors that heavily underapproximate the outlier class when utilizing only
the FPR detector feature, e.g., by choosing close to zero values for the hyperparameter used to define a threshold on the decision function (typically denoted
as *contamination*). Analogously, the hypervolume-based approach would favor
detectors that heavily overapproximate the outlier class. Thus, the exploration
of the configuration space of anomaly detectors done by the AutoML system
would almost exclusively explore such bad detectors. Generally, the FPR-based

Table 1. Correlation analysis results. Small arrows indicate whether a metric is to be maximized or minimized where the best performance per dataset and metric is bolded. For datasets marked with a †, anomaly detectors considered for prediction have a mean scaled MCC score of at least 0.6, meaning they actually are capable of identifying anomalies reasonably. We use the following abbreviations for the ranking approaches: R = random, FPR = false positive rate, HV = hypervolume, L = linear combination, M = meta-model, and Mc = cherry-picked variant of the meta-model.

Meta-test Dataset	scaled MCC Max	Mean	Min	regret@1↓ R	FPR	HV	L	M	Mc	regret@5↓ R	FPR	HV	L	M	Mc	NDCG↑ R	FPR	HV	L	M	Mc	τ↑ R	FPR	HV	L	M	Mc
Balance Middle	.59	.47	.40	.12	.11	**.07**	.12	.12		.05	.10	.05	**.03**	.04		.96	**.97**	.97	.96	.96		-.01	**.15**	-.16	-.01	-.14	
Biomed Healthy	.84	.45	.31	.35	.38	**.30**	.42	.38		**.22**	.38	.30	.42	.38		**.89**	.89	.89	.86	.89		.01	**.22**	-.26	-.22	.06	
Diabetes Absent	.67	.56	.45	.07	.18	.13	.19	**.04**		.03	.12	**.02**	.12	.03		.96	.93	**.97**	.94	.97		.01	-.38	**.26**	-.23	-.06	
Diabetes Present	.67	.51	.41	.16	.19	.17	.22	**.14**		.09	.19	**.07**	.22	.14		.95	.94	.95	.93	**.96**		0	-.11	-.26	-.36	**.03**	
Ecoli Periplasm	.73	.57	.40	.15	**.01**	.17	.17	.20		.07	**.01**	.15	.11	.05		.95	**.96**	.95	.95	.94		-.01	.03	.13	**.16**	-.01	
Iris Versicolor	.73	.49	.36	.26	.26	**.14**	.26	.30		.13	**0**	.10	.12	.12		.93	.93	**.96**	.94	.92		-.02	-.19	**.28**	.13	-.06	
Iris Virginica	.84	.53	.35	.24	.37	.21	**0**	.37		.14	**0**	.13	0	0		.91	.89	**.96**	.96	.90		.01	-.11	**.45**	.39	.17	
Liver 1	.64	.49	.40	**.14**	.16	.16	.19	.16		.08	.16	.10	.10	**.03**		.95	.95	.94	.94	**.96**		0	**0**	-.20	-.09	-.04	
Survival G5	.85	.54	.40	.31	.39	.39	**.19**	.34		.22	**0**	.25	0	.23		.94	.95	.93	**.97**	.93		-.02	**.30**	.05	-.41	-.11	
Survival S5	.72	.50	.37	**.21**	.26	.27	.27	.27		.11	.24	**.08**	.17	.26		**.94**	.92	.94	.92	.92		**.03**	-.15	-.18	0	-.05	
Biomed Diseased†	1	.64	.31	.15	**0**	.50	0	0	0	.09	0	.37	**0**	0	0	.90	.97	.89	.96	**.98**	.98	0	.44	-.02	.41	**.54**	.54
Balance Left†	.92	.62	.44	.30	.33	.31	**0**	.33	.13	.18	.13	.30	**0**	.16	.13	.94	.94	.94	**.98**	.95	.96	0	.07	.24	**.48**	.29	.39
Balance Right†	.86	.62	.42	.25	.39	.23	**0**	.25	.08	.11	.08	.23	**0**	.23	0	.94	.92	.94	.98	.93	**.99**	0	.12	.12	.53	.01	**.59**
Iris Setosa†	1	.68	.41	.22	**0**	.42	.27	.35	.27	.16	**0**	.34	0	.27	0	.93	**.98**	.91	.95	.92	.97	0	**.46**	-.24	.28	-.18	.28
Liver 2†	.85	.60	.35	.12	**0**	.32	0	.32	0	.05	**0**	.18	0	.18	0	.93	**.99**	.91	.96	.91	.97	.02	**.49**	-.17	.21	-.18	.23
Mean	.79	.55	.39	.20	.20	.25	.15	.24	**.10**	.12	.09	.18	**0**	.14	.03	.93	.94	.94	.95	.94	**.97**	0	.09	.03	.14	.02	**.41**
Median	.84	.54	.40	.21	.19	.23	.19	.27	**.08**	.11	.08	.15	**0**	.14	0	.94	.94	.94	.95	.93	**.97**	0	.07	.05	.16	-.01	**.39**

approach performs much better than we intuitively expected. A reason for this might be that there are only relatively few anomaly detectors that heavily under-approximate the outlier class in the 50 randomly configured anomaly detectors per dataset. Another reason is that the FPR directly influences the MCC as described before.

Referring to **RQ2** (cf. Sect. 5), ranking according to the linear combination of hypervolume and FPR detector features yields decent results. The linear combination of the hypervolume and the FPR seems to be a suitable means to compare different anomaly detectors to each other and rank them accordingly. However, investigating τ for the separate datasets yields that 5 datasets show negative τ values, which is also the reason for the relatively low mean and median τ of .14 and .16, respectively. Cross-checking these results with the mean MCC values of the 50 randomly configured anomaly detectors per dataset indicates a strong relation. If we restrict the evaluation to those datasets with a mean scaled MCC of at least .6, we end up with a mean regret@1 of .05, regret@5 of 0, NDCG of .97, and τ of .38 (not shown in Table 1). Hence, the features seem to be more informative for those datasets for which we randomly find better anomaly detectors on average.

The meta-model-based approach, which uses the hypervolume and FPR as features for both datasets and the anomaly detector, is clearly worse than the linear combination of the detector features alone. Overall, it can be regarded as being on par with the randomized approach. Given the decent performance of the linear ranking approach, we assume that the FPR and hypervolume landmark features are not informative and simply introduce noise into the training process of the meta-model when considering all datasets.

However, a leave-one-out evaluation on cherry-picked datasets – analogously to what we describe in the context of the linear ranking – leads to the meta-model slightly outperforming the cherry-picked linear ranking in terms of the mean NDCG and τ. Thus, we once more see a strong relation to particularities of the corresponding meta-learning datasets and the performance of the encompassed anomaly detectors, which – obviously – the landmarking features cannot sufficiently express. Additionally, the hypervolume estimation is very sensitive to the data distribution of the target class. On the one hand, for distributions well approximated by the sampled hypersphere, a good anomaly detector has a hypervolume close to 1. On the other hand, for a distribution on a submanifold, the hypervolume of a good detector tends towards 0. This relation might be hard to learn.

Still, we find the results to be promising, particularly considering that we do not assume any information regarding the outlier class of the unseen dataset. Thus, they form a potential base from which interesting future work may emerge. Especially whether additional meta-features, other types of meta-models, or improved sampling strategies for the hypervolume estimation may help to improve the accuracy of the predictions for the MCC.

6   Conclusion

In this paper, we have considered the learning problem of anomaly detection where during training, only a dataset with normal data points is available. While this impedes the use of performance measures explicitly quantifying how well an anomaly detector is able to identify anomalies for the task of automatically selecting and configuring anomaly detectors, we proposed to employ meta-learning to predict measures of interest. In this regard, we considered mainly two types of meta-features. One that is based on the hypervolume covered by trained anomaly detectors and the other one considering the FPRs of anomaly detectors. While a lower value for the former seems favorable as the anomaly detector fits the training data more tightly, this is usually in conflict with minimizing the FPR since the smaller the hypervolume, the more training data points may be classified as positive. Used in combination, the two features have shown promising performance to be used directly for ranking anomaly detectors and as meta-features for a meta-model to predict a measure of interest that would actually require anomalies in the training data to be evaluated. Moreover, results of the corresponding experiments, which can rather be considered a proof of concept, indicate that AutoML for anomaly detectors might be feasible using such surrogate measures for performance evaluation.

Whether AutoML systems can really work well with such surrogate models for selecting and configuring anomaly detectors, however, is still an open question and also outlines interesting future work. Thus, we aim to extend our approach by augmenting the set of meta-features with more types of informative features describing the data or the anomaly detector to improve the quality of the meta-model. Other research directions are to improve the hypervolume estimation and to formalize the problem directly as ranking problem.

Acknowledgments. This work has been partially supported by the BaSys überProd project funded by the German Federal Ministry of Education and Research (No. 01IS20094D). Furthermore, we want to thank Jörg Holtmann and Eyke Hüllermeier for their valuable feedback.

References

1. Hutter, F., Kotthoff, L., Vanschoren, J.: Automated Machine Learning: Methods, Systems, Challenges. Springer Nature, Cham (2019). https://doi.org/10.1007/978-3-030-05318-5
2. Zöller, M., Huber, M.: Benchmark and survey of automated machine learning frameworks. J. Artif. Intell. Res. **70** (2021)
3. Thornton, C., Hutter, F., Hoos, H.H., Leyton-Brown, K.: Auto-WEKA: combined selection and hyperparameter optimization of classification algorithms. In: Proceedings of the 19th International Conference on Knowledge Discovery and Data Mining (2013)
4. Feurer, M., Klein, A., Eggensperger, K., Springenberg, J., Blum, M., Hutter, F.: Efficient and robust automated machine learning. Adv. Neural Inf. Process. Syst. **28** (2015)

5. Mohr, F., Wever, M., Hüllermeier, E.: ML-Plan: automated machine learning via hierarchical planning. Mach. Learn. **107**(8) (2018)
6. Ren, P., et al.: A comprehensive survey of neural architecture search: challenges and solutions. ACM Comput. Surv. **54**(4) (2021)
7. Wever, M., Tornede, A., Mohr, F., Hüllermeier, E.: Automl for multi-label classification: Overview and empirical evaluation. IEEE Trans. Pattern Anal. Mach. Intell. **43**(9) (2021)
8. Tornede, T., Tornede, A., Wever, M., Hüllermeier, E.: Coevolution of remaining useful lifetime estimation pipelines for automated predictive maintenance. In: Proceedings of the Genetic and Evolutionary Computation Conference (2021)
9. Celik, B., Singh, P., Vanschoren, J.: Online automl: an adaptive automl framework for online learning. arXiv preprint arXiv:2201.09750 (2022)
10. Bisong, E.: Google AutoML: cloud natural language processing. In: Building Machine Learning and Deep Learning Models on Google Cloud Platform, pp. 599–612. Apress, Berkeley, CA (2019). https://doi.org/10.1007/978-1-4842-4470-8_43
11. Mueller, J., Shi, X., Smola, A.: Faster, simpler, more accurate: practical automated machine learning with tabular, text, and image data. In: Proceedings of the 26th International Conference on Knowledge Discovery & Data Mining (2020)
12. Chandola, V., Banerjee, A., Kumar, V.: Anomaly detection: a survey. ACM Comput. Surv. **41**(3) (2009)
13. Ahmed, M., Mahmood, A.N., Hu, J.: A survey of network anomaly detection techniques. J. Netw. Comput. Appl. **60** (2016)
14. Zhao, Y., Rossi, R., Akoglu, L.: Automatic unsupervised outlier model selection. Adv. Neural Inf. Process. Syst. **34**. Curran Associates Inc. (2021)
15. Putina, A., Bahri, M., Salutari, F., Sozio, M.: AutoAD: an automated framework for unsupervised anomaly detection. In: IEEE International Conference on Data Science and Advanced Analytics., Paris / Virtual Event, France. IEEE (2022)
16. Li, Y., Zha, D., Venugopal, P., Zou, N., Hu, X.: Pyodds: an end-to-end outlier detection system with automated machine learning. In: Companion Proceedings of the Web Conference 2020, New York, NY, USA, 2020. Association for Computing Machinery (2020)
17. Lai, K., et al.: Tods: an automated time series outlier detection system. In: Proceedings of the AAAI Conference on Artificial Intelligence, vol. 35, no. 18 (2021)
18. Li, Y., et al.: Automated anomaly detection via curiosity-guided search and self-imitation learning. IEEE Trans. Neural Netw. Learn. Syst. **33**(6) (2022)
19. Zhao, Y., Nasrullah, Z., Hryniewicki, M.K., Li, Z.: LSCP: locally selective combination in parallel outlier ensembles. In: Proceedings of the 2019 International Conference on Data Mining. SIAM (2019)
20. Tax, D., Muller, K.: A consistency-based model selection for one-class classification. In: Proceedings of the 17th International Conference on Pattern Recognition, 2004, vol. 3 (2004)
21. Tax, D., Duin, R.: Outliers and data descriptions. In: Proceedings of the 7th Annual Conference of the Advanced School for Computing and Imaging, 2001. ASCI (2001)
22. Tax, D., Duin, R.: Uniform object generation for optimizing one-class classifiers. J. Mach. Learn. Res. **2**(2) (2001). Special Issue on Kernel Methods
23. Object Management Group (OMG). Business process model and notation (bpmn), version 2.0.2. Technical report (2014)
24. Tax - One-class Datasets. https://homepage.tudelft.nl/n9d04/occ/index.html. Accessed 16 Nov 2022
25. Zhao, Y., Nasrullah, Z., Li, Z.: Pyod: a python toolbox for scalable outlier detection. J. Mach. Learn. Res. **20**(96) (2019)

26. Knight, W.: A computer method for calculating Kendall's tau with ungrouped data. J. Am. Stat. Assoc. **61**(314) (1966)
27. Järvelin, K., Kekäläinen, J.: Cumulated gain-based evaluation of IR techniques. ACM Trans. Inf. Syst. **20**(4) (2002)

Should We Consider On-Demand Analysis in Scale-Free Networks?

Arnaud Soulet[(✉)]

Université de Tours, LIFAT, Blois, France
`arnaud.soulet@univ-tours.fr`

Abstract. Networks are structures used in many fields for which it is necessary to have analytical systems. Often, the size of networks increases over the time so that the connectivity of the nodes follows a power law. This scale-free nature also causes analytical queries to be concentrated on nodes with higher connectivity. Rather than computing the query results for each node in advance, this paper considers an on-demand approach to evaluate its potential gain. To this end, we propose a cost model dedicated to scale-free networks for which we compute the cost for both the offline and on-demand systems. It is reasonable in an on-demand approach to cache part of the results on the fly. We study theoretically and on real-world networks three policies: caching nothing, caching everything and minimizing the total cost. Experiments show that the on-demand approach is relevant if some of the results are cached, especially when the query load is low and the query complexity is reasonable.

Keywords: Decision support system · Analytical queries · On-demand system

1 Introduction

Networks are complex structures often used to represent information where vertices are entities and edges are their relationships. For example, in citation networks, articles are connected by their references. The Web is a set of pages connected by their links. Social networks connect people through directed ("follower" relationship) or non-directed links ("frienship"). Finally, knowledge graphs connect entities by directed and labeled relations. Most often these networks are characterized by a rapid growth of the number of nodes where the connectivity follows a preferential attachment mechanism leading to nodes that concentrate links – see the Web [2], social networks [5] or knowledge graphs [7].

In many fields including bibliometrics [4] or webometrics [3], network analytics aims to analyze and extract insights from networks. Many analytical systems produce rich indicators for each node of the network. Typically, Google Scholar offers the same page for each author (with citation number, h-index, and so on). Producing these analyses is a challenge because of the volume of data and the complexity of some indicators. In the era of Big Data, these challenges have mostly been met by resorting to cluster computing frameworks (e.g., MapReduce or Spark) [21]. Scaling up is not a difficulty for this kind of approach where it is always possible to add computer nodes if the volume of data increases or if the difficulty of the analysis requires it. With this approach, the results once computed are kept and then, when the system is online, the results are returned instantly to the users following their queries. However, since networks evolve rapidly, it is necessary to repeat the costly offline processing regularly. Moreover, with the preferential attachment mechanism, the most popular nodes are also the most accessed. For instance, Yann LeCun's page will be more accessed than those of the PhD students in deep learning publishing their first paper. The undifferentiated processing of nodes thus leads to the generation of analyses that will be computed and stored uselessly because they will never be consulted before the next refresh. It would be possible not to analyze the less popular nodes of the network to reduce costs, or even forget them [6,13]. But, such an approach would be detrimental to the diversity of the analytical system, especially since some rare entities are sometimes the most important (e.g., some little-known entries in a dictionary like DBnary studied in Sect. 6). In order to reduce costs while preserving the diversity of the network, this paper aims at determining whether it would not be preferable to produce the analyses on-demand.

In this paper, we propose a generic model to understand the impact of different computational and storage strategies for analytical systems in scale-free networks. More specifically, our contributions are as follows:

- We propose a cost model specific to analytical systems in scale-free heterogeneous networks based on types (source entities) and their items (produced entities).
- We theoretically compare offline systems with on-demand systems by distinguishing several caching policies (all, nothing, compromise). In particular, we study the utility ratio of cached results.
- We evaluate our approach on several real-world graphs showing the interest of using on-demand systems in certain scenarii.

The outline of this paper is as follows. Section 2 reviews some related work about decision support systems. Section 3 introduces basic definitions and the cost model framework. We compute the cost of offline and on-demand systems in Sects. 4 and 5 respectively. We apply these models on real-world networks in Sect. 6 and conclude in Sect. 7.

2 Related Work

In the introduction, we have already mentioned approaches based on the MapReduce paradigm to implement distributed and parallel algorithms on clusters. Typically, [16] proposes a method for large-scale social network analysis based on MapReduce. There are generalist frameworks based on MapReduce to analyze data like Apache Pig or Apache Hive [8]. In-memory analytics frameworks, such as Apache Spark, that also well-adapted for graph-based analytics [1] have also been designed to better handle iterative processes. [15] shows how to exploit in-memory frameworks to analyze knowledge graphs. All these approaches process the entirety of the data indiscriminately to produce analyses for each entity. For simplicity, we will refer to this type of approach as offline analytical systems.

Many works in the literature have been interested in analytical queries in the database field (where by nature, the system returns the result of its query on the fly). In particular, they have proposed caching policies and cost models dedicated to this type of queries in relational model [11,22] and non-relational models [10,17]. However, the goal of these decision support systems is to be able to answer very diverse queries, whereas in our case we are always interested in the same query applied to a large number of distinct entities. Consequently, most of these works seek to minimize the cost of executing a query while we seek to minimize the cost of a set of queries. More specifically, it is our query set that is unbalanced (few frequent queries, many rare ones) rather than the data associated with a query (which may require the use of histograms [12]). Moreover, in the following, the cost function of a query is an input to our problem. Another consequence of the variety of queries is that database systems cannot store all the results, contrary to our setup. They therefore implement cache replacement policies [14] such as least recent used or least frequently used. In contrast, in this work, we implement a global cache policy that avoids any replacement.

3   Preliminaries

Scale-Free Heterogeneous Network. Let us consider a (heterogeneous) network $\mathcal{N} = \langle I, T, \tau \rangle$ made of a set of items I, a set of types T and a membership relation $\tau \subseteq I \times T$. $(i, t) \in \tau$ means that the item i belongs to the type t. Considering that $I \subseteq V$ and $T \subseteq V$, we can also consider non-heterogeneous network. The number of types is denoted by s: $s = |T|$. The degree of the type $t \in T$ is the number of items in relation with that t: $\deg_\tau(t) = |\{i \in I \; : \; (i, t) \in \tau\}|$. The frequency $n_\tau(k)$ in τ counts how many types $t \in T$ are exactly in relation with k items (i.e., its degree $\deg_\tau(t) = k$): $n_\tau(k) = |\{t \in T \; : \; \deg_\tau(t) = k\}|$. The total degree is defined as $m_\tau = \sum_{k=1}^{\infty} k \times n_\tau(k) = |\tau|$. When the membership relation τ is clear, we omit it: $\deg(t)$ refers to $\deg_\tau(t)$, $n(t)$ refers to $n_\tau(t)$ and so on. A scale-free network is a network whose frequencies $n(k)$ follow a power law (at least asymptotically): $n(k) \sim k^{-\gamma}$ with $\gamma > 2$. For instance, Fig. 1 shows on the left the in-degree distribution of entities in Wikidata (see Sect. 6 for more details) with the magenta dots. The distribution can be approximated by a power law of exponent $\gamma = 2.058$ (see the dash lines).

Analytical System. The end users are interested in an analytical query Q on the network $\mathcal{N}$ for different types t. This query Q may lead to complex manipulations of the items of t. For instance, if the types are authors and the items are their publications (with years and citation numbers), a query may compute at the same time the total number of citations, the number of papers per year, the h-index, and so on. An analytical system $\mathcal{S}_Q$ (or simple $\mathcal{S}$) is a decision support

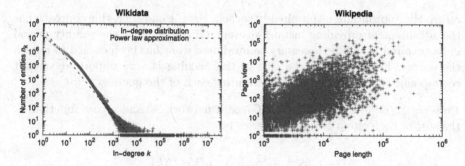

Fig. 1. Rationale for our cost model

system that efficiently evaluates the analytical query Q on the network $\mathcal{N}$ to obtain the result $Q(t, \mathcal{N})$. In the following, we consider that the network evolves and that a result remains valid for a period Δ.

Cost Model. Given a type t and a network $\mathcal{N}$, we assume that the cost of the query Q for t in $\mathcal{N}$ only depends on its degree $\deg(t)$: $C(\deg(t))$ where C is a cost function. In the following, our goal is to evaluate the (average) cost $C(\mathcal{S}, q, \mathcal{N})$ of the system $\mathcal{S}$ when it receives q queries during the validity period Δ. Naively, we could think that this average cost of executing q queries is equal to the average query cost repeated q times:

$$C(\mathcal{S}, q, \mathcal{N}) = q \cdot \frac{1}{s} \cdot \sum_{t \in T} C(\deg(t))$$

This formula is wrong for two reasons: 1) types do not have the same probability of being queried and 2) the system $\mathcal{S}$ (whether offline or on-demand) can store query results avoiding the repetition of some queries. For the first point, it is clear that some types will be queried more because of their popularity (as illustrated in the introduction with the Yann LeCun's page). For this purpose, we make the important and realistic assumption that the probability of querying a type t is proportional to its degree $\deg(t)$. It is difficult to find data to justify this assumption. Nevertheless, Fig. 1 on the right illustrates this phenomenon with the pages of Wikipedia where we see that globally the longest pages are also the most viewed. For the second point, we will see the impact of storing all query results in advance (see offline analytical system in Sect. 4) or caching a part of the query results on the fly (see on-demand analytical system in Sect. 5).

4　Offline Analytical Systems

The principle of an offline system is to compute in advance the queries for all types in order to cache them. When the system is online, it will be enough to return the appropriate result already cached. For the end-user, this approach has the advantage of providing instant answers. Of course, once the validity period Δ has expired, it will be necessary to invalidate what has been cached to refresh the results. Unfortunately, the cost of this caching is very important since it corresponds exactly to the cost of executing each of the queries:

Property 1 (Offline system cost). Given a network $\mathcal{N}$ and a cost function C, the cost of the system $\mathcal{S}^{\text{off}}$ for q queries is:

$$C(\mathcal{S}^{\text{off}}, q, \mathcal{N}) = \sum_{k=1}^{\infty} n(k) \cdot C(k)$$

Due to the lack of space, we omit most of the proofs. For simplicity, Property 1 ignores the parallelization costs which can be significant in some cluster computing architectures. Nonetheless, this cost remains very high because all network types in T are considered without distinction. Unfortunately, a large part of the cached results will never be used. Considering a linear cost $C(k) = \alpha k$ and a scale-free network with exponent $\gamma > 2$, we can demonstrate that the cost of the system $\mathcal{S}^{\text{off}}$ is simply $s \times \alpha \times \frac{\gamma-1}{\gamma-2}$ as the mean degree is given by $(\gamma-1)/(\gamma-2)$ [18].

To evaluate the quality of a system, we calculate the utility ratio (denoted by UR) that is the proportion of the caching cost that was reused. The idea is to evaluate how much of the caching effort was worthwhile. The utility ratio is close to 1 when each caching cost involved a query that was queried again. Conversely, it is close to 0 when what was cached was not queried again resulting in unnecessary caching costs. The following property gives this measure for an offline system:

Property 2 (Utility ratio). Given a network $\mathcal{N}$ and a cost function C, the utility ratio of the system $\mathcal{S}^{\text{off}}$ for q queries is:

$$UR(\mathcal{S}^{\text{off}}, q, \mathcal{N}) = \frac{1}{C(\mathcal{S}^{\text{off}}, q, \mathcal{N})} \cdot \sum_{k=1}^{\infty} \left(1 - \left(1 - \frac{k}{m}\right)^q\right) \cdot n(k) \cdot C(k)$$

This property calculates the average probability that a type is queried at least once (i.e., $1 - (1 - k/m)^q$) in order to obtain the average number of types queried at least once. Only the costs corresponding to these queries are really useful. It is easy to see that the utility ratio tends towards 1 when the number of queries q becomes large. In contrast, offline systems may not be very relevant in very high velocity networks where the refresh are numerous leading to low query number q.

5   On-Demand Analytical Systems

The principle of an on-demand system is to evaluate the query for a type t at the time the user requests it. However, it may make sense to cache some frequently requested queries. The challenge is to choose which queries to cache to minimize the overall cost of the query set. To address this problem, we first compute the on-demand system cost (see Sect. 5.1) and then, we study several caching policies (see Sect. 5.2).

5.1 Cost of On-Demand Systems

With our cost model, the higher the degree of a type, the more likely it is to be queried. Intuitively, we must therefore determine a degree k_{cache} beyond which we must cache all queries. Below this threshold k_{cache}, the queries will be systematically computed on-demand without any storage. Above this threshold k_{cache}, we check if the result is already cached for returning it directly. If the result is not cached, it is computed and cached. The following property provides the theoretical cost of the system S^{on}:

Property 3 (On-demand system cost). Given a network $\mathcal{N}$ and a cost function C, the cost of the system S^{on} for q queries and the degree k_{cache} is:

$$C(S^{on}, q, \mathcal{N}) = \sum_{k < k_{cache}} \frac{q \cdot k}{m} \cdot n(k) \cdot C(k) + \sum_{k \geq k_{cache}} \left(1 - \left(1 - \frac{k}{m}\right)^q\right) \cdot n(k) \cdot C(k)$$

This property sums the costs degree by degree by separating them into two parts with respect to the threshold k_{cache}. For degrees less than k_{cache}, the cost for a degree k is the product of the number of performed queries $q \times k/m \times n(k)$ and the cost of a query $C(k)$. For degrees greater than k_{cache}, only the types queried incur a cost and this cost is unique (because a second query exploits the cache).

It is important to note that the part without cache increases linearly with the number of queries q while the part with cache is upper bounded by $\sum_{k \geq k_{cache}} n(k) \times C(k)$. To estimate the computation cost C_c, these two parts are useful and we have, in the linear case, the following cost per degree: $C_c(k) = \alpha_c k$. Conversely, for the storage cost C_s, the part without cache has no cost which can be modeled with the following cost per degree:

$$C_s(k) = \begin{cases} 0 & \text{if } k < k_{cache} \\ \alpha_s k & \text{otherwise} \end{cases}$$

As in the previous section, we compute the utility ratio:

Property 4 (Utility ratio). Given a network $\mathcal{N}$ and a cost function C, the utility ratio of the system S^{on} for q queries (assuming that $q \gg 1$) is:

$$UR(S^{on}, q, \mathcal{N}) \approx \frac{1}{C(S^{on}, q, \mathcal{N})} \cdot \sum_{k \geq k_{cache}} \left(1 - \left(1 - \frac{k}{m}\right)^q \left(1 + \frac{q \cdot k}{m}\right)\right) \cdot n(k) \cdot C(k)$$

It is again necessary to calculate the cost of what is really useful by normalizing it by the cost of what has been cached. Caching is useful if a type of degree greater than degree k_{cache} is queried at least twice (1 time for caching and 1 time for use). The probability of querying a type at least twice is $1 - \sum_{i=0}^{1} \binom{q}{i} (1 - k/m)^{q-i} (k/m)^i$ which simplifies if q is large by assuming that $(1 - k/m)^{q-1} \approx (1 - k/m)^q$. As previously, this utility ratio tends towards 1 when the number of queries q becomes large.

5.2 Cache Policies

Based on the cost model for the on-demand system, we can now consider different policies: caching nothing, caching everything or minimizing the overall cost. Other policies could be considered such as having a maximum cache size.

Let us first consider the extreme cases. The policy of caching nothing (i.e., $k_{\text{cache}} = +\infty$) boils down to a linear cost with the number of queries. Therefore, as soon as the number of queries is very large, this strategy has a high cost making it inefficient. On the opposite, the policy of caching everything (i.e., $k_{\text{cache}} = 0$) appears interesting if you have an unlimited amount of storage space. However, the results cached for queries concerning types with a low degree have little chance of being used again. Even if they have a low storage cost, it is a waste of storage.

For this reason, we propose to choose the degree k_{best} that minimizes the total linear cost $C_t(k) = C_c(k) + C_s(k)$ (see above for the definition of $C_c(k)$ and $C_s(k)$). Intuitively, the idea is to choose the degree where the costs with cache and without are in equilibrium. In the case of a linear complexity, the below property approximates k_{best}:

Property 5 (Theoretical result). Given a network $\mathcal{N}$ and the total linear cost function $C_t(k)$, the degree k_{best} minimizing the overall cost is given by:

$$k_{\text{best}} \approx \frac{m}{q} \cdot \frac{\alpha_c \cdot W\left(-\frac{(\alpha_c + \alpha_s)e^{-\alpha_s/\alpha_c - 1}}{\alpha_c}\right) + \alpha_c + \alpha_s}{\alpha_c}$$

where W is the Lambert W function.

Proof. We give the main steps of the proof. First, we look for the degree k such that the cost without cache is equal to that with cache (see Property 3) by injecting the costs (i.e., $C_t(k) = \alpha_c k$ for "with cache" and $C_t(k) = (\alpha_c + \alpha_s)k$ for "without cache"):

$$\frac{q \cdot k}{m} \cdot n(k) \cdot \alpha_c k = \left(1 - \left(1 - \frac{k}{m}\right)^q\right) \cdot n(k) \cdot (\alpha_c + \alpha_s)k$$

It is possible to approximate $(1 - k/m)^q$ by $\exp(-q \cdot k/m)$ as $k/m \ll 1$ (that explains the approximation $\approx$ in the final result). After simplification, we obtain the following equation:

$$\frac{q \cdot k}{m} \cdot \alpha_c = \left(1 - \exp\left(-\frac{q \cdot k}{m}\right)\right) \cdot (\alpha_c + \alpha_s)$$

Solving this equation gives the right-hand side of the result in Property 5.   □

This property gives a theoretical approximation of the degree k_{cache} which is based on the Lambert W function, quite complex to compute in practice. Therefore, it is simpler to determine k_{cache} by performing a dichotomic search between 0 and k_{max} looking for the minimum cost $C(\mathcal{S}^{on}, q, \mathcal{N})$. Interestingly, this algorithmic approach works for any convex cost function. Furthermore, considering the total linear cost $C_t(k)$ and a scale-free network with exponent $\gamma > 2$, we simplify the approximation k_{best} by neglecting the Lambert function term and injecting the mean:

$$k_{approx} = \frac{m}{q} \cdot \frac{\alpha_c + \alpha_s}{\alpha_c} = \frac{s}{q} \cdot \frac{\gamma - 1}{\gamma - 2} \cdot \frac{\alpha_c + \alpha_s}{\alpha_c}$$

6 Experimental Study

This experimental study applies our cost models on real-world networks in order to identify the evolution of the total cost and the utility ratio with the number of queries and the complexity of the query.

Table 1. Main characteristics of networks

Network	Vertices number	Edge number	Maximum degree	Exponent γ
Wikidata	37,256,044	675,226,687	37,656,116	2.058
DBnary	38,069,118	198,355,239	19,310,138	2.237
Cell-Cell	1,018,524	49,471,006	848	2.021
Twitch	168,116	13,595,116	35,279	2.012

Experimental Setting. We prepare two knowledge graph benchmarks (denoted by Wikidata and DBnary) based on crowdsourcing projects of Wikimedia Foundation: Wikidata [24] and DBnary [20]. For Wikidata, we used a truthy dump (February 2022)[1] and we filtered each dump to remove literals and external entities whose Uniform Resource Identifier (URI) is not prefixed by http://www.wikidata.org/. For DBnary, we simply used a dump in turtle format (May 2022)[2]. We also use two existing networks from the SNAP repository: Cell-Cell [25] and Twitch [19]. Table 1 provides the main characteristics of these four networks. We set the same complexity for storage and computation (e.g., $C_c(k) = C_s(k) = k^2$ for quadratic complexity). For reproducibility, the source code and frequency distributions corresponding to the four networks are available online: https://github.com/asoulet/ida2023ondemand

---

[1] https://dumps.wikimedia.org/other/incr/wikidatawiki/.
[2] http://kaiko.getalp.org/static/ontolex/latest/.

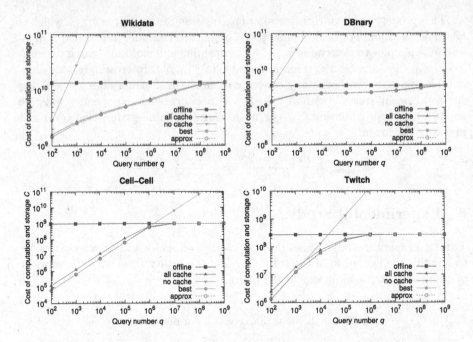

Fig. 2. Total cost of offline/on-demand systems with different cache policies

Total Cost Study. We will start by evaluating which strategy is the most parsi-monious i.e., with the lowest total cost (computation plus storage). Considering a linear complexity, Fig. 2 presents the evolution of the cost with the number of queries for 5 approaches: the offline system, the on-demand system with every-thing cached (i.e., $k_{cache} = 0$), nothing cached (i.e., $k_{cache} = +\infty$), best (using k_{best}) and scale-free approximation (using k_{approx}). Note that the scales are log-arithmic. The offline system (denoted by offline) is obviously independent of the number of queries. Therefore, the more the load increases, the more relevant this approach is. But for the largest datasets, a very high number of queries must be reached, which is unlikely to be achieved due to the data velocity of most net-works. Unsurprisingly, the no-cache system (denoted by no cache) is unattractive because its linear cost grows rapidly with the number of queries, exceeding the offline system. The on-demand system caching all queries (denoted by all cache) is not far from te best policy. It is therefore probably the most reasonable policy if the distribution over the whole network is not known. Of course, the on-demand system with cost minimization (denoted by best) is the least expensive approach whatever the number of queries even if its gain is significant for low volumes of queries. Finally, the proposed approximation (denoted by approx) works particu-larly well since the deviation is not large enough to distinguish the approx curve from the best curve.

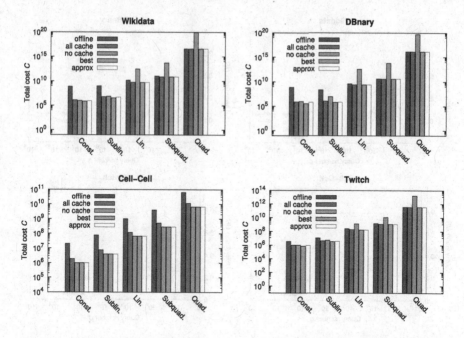

Fig. 3. Total cost of offline/on-demand systems w.r.t. cost function complexity

Now considering 100,000 queries, Fig. 3 gives the total cost with different cost function complexity for the same 5 approaches as above. Of course, the higher the complexity of the cost function, the higher the total cost. Typically, there is an order of magnitude increase from constant to linear and from linear to quadratic. The gain of the on-demand system decreases when the complexity is higher. Nevertheless, depending on the network, the on-demand system can remain relevant as it is the case for Cell-Cell with 100,000 queries. As in the previous experiment, the no-cache policy appears to be of little relevance (except for low complexity). Finally, the policy where everything is cached remains competitive with the best and approx policies which, although more subtle, are not significantly better.

Utility Ratio Study. We now study the interest of what has been cached. Considering a linear complexity, Fig. 4 presents the evolution of the utility ratio with the number of queries for 4 approaches: the offline system, the on-demand system with three policies: all cache, best and approx. The no-cache policy is not relevant for this study. It clearly appears that the least good approaches are offline and all-cache where a large part of the stored information is never queried. The utility ratio for Cell-Cell is zero up to 10,000 queries because the best solution is to cache nothing. In this experiment, it is visible that the best and approx policies are slightly different. Most often, the best approach stores a little less data explaining a slightly higher utility ratio.

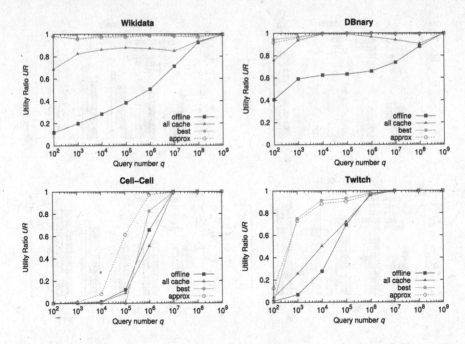

Fig. 4. Utility ratio of offline/on-demand systems w.r.t. load of queries

To sum up, the on-demand system is always the most parsimonious if a cache is used. Its gain is stronger when the query load is low and the cost function complexity per query is low. With the best and approx policies, on-the-fly caching is efficient guaranteeing high reuse.

7 Concluding Remarks

Our study based on a cost model shows the importance of considering an analytical system as a whole. It determines the advantages and weaknesses of different strategies depending on the query complexity and the number of queries:

On-Demand System Interest. When the query load is low or network velocity quickly invalidates the cache, on-demand systems are preferred. This avoids performing computation and storing data for types that will never be queried. It should be noted, however, that in our study we did not consider a mechanism for updating the cache of the query result. For example, when a new publication arrives, it is easy to update the different bibliometric indicators without recalculating everything from scratch. Such mechanisms are possible for certain queries, as it is the case with materialized views [9] in the database field.

Cache Policy Recommendations. The use of a cache is absolutely mandatory for on-demand systems in order to store the results of the types with the highest degrees (that are the most queried). When one knows the distribution of the data, it is easy to determine the degree above which query results should be kept. Otherwise, caching everything is still a reasonable policy because types with lower degree have lower storage complexity. However, our study has implicitly focused on a centralized context where caching and retrieval are negligible. In a decentralized context, the situation can be reversed with high storage complexity (due to network communications) while computational costs at the client can be neglected.

Interactivity Challenge. The advantage of offline systems is that they guarantee excellent interactivity when their results are used online. For a query on a given type, it is fast to return to the user the result already pre-computed. For on-demand systems, this interactivity is more complicated to guarantee especially for queries with high complexity on a high degree type. We think that two main workarounds can be used. First, systems with low response time (e.g., based on anytime algorithm [23]) should be preferred to those with low execution time. Indeed, it is often possible to propose a partial result quickly that will be refined later. Second, a hybrid strategy could be considered by pre-computing the answers for all types above a certain degree.

Acknowledgments. This work was partially supported by the grant ANR-21-CE23-0033 ("SELEXINI").

References

1. Andersen, J.S., Zukunft, O.: Evaluating the scaling of graph-algorithms for big data using GraphX. In: 2016 2nd International Conference on Open and Big Data (OBD), pp. 1–8. IEEE (2016)
2. Barabási, A.L., Albert, R.: Emergence of scaling in random networks. Science **286**(5439), 509–512 (1999)
3. Björneborn, L., Ingwersen, P.: Toward a basic framework for webometrics. J. Am. Soc. Inform. Sci. Technol. **55**(14), 1216–1227 (2004)
4. Broadus, R.N.: Toward a definition of bibliometrics. Scientometrics **12**, 373–379 (1987)
5. Csányi, G., Szendrői, B.: Structure of a large social network. Phys. Rev. E **69**(3), 036131 (2004)
6. Davidson, S.B., Gershtein, S., Milo, T., Novgorodov, S.: Disposal by design. Data Engineering, p. 10 (2022)
7. Ding, L., Finin, T.: Characterizing the semantic web on the web. In: The Semantic Web - ISWC 2006. ISWC 2006. LNCS, vol. 4273, pp. 242–257. Springer, Berlin, Heidelberg (2006). https://doi.org/10.1007/11926078_18
8. Fuad, A., Erwin, A., Ipung, H.P.: Processing performance on Apache Pig, Apache Hive and MySQL cluster. In: Proceedings of the International Conference on Information, Communication Technology and System (ICTS) 2014, pp. 297–302. IEEE (2014)
9. Gupta, A., Mumick, I.S., et al.: Maintenance of materialized views: problems, techniques, and applications. IEEE Data Eng. Bull. **18**(2), 3–18 (1995)

10. Hewasinghage, M., Abelló, A., Varga, J., Zimányi, E.: A cost model for random access queries in document stores. VLDB J. **30**(4), 559–578 (2021). https://doi. org/10.1007/s00778-021-00660-x

11. Ioannidis, Y.E.: Query optimization. ACM Comput. Surv. (CSUR) **28**(1), 121–123 (1996)

12. Ioannidis, Y.E., Poosala, V.: Balancing histogram optimality and practicality for query result size estimation. ACM SIGMOD Rec. **24**(2), 233–244 (1995)

13. Kersten, M.L., Sidirourgos, L.: A database system with amnesia. In: CIDR (2017)

14. Lee, D., et al.: On the existence of a spectrum of policies that subsumes the least recently used (LRU) and least frequently used (LFU) policies. In: Proceedings of the 1999 ACM SIGMETRICS International Conference on Measurement and Modeling of Computer Systems, pp. 134–143 (1999)

15. Lehmann, J., et al.: Distributed semantic analytics using the SANSA stack. In: d'Amato, C., et al. (eds.) ISWC 2017. LNCS, vol. 10588, pp. 147–155. Springer, Cham (2017). https://doi.org/10.1007/978-3-319-68204-4_15

16. Liu, G., Zhang, M., Yan, F.: Large-scale social network analysis based on MapReduce. In: 2010 International Conference on Computational Aspects of Social Networks, pp. 487–490. IEEE (2010)

17. Müller, S., Plattner, H.: Aggregates caching in columnar in-memory databases. In: Jagatheesan, A., Levandoski, J., Neumann, T., Pavlo, A. (eds.) IMDM 2013-2014. LNCS, vol. 8921, pp. 69–81. Springer, Cham (2015). https://doi.org/10.1007/978-3-319-13960-9_6

18. Newman, M.E.: Power laws, pareto distributions and Zipf's law. Contemp. Phys. **46**(5), 323–351 (2005)

19. Rozemberczki, B., Sarkar, R.: Twitch gamers: a dataset for evaluating proximity preserving and structural role-based node embeddings (2021)

20. Sérasset, G.: DBnary: wiktionary as a lemon-based multilingual lexical resource in RDF. Semant. Web **6**(4), 355–361 (2015)

21. Shi, J., et al.: Clash of the titans: MapReduce vs. Spark for large scale data analytics. Proc. VLDB Endow. **8**(13), 2110–2121 (2015)

22. Shim, J., Scheuermann, P., Vingralek, R.: Dynamic caching of query results for decision support systems. In: Proceedings of the Eleventh International Conference on Scientific and Statistical Database Management, pp. 254–263. IEEE (1999)

23. Soulet, A., Suchanek, F.M.: Anytime large-scale analytics of linked open data. In: Ghidini, C., et al. (eds.) ISWC 2019. LNCS, vol. 11778, pp. 576–592. Springer, Cham (2019). https://doi.org/10.1007/978-3-030-30793-6_33

24. Vrandečić, D., Krötzsch, M.: Wikidata: a free collaborative knowledgebase. Commun. ACM **57**(10), 78–85 (2014)

25. Zheng, G.X., et al.: Massively parallel digital transcriptional profiling of single cells. Nat. Commun. **8**(1), 1–12 (2017)

ROCKAD: Transferring ROCKET to Whole Time Series Anomaly Detection

Andreas Theissler[✉][iD], Manuel Wengert, and Felix Gerschner[iD]

Aalen University of Applied Sciences, 73430 Aalen, Germany
andreas.theissler@hs-aalen.de

Abstract. The analysis of time series data is of high relevance in fields like manufacturing, health, automotive, or science. In this paper, we propose ROCKAD, a kernel-based approach for semi-supervised whole time series anomaly detection, i.e. the assignment of a single anomaly score to an entire time series. Our key idea is to use ROCKET as an unsupervised feature extractor and to train a single as well as an ensemble of k-nearest neighbors anomaly detectors to deduce an anomaly score. To the best of our knowledge, this is the first approach to transfer the ideas of ROCKET to the task of anomaly detection. We systematically evaluate ROCKAD for univariate time series and show it is statistically significantly better compared to baseline methods. Additionally, we show in a case study that ROCKAD is also applicable to multivariate time series.

Keywords: Machine learning · Anomaly detection · Time series

1 Introduction

Time series data are omnipresent, examples being manufacturing data [3,17], recordings from automotive systems [32,34], medical data [23,27], or environmental data [1,35]. Machine learning-based classification, forecasting, or anomaly detection has enabled new applications in the aforementioned fields.

In this paper, we address the problem of *whole time series anomaly detection*, i.e. assigning a single anomaly score to an entire time series [10,30]. The majority of research on time series anomaly detection (AD) aims to detect anomalous data points or subsequences. Whole time series anomaly detection [10], on the other hand, is used (a) when the entire time series is expected to have anomalous behavior, or (b) when time series are segmented prior to the analysis either by fixed interval sizes or based on change-point detection. Examples are time series from manufacturing [2], from vehicle tests [32], or medical data like ECG [10].

We address the question, if the unsupervised feature extraction of the time series classifier ROCKET (RandOm Convolutional KErnel Transform), proposed by Dempster et al. [14], can be exploited for anomaly detection. We propose the approach **ROCKAD** (**ROCKET** **A**nomaly **D**etector), using ROCKET, enhanced by an anomaly detection pipeline. To the best of our knowledge, this is the *first attempt to transfer ROCKET to the task of anomaly detection*.

A. Theissler, M. Wengert and F. Gerschner—contributed equally.

© The Author(s), under exclusive license to Springer Nature Switzerland AG 2023
B. Crémilleux et al. (Eds.): IDA 2023, LNCS 13876, pp. 419–432, 2023.
https://doi.org/10.1007/978-3-031-30047-9_33

Since anomalies correspond to rare and unusual events, it is typically not possible to obtain a representative set of anomalous time series. A set of normal data, on the other hand, can be obtained easily, for example by monitoring a system in its normal operation mode. Hence, unsupervised and semi-supervised AD [9] are common settings. We address *semi-supervised AD* [9] where an anomaly detector M_{AD} is trained on a training set $\mathcal{X}_{tr}$ solely containing data from the normal class $\mathcal{C}_N$.

While semi-supervised settings are more likely to be applicable to real problem settings, they suffer from the inherent problem that model tuning is based on the normal class only. This has the following implications: (1) The discriminative power of some feature space $\mathcal{F}$ cannot be determined with approaches like recursive feature elimination. The same is valid for implicit features in some latent space of Deep Learning models, since these are found w.r.t. tuneable hyperparameters. (2) A model's hyperparameters cannot be tuned with standard loss functions that incorporate the trade-off between the different per-class errors.

As a solution, existing approaches use data heuristics, create artificial anomalies [26], incorporate data from the anomaly class [28], use the fraction of expected anomalies [28], define some cost function based on assumptions [28,31], or allow users to interactively tune the model [33,37].

Our key idea is, to use the original ROCKET as an unsupervised feature extractor, yielding a high-dimensional feature space $\mathcal{F}_\mathcal{R}$ without parameter tuning. In the feature space, we evaluate anomaly detectors, which for a time series x_T output an anomaly score α. Specifically designed for time series, ROCKET uses convolutional kernels known from 1D-convolutional neural networks. However, as the key difference which we exploit in our paper, rather than learning the kernel parameters, ROCKET creates a high number of kernels with randomly sampled parameters. Moving these kernels over a time series x_T, ROCKET transforms x_T into a feature space where a small number of features is relevant for class discrimination. The original ROCKET uses a subsequent classifier to separate the classes. We replace this by an AD pipeline (preprocessing and anomaly detectors).

This paper makes the following contributions:

1. We transfer the ideas of the ROCKET time series classifier to the task of anomaly detection (AD), hoping to inspire research in AD to exploit ROCKET's success for classification tasks.
2. We evaluate the approach w.r.t. further anomaly detectors on the UCR time series repository, allowing for reproducible and comparable research.

2   Related Work

The general field of anomaly detection is surveyed and structured in [9]. Current surveys on AD for time series are presented in [7,24]. In [7,10], the problem setting of *whole time series anomaly detection* is defined.

In [10], anomalous segments in univariate ECG time series are detected in a semi-supervised setting using nearest neighbors. In [32], anomalies in multivariate time series are detected with an ensemble. With a sliding window, fixed-sized segments are extracted and evaluated by an ensemble of different anomaly detectors. In contrast to our approach, [10,32] work on the raw time series representations.

As opposed to working on the time series representation, a variety of methods for feature extraction were proposed in literature. Examples, which were used for AD, are briefly reviewed in the following: In [29], tsfresh [11] is used to extract a set of features which is then reduced to a small set of discriminative features. The number of those features is user-defined. AD is achieved in the reduced feature space using PCA. In contrast to [29], our approach uses ROCKET as a feature extractor. A different approach is to learn shapelets [38], which are subsequences that are most representative for a class membership. Hence, shapelets can be used to classify time series and have also been used for AD [5,41]. While the extraction of shapelets in [5,41] requires a learning procedure, our approach uses the computationally efficient idea to use a randomly generated, overparmeterized features space.

3   ROCKAD: Kernel-Based Whole Series Anomaly Detection

We introduce an anomaly detector for time series, the Random Convolutional Kernel Transform Anomaly Detector (ROCKAD)[1]. In short, ROCKAD transforms a time series to a more abstract feature space, capturing its temporal information. Using a nearest neighbor approach, the transformed time series are compared by their distances to the transformed normal time series in $\mathcal{X}_{tr}$ and an anomaly score α is deduced. We systematically evaluate ROCKAD for univariate time series, however, since the underlying ROCKET also works for multivariate time series, ROCKAD is also applicable to these. The approach was inspired by ROCKET's strong performance in time series classification and was developed on the same data sets as the classification counterpart [13].

3.1   Definitions

Time series can be univariate or multivariate, where observations of one variable are referred to as univariate time series, defined as follows:

Definition 1. *A **univariate time series** $x_T : \overrightarrow{T} \mapsto \overrightarrow{X}$ is a finite sequence of N data points ordered by time. For every time point t_i in $\overrightarrow{T} = \{t_1, ..., t_N\}$, there exists one data point x_i in $\overrightarrow{X} = \{x_1, ..., x_N\}$. The time points t_i are equidistant.*

For time series, we address the problem of semi-supervised anomaly detection. As opposed to the detection of single abnormal data points or the detection of abnormal subsequences, our focus is whole series anomaly detection which we define as follows (in accordance with [10]):

---

[1] ROCKAD source code and further information: https://ml-and-vis.org/rockad.

Definition 2. *Semi-supervised whole series anomaly detection* is the assignment of a single anomaly score α to an entire time series x_T by an anomaly detector M_{AD}, i.e. "whole series". M_{AD} is trained on a training set $\mathcal{X}_{tr}$ solely containing instances from the normal class $\mathcal{C}_N$, i.e. "semi-supervised".

3.2 Architecture of ROCKAD

ROCKAD consists of three stages, namely feature extraction, transformation and AD: (1) on each input time series x_T, ROCKET is used as an unsupervised feature extractor, (2) on the ROCKET-induced feature space, a power transformer is applied, and (3) k-nearest neighbors is used to deduce an anomaly score α for x_T. The components are depicted in Fig. 1 and described in the following:

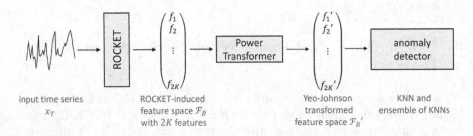

Fig. 1. Steps of ROCKAD (Power-transformed **ROCK**ET **A**nomaly **D**etector).

1. **ROCKET:** Our approach builds on the time series classifier ROCKET [14] which generates a set of K convolutional kernels (default: $K = 10,000$ [14]) that are moved over the time series x_T. The kernel parameters (weights, bias, dilation and padding) are randomly sampled from predefined ranges or distributions. Sliding the kernels over x_T, two features are extracted for each kernel: (1) the greatest weighted sum of the kernels' elements from all convolutional operations (*max*) and (2) the proportion of weighted sums that were positive over all convolutional operations for the kernel (*ppv*). ROCKET reaches SOTA accuracy for time series classification while demanding only a fraction of time [14]. We use ROCKET as an unsupervised feature extractor, yielding a high-dimensional feature space without parameter tuning. We denote this *ROCKET-induced feature space* by $\mathcal{F}_{\mathcal{R}}$.

2. **Power transformer:** We found that transforming $\mathcal{F}_{\mathcal{R}}$ to a more Gaussian-like distribution, improves the separation of $\mathcal{C}_N$ and $\mathcal{C}_A$. We use the power transformer by Yeo and Johnson [39], which can deal with negative values. The resulting features are scaled using z-score, i.e. mean $= 0$, std $= 1$. We denote the transformed feature space by $\mathcal{F}_{\mathcal{R}}'$.

3. **Anomaly detector (k-nearest neighbors, KNN):** We use KNN to determine an anomaly score α for a time series $x_T \in \mathcal{X}_{te}$, by calculating the average

distance of $\mathcal{F_R}'(x_T)$ to its k nearest neighbors in $\mathcal{F_R}'(\mathcal{X}_{tr})$. We use two variants: a single KNN and a bagging ensemble of n KNN estimators.

We created two variants of ROCKAD (see Table 1): ROCKAD$_{(1)\_default}$ uses a single KNN and ROCKAD$_{(n)\_default}$ uses an ensemble of n KNNs. In addition, we evaluate tuned versions denoted as ROCKAD$_{(1)\_tuned}$ and ROCKAD$_{(n)\_tuned}$ (see Table 1). We tuned the hyperparameters $\{K, n, k\}$ of ROCKAD$_{(n)}$ using the "development data sets" from the ROCKET paper [13]. These data sets are also included in the evaluation, hence, we acknowledge that the tuned versions could have a tendency to be optimized towards the UCR evaluation data.

Table 1. Evaluated ROCKAD variants

variant	configuration	parameters
ROCKAD$_{(1)\_default}$	single KNN; default parameters	$K = 10000, k = 5$
ROCKAD$_{(n)\_default}$	KNN-ensemble; default parameters	$K = 10000, k = 5, n = 10$
ROCKAD$_{(1)\_tuned}$	single KNN; tuned parameters	$K = 600, k = 3$
ROCKAD$_{(n)\_tuned}$	KNN-ensemble; tuned parameters	$K = 600, k = 3, n = 21$

3.3 Training and Detection Procedure

During training of ROCKAD$_{(n)}$, ROCKET generates a set of K kernels and extracts the *max* and *ppv* features (see [14]) from the time series in $\mathcal{X}_{tr}$ which are then transformed as described in Sect. 3.2. The transformed series $\mathcal{F_R}'(\mathcal{X}_{tr})$ are passed to the KNN-ensemble, whereas each estimator $j \in [1; n]$ is built on a bootstrapped subset of the transformed training data, denoted as $\mathcal{F_R}'(\mathcal{X}_{tr})_j$. During detection, the kernels obtained from training are applied on $x_T \in \mathcal{X}_{te}$ and extract the features $\mathcal{F_R}(x_T)$. The power transformer maps $\mathcal{F_R}(x_T)$ with the parameters found in the training phase. Afterwards, z-score is applied using the parameters obtained from $\mathcal{F_R}(\mathcal{X}_{tr})$. Eventually, the KNN-ensemble queries the resulting $\mathcal{F_R}'(x_T)$ through each estimator j which calculates the average Euclidean distance of $\mathcal{F_R}'(x_T)$ to the k nearest neighbors of the estimators' training subset $\mathcal{F_R}'(\mathcal{X}_{tr})_j$. The resulting distances are averaged over all estimators resulting in α:

$$\alpha = \frac{1}{n} \sum_{j=1}^{n} \frac{\sum_{i=1}^{k} dist(\mathcal{F_R}'(x_T), NN_i(\mathcal{F_R}'(X_{tr})_j))}{k} \tag{1}$$

ROCKAD$_{(1)}$ adopts these procedures but trains a single KNN on the entire training data. The anomaly score for ROCKAD$_{(1)}$ corresponds to (1) with $n = 1$ and $\mathcal{F_R}'(X_{tr})_j$ replaced by $\mathcal{F_R}'(X_{tr})$.

4   Experimental Results

This section describes the evaluation data sets and data preparation steps, followed by an introduction of the baseline models, ROCKAD is compared with. Following that, the results are reported in critical difference diagrams. More details and the raw results can be found at https://ml-and-vis.org/rockad.

4.1   Evaluation Data Sets

Data sets for whole time series AD are rarely available. Therefore, the widely used UCR time series classification data sets [12], were used for development and evaluation. In order to transform the problem setting from classification to AD, each data set was prepared as follows:

1. The class containing the most time series is used as the normal class C_N, the class with the least time series as the anomaly class C_A.
2. $\mathcal{X}_{tr}$ is created using 80% of the data in C_N.
3. The test set $\mathcal{X}_{te}$ contains the remaining 20% of C_N and is enriched with time series from C_A, such that $\lfloor 10\% \rfloor$ of $\mathcal{X}_{te}$ are anomalies.
4. Data sets, where 10% of $\mathcal{X}_{te} > |C_A|$ or sampling 10% anomalies for $\mathcal{X}_{te}$ leads to $|C_A| < 1$, were excluded. As a consequence, 93 UCR data sets were used in the evaluation.
5. The above procedure was repeated such that 10 resamples were drawn. The 10 results are averaged to obtain robust results, as suggested in [12].

4.2   Baseline Anomaly Detectors

As baselines to compare ROCKAD to, we first use a number of feature space models that do not capture the temporal relations in time series, i.e. each data point x_{t_i} is interpreted as a feature:

1. **Isolation Forest (iForest)**: An isolation forest [19] is an ensemble of isolation trees. Anomalies are detected by separating rare and unusual data points from the rest of the data. The underlying idea is that anomalous data points are easier to separate, creating a tree path that is less deep.
2. **Local Outlier Factor (LOF)**: LOF [8] assigns outlier scores to data points based on the idea that anomalies are in low density regions of the feature space. We use the LOF-variant for novelty detection, implemented in sklearn [22].
3. **One Class Support Vector Machine (OCSVM)**: An OCSVM [25] separates the classes C_N and C_A with a hyperplane in a kernel-induced feature space. We set the hyperparameter $\nu = 0.05$ and use the RBF-Kernel with $\gamma = 1/(|\mathcal{F}| \times \sigma^2)$.
4. **k-nearest neighbors (KNN)**: KNN is adapted to AD by calculating a data point's mean distance to its k nearest neighbors in $\mathcal{X}_{tr}$ and reporting this distance as anomaly score α.

As further baseline models, we use a number of time series models, modeling the temporal behaviour – two of them adapted to whole series AD:

5. **LSTM Encoder Decoder (LSTMED)**: An LSTMED consists of LSTMs [16] for encoding and decoding of a time series. Based on the reconstruction error, anomalies are detected – similar to the idea of autoencoders. We adapted LSTMED for whole series AD by reporting the sum of reconstruction errors as anomaly score.

6. **Temporal Convolutional Network Autoencoder (TCNAE)**: Based on the temporal convolutional network (TCN) [4], an enhancement with an autoencoder for AD was proposed in [36]. Analogously to LSTMED, the sum of reconstruction errors is used as anomaly score.

7. **LSTM-Deep Autoencoding Gaussian Mixture Model (L-DAGMM)**: The DAGMM [42] is a sequence of a deep autoencoder and a Gaussian mixture model, jointly optimizing the parameters of both models. We use it in the variant replacing the deep autoencoder with an LSTMED [18].

8. **Recurrent Energy Based Model (REBM)**: The REBM [40] combines the ideas of Restricted Boltzman Machines and recurrent networks.

Furthermore, as an alternative feature extractor we use `tsfresh` [11] in combination with the feature space models 1–4 (denoted as tsfresh_OCSVM, tsfresh_iForest, etc.). We use the full `tsfresh`-feature space, but we acknowledge that in [29] it was shown that reducing the number of `tsfresh`-features yields higher detection rates. However, this requires a user-defined number of features.

4.3   Evaluation Metrics

For each anomaly detector M_{AD} we obtain anomaly scores α instead of crisp decisions, for example distances or inherent outlier scores. From α, we calculcate the ROC curve showing the trade-off between TPR and FPR, where we define C_A as the positive class. As evaluation metric, we report the AUROC (area under the ROC curve) which expresses the overall performance of each M_{AD} independent of a threshold. The AUROC is considered a common metric for evaluating anomaly detectors as shown e.g. in [21]. To compare the M_{AD}, we use the Friedman-posthoc-Wilcoxon-Holm test. As recommended in [15], Friedman's test compares the anomaly detectors by their average rank on the evaluation data sets and tests the differences for statistical significance. In accordance with [6], a Wilcoxon signed rank test with Holm's alpha correction is applied. The results are shown as a critical difference (CD) diagram showing the M_{AD} sorted by their rank averaged over the evaluation data sets. Those M_{AD}, with differences in the ranks not being statistically significant, are connected by a horizontal line.

4.4   Results for ROCKAD Compared to Baseline Models

The evaluation shows both ROCKAD models compared to the baselines. The primary result of our evaluation is that ROCKAD significantly outperformed the baseline models for both versions on the 93 UCR data sets, as shown in Fig. 2. The temporal models yielded a significantly lower mean rank than the others. We acknowledge, that we had to adapt two of these models to whole series AD. Furthermore, the models' performances depend on hyperparameter tuning using validation sets, which was not conducted throughout the evaluation.

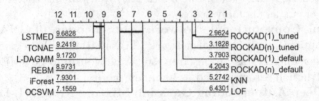

Fig. 2. CD-Diagram of the anomaly detection evaluation results (AUROC) for the baseline anomaly detectors and ROCKAD (small ranks = better performance).

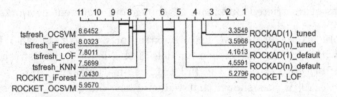

Fig. 3. CD-Diagram (AUROC) of different M_{AD} attached to a sequence of ROCKET ($K = 10,000$) and power transformer, compared to ROCKAD.

Figure 3 shows the performance for different estimators applied to ROCKET and `tsfresh` with a power transformer. The diagram displays a significant difference between the `tsfresh` and ROCKET models, except for ROCKET_iForest. ROCKAD$_{(1)\_default}$ outperformed the ensemble on the default setting. However, if the number of kernels K has to be reduced due to computational limitations, we suggest to try both versions since the tuned models with a smaller K had no statistical difference.

4.5   Sensitivity Analysis

We examine the effects of varying the number of kernels, estimators, and neighbors with a sensitivity analysis. The analysis was conducted on 31 of ROCKET's "development data sets" [13] that were in accordance with steps 1–5 in Sect. 4.1. Starting with ROCKAD's default settings, the hyperparameters $\{K, n, k\}$ were tuned sequentially. Therefore, the best number of kernels K was determined and used to evaluate the number of estimators n and the best $\{K, n\}$ were taken to assess the number of neighbors k per estimator. The AUROCs of each model were subtracted from the model with the default parameter of its origin. The

AUROC changes are shown by the differences for the data sets, where a difference of 0 corresponds to the origin model. A negative difference implies a worse performance than the origin. To select the best hyperparameter, the differences' mean ranks are given in parentheses (smaller = better). The best hyperparameters are used for the tuned versions $ROCKAD_{(1)\_tuned}$ and $ROCKAD_{(n)\_tuned}$.

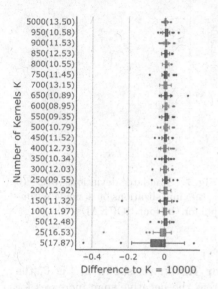 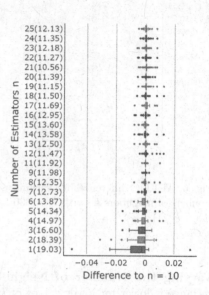

Fig. 4. AUROC differences for varying numbers of kernels K. On the development data, $K = 600$ kernels yielded the best results.

Fig. 5. AUROC differences for numbers of estimators n with $K = 600$, where $n = 21$ yielded the best results on the development data.

Varying the Number of Kernels: First, the impact of the number of kernels K is evaluated. As shown in Fig. 4, a small K can lead to slightly better results compared to $K = 10,000$. However, $K = 500$ and $K = 100$ show that a smaller K may also produce outstanding worse or better results for individual sets. Moreover, a high K reduces the variance of AUROC over multiple data sets, as visible for $K = 5,000$. Summarizing, a small K can increase the AUROC, if a few labels are available for a sanity check. Otherwise, a high K seems the safer choice. $K = 600$ had the best mean rank in this experiment and was chosen for the tuned versions of ROCKAD.

Varying the Number of Estimators: Figure 5 shows the benefit of ensembling, as the model with $n = 1$ yields the highest variance over multiple data sets. Additionally, a leveling trend towards 0 can be seen for an increasing n. The slope of increasing AUROCs decreases and flattens at approximately $n = 10$. The best mean rank was achieved at $n = 21$.

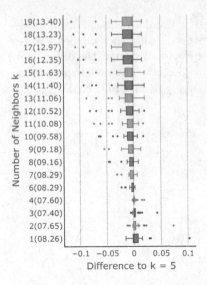

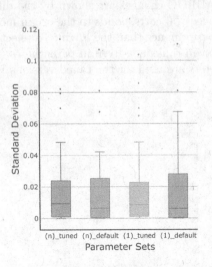

Fig. 6. AUROC differences for varying numbers of neighbors k (with $K = 600$ and $n = 21$).

Fig. 7. Standard deviations over 10 kernel initializations on a single resample for the four ROCKADs.

Varying the Number of Neighbors for KNN: As shown in Fig. 6, the variance is lower for lower values of k, whereas the negative span increases for $k > 6$. The default parameter $k = 5$ is near-optimal, the best mean rank was obtained with $k = 3$. Furthermore, [14] pointed out that a small K leads to a high standard deviation for different kernel initializations. As the tuned model suggests a small K, the robustness was investigated by measuring the standard deviation of the four models over 10 different initializations on a single resample. The results, shown in Fig. 7, could not confirm this behavior with KNNs.

5   Case Study: Detecting Anomalous Textures

To show the applicability of ROCKAD, a texture recognition case study was conducted with a data set of recordings from a 6-axis force/torque sensor, resulting in a 6D-multivariate time series (MTS) (taken from [20]). Transforming the classification data set to AD, the procedure described in Sect. 4.1 was used. The task was to train on time series of the material wood ($\mathcal{C}_N = \{\text{"wood"}, \text{"osb"}\}$) and detect the remaining 19 materials as anomalies. In order to obtain a crisp classification result for the nearest neighbor-based anomaly score α, i.e. $\{\mathcal{C}_N, \mathcal{C}_A\}$, we use the method proposed in [28], turning ROCKAD into a one-class classifier. We report the AUROC and the F1-score in Table 2.

We evaluated $\text{ROCKAD}_{(n)\_\text{tuned}}$, $\text{ROCKAD}_{(1)\_\text{tuned}}$ and ROCKET+ OCSVM. First, ROCKAD was analyzed for univariate whole series AD using the measured horizontal force F_z applied on the texture. Further, we show

Table 2. Results for the texture recognition case study. F_z denotes the univariate and $\vec{F}\&\vec{T}$ the multivariate time series.

	F_z (univariate)		$\vec{F}\&\vec{T}$ (multivariate)	
	F1-score	AUROC	F1-score	AUROC
$ROCKAD_{(n)\_tuned}$	**0.916**	**0.936**	**0.864**	**0.977**
$ROCKAD_{(1)\_tuned}$	0.739	0.754	0.740	0.778
ROCKET_OCSVM	0.907	0.915	0.870	0.884

that ROCKAD can be applied to multivariate whole time series AD. The 6-dimensional MTS composed of measurements of the applied force and torque in x, y and z direction ($\vec{F}\&\vec{T}$) is used. Analogously to the univariate case, this experiment was done using ROCKAD and ROCKET+OCSVM. Applying the proposed model to MTS anomaly detection is possible without further adjustments since ROCKET works for both cases (i.e. univariate and multivariate time series).

6   Conclusion

The novel approach ROCKAD for whole series anomaly detection was proposed, exploiting the time series classifier ROCKET [14]. Using ROCKET as a feature extractor, ROCKAD transforms the data and trains a single KNN and an ensemble of KNN anomaly detectors. The results on the univariate UCR time series showed the efficacy of the approach. Furthermore, a case study showed promising results regarding ROCKAD's generalizability as well as its use on multivariate time series. Yet, our research is not without limitations: (a) We proposed ROCKAD in a default setting and a setting tuned on a subset of the UCR data. While this procedure is in accordance with [14], we acknowledge that the tuned variant might have been implicitly optimized towards the UCR data. Yet, we could report that ROCKAD with default settings is also statistically significantly better than the baseline models. (b) We had to adapt two of the time series baseline models to the problem of whole series anomaly detection. (c) In order to be comparable, we used the UCR data set. However, these data sets are relatively small, i.e. the neural network baseline models might perform better than reported on larger data sets. Future work could be (1) a systematic evaluation, possibly an advancement, of ROCKAD for multivariate time series and (2) the development of a novel one-class classifier based on ROCKAD, i.e. the introduction of crisp decisions as opposed to an anomaly score, as we have preliminarily shown in the case study.

References

1. Abulibdeh, A.: Time series analysis of environmental quality in the state of Qatar. Energy Policy **168**, 113089 (2022)

2. Ahmad, A., Song, C., Tan, R., Gärtler, M., Klöpper, B.: Active learning application for recognizing steps in chemical batch production. In: 2022 IEEE 27th International Conference on Emerging Technologies and Factory Automation (ETFA), pp. 1–4. IEEE (2022)
3. Atzmueller, M., Hayat, N., Schmidt, A., Klöpper, B.: Explanation-aware feature selection using symbolic time series abstraction: approaches and experiences in a petro-chemical production context. In: 2017 IEEE 15th International Conference on Industrial Informatics (INDIN), pp. 799–804. IEEE (2017)
4. Bai, S., Kolter, J.Z., Koltun, V.: An empirical evaluation of generic convolutional and recurrent networks for sequence modeling. ArXiv (2018)
5. Beggel, L., Kausler, B.X., Schiegg, M., Pfeiffer, M., Bischl, B.: Time series anomaly detection based on shapelet learning. Comput. Stat. **34**, 945–976 (2019)
6. Benavoli, A., Corani, G., Mangili, F.: Should we really use post-hoc tests based on mean-ranks? J. Mach. Learn. Res. **17**, 1–10 (2016)
7. Blázquez-García, A., Conde, A., Mori, U., Lozano, J.A.: A review on outlier/anomaly detection in time series data. ACM Comput. Surv. (CSUR) **54**(3), 1–33 (2021)
8. Breunig, M.M., Kriegel, H.P., Ng, R.T., Sander, J.: LOF: identifying density-based local outliers. In: ACM SIGMOD International conference on Management of data, pp. 93–104 (2000)
9. Chandola, V., Banerjee, A., Kumar, V.: Anomaly detection: a survey. ACM Comput. Surv. **41**(3), 15:1–15:58 (2009)
10. Chandola, V., Cheboli, D., Kumar, V.: Detecting anomalies in a time series database (2009)
11. Christ, M., Braun, N., Neuffer, J., Kempa-Liehr, A.W.: Time series FeatuRe extraction on basis of scalable hypothesis tests (tsfresh – a python package). Neurocomputing **307**, 72–77 (2018)
12. Dau, H.A., et al.: The UCR time series archive. IEEE/CAA J. Automatica Sinica **6**(6), 1293–1305 (2019)
13. Dempster, A., Petitjean, F., Webb, G.I.: ROCKET: exceptionally fast and accurate time series classification using random convolutional kernels. CoRR abs/1910.13051 (2019)
14. Dempster, A., Petitjean, F., Webb, G.I.: ROCKET: exceptionally fast and accurate time series classification using random convolutional kernels. Data Min. Knowl. Disc. **34**(5), 1454–1495 (2020). https://doi.org/10.1007/s10618-020-00701-z
15. Demšar, J.: Statistical comparisons of classifiers over multiple data sets. J. Mach. Learn. Res. **7**, 1–30 (2006)
16. Hochreiter, S., Schmidhuber, J.: Long short-term memory. Neural Comput. **9**(8), 1735–1780 (1997)
17. Hsu, C.Y., Liu, W.C.: Multiple time-series convolutional neural network for fault detection and diagnosis and empirical study in semiconductor manufacturing. J. Intell. Manuf. **32**(3), 823–836 (2021). https://doi.org/10.1007/s10845-020-01591-0
18. Li, Y., Zha, D., Zou, N., Hu, X.: PyODDS: an end-to-end outlier detection system (2019)
19. Liu, F.T., Ting, K.M., Zhou, Z.H.: Isolation forest. In: 2008 Eighth IEEE International Conference on Data Mining, pp. 413–422 (2008)
20. Markert, T., Matich, S., Hoerner, E., Theissler, A., Atzmueller, M.: Fingertip 6-axis force/torque sensing for texture recognition in robotic manipulation. In: International Conference on Emerging Technologies and Factory Automation (ETFA). IEEE (2021)

21. Seliya, N., Abdollah Zadeh, A., Khoshgoftaar, T.M.: A literature review on one-class classification and its potential applications in big data. J. Big Data **8**(1), 1–31 (2021). https://doi.org/10.1186/s40537-021-00514-x
22. Pedregosa, F., et al.: Scikit-learn: machine learning in python. J. Mach. Learn. Res. **12**, 2825–2830 (2011)
23. Raab, D., Theissler, A., Spiliopoulou, M.: XAI4EEG: spectral and spatio-temporal explanation of deep learning-based seizure detection in EEG time series. Neural Comput. Appl., pp. 1–18 (2022). https://doi.org/10.1007/s00521-022-07809-x
24. Schmidl, S., Wenig, P., Papenbrock, T.: Anomaly detection in time series: a comprehensive evaluation. Proc. VLDB Endowment **15**(9), 1779–1797 (2022)
25. Schölkopf, B., Platt, J.C., Shawe-Taylor, J.C., Smola, A.J., Williamson, R.C.: Estimating the support of a high-dimensional distribution. Neural Comput. **13**, 1443–1471 (2001)
26. Steinbuss, G., Böhm, K.: Generating artificial outliers in the absence of genuine ones - a survey. ACM Trans. Knowl. Disc. Data **15**(2), 1–37 (2021)
27. Sun, J., et al.: A hybrid deep neural network for classification of schizophrenia using EEG Data. Sci. Rep. **11**(1), 4706 (2021). https://doi.org/10.1038/s41598-021-83350-6
28. Tax, D.M.: One-class classification. Concept-learning in the absence of counter-examples. Ph.D. thesis, Delft University of Technology (2001)
29. Teh, H.Y., Kevin, I., Wang, K., Kempa-Liehr, A.W.: Expect the unexpected: unsupervised feature selection for automated sensor anomaly detection. IEEE Sens. J. **21**(16), 18033–18046 (2021)
30. Teng, M.: Anomaly detection on time series. In: 2010 IEEE International Conference on Progress in Informatics and Computing, vol. 1, pp. 603–608 (2010)
31. Theissler, A.: Detecting anomalies in multivariate time series from automotive systems. Ph.D. thesis, Brunel University London (2013)
32. Theissler, A.: Detecting known and unknown faults in automotive systems using ensemble-based anomaly detection. Knowl.-Based Syst. **123**(C), 163–173 (2017)
33. Theissler, A., Kraft, A.L., Rudeck, M., Erlenbusch, F.: VIAL-AD: visual interactive labelling for anomaly detection - an approach and open research questions. In: International Workshop on Interactive Adaptive Learning (IAL). CEUR-WS (2020)
34. Theissler, A., Pérez-Velázquez, J., Kettelgerdes, M., Elger, G.: Predictive maintenance enabled by machine learning: use cases and challenges in the automotive industry. Reliab. Eng. Syst. Saf. **215**, 107864 (2021)
35. Theissler, A., Thomas, M., Burch, M., Gerschner, F.: ConfusionVis: comparative evaluation and selection of multi-class classifiers based on confusion matrices. Knowl.-Based Syst. **247**, 108651 (2022)
36. Thill, M., Konen, W., Bäck, T.: Time series encodings with temporal convolutional networks. In: Filipič, B., Minisci, E., Vasile, M. (eds.) BIOMA 2020. LNCS, vol. 12438, pp. 161–173. Springer, Cham (2020). https://doi.org/10.1007/978-3-030-63710-1_13
37. Trittenbach, H., Böhm, K., Assent, I.: Active learning of SVDD hyperparameter values. In: 2020 IEEE 7th International Conference on Data Science and Advanced Analytics (DSAA), pp. 109–117 (2020)
38. Ye, L., Keogh, E.: Time series shapelets. In: Proceedings of the 15th ACM SIGKDD international conference on Knowledge discovery and data mining. ACM (2009)
39. Yeo, I.K., Johnson, R.: A new family of power transformations to improve normality or symmetry. Biometrika **87**, 954–959 (2000)

40. Zhai, S., Cheng, Y., Lu, W., Zhang, Z.: Deep structured energy based models for anomaly detection. In: International Conference on Machine Learning, pp. 1100–1109. PMLR (2016)
41. Zhang, J., Zeng, B., Shen, W., Gao, L.: A one-class Shapelet dictionary learning method for wind turbine bearing anomaly detection. Measurement **197**, 111318 (2022)
42. Zong, B., et al.: Deep autoencoding gaussian mixture model for unsupervised anomaly detection. In: International Conference on Learning Representations (2018)

Out-of-Distribution Generalisation with Symmetry-Based Disentangled Representations

Loek Tonnaer$^{(\boxtimes)}$, Mike Holenderski, and Vlado Menkovski

Eindhoven University of Technology, Eindhoven, The Netherlands
{l.m.a.tonnaer,m.holenderski,v.menkovski}@tue.nl

Abstract. Learning disentangled representations is suggested to help with generalisation in AI models. This is particularly obvious for combinatorial generalisation, the ability to combine familiar factors to produce new unseen combinations. Disentangling such factors should provide a clear method to generalise to novel combinations, but recent empirical studies suggest that this does not really happen in practice. Disentanglement methods typically assume *i.i.d.* training and test data, but for combinatorial generalisation we want to generalise towards factor combinations that can be considered out-of-distribution (OOD). There is a misalignment between the distribution of the observed data and the structure that is induced by the underlying factors.

A promising direction to address this misalignment is symmetry-based disentanglement, which is defined as disentangling symmetry transformations that induce a group structure underlying the data. Such a structure is independent of the (observed) distribution of the data and thus provides a sensible language to model OOD factor combinations as well. We investigate the combinatorial generalisation capabilities of a symmetry-based disentanglement model (LSBD-VAE) compared to traditional VAE-based disentanglement models. We observe that both types of models struggle with generalisation in more challenging settings, and that symmetry-based disentanglement appears to show no obvious improvement over traditional disentanglement. However, we also observe that even if LSBD-VAE assigns low likelihood to OOD combinations, the encoder may still generalise well by learning a meaningful mapping reflecting the underlying group structure.

1 Introduction

It is suggested that learning representations that disentangle underlying factors of variation in the data is an important goal towards better generalisation [1]. This is particularly obvious if we consider combinatorial generalisation, the ability to generalise to novel combinations of previously seen factors. Ideally a model should be able to disentangle the underlying factors of a datapoint, even if that particular combination of factors was never observed during training, as long as each individual factor value has been seen before.

© The Author(s), under exclusive license to Springer Nature Switzerland AG 2023
B. Crémilleux et al. (Eds.): IDA 2023, LNCS 13876, pp. 433–445, 2023.
https://doi.org/10.1007/978-3-031-30047-9_34

Disentanglement models are typically trained on data that covers most factor combinations [10], but the number of combinations scales exponentially with the number of factors, which quickly becomes unmanageable for realistic scenarios with more than a few factors. Thus, it is beneficial if models can learn the underlying mechanisms behind the factors without seeing all possible combinations.

However, recent studies have shown that current disentanglement methods do not deliver on their promise in this so-called out-of-distribution (OOD) generalisation setting [12,17]. Correlations between the factors of variation in the observed data are reflected in the learned latent representations of disentanglement models [20], since the training methods are designed for independent and identically distributed (*i.i.d.*) data. For example, two factors that correlate strongly with each other may be represented in a single latent dimension, even if they represent two fundamentally different properties.

There is clearly a misalignment between the concept of disentangling underlying factors of variation, which may not be independently distributed in the data, and learning to model the distribution of the data. Whereas most disentanglement methods aim to uncover the former, their methodology mostly focuses on the latter.

One promising direction to resolve this misalignment is symmetry-based disentanglement (SBD) [5], which provides a formal language (using group theory) to reason about the structure of underlying factors. This allows to model this structure separate from the distribution of the data. SBD focuses on disentangling symmetry transformations that act on the data, reasoning that these transformations induce the underlying factors of variation. The idea is to learn representations that are equivariant to such transformations. By focusing on these transformations during training, we hope to learn a model that can generalise well to unseen factor combinations, since such combinations are still the result of applying transformations that were seen during training.

In this paper, we evaluate how well disentangled representations generalise to unseen factor combinations, with a particular focus on linear SBD (LSBD) representations, which provide a fixed formulation of how transformations affect representations. We investigate how much coverage of factor combinations is needed for current methods to generalise well to unseen combinations, exposing the limits of these methods. We confirm previous findings that disentanglement does not seem to improve OOD generalisation much, even in relatively simple settings. Moreover, we show that LSBD representations also generalise poorly to unseen factor combinations, despite the more suitable perspective of modelling transformations. However, we also observe that this partially depends on how we measure OOD generalisation: for VAE-based models, unseen factors may be encoded fairly well (reflecting the underlying factor structure) even if they are decoded poorly.

Our results suggest that more work is needed to learn representations that can generalise better to unseen factor combinations, even if equivariance with respect to disentangled transformations is used as a learning signal. We expose the limitations of LSBD representations to generalise to unseen factor combinations, even if the transformation mechanisms are captured well for the observed data. We hope that our results provide a basis for further research on how to design methods that generalise better to unseen factor combinations, and on how to evaluate such generalisation.

2   Background

Disentanglement. Since the suggestion that disentangling underlying factors of variation in data is important for better generalisation [1], various methods and metrics to learn and evaluate disentangled representations have been proposed [3,6,7,9]. Most methods focus on expanding a Variational Autoencoder (VAE) [8,16] with some regularisation term to encourage disentanglement. They assume the data is independent and identically distributed. However, these methods showed only limited success, mostly on toy problems. Moreover, disentanglement in general is shown to be impossible without some sort of inductive bias [10].

Symmetry-Based Disentanglement. Disentanglement is typically explained from the perspective of independent factors of variation, but this doesn't provide a formal definition of what constitutes a disentangled representation. To address this, symmetry-based disentanglement (SBD) [5] gives a formal definition by means of group theory. The motivation is that symmetries in the real world are what leads to variability in data observations. A symmetry is a transformation that affects some factor of variation, but leaves all others invariant. Such symmetry transformations can be described with the formal language of group theory, as a decomposable transformation group acting on both the data and the learned representations. The goal of SBD is then to learn representations that are invariant to such symmetries, resulting in disentangled subspaces where one symmetry (or subgroup) affects only one subspace. Several recent works focus on learning SBD representations [4,13,15,19], with varying degrees of supervision on transformations.

We focus on Linear SBD (LSBD), where an additional requirement is that transformations act on the representations as linear operators (i.e. matrix multiplications). This allows for a simple and consistent description of transformations in latent space, which is useful for generalising beyond factor combinations seen during training. We provide more detail in Sect. 3.

OOD Generalisation. Generalisation has always been one of the main goals in machine learning, and representation learning in particular. Machine learning methods are however mostly based on the *i.i.d.* assumption that train and test data are independent and identically distributed. In realistic scenarios this is hard to realise, so there is significant interest in dropping this assumption and investigating out-of-distribution (OOD) generalisation (see [18] for an overview).

Disentanglement seems a sensible solution to improve OOD generalisation, especially from the perspective of combinatorial generalisation where the goal is to generalise to unseen combinations of factor values observed during training.

However, recent work [12] shows somewhat surprisingly that disentanglement does not seem to help with combinatorial generalisation. In this paper we follow their evaluation protocol and expand on it, focusing mostly on LSBD instead of traditional disentanglement. Our setting is similar to the "composition" setting in [17], which provides a benchmark to evaluate OOD detection for visual representation learning. Related work [20] shows that disentanglement methods trained on data with correlations between the factors of variation learn representations that reflect these correlations, which further emphasises the need for better disentanglement methods that can generalise to OOD data.

3   Symmetry-Based Disentanglement Should Enable Out-of-Distribution Generalisation

Linear Symmetry-Based Disentanglement (LSBD). We are mostly interested in a specific definition of disentanglement, LSBD [5]. The idea is to disentangle symmetry transformations that act on a set of real-world factors W. Such transformations can be modelled by a group G acting on W. The key assumption is that this group can be decomposed into subgroups $G = G_1 \times \ldots \times G_K$, such that each subgroup only affects one aspect of the real world W. Data observations are assumed to come from a generative process $b : W \to X$. A learned inference map (i.e. encoder model) $h : X \to Z$ to a representation space (or latent space) Z is defined to be LSBD if the following holds:

1. there is a decomposition of the latent space into subspaces $Z_1 \oplus \ldots \oplus Z_K$,
2. there is a group representation for each subgroup in the corresponding vector subspace $\rho_k : G_k \to \mathrm{GL}(Z_k)$ for $k = 1, \ldots, K$,
3. the group representation $\rho : G \to \mathrm{GL}(Z)$ acts on Z as $\rho(g) \cdot z = (\rho_1(g_1) \cdot z_1, \ldots, \rho_K(g_K) \cdot z_K)$, and
4. the composition $f = h \circ b$ is *equivariant* with respect to the actions of G on W and Z, i.e. for all $w \in W$ and $g \in G$ it holds that $f(g \cdot w) = \rho(g) \cdot f(w)$.

There is also a non-linear definition of symmetry-based disentanglement [5], but we focus on the linear version since it allows for a simple and explicit way to express transformations acting in the latent space, namely as linear operators.

We use the D_{LSBD} metric proposed by [19] to quantify disentanglement with respect to the LSBD definition.

OOD Generalisation Through Equivariance. LSBD is based on the idea that transformations describe the sources of variability in observed data. From this perspective, OOD generalisation means that we want to generalise to any observation that is the result of applying some transformation to a previously seen data observation, even if the resulting observation has never been seen before. Since LSBD provides a clear formulation of how a transformation should act on a latent representation, a well-trained LSBD model should also be able to represent unseen data that is the result of such a transformation, because of the equivariance of the model with respect to this transformation. LSBD thus provides a suitable framework for this kind of OOD generalisation in theory.

4 Experimental Setup

We investigate the out-of-distribution (OOD) generalisation of a number of VAE-based disentanglement models for datasets with known factorised generative factors, by splitting off data with certain factor combinations into an OOD set and using the remaining data as training set. We then evaluate those models on their performance on the left-out OOD data using various metrics in Sect. 5.

Datasets. We consider two datasets with an underlying $SO(2) \times SO(2)$ group structure (i.e. having 2 cyclic factors), Square and Arrow; as well as two popular disentanglement datasets, dSprites and 3D Shapes. See Fig. 1 for some examples. The datasets can be fully generated by known factors of variation. All datasets contain images with 64×64 pixels (black-and-white or RGB).

Square has factors *x-position* and *y-position*, squares wrap around the edges of the canvas so the factors are cyclic. Arrow has factors *orientation* and *hue*. Each factor can attain 64 values, yielding 4096 datapoints in each dataset.

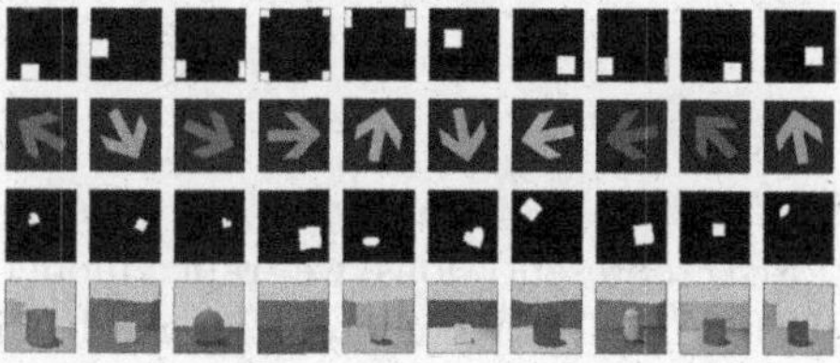

Fig. 1. Example images of Square, Arrow, dSprites, and 3D Shapes.

dSprites [11] has factors *shape, scale, orientation, x-position*, and *y-position*, with 3, 6, 40, 32, and 32 values each, for a total of 737,280 datapoints. The *shape* factor is categorical, whereas the other factors describe continuous properties, of which *orientation* is cyclic. 3D Shapes [2] has factors *floor hue, wall hue, object hue, scale, shape*, and *orientation*, with 10, 10, 10, 8, 4, and 15 values each, for a total of 480,000 datapoints. The *shape* factor is categorical, whereas the other factors describe continuous properties, of which all *hue* values are cyclic.

OOD Splits: Left-Out Factor Combinations. For Arrow and Square we define different OOD splits by leaving out images where both factors are within a certain range. If we represent the factors as values f_1 and f_2 on a scale from 0 to 1, we split off images with factor values $f_1 < r$ and $f_2 < r$ simultaneously, see Fig. 2. We do this for various values of r, namely 0.125, 0.25, 0.375, 0.5, 0.625, 0.750, and 0.875, leading to 7 different OOD splits. Note that the ratio of the number of datapoints we split off from the full dataset grows quadratically as r^2, i.e. each OOD split contains roughly 1.5%, 6%, 14%, 25%, 39%, 56%, and 77% of the full dataset, respectively.

For the dSprites and 3D Shapes dataset, we follow the experimental setup of [12], defining three experiments for each dataset: recombination-to-element (RTE), recombination-to-range (RTR) and extrapolation (EXTR). Table 1 summarises the factor combinations that are left out as OOD combinations for each of these settings on both datasets. Note that factors *scale* and *x-position* for dSprites and *object scale, orientation* and all hues for 3D Shapes are given as values from 0 to 1. Hue values above 0.5 correspond to cyan, blue, and purple.

Table 1. OOD splits for dSprites and 3D Shapes.

	RTE	RTR	EXTR
dSprites	*shape = ellipsis, scale < 0.6,* *120° < orientation < 240°,* *x-position ≥ 0.6*	*shape = square,* *x-position ≥ 0.5*	*x-position > 0.5*
3D Shapes	*floor hue ≥ 0.5, wall hue ≥ 0.5,* *object hue ≥ 0.5,* *object shape = cylinder,* *object scale = 1, orientation = 0*	*object hue ≥ 0.5,* *object shape = oblong*	*floor hue ≥ 0.5*

LSBD-VAE. To investigate the generalisation of LSBD representations, we train a model called LSBD-VAE [19]. We choose this model since it can be trained with supervision on the underlying transformations in a batch of datapoints, such that we can easily and reliably learn LSBD representations for the training data even with challenging OOD splits.

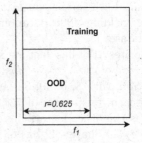

Fig. 2. OOD splits for datasets with 2 factors.

Like a regular Variational Autoencoder (VAE) [8,16], LSBD-VAE consists of an encoder (or approximate posterior) $q(Z|X)$, a prior $p(Z)$, and a decoder $p(X|Z)$, where X and Z are the data and latent space, respectively. Given a group decomposition $G = G_1 \times \ldots \times G_K$ that represents the symmetry structure underlying the data, LSBD-VAE defines suitable topologies for the corresponding latent subspaces $Z = Z_1 \times \ldots \times Z_K$ and a matching linearly disentangled group representation ρ. The factors in our Square and Arrow datasets can be described with subgroups $G_k = SO(2)$, i.e. the Special Orthogonal group of 2D rotations. As matching latent subspaces we use unit circles (or 1-spheres) $Z_k = S^1 = \{z \in \mathbb{R}^2 : ||z|| = 1\}$. Group representations ρ_k are then 2D rotation matrices.

LSBD-VAE uses a ΔVAE [14] to learn encoded (posterior) distributions on these latent subspaces. Like a regular VAE, the unsupervised ΔVAE model is trained by minimising the negative Evidence Lower Bound (ELBO), but the prior and approximate posterior (encoder) are defined on a (typically non-Euclidean) Riemannian manifold Z. The prior is uniform over this manifold, whereas the approximate posterior is defined by a location and scaling parameter. To estimate the intractable terms of the negative ELBO, the reparameterisation trick is implemented via a random walk.

Transformation-Supervised Batches. The LSBD-VAE can be trained both on unsupervised images and transformation-supervised batches $(x_1, x_2, \ldots, x_M)$ where all samples can be expressed as a known transformation of the first, i.e. $x_m = g_m \cdot x_1$ for $m = 2, \ldots, M$. Each transformation g corresponds to changes in each of the factor values, and can be represented with rotation matrices acting on each of the latent subspaces.

Given such transformation-supervised batches, LSBD-VAE includes an additional loss term $\mathcal{L}_{LSBD}$ that encourages learning LSBD representations. $\mathcal{L}_{LSBD}$ measures the dispersion of the points $\rho(g_m^{-1}) \cdot z_m$ for $m = 1, \ldots, M$, where z_m is the model's encoding of data point x_m. Ideally, since $x_1 = g_m^{-1} \cdot x_m$, all these points should be equal to achieve LSBD, so the dispersion provides a term to encourage LSBD. Formally, it is defined as

$$
\mathcal{L}_{LSBD} = \frac{1}{M} \sum_{m=1}^{M} \left\| \rho(g_m^{-1}) \cdot z_m - \Pi \left(\frac{1}{M} \sum_{m=1}^{M} \rho(g_m^{-1}) \cdot z_m \right) \right\|^2,
$$

where $g_1 = e$, the group identity.

In this paper we train only on transformation-supervised batches without unsupervised training. This is a rather strong type of supervision, but our goal is to investigate how well-trained LSBD representations perform in OOD generalisation, not to find the most efficient way to train an LSBD-VAE model. We split up the training set into transformation-supervised batches of size $M = 32$.

Non-Cyclic Factors. Although dSprites and 3D Shapes contain factors that don't really have an underlying SO(2) structure, we can still use this formulation to train an LSBD-VAE on these datasets, by mapping the factor values to suitable angle values from 0 to 2π radians. The *shape* factors in both datasets can be represented as equally spaced values with an in-between angle of $\frac{2\pi}{n_{classes}}$. Since there are only a few classes, this naturally encourages a class-based clustering in the corresponding latent subspaces. Factors *scale*, *x-position* and *y-position* in dSprites as well as *scale* and *orientation* in 3D Shapes, are essentially continuous but not cyclic. Thus, we map them to a range of angle values from 0 to $0.9 \cdot 2\pi$ radians, to introduce a discontinuity in the lowest and highest observed factor value. The remaining factors are cyclic so we represent them with regular angle values between 0 and 2π.

Architecture and Hyperparameters. For the encoder we use a convolutional architecture as in [10], with 4 convolutional layers (4×4 kernels and 2×2 strides, each layer has 32, 32, 64, and 64 filters, respectively), followed by a fully-connected layer with 256 units. The output of this is connected by fully-connected layers to the parameters of the LSBD-VAE latent subspaces. The decoder is the reverse of this, using strided transposed convolutions. All hidden layers use "ReLU" non-linearity, the output layer uses a sigmoid activation to predict pixel values. Each model is trained with the Adam optimiser, using an early stopping criterion. We use a mini-batch size of 8, where each element in the mini-batch is in fact a transformation-supervised batch of $M = 32$ images, thus each mini-batch consists of 256 images with batch shape (8, 32). We train each dataset and OOD split combination 3 times, and report the mean scores over these 3 iterations in our evaluations.

Traditional Disentanglement Models. For comparison, we train a regular VAE [8,16] as well as 5 traditional unsupervised disentanglement models: BetaVAE [6], DIP-VAE-I and II [9], FactorVAE [7], and cc-VAE [3], as implemented in `disentanglement_lib` [10], which use the same architecture as described above for LSBD-VAE. These models do not address disentanglement from the perspective of LSBD, but are instead based on statistical properties of the data. We train each model-dataset combination only once, due to the large number of such combinations and limited computing power. Each model is trained for 30,000 training steps with a batch size of 64.

For the Square and Arrow datasets, which have 2 cyclic factors, we trained each model-dataset combination with 2, 4 and 7 latent dimensions. Although 2 factors could technically be disentangled in 2 latent dimensions, at least 4 dimensions are needed to represent the cyclic topology as well (see [5] and [14]). Nevertheless, we observed that the extra capacity of 7 latent dimensions led to better OOD generalisation results, thus we report only those experiments in Sect. 5. For dSprites and 3D Shapes (with 5 and 6 factors, respectively) we only trained with 7 latent dimensions.

5   Experiments and Results

Likelihood Ratio: Training vs. OOD ELBO. We follow a similar evaluation protocol to [12], where we compute the mean negative log-likelihood (approximated with the negative Evidence Lower Bound or ELBO) for the training data as well as the OOD test data. A small difference between the training and OOD ELBOs indicates good generalisation. Note that since the ELBO is an approximation for the log-likelihood, the difference between ELBOs represents a likelihood ratio.

Figure 3 shows the differences for all models and datasets. We observe that LSBD-VAE only shows improved generalisation for the Square dataset, whereas for other datasets it mostly gets outperformed by the traditional methods.

This is particularly surprising for the Arrow dataset, where the symmetry-based paradigm is most suitable. We suspect that the increased model capacity of the traditional models with 7 latent dimensions helps more for generalisation than the strong regularisation of the LSBD-VAE. To illustrate this, we compare the reconstructions of OOD samples from the Arrow 0.625 split by DIP-VAE and LSBD-VAE, see Fig. 4. Both models reconstruct training data similarly well, but their behaviour on OOD data is quite different. DIP-VAE reconstructs OOD samples to darkened images with an uneven colour that nevertheless capture the underlying factors (orientation and hue) fairly well, whereas LSBD-VAE reconstructs into clearer images with an incorrect factor combination (one that

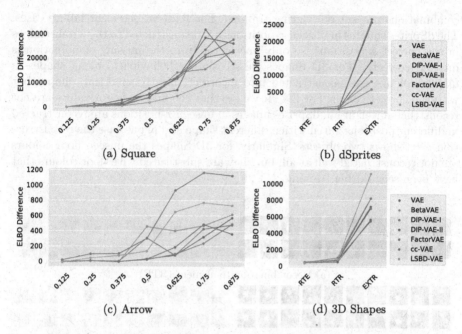

(a) Square

(b) dSprites

(c) Arrow

(d) 3D Shapes

Fig. 3. Differences between train and OOD ELBO for all datasets and models. The horizontal axis shows different OOD splits.

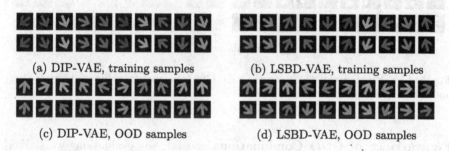

(a) DIP-VAE, training samples   (b) LSBD-VAE, training samples

(c) DIP-VAE, OOD samples   (d) LSBD-VAE, OOD samples

Fig. 4. Examples of training and OOD samples (top lines) and their reconstructions (bottom lines) by two different models, for the Arrow 0.625 split.

has been seen during training). Therefore, DIP-VAE achieves a lower pixel-wise reconstruction error, which is a main component of the ELBO computation.

Reconstructions of OOD Combinations. We can better understand the generalisation of LSBD-VAE on dSprites and 3D Shapes by inspecting reconstructions of OOD data from the different splits, as shown in Fig. 5. For RTE we see that OOD data is reconstructed fairly well for both datasets, although dSprites images are sometimes reconstructed with the wrong shape. This seems mostly the effect of good interpolation, since for RTE only a limited number of combinations are left out during training. For RTR we see clear failure cases. The dSprites squares in unseen x-positions are reconstructed to the wrong shapes in the correct x-positions, so only one factor from the missing combination is inferred correctly. For 3D Shapes we see similar behaviour; oblong shapes in unseen colours are reconstructed into incorrect (mostly spherical) shapes but with the correct colour. For EXTR we see that the unseen factor values are not reconstructed well at all, dSprites images in unseen x-positions are reconstructed in different positions, often with a different shape and in one case the reconstruction even shows two objects. Similarly, for 3D Shapes the unseen floor colours are not reconstructed well at all, but they are substituted with floor colours that have been seen during training.

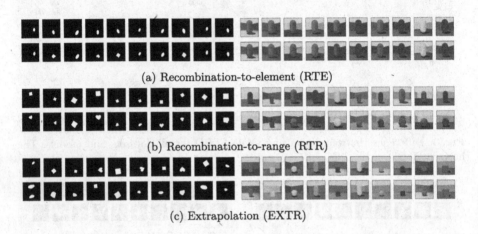

(a) Recombination-to-element (RTE)

(b) Recombination-to-range (RTR)

(c) Extrapolation (EXTR)

Fig. 5. LSBD-VAE reconstructions of OOD data from various splits of dSprites (left) and 3D Shapes (right).

Equivariance of OOD Combinations. So far, the evaluations we showed rely mostly on reconstruction performance, which is the main component of the ELBO in a VAE-based model. Such evaluations are heavily focused on the generalisation of the decoder. Yet, in a representation learning setting we are typically mostly interested in the behaviour of the encoder, which is the model that actually learns representations.

In symmetry-based disentanglement (SBD), we can use the notion of equivariance to evaluate the generalisation of the encoder. In particular, for Linear SBD (LSBD) we can use the D_{LSBD} metric from [19] to quantify the equivariance with respect to the transformations in the full dataset including the left-out OOD combinations. This gives us a measure of how well a model can represent (in a linear manner) the underlying structure of the data, even if it hasn't observed certain parts of this structure.

Figure 6 shows the D_{LSBD} scores for all models on the Square and Arrow datasets, lower scores are better (0 is optimal). dSprites and 3D Shapes contain factors that cannot clearly be mapped to LSBD symmetries, so we do not evaluate D_{LSBD} on those datasets. We emphasise that LSBD-VAE is the only model that attempts to disentangle from an LSBD point of view, so the traditional models are included not for fair comparison but as indicative results. Indeed we observe that LSBD-VAE achieves better D_{LSBD} scores overall, though some traditional models perform fairly well on the Square dataset. LSBD-VAE performance on D_{LSBD} hardly suffers for the smaller OOD splits (up until the 0.5 split), and still performs fairly well for the larger OOD splits. This indicates that even though OOD generalisation seems poor when inspecting ELBO values, the encoder appears to represent the underlying structure of the data quite well.

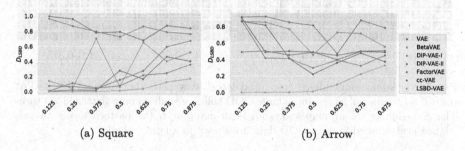

(a) Square   (b) Arrow

Fig. 6. D_{LSBD} scores (lower is better) for various OOD splits.

To visualise this more clearly, we show 2D latent embeddings and traversals for LSBD-VAE on the Arrow datasets in Fig. 7, for increasingly large OOD splits. From the top row we see that until the 0.5 split, the underlying structure is captured quite well even for unseen OOD combinations. For the 0.5 split we can clearly identify where the unseen OOD combinations are encoded since they break the axis-alignment, but the overall topology is still intact. For larger OOD splits we see that this topology starts to break, and that OOD encodings start overlapping with training encodings, thus the model starts failing to represent these OOD factor combinations.

The bottom row of Fig. 7 shows how the decoder fails for OOD combinations. For the 0.5 split, orientation and hue are easily recognisable and disentangled in the generated images. For larger splits, shapes become more disfigured, and eventually orientation and hue are no longer well-represented.

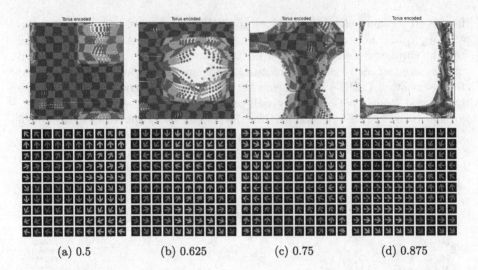

Fig. 7. 2D latent embeddings (top) and latent traversals (bottom) for LSBD-VAE trained on Arrow for increasingly large OOD splits, visualised on a flattened 2D torus. The embeddings colour map shows an ideal mapping if the pattern forms an axis-aligned grid. Embeddings of OOD data are shown in a lighter shade.

6   Conclusion

In this work we investigated the out-of-distribution (OOD) generalisation of disentangled representations, in particular from the perspective of linear symmetry-based disentanglement (LSBD). We reason why such representations should in theory generalise well to unseen (OOD) factor combinations, but in practice we observe that disentanglement models struggle to generalise well. We provide empirical results that showcase for which settings OOD generalisation seems to work, and where and how models fail. Overall our results imply that there is still work to be done to achieve OOD generalisation with disentangled models. The results, however, also show some promise that LSBD models can learn to express unseen factor combinations if there is sufficient coverage of combinations, although in practice decoders seem to struggle more with correctly representing the unseen combinations even if their encodings satisfy equivariance quite well.

References

1. Bengio, Y., Courville, A., Vincent, P.: Representation learning: a review and new perspectives. IEEE Trans. Pattern Anal. Mach. Intell. **35**(8), 1798–1828 (2013)
2. Burgess, C., Kim, H.: 3D shapes dataset (2018). https://github.com/deepmind/3dshapes-dataset/
3. Burgess, C.P., et al.: Understanding disentangling in β-VAE. arXiv preprint arXiv:1804.03599 (2018)

4. Caselles-Dupré, H., Ortiz, M.G., Filliat, D.: Symmetry-based disentangled representation learning requires interaction with environments. In: Advances in Neural Information Processing Systems, pp. 4606–4615 (2019)
5. Higgins, I., et al.: Towards a definition of disentangled representations. arXiv preprint arXiv:1812.02230 (2018)
6. Higgins, I., et al.: β-VAE: learning basic visual concepts with a constrained variational framework. In: International Conference on Learning Representations (2017)
7. Kim, H., Mnih, A.: Disentangling by factorising. In: International Conference on Machine Learning, pp. 2649–2658 (2018)
8. Kingma, D.P., Welling, M.: Auto-encoding variational Bayes. In: International Conference on Learning Representations (2014)
9. Kumar, A., Sattigeri, P., Balakrishnan, A.: Variational inference of disentangled latent concepts from unlabeled observations. In: International Conference on Learning Representations (2018)
10. Locatello, F., Bauer, S., Lucic, M., Gelly, S., Schölkopf, B., Bachem, O.: Challenging common assumptions in the unsupervised learning of disentangled representations. In: International Conference on Machine Learning (2019)
11. Matthey, L., Higgins, I., Hassabis, D., Lerchner, A.: dSprites: disentanglement testing sprites dataset (2017). https://github.com/deepmind/dsprites-dataset/
12. Montero, M.L., Ludwig, C.J., Costa, R.P., Malhotra, G., Bowers, J.: The role of disentanglement in generalisation. In: International Conference on Learning Representations (2020)
13. Painter, M., Prugel-Bennett, A., Hare, J.: Linear disentangled representations and unsupervised action estimation. In: Advances in Neural Information Processing Systems, vol. 33 (2020)
14. Perez Rey, L.A., Menkovski, V., Portegies, J.: Diffusion variational autoencoders. In: Proceedings of the Twenty-Ninth International Joint Conference on Artificial Intelligence, IJCAI-20, pp. 2704–2710 (2020)
15. Quessard, R., Barrett, T.D., Clements, W.R.: Learning group structure and disentangled representations of dynamical environments. In: Advances in Neural Information Processing Systems, vol. 33 (2020)
16. Rezende, D.J., Mohamed, S., Wierstra, D.: Stochastic backpropagation and approximate inference in deep generative models. In: Proceedings of the 31st International Conference on International Conference on Machine Learning, vol. 32 (2014)
17. Schott, L., et al.: Visual representation learning does not generalize strongly within the same domain. arXiv preprint arXiv:2107.08221 (2021)
18. Shen, Z., et al.: Towards out-of-distribution generalization: a survey. arXiv preprint arXiv:2108.13624 (2021)
19. Tonnaer, L., Pérez Rey, L.A., Menkovski, V., Holenderski, M., Portegies, J.W.: Quantifying and learning linear symmetry-based disentanglement. In: Proceedings of the 39th International Conference on Machine Learning (ICML) (2022). http://arxiv.org/abs/2011.06070
20. Träuble, F., et al.: On disentangled representations learned from correlated data. In: International Conference on Machine Learning, pp. 10401–10412. PMLR (2021)

Forecasting Electricity Prices: An Optimize Then Predict-Based Approach

Léonard Tschora[1,2(✉)], Erwan Pierre[2], Marc Plantevit[3], and Céline Robardet[1]

[1] Univ Lyon, INSA Lyon, LIRIS, UMR5205, 69621 Villeurbanne, France
[2] BCM Energy, 69003 Lyon, France
`leonard.tschora@bcmenergy.fr`
[3] EPITA Research Laboratory (LRE), 94276 Le Kremlin-Bicêtre, France

Abstract. We are interested in electricity price forecasting at the European scale. The electricity market is ruled by price regulation mechanisms that make it possible to adjust production to demand, as electricity is difficult to store. These mechanisms ensure the highest price for producers, the lowest price for consumers and a zero energy balance by setting day-ahead prices, i.e. prices for the next 24 h. Most studies have focused on learning increasingly sophisticated models to predict the next day's 24 hourly prices for a given zone. However, the zones are interdependent and this last point has hitherto been largely underestimated. In the following, we show that estimating the energy cross-border transfer by solving an optimization problem and integrating it as input of a model improves the performance of the price forecasting for several zones together.

Keywords: Electricity Price Forecasting · Optimization-based data augmentation · Machine learning

1 Introduction

Energy challenges are even more important as our societies have become extremely dependent on it. However, the production of energy, and in particular electricity, is linked to many intricate factors, based on different estimates such as weather forecasts (influencing both production and consumption) or production capacities for various means. Added to this complexity is a tariff regulation mechanism [13] used to balance production and consumption, as electricity is hard to store. This algorithm maximizes social welfare defined as the sum of consumer surplus, supplier surplus and congestion rents from cross-border exchanges. It ensures the highest price for producers, the lowest price for suppliers and a constant energy balance by setting day-ahead prices, i.e., 24-hourly prices for the next day.

Being able to forecast day-ahead energy prices is crucial to control energy production and for a successful energy transition. Thus, many works [5,9,15,16] have sought to produce the most accurate price prediction models possible. In [14], we have shown that approaches based on machine learning models are

B. Crémilleux et al. (Eds.): IDA 2023, LNCS 13876, pp. 446–458, 2023.
https://doi.org/10.1007/978-3-031-30047-9_35

superior to benchmark auto-regressive models. They provide much more accurate predictions and are fast enough to be used operationally. We also strove for predicting the prices of different zones jointly. Although we did not obtain a significant improvement in the forecasts, the analysis of the contributions of the variables highlighted the importance of integrating data from foreign countries for the price forecast. For example, we have shown that Swiss prices contribute significantly to increasing the accuracy of French, Belgian and German price forecasts. We concluded that we had not used enough information to correctly model the European network, in particular that we had not sufficiently taken into account transfer capacities and cross-border energy flows in our models.

We propose to overcome these limitations by putting forward different ways to integrate cross-border flows into predictive models. Cross-border flows are constrained by the Available Transfer Capacity (ATC) between two countries that share a border. However, this maximum capacity is not fully used continuously and knowing the flows between countries would undoubtedly improve the prediction models. For this, we propose to take advantage of domain knowledge to estimate cross-border flows by a combinatorial optimization model.

The proposed approach is reversed from the predict-then-optimize approaches [3, 11] used to solve many decision-making problems by combining machine learning and combinatorial optimization. In this framework, some parameters of a combinatorial optimization problem are estimated from other features based on historical data. Our approach use a combinatorial optimization model to estimate features that are then used to train a machine learning model.

In this paper, we introduce the problem of electricity price forecasting on the European market (Sect. 2). Our research hypothesis is that we can improve the model prediction by enriching the input data thanks to domain knowledge. Especially, we introduce the problem of estimating the cross-border flows (Sect. 3). We design two distinct combinatorial optimization problems and their combination. Then, we use the results of these optimization problems in a multi-zone forecasting model that predicts prices for 35 distinct zones of the European market (Sect. 4). The experimental evaluation (Sect. 5) confirms that the cross-border flows estimation makes it possible to improve the model performance. We then conclude with a broader discussion and a forward look (Sect. 6).

2   Electricity Price Forecasting Problem on the European Market

Unlike other commodities (e.g., cereals, oil), electricity cannot be efficiently stored. To prevent failures on the electricity network, balancing algorithms are used. On the European market, the EUPHEMIA [13] algorithm fixes hourly prices by matching demand, production and exchanges across Europe in a way to maximize the social welfare while taking into account the market and network constraints: (1) The energy balance must be zero for all zones at all times. (2) The flow of energy between two zones must not exceed the maximum transfer

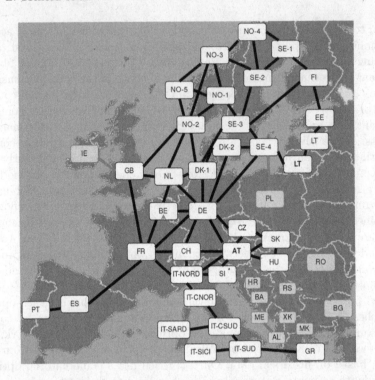

Fig. 1. European electricity market map: Some countries are divided into several zones (e.g., Italy, Norway). Prices are established for each zone. Energy can flow between connected zones. Areas or connections colored in red are excluded from our dataset due to lack of data.

capacity between these two zones. **(3)** Where possible, the energy flow between two areas is maximized to generate more profit from congestion rents. This algorithm runs daily at noon and determines the day-ahead prices, matched demand and supply and energy flows of the 46 European zones (see Fig. 1). The electricity price forecasting problem (EPF) consists in predicting the prices over 24 h before their settlement. Electricity prices are constrained by fundamentals variables: consumption, generation, transfer capacities. More precisely, pricing algorithms use a forecast of those variables for the next day.

To solve the EPF problem, we represent the European market on day d using a graph. Each zone is represented by a node z for which day-ahead prices $\mathbf{D}_z \in \mathbb{R}^{24}$ must be predicted. For some problems, the required amount of energy to be produced $\mathbf{E}_z \in \mathbb{R}^{24}$ also has to be predicted. Connected zones on the market are linked in the graph by edges (z, z'), associated with day-ahead flows $\mathbf{F}_{z,z'} \in \mathbb{R}^{24}$. The features used by the pricing algorithm are **(1)** Consumption forecast for the next day $\mathbf{C}_z \in \mathbb{R}^{24}$ **(2)** Renewable generation forecast for the next day $\mathbf{R}_z \in \mathbb{R}^{24}$, **(3)** Programmable generation forecast for the next day $\mathbf{G}_z \in \mathbb{R}^{24}$, **(4)** Maximal generation capacity for the next day $\mathbf{V}_z \in \mathbb{R}^{24}$, **(5)** Current Prices $\mathbf{P}_z \in \mathbb{R}^{24}$, **(6)** Available Transfer Capacities for the next day

$\mathbf{A}_{z,z'} \in \mathbb{R}^{24}$ which is the maximum amount of energy that can be sent from z to z'. Since renewable energy production is subject to external factors that are not controllable (wind speed, solar radiation, etc.), we distinguished the two types of source by $\mathbf{R}_z$ and $\mathbf{G}_z$.

Hence, $\mathbf{C}, \mathbf{R}, \mathbf{G}, \mathbf{P}, \mathbf{V}$ and $\mathbf{A}$ are known at prediction time, while $\mathbf{D}, \mathbf{E}$ and $\mathbf{F}$ are unknown. In what follows, we propose to take advantage of knowledge from the field of electricity pricing to estimate the flows $\mathbf{F}$ between zones by combinatorial optimization, before using those results to forecast the day-ahead prices $\mathbf{D}$.

3 Estimate Cross-Border Flows by Combinatorial Optimization

The EUPHEMIA algorithm sets electricity prices on the European market by satisfying the constraints listed in Sect. 2. These constraints lead to sophisticated and counter-intuitive flows between zones, some zones playing the role of transit zones to make possible energy exchanges between two other zones. To better model these dynamics, we use domain knowledge to approximate day-ahead flows $\mathbf{F}$ and use them as predictive variables into EPF models. In this section, we describe four different methods for predicting $\mathbf{F}$.

3.1 A Formalization by Linear Programming

The most natural way to formulate the flow optimization problem is to write the EUPHEMIA algorithm as a linear programming problem. In doing so, the network constraints are explicitly enforced, and the Day-Ahead Flow $\mathbf{F}$ and required energy generation $\mathbf{E}$ are computed:

$$\text{Flin} = \underset{\mathbf{F}_{z,z'} \text{ and } \mathbf{E}_z}{\arg\max} \sum_{z,z'} \mathbf{F}_{z,z'}(\mathbf{P}_{z'} - \mathbf{P}_z)$$

$$\textit{under const.} \begin{cases} \mathbf{C}_z - \mathbf{R}_z - \mathbf{E}_z + \sum_{z'} \mathbf{F}_{z,z'} - \sum_{z'} \mathbf{F}_{z',z} = 0 & \forall z \\ \mathbf{E}_z \leq \mathbf{V}_z & \forall z \\ \mathbf{F}_{z,z'} \leq \mathbf{A}_{z,z'} & \forall z, z' \end{cases}$$

Flow-related profit is maximized under three constraints. The first constraint ensures a zero energy balance, the second stipulates that the planned production must not exceed its maximum capacity, and the third imposes that the flows do not exceed the capacities of the lines. The cost aims to maximize congestion rents by maximizing potentially valuable flows: flows from a zone with lower prices to a zone with higher prices. In this set-up, we consider the consumption $\mathbf{C}_z$ and renewable generation forecasts $\mathbf{R}_z$ as fixed. The required generation $\mathbf{E}_z$ is determined to match $\mathbf{C}_z - \mathbf{R}_z$.

Table 1. CC, MAE and SMAPE metrics between **Flin** and **Flsq** optimized flows and actual day-ahead flow values **F** on the train dataset.

Problem	CC	MAE (MWh	SMAPE (%)
Flin	0.153	944.63	120.32
Flsq	**0.389**	**418.05**	**105.93**

3.2   Formalizing the Problem by a Least-Squares Loss

The formulation of the problem by linear programming has a major drawback. We allow the generation of zone to expand to its maximum capacity $E_z \leq V_z$ without penalty to the cost. In practice, switching power plants on or off has a cost that is not linear with respect to the generated volume. We thus propose to rewrite the problem by transforming the energy balance constraint into a cost to be minimized.

$$\mathbf{Flsq} = \underset{\mathbf{F}_{z,z'}}{\arg\min} \sum_z \left(\sum_{z'} \mathbf{F}_{z,z'} - \sum_{z'} \mathbf{F}_{z',z} + \mathbf{C}_z - (\mathbf{R}_z + \mathbf{G}_z) \right)^2$$

under constraint $0 \leq \mathbf{F}_{z,z'} \leq \mathbf{A}_{z,z'} \ \forall_{z,z'}$

The squared loss ensures that unbalanced zones are heavily penalized. Thus, we do not have to penalize the objective by the price difference and, we can also remove the determination of $\mathbf{E}_z$ from the problem and we use the programmable generation forecast $\mathbf{G}_z$ instead.

3.3   Combining the Two Formalizations

To study the estimation quality of these two models, we solved the two optimization problems for each hour of our train dataset (see Sect. 4.1) using scipy[1]. As flow values are known a posteriori, we can evaluate the quality of the estimation on the train set using standard measures (see description in Sect. 5). The metrics obtained are reported in Table 1. It is obvious that **Flsq** outperforms **Flin** on the dataset. However, by analyzing the estimations with a lower granularity, we observe that the performances vary according to graph edges. For example, the flow on the edge between Norway-5 and Norway-1 is well handled by problem **Flsq** as shown in Fig. 2 (left) while the flow on edge between France and Germany is better handled by problem **Flin** (see Fig. 2 right). To take advantage of these two models, we sought to identify the market conditions allowing to differentiate these two scenarios. For this, we first define the loss difference between the results of the two problems as

$$\mathbf{L}^{(t)}(z, z') = |\mathbf{F}_{z,z'}^{(t)} - \mathbf{Flsq}_{z,z'}^{(t)}| - |\mathbf{F}_{z,z'}^{(t)} - \mathbf{Flin}_{z,z'}^{(t)}|$$

---

[1] https://scipy.org/.

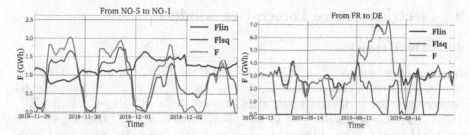

Fig. 2. Optimized flows **Flin** (blue), **Flsq** (green) and the actual Day-Ahead flows **F** (red) between the Norway-5 and Norway-1 zones (left) and the France and Germany zones (right). On the left, we observe that **Flsq** comes close enough to the Day-Ahead flow **F**, while **Flin** does not. On the right, we observe the opposite.

where $t \doteq (d, h)$ is one of the N possible time-steps. We analyze the relationship between $\mathbf{L}^{(t)}(z, z')$ and the characteristics of the market

$$x \in (\mathbf{C}_z, \mathbf{C}_{z'}, \mathbf{R}_z, \mathbf{R}_{z'}, \mathbf{P}_z, \mathbf{P}_{z'}).$$

We break down x into 100 quantiles x_q and compute the average loss for each (x, q):

$$\mathbf{L}(z, z', x, q) = \frac{1}{N} \sum_{t \in T(x,q)} \mathbf{L}^{(t)}(z, z')$$

with $T(x, q) = \{t \mid x^{(t)} \in [x_q, x_{q+1}]\}$. Market conditions where $\mathbf{L}(z, z', x, q) > 0$ correspond to situations where it is preferable to use **Flin** instead of **Flsq**. We name the results of this combination **Fcmb**. To generate **Fcmb** on the test dataset, we keep the same market conditions (z, z', x, q) as found on the train dataset. This prevents data leaks related to the use of posterior data for a prediction.

3.4 One-Sided Flows

In the above formalization, we enable bilateral flows between two zones, i.e. $\mathbf{F}_{z,z'} > 0$ and $\mathbf{F}_{z',z} > 0$ can both occur, which matches the logic of EUPHEMIA. However, in practice most connections never have two-sided flows. To further improve our flow modeling, we identify one-sided connections and apply one-sidedness in our flow estimations. For each link (z, z'), we count the number of times on the train dataset when the flow is one-sided i.e. when we have $\mathbf{F}_{z,z'} \geq 0$ and $\mathbf{F}_{z',z} = 0$. If this occurs more than 75% of the time, we consider the edge (z, z') as always one-sided. For this, we keep the most important predicted flow from which we subtract the least important flow. We set the latter to 0. In this way, the energy balance in the two zones remains the same. We apply this transformation to **Fcmb** and call the result **Fos**.

4   Electricity Price Forecasting Models

4.1   The Dataset

In this section, we tackle the EPF problem on the European market. The data is available free of charge[2] and we collected 35 out of 46 zones, linked by 63 connections. For each zone z, the attributes are $\mathbf{X}_z = (\mathbf{C}_z, \mathbf{R}_z, \mathbf{G}_z, \mathbf{P}_z) \in \mathbb{R}^{96}$. Hence, each day is described by 35×96 predictive features and the targets to be predicted are the 24 hourly prices for each zone. We exclude the Swiss and Great-Britain prices from the prediction task. Although they are part of the network, their prices are determined prior to the closing of EUPHEMIA and we prefer to use them as predictive variables. We predict the 24 prices of the remaining 33 zones every day: $\mathbf{Y} \in \mathbb{R}^{792}$. Our dataset spans from 01/01/2016 to 31/12/2021. We use the last two years (2020, 2021) as test set. Two years is a good duration because the prices show a strong seasonality. The year 2019 is kept as a validation set for hyper-parameter search.

In addition to the 35×96 predictive features cited above, we consider the Available Transfer Capacities for each connection $\mathbf{A}$ or instead one of the flow estimates $\mathbf{Flin}$, $\mathbf{Flsq}$, $\mathbf{Fcmb}$, or $\mathbf{Fos}$ for each link, leading to 126×24 additional predictive variables. Each line of our dataset corresponds to a day and has 6384 values.

4.2   The Machine Learning Models

We use Deep Neural Network and Convolutional Neural Network to predict the electricity prices. Deep Neural Networks (**DNN**) [6–8,12] are the most commonly used models in EPF. Its training samples are vectors $s \in \mathbb{R}^{6384}$. Convolutional Neural Networks (**CNN**) have also seen a growing interest in EPF over the past years [1,4,7]. We compute the convolutions along time and each sample is a vector $s \in \mathbb{R}^{(35+126) \times 24}$. Finally, we propose to use a Graph Neural Network (**GNN**), which is new for the EPF domain. **GNNs** make it possible to exploit data structured as graphs as described in Sect. 2. We train our **GNN** for the node prediction problem by stacking graph convolution layers that update the node embeddings. This is followed by linear layers that map node embeddings to their predicted values. We use tensorflow and pytorch-geometric libraries[3]. Each model (**DNN, CNN, GNN**) is trained on 5 different versions of our dataset according to the method use to estimate $\mathbf{F}$: $\mathbf{A}$, $\mathbf{Flin}$, $\mathbf{Flsq}$, $\mathbf{Fcmb}$, $\mathbf{Fos}$. To be fair in our experiments, we set a time limit for the hyper-parameter search. More precisely, we let our program explore the hyper-parameter grid for 24 h for each model with $\mathbf{F} = \mathbf{A}$ on a 20cpus computer and use the same configuration for all variants of $\mathbf{F}$. This introduces a slight bias as the resulting best configuration is chosen for its performance on the $\mathbf{A}$ dataset. After finding the optimal configuration, we calculate forecasts on the test dataset using recalibration. It consists in re-training the model using the most recent data before making forecasts. Once a

---

[2] https://transparency.entsoe.eu.
[3] https://www.tensorflow.org/, https://pytorch-geometric.readthedocs.io.

Table 2. Metrics for flow estimation on the test dataset for the different methods. The **Flsq** method outperforms the **Flin** methods. The **Fcmb** method does not improves the metrics, while the **Fos** method improves performances.

Problem	CC	MAE (MWh)	SMAPE (%)
A	0.14	917.59	111.43
Flin	0.116	876.51	111.12
Flsq	**0.380**	388.95	105.35
Fcmb	0.367	396.46	102.52
Fos	0.375	**314.5**	**81.19**

test set sample is predicted, we can integrate its predictions into the training dataset and retrain the model. We recalibrate our models every 30 days.

5   Experiments

We compare the values between the predicted $\hat{Y}_{z,h}$ and the real $Y_{z,h}$ target variables for the different zones z and hours h. We use standard measures as $MAE(Y, \hat{Y})$ (the average of the absolute difference between the values over the target variables), $SMAPE(Y, \hat{Y})$ (the symmetric mean absolute percentage error over the target variables), and CC (the average correlation coefficient over the target variables). To check the statistical significance of the results, we use the Diebold & Mariano (DM) test [2] that compares two models M_1 and M_2. The null hypothesis H_0 is that $Loss(M_1) > Loss(M_2)$, i.e.the first model is less efficient than the second. We can reject H_0 and conclude that M_1 outperforms M_2 if the resulting P-value is lower than a fixed threshold of 0.05. We use $SMAPE$ as $Loss$ to better account for the different price scales. To make the experiments reproducible, the source code and the data are made available[4].

5.1   Results

Flow Estimate. The results of the flow estimation problems on the test set are first presented in Table 2. For comparison, we also calculated the error between the network constraints **A** and the actual flows. We make the same observation as for the train set: **Flin** barely improves the quality of the flows while **Flsq** dramatically reduces the error. Then, their combination **Fcmb** does not shows notable metric improvement while setting up one-sided flows **Fos** does. We perform DM tests that confirms that the flow estimate quality increases with the complexity of the estimation method i.e. **Fos** outperforms every method, **Fcmb** outperforms every method except **Fos** and **Flsq** is better than **Flin**.

---

[4] https://github.com/Leonardbcm/OPALE.git.

Price Forecast. The results of the EPF problem on the test period are presented in Table 3. The left part display the metrics, while the right part of details the P-values of the DM tests. On each line, we first compare the model on the line with the same model using other flow estimates (first 5 columns), then we compare it to other models using the same flow estimate (last 3 columns). We can for instance confirm that the **DNN** model using the network constraints **A** is significantly more efficient than the **CNN** using **A** (first line).

The **CNN** models are less competitive. They obtain the worst metrics and the DM test confirms that they are significantly less efficient than other models using the same flows (penultimate column). The **GNN** models are the most adequate models for this problem. Their metrics are better and the DM test statistically confirms that they outperform other models using the same flows (last column). The **DNN** models thus stand in between. We now analyse the performance variations with respect to the flow estimation method. We compare results obtained using the network constraints **A** and those using estimation methods **F** (4th column). We notice that, except for **Flin**, estimating the flows significantly improves performances for all models. Moreover, the **Flin** method is significantly less efficient than every other (5th column). However, which flow estimation method is better for all models remains unclear. The best flow estimate for the **DNN** model is the **Flsq** (3rd row), **Fcmb** (9th row) for the **CNN**, while for the **GNN**, it is impossible to statistically decide between **Fos** and **Flsq**, despite metric differences.

Detailing the DM tests by zone, we observe that replacing **A** by **Flsq**, **Fcmb** or **Fos** leads to overall improvements, even though local decrease can occur (FR, HU, CZ, SK, SI, NO-5, DE, AT). Using **Flin** improves performances less often than other methods and can degrade forecasts on multiple neighboring areas (Italy for the CNN). **Fcmb** shows the biggest improvements and the lowest decrease for all models. Using **Fcmb** in a EPF model seems to be a reasonable default choice. Lastly, almost all zones profit from using **F** for the **DNN**.

5.2   SHAP Values

It is possible to further analyze our models and determine the impact of the different groups of features on the predictions. To that end, we consider the SHAP value approach [10], a feature attribution method that assigns to each feature a value that reflects its contribution in the prediction process. We denote the contribution of a column c to the target o on day d as $\Phi_c^{d,o}$. A column $c = (f, h, z)$ refers to the feature f at hour h for zone z or pair of zones (z, z') if f is an edge attribute. Hence, the contribution tensor $\Phi \in \mathbb{R}^{731 \times 792 \times 6385}$ is made of 3.7 billion values. For computational issues, we only compute 500 SHAP values on the first 30 days of the test dataset. We normalize the results so that the sum of each contribution equals 1 for each target of a given day to obtain $\bar{\Phi}_c^{d,o}$, and the sum of the contributions for each feature f is denoted $\bar{\Phi}_f$.

We compute $\bar{\Phi}_f$ for each $f \in (\mathbf{C}, \mathbf{G}, \mathbf{R}, \mathbf{P}, \mathbf{F})$ and display them in Table 4. First, we observe that the **GNN**'s top contributing features are the prices that explain 30% of the forecasts, against approximately 20% for the other models.

Table 3. (Left) Metrics on the test period. (Right) DM test's P-values. For each trained model (line), the P-value is computed against the same model with other flows (first 5 columns) and against other models with the same flows (last 3 columns). The null hypothesis states that the column model outperforms the row model. With a threshold of 0.05, the bold values indicate that the row model outperform the column model.

Models	CC	MAE (€/MWh)	SMAPE (%)	A	Flin	Flsq	Fcmb	Fos	DNN	CNN	GNN
DNN_A	0.893	13.97	29.76	-	1.0	1.0	1.0	1.0	-	**0.008**	1.0
DNN_Flin	0.903	13.44	28.51	**0.0**	-	1.0	1.0	0.998	-	**0.0**	1.0
DNN_Flsq	0.904	12.96	28.26	**0.0**	**0.0**	-	**0.001**	**0.0**	-	**0.0**	1.0
DNN_Fcmb	0.906	13.07	28.38	**0.0**	**0.0**	0.999	-	**0.001**	-	**0.0**	1.0
DNN_Fos	0.909	13.2	28.84	**0.0**	**0.002**	1.0	0.999	-	-	**0.0**	1.0
CNN_A	0.866	14.41	32.17	-	**0.035**	0.992	1.0	0.958	0.992	-	1.0
CNN_Flin	0.865	14.54	32.23	0.965	-	1.0	1.0	0.998	1.0	-	1.0
CNN_Flsq	0.875	14.19	32.01	**0.008**	**0.0**	-	0.996	0.175	1.0	-	1.0
CNN_Fcmb	0.867	14.04	31.81	**0.0**	**0.0**	**0.004**	-	**0.002**	1.0	-	1.0
CNN_Fos	0.872	14.26	31.87	**0.042**	**0.002**	0.825	0.998	-	1.0	-	1.0
GNN_A	0.925	10.23	24.59	-	0.819	0.981	0.928	1.0	**0.0**	**0.0**	-
GNN_Flin	0.926	10.22	24.6	0.181	-	0.957	0.884	0.996	**0.0**	**0.0**	-
GNN_Flsq	0.925	10.17	24.6	**0.019**	**0.043**	-	0.31	0.942	**0.0**	**0.0**	-
GNN_Fcmb	0.926	10.18	**24.46**	0.072	0.116	0.69	-	0.937	**0.0**	**0.0**	-
GNN_Fos	**0.926**	**10.14**	24.52	**0.0**	**0.004**	0.058	0.063	-	**0.0**	**0.0**	-

Table 4. Average contribution (%) for the predictions grouped by feature. For the **DNN** and **CNN** models, we observe that the average contribution of the flows **F** increases as we use more sophisticated estimation methods.

Model	DNN					CNN					GNN				
	FA	Flin	Flsq	Fcmb	Fos	FA	Flin	Flsq	Fcmb	Fos	FA	Flin	Flsq	Fcmb	Fos
C	19.4	19.2	19.3	19.3	18.8	19.3	19.2	18.9	18.8	18.6	16.7	17.0	16.7	16.7	17.0
G	20.4	20.5	20.0	20.0	19.5	20.1	20.6	10.7	20.5	20.0	18.1	18.0	18.0	18.0	18.3
R	21.5	21.6	20.7	20.7	20.1	22.1	21.5	21.0	20.9	22.3	19.4	19.1	19.5	19.6	19.5
P	20.5	20.5	20.7	20.5	20.3	20.1	20.5	21.3	21.7	20.4	31.6	31.5	31.0	30.9	30.6
F	18.5	18.2	19.3	19.4	21.2	18.4	18.1	18.1	18.2	18.7	14.2	14.4	14.8	14.8	14.5

The **GNN** also uses **F** the less (14% against 18–20%). Next, we observe that the **DNN** model favors the use of **F** at the expense of **C**, **G** and **R** as we use more sophisticated flow estimate (**Flsq, Fcmb, Fos**). In contrast, the average contribution of **F** in the **CNN** and **GNN** does not show a clear pattern. To detail these observations, we display in Fig. 3 the differences of contribution between **A** and the used estimate **F**. Green squares on coordinate (i, j) indicate that the contribution of **F** is more important than the contribution of **A** for predicting the zone i for model j. We observe that the **Flin** contribution differences are mostly negative i.e. models rely less on **Flin** than on **A** for forecasting prices. Next, we see that the **DNN** increases the contribution of **F** for almost all zones. Finally, the **CNN** always lowers the contribution of Spain (ES), Portugal (PT), and Italy (CNOR, CSUD and SARD). These zones are characterized by having few (1 or 2) connections. Latvia (LT) has a similar behavior for the **GNN** model.

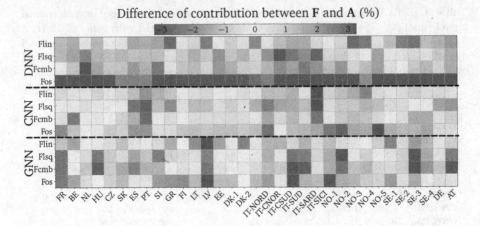

Fig. 3. Difference in contribution made by the flow estimates **F** compared to the available transfer capacity **A** for the different models and zones. The green (resp. red) squares indicate that **F** contributes more (resp. less) than **A**.

5.3   Discussion

The joint analysis of the model's performances and SHAP values of **Flin** shows us significant degradation of the forecasts and less contribution for the forecast than **A**. This leads us to conclude that the **Flin** method is not a good flow estimation method. Apart from **Flin**, other flow estimation methods are all beneficial for the EPF task, without being able to select the best one overall. The **DNN** is the less sophisticated model and cannot model the network. However, observing both a significant performance improvement and an increase of the average contribution of **F** over **A** for almost all zones, we infer that the **DNN** model takes benefit from using flow estimation methods. Next, the **CNN** use a matrix of arbitrary-arranged input features and a convolution kernel and dilation rate inconsistent with the European network. Consequently, zones and flows are not associated. Hence, **CNN** is the less tailored model for EPF. Lastly, the GNN model uses the graph representation of the network, with connections modeled as edges. This ability lowers the contribution of **A** or **F** in the forecast: node embeddings are already updated using their neighbors even with no flows. This is even more the case for isolated zones as their relationships with other zones are simpler. Another consequence is that **GNN** is the best model for EPF at the European scale.

6   Conclusion

In this paper, we introduce the problem of day-ahead electricity price forecasting considering many zones together and their interdependence due to price regulation mechanisms. While many works have focused on the construction of increasingly sophisticated models for specific regions of the European market, we propose new ways of estimating features based on domain knowledge, and this upstream of learning. We show that an *optimize then predict* strategy makes it possible to improve the learned models by fully considering cross-border energy flows estimated by several optimization problems. A SHAP-value analysis confirms that the estimated flows contribute more to the prediction than the Available Transfer Capacities, especially when the model is simple (DNN). For more sophisticated models such GNN, flows better influence predictions at the center of the European market while being less important for the zones at the periphery of the market.

Two main directions can be considered as future work. First, we could replace the generation forecast used in our models by a start-up/shut-down cost model for power plants. It would better capture dynamics between generation and day-ahead prices. Going further, we could also model part of the EUPHEMIA algorithm. Then, our work brings forward the question of mixing optimization problems and Machine Learning. Integrating the optimization problem as a layer in our Neural Network to achieve a *Optimize and Predict* framework would directly link the task loss (day-ahead price forecast) to the sub-task (flow estimation).

Acknowledgment. This research has partially been funded by the ANRT (French National Association for Research and Technology).

References

1. Cheng, H.-Y., Kuo, P.-H., Shen, Y., Huang, C.-J.: Deep convolutional neural network model for short-term electricity price forecasting (2020)
2. Diebold, F., Mariano, R.: Comparing predictive accuracy. J. Bus. Econ. Stat. **20**, 134–144 (1992)
3. El Balghiti, O., Elmachtoub, A. N., Grigas, P., Tewari, A.: Generalization bounds in the predict-then-optimize framework. In: Neurips, vol. 32 (2019)
4. Khan, Z.A., et al.: Short term electricity price forecasting through convolutional neural network (CNN). In: Barolli, L., Amato, F., Moscato, F., Enokido, T., Takizawa, M. (eds.) WAINA 2020. AISC, vol. 1150, pp. 1181–1188. Springer, Cham (2020). https://doi.org/10.1007/978-3-030-44038-1_108
5. Krizhevsky, A. Sutskever, L., Hinto, G.E.: Imagenet classification with deep CNNs. Technical report, University of Toronto (2012)
6. Lago, J., Marcjasz, G., De Schutter, B., Weron, R.: Forecasting day-ahead electricity prices: a review of state-of-the-art algorithms, best practices and an open-access benchmark. Appl. Energy **293**, 116983 (2021)
7. Lago, J., Ridder, F.D., Schutter, B. D.: Forecasting day-ahead electricity prices deep learning approaches and empirical comparison of traditional algorithms. Technical report, Delft University of Technology (2018)
8. Lago, J., Ridder, F.D., Vrancx, P., Schutter, B.D.: Forecasting day-ahead electricity prices in Europe: the importance of considering market integration. Appl. Energy **211**, 890–903 (2018)
9. Li, Y., Yu, R., Shahabi, C., Liu, Y.: Graph convolutional recurrent neural network: Data-driven traffic forecasting. CoRR, abs/1707.01926 (2017)

10. Lundberg, S., Lee, S.: A unified approach to interpreting model predictions. CoRR, abs/1705.07874 (2017)
11. Mandi, J., Bucarey, V., Mulamba, M., Guns, T.: Predict and optimize: through the lens of learning to rank. CoRR, abs/2112.03609 (2021)
12. Mosbah, H., El-Hawary, M.: Hourly electricity price forecasting for the next month using multilayer neural network. Can. J. Electr. Comput. Eng. **39**, 283–291 (2015)
13. PCR. Euphemia public description. Technical report, Price Coupling of Region (2016)
14. Tschora, L., Pierre, E., Plantevit, M., Robardet, C.: Electricity price forecasting on the day-ahead market using machine learning. Appl. Energy **313**, 118752 (2022)
15. Vlahogianni, E.I., Karlaftis, M.G., Golias, J.C.: Short-term traffic forecasting: where we are and where we're going. Transp. Res. Part C: Emerg. Technol. **43**, 3–19 (2014)
16. Zheng, Y., Liu, Q., Chen, E., Ge, Y., Zhao, J.L.: Exploiting multi-channels deep CNNs for multivariate time series classification. University of China Hefei, Technical report (2015)

A Similarity-Guided Framework for Error-Driven Discovery of Patient Neighbourhoods in EMA Data

Vishnu Unnikrishnan[1]([🖂])[iD], Miro Schleicher[1][iD], Clara Puga[1][iD],
Ruediger Pryss[2][iD], Carsten Vogel[2][iD], Winfried Schlee[2,3][iD],
and Myra Spiliopoulou[1][iD]

[1] Otto-von-Guericke University Magdeburg, Magdeburg, Germany
{vishnu.unnikrishnan,miro.schleicher,clara.puga,myra}@ovgu.de
[2] Institute for Clinical Epidemiology and Biometry,
University of Würzburg, Würzburg, Germany
{ruediger.pryss,carsten.vogel}@uni-wuerzburg.de
[3] Department of Psychiatry and Psychotherapy, University of Regensburg,
Regensburg, Germany
winfried.schlee@tinnitusresearch.com

Abstract. Recent advances in technology and societal changes have increased the amount of patient data that is being collected remotely, outside of hospitals. As technology enables the ability to collect Ecological Momentary Assessments (EMAs) of patient symptoms remotely, personalised predictors have become especially relevant in the field of medicine. However, focusing a predictive model on a single patient's data comes with sometimes extreme trade-offs on the amount of data available for training. While it is possible to mitigate this loss of data by including data from similar patients, the concept of similarity itself may be poorly defined in cases where patient data are available in two modalities - one that is fixed and relatively static (for e.g.: age, gender, etc.), and those that are more dynamic (instantaneous symptom severity). Including data from users with similar EMA data and disease characteristics has been explored with respect to building personalised predictors of the near future of a patient. We propose a method to build personalised predictors by discovering a neighbourhood for each user that decreases the prediction error of a model over that user's data. This method is useful not just for building better personalised predictors, but may also serve as a starting point for future investigations into what properties are shared by patients whose EMA data predict each other. We test our method on two EMA datasets, and show that our proposed method achieves significantly better RMSE than a single non-personalised global model, and that our framework provides better predictions for 82%–89% of the users compared to the global model for two datasets.

This work has received funding from the European Union's Horizon 2020 Research and Innovation Programme, Grant Agreement 848261 "Unification of treatments and Interventions for Tinnitus patients" (UNITI).

Keywords: mHealth · Tinnitus · Ecological Momentary Assessments · Personalised Predictors

1   Introduction

Along with rising interest and recent developments in extracting knowledge from healthcare data, the digital revolution and the ubiquity of smartphones is enabling clinicians, through the use of mobile health (mHealth) tools, to monitor the disease of the patient outside the hospital. Data can be collected to assess the momentary state of a variable of interest, and since the measurement happens in the patient's natural surroundings, these measurements are called "Ecological Momentary Assessments" (EMAs). EMAs are widely applied in the study of mental health and psychological affect, and are expected to reduce phenomena like the 'recall effect' [13], where the patient's recollection of an event is affected by events that have occurred later.

This work uses EMA data from two mHealth smartphone applications: Track-YourTinnitus, and the UNITI app. Both apps collect EMA data pertaining to tinnitus, a psychoacoustic disorder characterised frequently by the perception of a phantom sound in the absence of external stimuli. Each app caters to different demographics. The UNITI app was developed as a part of the UNITI project [12], with the main purpose of studying the changes in the dynamic presentation of the disease as the patient takes part in a randomised controlled trial where they receive a randomly-selected treatment, while the TrackYourTinnitus app is aimed at the general public. The patients in the UNITI app are also monitored for their participation in the app, strongly affecting the amount of data available for learning. It is important to note that 'typical' user engagement is more likely to approximate that observed in the TrackYourTinnitus app. Since the users of both apps are assumed to suffer from tinnitus, we will use the word 'patient' and 'user' interchangeably, although only a subset of users of the UNITI app are confirmed patients in a hospital.

Previous work on developing personalised predictors for similar data has already shown that data from other similar patients/users helps in improving the predictive performance of the models. However, this presents two problems: (a) The exact features that contribute in measuring the 'similarity' between two patients is sometimes unknown, and (b) the exact number of users whose data are used to develop the personalised predictor (i.e., the 'neighbourhood size') is fixed for all users. Given that each user may contribute different amounts of EMA data, a user with neighbours that contributed little data will have less data for learning compared to another user with neighbours that contributed heavily.

Our proposed neighbourhood discovery framework tackles this problem by optimising for the predictive performance of the personalised neighbourhood, instead of finding a one-size-fits-all neighbourhood size. i.e., all users who improve the predictive performance of the model are included in the neighbourhood. This approach is useful in cases where the notion of similarity is not clearly defined, and in cases where the neighbourhood as well as the size of the neighbourhood needs personalisation. The amount of data available from different

EMA users can vary over orders of magnitude (from a few tens of rows to the thousands), making traditional time series distance metrics like dynamic time warping (DTW) inapplicable. However, since a brute-force exploration of every possible combination of users for every possible number of user neighbourhoods is computationally intractable, we order the search process to include users in decreasing order of their similarity to the user for whom a predictor is being trained. Towards this end, we apply a simple cosine similarity between the first sim (in our case, $sim = 7$) observations of a target variable from two users. Using a small number of observations from the beginning of the time series makes it possible to order the brute-force search process, decreasing the number of models trained to a large extent. Additionally, the small number of observations used to compute the similarity decreases the 'leakage' from of testing data information, since they are sufficiently temporally distant from both the validation and test data. Each user's observations are also scaled separately, removing the effect of the mean values from each user, and focusing instead on the variations.

To summarise, this work proposes a neighbourhood discovery framework that· answers the following questions:

- How to discover an 'optimal' neighbourhood for a user in the context of building personalised predictors?
- How does the discovered neighbourhood perform compared to a non-personalised global model that is trained on data from all users?
- Which users benefit from a personalised neighbourhood?

This paper is structured as follows: Sect. 2 presents some related literature, and Sects. 3 and 4 introduce the proposed method and details of the dataset that we apply it on, followed by a discussion of the results. We close with some remarks on threats to validity and some possible extensions in Sect. 5.

2   Related Work

The idea of using machine learning to develop personalised predictors is receiving increasing attention, as healthcare information from electronic health records is presented with the opportunity to be integrated with additional data sources like wearable sensors, etc. [1]. While many studies recommend caution when it comes to promising better outcomes through personalised medicine [5,17], the authors also suggest that the focus should be on how to best combine information from machine learning models to assist clinicians understand the disease better. This has been shown in [7], where the predictive power of a model that incorporates newly identified proteins and genes in colorectal cancer patients is found to be better than current state of the art models without these novel variables. In our work, since there is no clearly established positive or negative outcome in the EMA domain for tinnitus, we focus instead on predicting the EMA data in the immediate future.

Developing personalised predictors for EMA data has been explored in [14], but it has been found that the relevant neighbourhood of a patient can be difficult to infer based on the relatively 'static' information (age, gender, sociodemographics, and standardised questionnaire assessments collected in a hospital) available on the patient. While such 'neighbourhoods' may be improved with expert information, there has still been no exploration on how the data from the EMA domain may be used to improve the prediction methods. The idea that the time series information may be accompanied by additional exogenous variables is only recently gaining attention, as only the most recent version of the M-5 time series forecasting competition has included such information alongside time series [4]. The M-5 competition also acknowledges other properties of collections of time series, like that they might be intermittent (like for time series that capture demand that may be sporadic), and include many zeros. The intermittent nature of data arrival is a property that is of high relevance to EMA data, although unlike in the case discussed in [4], the intermittence does not suggest the absence of data, only that it has not been observed.

Although the idea that a single model may be trained from multiple time series as inputs has been proposed early [10], the topic is still a matter of active research, with interest not just in developing global time series models, but also explanations for forecasts [11]. The latter point is especially relevant because practitioners have found that well tuned statistical models are often competitive when compared to modern machine learning methods, although a well tuned machine learning model can be expected to outperform their statistical counterparts [8]. It is further acknowledged in [2] that more research is needed to understand the relationships between individual time series that make a global time series model built on them more accurate than if the models were trained on the separate time series. Although not explored in this work, this is a strong motivation for pursuing our proposed methodology of discovering patients with similar symptom presentation, since a global time series that benefits from the shared data of multiple patients may reveal hitherto unknown clinically relevant similarities between these patients.

Another difference of the proposed workflow is that unlike [14] which develops neighbourhood based predictors of the current state of a target variable given all other EMA variables, but we extend this to a basic one-step forecaster, which forecasts the target variable for the next day given all EMAs of today. This extension follows naturally since it has already been seen in [3] that there are Granger-causal relationships between the EMA variables - i.e., the past values of some variables are predictive of the future values of others. Predicting the immediate future is also expected to be of more interest to the clinician, since advance warning of short-term symptom deterioration can facilitate early intervention.

The notion of discovering the neighbourhood for a user has also been explored in [15], but in the context of learning models for users with very little data. We generalise the framework to search for a neighbourhood for all users in the dataset, not just the ones that have too little data for learning.

Our workflow's error-driven adjustments are similar to boosting methods like localised boosting, [6] - but different in a few ways: firstly, that we deal with a regression, not a classification problem, but also in that the addition of a user adds all data from taht user's training data to the data accumulated for learning, as opposed only to localised weighting of the accumulated data from different users. This is bacause of the third and arguably the most important difference - that our proposed method provides us a set of useful neighbours for each user, which in turn can be useful in guiding further clinical analyses, since it is possible to explore the data of those users' static data to investigate what similarities exist between them. This is useful because understanding what features contribute to a meaningful similarity between users can help understand the dynamics of the disease better.

3 Neighbourhood Discovery Framework

3.1 Definitions

For all patients $p \in P$, our approach discovers a $neigbourhood(p) \subseteq P$ such that a predictive model trained on the data of all patients in $neighbourhood(p)$ achieves the best performance. It is clear that as $|P|$ increases, the number of unique subsets of P increases exponentially. Therefore, we discover the neighbourhood of patient p incrementally using a greedy approach, which uses a similarity function $S(p_i, p_j), \forall p_i, p_j \in P$ and $i \neq j$.

EMA Data: Our proposed approach trains a personalised predictor for the time series of each patient p in the dataset D. The n dimensional multivariate time series of each patient p's EMA data is denoted by $EMAs(p) = X_0^p \ldots X_T^p$, where T denotes time, and $X_{T=t}^p = \{x_1^t \ldots x_n^t\}$ are the n dimensions (questions) captured at time step t, where $0 \leq t \leq T$. The superscript 't' is omitted when referring to the entire series of observations collected for a particular dimension over all time points.

Train, Test and Validation Data: The data of each patient $EMAs(p)$ is split into $train\_data(p), val\_data(p)$ and $test\_data(p)$, where $train\_data(p) = X_0^p \ldots X_{t_v-1}^p$, $t_v < T$, $val\_data(p) = X_{t_v}^p \ldots X_{t_e-1}^p$, and $test\_data(p) = X_{t_e}^p \ldots X_T^p$. i.e., the full sequence of observations from a user p are split into contiguous chunks of train, validation and test data while preserving chronological ordering (validation data comes after test data, etc.). In this case, we reserve the last 30% of the data for testing, and use 70% of the rest for training, and 30% for validation. The *target variable* of prediction interest is $x_{target} \in x_1 \ldots x_n$.

Similarity Function $S(p_i, p_j)$: This function computes the similarity between two patients p_i and p_j. Any measure of similarity can be used here, but for the sake of simplicity, we use cosine similarity between the first sim observations of the target variable. $S(p_i, p_j)$ is computed as the cosine similarity between the EMA values for $p_i(x_{target}^{0\ldots sim})$ and $p_j(x_{target}^{0\ldots sim})$.

  Applying the similarity function S to all pairs of patients $(p_i, p_j) \forall i, j \in P$ provides a similarity matrix SM. For each p, we can select the row corresponding to that user and sort the values in descending order in order to find the nearest neighbours of user p. The framework searches this list in decreasing order of similarity in order to discover an optimal neighbourhood $DN(p) \subseteq P$ that maximises the predictive performance of a personalised model $PM(p)$ trained over the data of all users in $DN(p)$.

3.2 Error-Driven Neighbourhood Discovery Framework

Figure 1 shows an overview of the neighbourhood discovery framework, including the data used for training, validation, or testing by each component. The main goal of the framework is to train a personalised model for each patient p in the dataset, which is invoked only when predictions are required for patient p.

The neighbourhood discovery process begins with a model that includes data only from a single user, i.e., $DN = \{p\}$. The quality of this model is assessed against the validation data $val\_data(p)$. In the next step, the next-most-similar user from P is added to DN, and the performance of the model $PM(p)$ trained on all data in DN is assessed on $val\_data(p)$. If the performance of the model deteriorates, the last added user is removed from the discovered neighbourhood DN, and the search process continues until all users have been checked.

Models: Most users have too little EMA data, so building complex data-intensive models increases the likelihood of overfitting. Since the neighbourhood discovery process begins with just the user's own data, we stick to a simple linear regression model to predict the value of the target variable tomorrow given the EMA data of today. The framework design is, however, agnostic to the exact model used. More complex models can be used if there is more data available for learning.

Avoiding Local Minima: The search process has two ways to get stuck in local minima. One is due to a poorly defined similarity metric, which might compute a similarity on the basis of data that is not relevant to improving the prediction performance, and the other is due to small variations in the validation error of the personalised neighbourhood model as users are added. We mitigate the impact of wrong users added early in the search process by not terminating the search until all users have been iterated through. This is expected to give the model the chance to overcome a poor initialisation. In order to avoid strongly restricting amount of training data available to the model, we track the RMSE of the best-performing neighbourhood seen so far, and allow users to be added to $DN(p)$ even if the error increases by a small margin, as long as the model performance does not deteriorate beyond 10% of the error achieved by the best-performing neighbourhood. This way, we hope to overcome the local minima from overfit neighbourhoods, and expect that the additional data from a user that increases the error by a minor amount improves generalisation performance.

Reusing Models to Decrease Performance Overhead: The neighbourhood discovery process starting at different users may iterate through models trained on the same set of users multiple times. i.e., two users p_a and p_b with neighbourhoods $DN(p_a) = \{p_a, p_b, p_u, p_v, p_w\}$ and $DN(p_b) = \{p_b, p_a, p_u, p_v, p_w\}$ share the exact same neighbourhood. If all the models of user p_a were trained before the

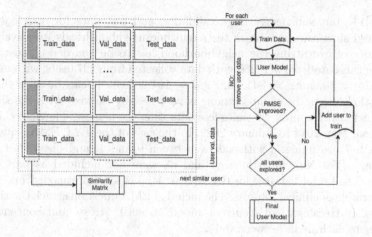

Fig. 1. An overview of the error-driven neighbourhood discovery process

neighbourhood discovery process for p_b began, then all the intermediate models $DN'(p_b) = \{p_b, p_a\}$, $DN'(p_b) = \{p_b, p_a, p_u\}$, $DN'(p_b) = \{p_b, p_a, p_u, p_v\}$ and $DN'(p_b) = \{p_b, p_a, p_u, p_v, p_w\}$ have already been trained once during the neighbourhood discovery process for user p_a, albeit evaluated on $val\_data(p_a)$. In order to avoid retraining models that have already been trained during the neighbourhood search process for other users, all models trained during the search process are saved in a way that they can be retrieved by a hash which is the set of users over which the model was learned. A model thus retrieved may be directly applied to compute the prediction error over the new user p_b's validation data.

4   Experiments and Discussion

4.1   Datasets

This study uses two datasets from two mHealth applications for tracking tinnitus: the TrackYourTinnitus (TYT) application [9], and the more advanced UNITI mHealth application [16]. The TrackYourTinnitus application is a mobile crowd sensing platform that is developed to enable experience sampling of the app user to better understand the dynamic presentation of tinnitus symptoms in the user's natural environment. The more recent UNITI app has been developed as part of the Horizon 2020 "UNITI" project [12], which aims at discovering treatment combinations that are particularly effective in treating tinnitus using a randomised controlled trial. Both apps periodically deliver notifications to the users on their mobile devices, prompting them to answer the EMA questionnaire. The user is also allowed to manually answer the EMA questionnaire at any time of their choosing.

TYT Dataset: For both the TYT and the UNITI datasets, all users with less than 30 days of data were excluded. This is higher than cut-offs applied in previous studies [14], but sufficiently large validation and test data for each patient p is required since the framework tests multiple neighbourhoods iteratively, and the chance of overfitting the neighbourhood has to be guarded against. After applying this cutoff, we were left with data collected from 227 users, with $mean = 90.0$, $min = 30$, $max = 841$, and $std = 104.9$ days. it can be seen that the distribution of lengths of interactions is very heavily skewed, with the shortest user contributing approx. 30 times less data than the longest. The data collected spans from May 2014 to January 2022. In the case of the TYT EMAs, questions 2 through 7 are numeric with values between 0 and 1, and question 1 and 8 are binary (Yes/No). Questions q1 and q8 were not included as part of the analysis. The target variable for the forecast step is set 'tinnitus distress', which is the variable of clinical interest. The included EMA questions track the tinnitus loudness, distress, patient's current mood, arousal, stress, and concentration (answers to each lie in between 0–1).

UNITI Dataset: Similarly to the TYT dataset, all users who contributed less than 30 days of data were excluded from the UNITI dataset. This left us with 222 patients, and on average 64.8 days of data per person, with a minimum of 30 days, a maximum of 263 days, and a standard deviation of 37.0 days. The data spans from Apr. 2021 to Apr. 2022, and hence does not allow for very long time series. The longest series is only approx. 9 times longer than the shortest, although the less skewed interaction patterns may also be a consequence of closer monitoring of these patients by physicians (the app has features for the physician to monitor patient state and provide feedback). The EMA questionnaire has 10 questions measuring momentary as well as daily symptoms: tinnitus loudness, distress, tension of the jaw and neck are measured at the moment, and daily values are also requested for how often the patient thought about tinnitus, extent to which the day was affected by tinnitus, maximum daily volume, amount of physical exercise, the stress level and general emotion (answers to each lie in between 0–100).

4.2 Experiments

In order to compare the performance of our proposed framework for both the TYT and UNITI datasets, we run experiments to answer the questions raised in Sect. 1.

We evaluate the neighbourhood discovery framework by comparing the prediction errors generated by the models trained over the discovered neighbourhood against those generated by a global model trained over the data of all users. The performance of the neighbourhood discovery framework is compared to the performance of a non-personalised global model trained on all the training data. Note: For most users, the model trained on the discovered neighbourhood has been trained on less data than the global model.

Apart from the root mean squared error (RMSE) over all predictions, we further investigate the user-level RMSEs to better understand the degree to which the personalised neighbourhood is more predictive than the RMSEs for those users when they are predicted by the global model. i.e., box plots of the user-level RMSEs are compared for the predictions from the global model and the personalised models. We further investigate the degree to which the personalised neighbourhood improves the predictions by comparing the number of users who were better predicted by their personalised neighbourhood than by the global model.

4.3   Results and Discussion

Overall Prediction Error: Figure 2 shows boxplots of the prediction errors from the personalised predictors and the global models for the TYT and UNITI datasets respectively. For both datasets, it can be seen that our neighbourhood discovery framework achieves lower errors than a single non-personalised global model.

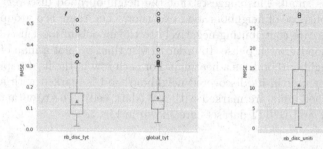

Fig. 2. User-level RMSEs for TYT dataset (left) and the UNITI dataset (right): Comparing predictions generated by the discovered neighbourhoods vs. the global model. (Green triangle: mean RMSE over all users, Green line: median RMSE over all users) (Color figure online)

For the TYT dataset, the mean RMSE over all users for the global model and the discovered neighbourhood were 0.1534 and 0.1335 for the TYT dataset. This translates to a 12.99% improvement when using the discovered neighbourhood instead of the global model for the TYT dataset.

For the UNITI dataset, the mean RMSE for the global model and the discovered neighbourhoods were 12.767 and 10.772. This translates to a 15.62% improvement.

User-Level RMSEs: The overall performance of the discovered neighbourhoods as measured by the RMSE for all predictions is shown as the green triangle in Fig. 2. The fact that the means are lower for the discovered neighbourhood shows that the average performance is indeed better. The medians in the boxplot indicate that the performance benefits are shared by all users, since the median user-level RMSEs for the discovered neighbourhoods are lower than the median user-level RMSEs when users are predicted by the global model.

A Wilcoxon signed-rank test was performed to check if the user-level RMSEs generated by the global and personalised models come from the same distribution. For both the TYT and the UNITI datasets, we can reject the null hypothesis that the user-level RMSEs generated by the global model and the personalised predictors come from the same distribution with p=4.55e-27, and p=3.16e-35 respectively.

Finally, we also check the number of users who are better predicted by their discovered neighbourhood compared to the global model (i.e., the number of times our proposed framework dominates the global model). For the TYT dataset, the neighbourhood discovered by our framework gave better predictions than the global model for **188/227** users (82.82% of the users), and for the UNITI dataset, our framework outperformed the global model for **199/222** users (89.64% of the users).

Though a majority of the users are better predicted by our approach, the difference in the mean RMSEs between the discovered neighbourhood and the global model (Fig. 2) is small. This suggests that the neighbourhood discovery might be serving the purpose of neighbourhood exclusion - i.e., there is a group of anomalous users who are contributing negatively to the models of most users during the neighbourhood growth phase. In order to test this, we plot a map of the discovered user neighbourhoods as a heatmap. For each row in the heatmap, the presence of another user in its discovered neighbourhood is marked with a 1 (bright cell), and other users are marked with a 0 (dark cell). The resultant heatmaps for the TYT and UNITI datasets are shown in Fig. 3.

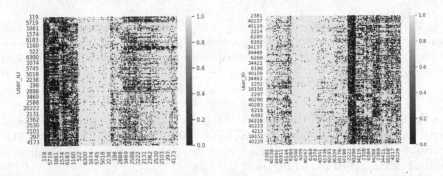

Fig. 3. Heatmap of discovered user neighbourhoods for TYT (left) and UNITI (right) datasets. Dark vertical bands (TYT - left and UNITI - mid-right) show 'unpopular' users who are not used in most users' neighbourhoods.

In both heatmaps, there is a black band of users (Fig. 3 left side of left figure for TYT, middle-right of right-figure for UNITI) who are appearing in the neighbourhoods of very few users, as well as a white band of users who are used very often in other's neighbourhoods. This result suggests that the benefit that most users get from the discovery framework is in the elimination of data that detracts from good predictive performance. The fact that there are users that are rarely used also has consequences for the likelihood of model reuse (discussed below), since the presence of a single rare user in the list of common users makes the use of a cached model much rarer.

Discussion on Model Reuse and Performance Considerations: The caching mechanishm described in Sect. 3 was designed to reduce the need to retrain models already built while exploring the neighbourhoods for another user. In order to assess the impact of this mechanism, we investigated the number of times a model was uncached vs. the number of times the cache was polled for a model trained over some particular set of users. Unfortunately, for both the TYT and UNITI datasets, very few of the trained models were requested by more than one user. For the TYT dataset, the hit rate for cached models was 105 out of 51529 requests (0.2%), and for the UNITI dataset there were 177 hits out of 49284 requests (0.3%). Given that the models we train were simple regressors, the training time even on commodity hardware was fast enough (<5 min) to not require performance optimisation using cached models. However, even small increases in the number of users might reverse this result, given the exponential increase in the number of trained models as users are added.

5   Closing Remarks

In this work, we proposed a method to build personalised neighbourhoods that maximise the predictive power over a user's EMA observations. We tackle the large number of user-neighbourhood permutations by searching the candidate neighbourhood of a user iteratively in decreasing order of similarity, and propose further memoisation-inspired techniques to reuse duplicate models trained over the same candidate set of users.

We tested our proposed framework on two EMA datasets, and showed that our framework achieves 13% and 15% better RMSE compared to a global model. We further show that the prediction errors on using the discovered neighbourhoods are statistically significantly lower with $p = 4.5e-27$ and $p = 3.1e-35$ when compared to the global model. **82.8%** and **89.6%** of the users are better predicted by our discovered neighbourhood compared to the global model (which in almost all cases is trained on more data).

Apart from our current results, several avenues remain open for further exploration. The current work only considers the errors for each user, where better predictions of the patient's near future can be of value especially when facilitating early interventions. It is expected, however, that apart from the prediction error, the predictive neighbourhood of a user can be of interest to the physician for diseases like tinnitus that have a very heterogeneous presentation. Our results suggest that there are some users that appear frequently in the neighbourhoods of others, and some users that are very rarely useful in predicting others. A closer analysis of the discovered neighbourhoods, and the users within those neighbourhoods that appear frequently and also infrequently may be useful in furthering the understanding of the disease itself. Graph mining methods may reveal interesting patterns or clusters of patients with similar disease development.

The current work also explores only linear regression models to predict tomorrow's EMA values given today's symptom severity. Since the framework is applicable for any data which is a collection of time series, it would be interesting to investigate if the framework is also applicable for more complex models from domains that are less strongly challenged by data availability. Larger datasets and more complicated models may also make caching the intermediate models useful again, a step that is not required for the datasets used in this study. It may also be possible to use the framework as a first step in learning global time series models over well-selected subsets of the time series.

References

1. Ahamed, F., Farid, F.: Applying internet of things and machine-learning for personalized healthcare: issues and challenges. In: 2018 International Conference on Machine Learning and Data Engineering (iCMLDE), pp. 19–21. IEEE (2018)
2. Hewamalage, H., Bergmeir, C., Bandara, K.: Global models for time series forecasting: a simulation study. Pattern Recognit. **124**, 108441 (2022). https://doi.org/10.1016/j.patcog.2021.108441
3. Jamaludeen, N., Unnikrishnan, V., Pryss, R., Schobel, J., Schlee, W., Spiliopoulou, M.: Circadian conditional granger causalities on ecological momentary assessment data from an mhealth app. In: 2021 IEEE 34th International Symposium on Computer-Based Medical Systems (CBMS), pp. 354–359. IEEE (2021)
4. Makridakis, S., Spiliotis, E., Assimakopoulos, V.: The M5 competition: background, organization, and implementation. Int. J. Forecast. **38**, 1325–1336 (2021)
5. Matheny, M.E., Whicher, D., Israni, S.T.: Artificial intelligence in health care: a report from the national academy of medicine. JAMA **323**(6), 509–510 (2020)
6. Meir, R., El-Yaniv, R., Ben-David, S.: Localized boosting. In: COLT, pp. 190–199. Citeseer (2000)
7. Nwaokorie, A., Fey, D.: Personalised medicine for colorectal cancer using mechanism-based machine learning models. Int. J. Mol. Sci. **22**(18), 9970 (2021)
8. Petropoulos, F., et al.: Forecasting: theory and practice. Int. J. Forecast. **38**(3), 705–871 (2022). https://doi.org/10.1016/j.ijforecast.2021.11.001
9. Pryss, R., Reichert, M., Langguth, B., Schlee, W.: Mobile crowd sensing services for tinnitus assessment, therapy, and research. In: 2015 IEEE International Conference on Mobile Services, pp. 352–359. IEEE (2015)
10. Roorda, B., Heij, C.: Global total least squares modeling of multivariable time series. IEEE Trans. Autom. Control **40**(1), 50–63 (1995)
11. Rožanec, J., Trajkova, E., Kenda, K., Fortuna, B., Mladenić, D.: Explaining bad forecasts in global time series models. Appl. Sci. **11**(19), 9243 (2021)
12. Schlee, W., et al.: Towards a unification of treatments and interventions for tinnitus patients: the EU research and innovation action UNITI. In: Progress in Brain Research, pp. 441–451. Elsevier BV (2021)
13. Sedgwick, P.: What is recall bias? BMJ **344** (2012)
14. Unnikrishnan, V., et al.: Entity-level stream classification: exploiting entity similarity to label the future observations referring to an entity. Int. J. Data Sci. Anal. **9**(1), 1–15 (2020)
15. Unnikrishnan, V., et al.: Love thy neighbours: a framework for error-driven discovery of useful neighbourhoods for one-step forecasts on EMA data. In: 2021 IEEE 34th International Symposium on Computer-Based Medical Systems (CBMS), pp. 295–300. IEEE (2021)

16. Vogel, C., Schobel, J., Schlee, W., Engelke, M., Pryss, R.: UNITI mobile-EMI-apps for a large-scale European study on tinnitus. In: 2021 43rd Annual International Conference of the IEEE Engineering in Medicine & Biology Society (EMBC), pp. 2358–2362. IEEE (2021)
17. Wilkinson, J., et al.: Time to reality check the promises of machine learning-powered precision medicine. Lancet Digit. Health **2**, e677–e680 (2020)

QBERT: Generalist Model for Processing Questions

Zhaozhen Xu[1]([✉]) and Nello Cristianini[2]

[1] Department of Computer Science, University of Bristol, Bristol, UK
zhaozhen.xu@bristol.ac.uk
[2] Department of Computer Science, University of Bath, Bath, UK
nc993@bath.ac.uk

Abstract. Using a single model across various tasks is beneficial for training and applying deep neural sequence models. We address the problem of developing generalist representations of text that can be used to perform a range of different tasks rather than being specialised to a single application. We focus on processing short questions and developing an embedding for these questions that is useful on a diverse set of problems, such as question topic classification, equivalent question recognition, and question answering. This paper introduces QBERT, a generalist model for processing questions. With QBERT, we demonstrate how we can train a multi-task network that performs all question-related tasks and has achieved similar performance compared to its corresponding single-task models.

Keywords: Multi-task Learning · Deep Learning · Text Processing

1 Introduction

There is increased attention to the problem of learning generalist agents (as opposite to specialist) in a way that the same representation can be used in a range of tasks, even if it does not excel at any specific task [14]. While a specialist should be expected to excel at its one task, a generalist is expected to be good at many problems. In this paper, we focus on building a generalist model for processing a special type of short text: Question.

The development of online communities produces a massive amount of text every day. For example, in the question domain, with the rise of commercial voice assistants such as Siri and Alexa and communities such as Quora, numerous questions are asked on a daily basis. Processing these questions can provide a new perspective on understanding communities and people's interests.

In this paper, we define the generalist model as a question-processing model that targets analysing the semantic and syntactic information in the question. More specifically, this generalist model can process the questions in terms of question topic classification, equivalent question recognition, and question answering, which will be explained in Sect. 2.

B. Crémilleux et al. (Eds.): IDA 2023, LNCS 13876, pp. 472–483, 2023.
https://doi.org/10.1007/978-3-031-30047-9_37

Some state-of-the-art deep learning models like Transformer [16] are widely used in natural language processing (NLP). They resulted in leading performance for various tasks [5, 10, 21]. Train a language model requires lots of training data. Therefore, researchers had to create pre-trained language models using large-scale unsupervised tasks and then fine-tune them with labelled task-specific data. However, labelled data for a specific task are always limited and hard to obtain. Besides, a language model can have a size of millions or billions of parameters. It is usually expensive to train and use a separate network for each task. A generalist model can help address these problems by applying multi-task learning, a learning approach that improves generalisation by adding inductive bias such as tasks and domain information [2].

There are two main strategies for multi-task learning. One standard approach is adding extra tasks, also referred to as auxiliary tasks, to improve the performance of the target task. Empirically, adding auxiliary tasks to a pre-trained network is more similar to transfer learning, which improves primary tasks with additional tasks. Another [9] is learning all the tasks jointly without identifying the primary task so that all the tasks can achieve balanced performance, which can be leveraged for training a generalist agent.

We fine-tune the pre-trained language model with all the tasks jointly without identifying primary and auxiliary tasks. These tasks share the same domain, which is referred to as inductive bias multi-task learning [13]. Research [8, 11] shows that multi-task learning and pre-trained language models are complementary and can be combined to generate better performance on learning text representations.

There are many different types of tasks included in multi-task natural language understanding. For example, single sentence classification like sentiment analysis, pairwise classification like natural language inference, and regression task like sentence similarity. MT-DNN [8] trains their multi-tasking model with the transformer encoder and task-specific layer so that it can apply to classification and regression tasks. To adapt to various tasks, some researchers re-frame all the datasets into the same format. MQAN [9] formulates all the datasets into question answering over context. T5 [11] creates a sequence-to-sequence format for the tasks. All these models focus on general language understanding tasks like GLUE [17], and decaNLP [9].

In contrast, we focus on a range of different tasks for processing questions. And we report here on a generalist network called QBERT to solve three processing tasks we defined in the question domain. QBERT intends to work as a "generalist" language model that can perform multiple question tasks rather than a "specialist" who is only trained to maximise the performance on one specific task.

QBERT is based on sentence-BERT (SBERT) [15], a Siamese BERT (Bidirectional Encoder Representations from Transformers) [5] that projects the sentences into high-dimensional vector space. This process is known as embedding. The sentence embeddings with similar semantic meanings are close to each other in the vector space. Note that our intention is not to design a new algorithm but to fine-tune SBERT in a multi-task way so that the same representations can be used for processing questions in multiple ways. After fine-tuning SBERT, the embeddings generated from the input sequence can be used for both classification and retrieval tasks.

A previous study [19] on the question-related multi-tasking model shows that the training curriculum is critical. They reported that one certain curriculum could obtain a balanced performance on all the tasks. However, one of the limitations of the previous study is that the model lacked consistency on different question tasks. Reference [19] performed topic classification with a single BERT structure, others with Siamese BERT. To improve this, we re-frame the single sentence classification into a retrieval task.

During inference, QBERT produces the representation of the input sequence without any task-specific modules. Instead, it contains a threshold filter to determine the cosine similarity of the embedding pairs. Compared to the standard multi-task structure, reducing task-specific layers simplifies the complexity of the network. The network shares all the weights between tasks, also known as hard parameter sharing. More details of QBERT will be explained in Sect. 3.

After that, we compare QBERT with SBERT and the single-task version of SBERT in Sect. 4. The results in Sect. 5 also show how the training curriculum affects the performance of QBERT.

2   Tasks

In this paper, we define task (T) by data (X), label (Y), and loss function (L) as follow.

$$T \doteq \{p(X), p(Y \mid X), L\} \tag{1}$$

where $p(X)$ is the distribution of the input data, $p(Y \mid X)$ is the distribution of label Y given data X, and L is the loss function.

QBERT combines 3 different types of tasks: **question topic classification**, **equivalent question recognition**, and **question answering**. These tasks target common natural language understanding problems such as single sentence classification, pairwise classification, and information retrieval.

Question Topic Classification (QT): Given a question, the model labels the topic of the question.

Equivalent Question Recognition (QE): There are two sub-tasks included in QE, classification and retrieval. In classification, the model aims at classifying if the question pairs are similar or not, and based on the outcome, retrieve all similar questions from a question corpus with the given question.

Question Answering (QA): Given a question, the model searches for the answer from lists of candidate sentences. We determine this task as an open-domain open-book QA in which the question has no limitation in domains; the model allows answering the question with the content provided.

3   QBERT: A Multi-task Question-Processing Version of BERT

Our model is inspired by SBERT [15], which projects input sequence (sentence in this case) in high dimensional space. In such a way, we can evaluate the similarity between input sequences in the vector space using cosine similarity and retrieve the most similar sequence within a given corpus. Additionally, we can apply our model in classification by introducing a similarity threshold.

The three question tasks we defined in the previous section include three kinds of machine learning tasks: single sentence multiclass classification, pairwise classification, and information retrieval. In the previous research [19], topic classification was performed with a separate network because it is a multiclass classification that requires single input instead of pairwise. To perform these three tasks with one Siamese model, we consider the topic classification as pairwise classification by taking the (Question, Topic) as the pairwise input. During inference, instead of categorising the topic of a question, we retrieve the closest topic to a question.

Figure 1 illustrates the architecture of QBERT. During training, all the tasks are trained as to minimise the cosine distance using the binary labels and update the shared BERT layer. Task-specific loss functions are introduced for different types of data. While inference, the model only requires shared layers without task-specific layers, which manages to simplify the model.

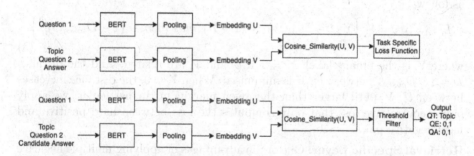

Fig. 1. QBERT architecture. The architecture is based on SBERT but trained to have a balanced performance on various tasks.Top: Training as binary classification, Bottom: Inference by calculating the cosine similarity between input sequences. All BERTs share the same parameters.

Input Layer: $S = (s_1, ..., s_n)$ is an input sequence with n words. The sequence can be either a topic, question, sentence, or paragraph. The model takes a pairwise input (S, S') such as a question pair, question-topic pair, or question-answer pair. The pairwise input is then passed to two identical BERTs.

BERT Layer: BERTs in this layer share all the parameters. The shared embedding layer following the setup of $BERT_{base}$ which takes the sequence input as word tokens and generates an output for each token as well as a [CLS] token at the beginning of the output sequence. $BERT_{base}$ uses the an encoder containing 12 layers and 110M parameters and is pre-trained with two unsupervised tasks: masked language model and next sentence prediction. The output of the BERT layer is in $\mathbb{R}^d$ vector space, and according to BERT, $d = 768$.

Pooling Layer: Similar to SBERT, the model leverages a mean pooling strategy that computes the mean of all output tokens (except [CLS]) of the sequence from BERT. According to SBERT, the mean pooling strategy outperforms using the [CLS] token as the embedding on capture sequence similarity. After the pooling function, the model generates a pair of embedding U and embedding V as Eq. 2, where $U \in \mathbb{R}^d$ and $V \in \mathbb{R}^d$.

$$Embedding = \frac{1}{n} \sum_{i=1}^{n} \Phi_{BERT}(s_i) \tag{2}$$

We apply two different loss functions for different types of data. Online contrastive loss is introduced for binary classification. And multiple negatives ranking loss is used for information-retrieving datasets (containing only positive samples) and multi-class classification. For multi-class classification, instead of classifying the category of the data, we retrieve the class label for the data. Adam optimiser [7] minimises the loss based on the cosine similarity $D_{cosine}(U, V)$.

Pairwise Classification Specific Layer: QBERT introduces the contrastive loss [6] for pairwise classification. It aims to gather positive pairs in the vector space while separating negative pairs. For embedding U, V, the loss is calculated as follows.

$$L_{contrastive} = \frac{1}{2} \left\{ Y \left(1 - D_{cosine}\right)^2 + (1 - Y) \left[max \left(0, m - (1 - D_{cosine})\right)\right]^2 \right\} \tag{3}$$

where Y is the binary label. $Y = 1$ if U and V are related. And the distance $D = 1 - D_{cosine}$ between U, V is minimised. When $Y = 0$, the distance increases between U, V until larger than the given margin m. In particular, we apply online contrastive loss that only computes the loss between hard positive and hard negative pairs.

Retrieval Specific Layer: One of the advantages of applying multiple negative ranking loss is that the training dataset no longer requires both positive and negative labels. For a given positive sequence pair (S_i, S_i'), the function assumes that any (S_i, S_j') is negative when $i \neq j$. For example, in QA, for question set $Q = \{q_1, ..., q_m\}$ and answer set $A = \{a_1, ..., a_m\}$, (q_i, a_i) is a positive pair given by the dataset, (q_i, a_j) is a negative pair randomly generated from the dataset. The cross-entropy loss of all the sequences pairs is calculated as follows.

$$L_{multiple\_negative} = -(Y log \left(D_{cosine}\right) + (1 - Y) log(1 - D_{cosine})) \tag{4}$$

During inference, QBERT no longer leverages a task specific layer. Instead, it introduces a threshold filter. QBERT calculates the $D_{cosine}(U, V)$, the cosine similarity between embeddings U and V, and applies different similarity thresholds for each task to determine if two sequences are related in terms of topic, equivalent question, or corresponding answer. The threshold of best performance is selected after training.

3.1   Training Curriculum

During training, the data in each dataset get divided into batches $B = \{b_1, ..., b_n\}$. In each step, one batch b_i is selected randomly, and the model parameters are updated by stochastic gradient descent. As shown in the previous research [19], the training curriculum was critical for multi-task question processing. In reference [19], the tasks were trained once at a time, from QE to QA to QT (QT was trained with different network architecture). However, the tasks learned in the earlier stage had a worse performance compared to the tasks learned in the later stage. To improve this, we train QBERT in a fixed-order round robin (RR) curriculum and compare the results with one by one (OBO) curriculum.

In the OBO approach, we train QBERT following QT, QE, and QA orders. Every dataset is divided into multiple batches, each with specific batch size, and is trained one at a time. The parameters are updated during the training and shared amongst all the datasets.

On the contrary, QBERT trains all the tasks simultaneously in the RR curriculum. The data in each task-specific layer are built as mini-batches and divided into two task-specific layers. During each step, the model is trained and updated by batches from both classification and retrieval tasks. QBERT-RR alternates between tasks during training which prevents the model from forgetting about the tasks learned at the beginning of the training.

3.2   Threshold Filter

In QE and QA, apart from classifying if the sequences are related, it is also crucial for the system to search all the related sequences (equivalent question or answer) for the given questions. The problem is how to quantify "related" with embeddings. A cosine similarity threshold is introduced in this model. Using a threshold simplifies the network structure of QBERT during inference by removing the task-specific layer. The threshold filter acts like a margin separating related (S, S') from others. Furthermore, with this threshold, the network can not only search the information that is closest to the query but can also identify if the information is related (close enough) to the query. For example, a question might be unique in the corpus so that the closest question to the given question is not equivalent to the given question if it has a smaller cosine similarity than the threshold; or a question might not have a high-confidence answer from the candidate corpus, the closest candidate with a cosine similarity smaller than the threshold will not be considered as the right answer.

To decide the threshold, first, all the sequences in the training set are embedded with the fine-tuned model. The sequence pairs are classified as positive if they have more similarity than the threshold. The similarity threshold with the best accuracy in the training set is found to quantify any question pairs during testing. With a threshold, the model is capable of searching and grouping all the related sequences in a given candidate corpus.

4   Experiments

Training QBERT includes pre-training and multi-task training. We follow the pre-training of BERT and SBERT. Then we perform multi-task learning on five question related datasets and evaluate on four of them.

4.1   Datasets

Quora Question Pair (QQP) [4] first released on Quora in 2017. It is a dataset that contains 404k question pairs collected and annotated by Quora. QQP labels if the questions are duplicated or not. There are 537k unique questions in the dataset. Training on QQP, we aim at improving the performance of QE tasks for QBERT. We then evaluate QQP in both pairwise classification and equivalent questions retrieval.

WikiQA [20] is a question-answering dataset which has the questions from query logs on Bing and answers from Wikipedia's summaries. The questions in WikiQA are factual questions that start with WH words like who, what, and when etc. The candidate answers are extracted sentences from the first paragraph of Wikipedia articles (also known as Wikipedia Summary). The dataset includes 3,047 questions and 26k candidate sentences, of which 1,239 questions contain a correct answer. We train WikiQA as a classification task and evaluate it as an answer selection task.

Yahoo! Answer [22] data were originally collected by Yahoo! Research Alliance Webscope program. Zhang et al., built up a corpus which contains 1.46M samples within 10 most popular topics on *Yahoo! Answer*. The sample includes the topic, question title, question content, and the best answer provided by the user. We apply this corpus in both QT and QA tasks. For QT, QBERT takes the question title and topic as the input sequence pairs. Question title and best answer are leveraged for training QA tasks. To distinguish the data used in different tasks, we use YT for the data applied in QT and YQA for data in QA.

Stanford Question Answering Dataset (SQuAD) [12] is a corpus that contains questions, answers and contexts for reading comprehension tasks. The contexts are extracted from Wikipedia. We use SQuAD 1.1, which all the questions have a corresponding answer phrase in the given context. There are 98,169 question-answer pairs in the dataset. To train QBERT, we take the question and the one sentence in the context that contains the answer phase as input.

4.2   Implementation Details

For each input sequence, the length is limited to 35 tokens because we use two BERTs to read the sequence pair instead of concatenating two sequences into one as the input. Besides, most questions in the datasets have less than 35 tokens. The sequence is truncated at the end if it is longer than the limitation (Table 1).

We train QBERT with the multiple negative ranking loss for QT and the online contrastive loss for QE. And we define the similarity threshold for QE based on the best accuracy on the training set. Then we evaluate the model on

Table 1. Statistics of the training datasets. Note that the test set of SQuAD is confidential from researcher. The number of test data states here is the validation set that is publicly accessed. The metrics for SQuAD in this paper is "exact match in sentence" which will defined in Sect. 5.

Dataset	#Train	#Test	Label	Metrics
YT	1,400,000	60,000	10	Accuracy
QQP	283,001	121,286	2	Accuracy/F1
WikiQA	23,080	6,116	2	Accuracy/F1
YQA	14,000,000	600,000	1	-
SQuAD	87,355	10,539	1	EM*

both QE classification and retrieval tasks. The QE retrieval candidate corpus is constructed by sampled queries in the QQP test set.

For QA, we train WikiQA with the online contractive loss and YQA and SQuAD with multiple negative ranking loss. This is because YQA and SQuAD only contain question answering pairs and do not come with negative samples. However, for WikiQA, there are questions with no answers in the dataset. Thus, a threshold is needed to identify if the closest candidate to the question is the high-confidence answer. The threshold is defined as the one that creates the best precision in the WikiQA training set.

The implementation of QBERT is based on PyTorch and SBERT. The margin for positive samples and negative samples is 0.5. We train the model for 5 epochs with a batch size of 32 and a learning rate of $2e-5$. 10% of the training data is used for warm-up.

We train QBERT with one GeForce GTX TITAN X GPU. To train QBERT-OBO, it takes 45.5 h, and 93 h for QBERT-RR. Even though training the model is time-consuming, once trained, the model is much faster during inference. It takes 1.5 ms, 5.44 ms, 19.62 ms, and 49.76 ms per question in YQT, QQP, WikiQA, and SQuAD, respectively.

5 Performance of QBERT

We evaluate QBERT with YT for QT classification, QQP for QE classification and QE retrieval, and WikiQA and SQuAD for QA retrieval.

For QE, we evaluate classification and retrieval task accuracy with the QQP dataset. If the question pair has a similarity larger than the threshold, it is categorised as equivalent in classification. To perform similar question mining, we create a question corpus based on QQP. First, all the relevant questions for the given query are included in the dataset, ensuring that there is always a relevant question in the corpus. Second, we fill the rest of the corpus with irrelevant questions. There are 104,033 samples in total. While mining the similar questions from the corpus, the candidate with the highest similarity larger than the threshold is defined as the duplicate question.

We assess the QA performance only on WikiQA and SQuAD, because the answers in the YQA are paragraphs provided by Yahoo! users, and it is hard to construct a candidate corpus used for single-sentence answer retrieval. We count the number of questions that can correctly identify the answer (or *None* for the questions without an answer) from the corpus while evaluating QA tasks.

In WikiQA, the question is not guaranteed to have an answer. Therefore, for each question, the model takes the sentence with the highest cosine similarity score in the candidate set and compares it with the threshold. If the similarity is above the threshold and the sentence is labelled as a correct answer, then the prediction is correct. For SQuAD, each query has a corresponding answer in the given context. Thus, we take the sentence with the highest cosine similarity score as the candidate. Note that for SQuAD, the ground truth answer is a short answer phrase extracted from the given context. Since QBERT retrieves one sentence as the answer, we evaluate the exact match phrase in the sentence, which depends on whether the answer phrase is in the selected sentence.

To understand the performance of multi-task learning, we use SBERT, which is fine-tuned with natural language inference dataset [1,18] and semantic textual similarity dataset [3] as our baseline. We also compare QBERT with the single-tasks model. The result is shown in Table 2.

Table 2. The performance of QBERT-RR and QBERT-OBO compares with single-task SBERT trained on QT, QA and QE. SBERT without training on any question dataset is used as our baseline. We evaluate the QQP dataset on both classification and retrieval. Note that for SQuAD, we measure the accuracy by checking if the exact match phrase is contained in the retrieved sentence.

Curriculum	YQT	QQP-C	QQP-R
	Acc.	Acc.	Acc./F1
Baseline	35.27 ± 0.58	74.80 ± 0.32	$54.53 \pm 0.89/53.01 \pm 0.82$
QT	72.44 ± 0.39		
QE		89.79 ± 0.23	$56.98 \pm 0.65/55.36 \pm 0.60$
QA			
OBO	59.84 ± 0.32	78.85 ± 0.44	$57.46 \pm 0.78/55.87 \pm 0.70$
RR	73.77 ± 0.58	90.13 ± 0.19	$58.22 \pm 0.78/56.53 \pm 0.75$

Curriculum	WikiQA	SQuAD
	Acc./F1	Acc.
Baseline	$77.46 \pm 2.82/58.24 \pm 12.48$	67.04 ± 0.93
QT		
QE		
QA	$79.05 \pm 5.89/72.50 \pm 11.08$	78.59 ± 0.82
OBO	$80.16 \pm 6.03/69.29 \pm 9.00$	76.09 ± 0.66
RR	$81.90 \pm 5.60/73.73 \pm 8.12$	71.42 ± 1.45

SBERT was only trained on natural language inference dataset and semantic textual similarity dataset containing sentence pairs with labels. It therefore, manages to detect similar question pairs albeit with poor performance. However, SBERT was not trained to group sentences with the same topic, and it is unable to identify the question topic. Since SBERT achieves a similar accuracy to other models on WikiQA dataset, it has a worse F1 score compared to others.

In Table 2, model SBERT-QT, SBERT-QE, SBERT-QA represent single-task training. It leverages the same architecture as QBERT. However, for each task, it has a separate model. While fine-tuning the single-task model, we update both the BERT layer and task-specific layer for each dataset. The results show that QBERT-RR achieves a similar performance on most question datasets, except for retrieving answer from SQuAD, with the generalist representation.

The previous research [19] proved that the training curriculum is important for training a multi-task network. Thus, we investigate two different training strategies. The QBERT-OBO shows better performance on WikiQA; on the other hand, it has worse performance on QT and QE compared to the single-task models. When training QBERT-OBO, we train one dataset after another. This causes the model to "forget" what it learnt during the early stages.

In contrast, while training with the RR curriculum, the model achieves a balanced performance on each task. Although QBERT-RR does not excel in any task compared to the single task model, it is able to generate a representation that can be used to perform a range of question tasks. Figure 2a shows the performance of QE and QA classification tasks.

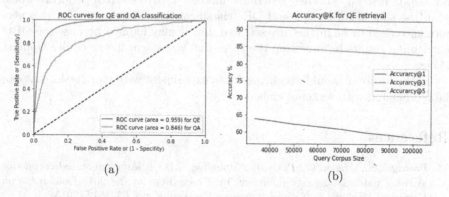

(a)   (b)

Fig. 2. (a): ROC curve for QE and QA classification from model QBERT-RR. The black dashed line represents the performance of a random classifier. (b): Accuracy@K of different corpus sizes in QE retrieval task.

We also evaluate the accuracy@k among different retrieval corpus sizes for QE using the QQP test set. Accuracy@k is a top-k accuracy classification score. In QE, it counts the number of times where the relevant question is contained in the top k candidates. According to the results illustrated in Fig. 2b, it is more challenging to retrieve among the larger corpus. When all the queries in the dataset are included in the retrieval corpus, the accuracy@1, accuracy@3, accuracy@5 are 58.24%, 82.72%, and 89.26%, respectively. More than 80% of the related questions are located in the top 3 candidates. However, only 58.150% of them are the closest to the given query, which can be improved in the future.

Lastly, we notice one limitation while evaluating QA retrieval with the SQuAD dataset. When creating the candidate corpus, we leverage a sentence tokenizer to split the paragraph into sentences. However, the sentence tokenizer split the sentence based on the punctuation. For example, "Washington, D.C." is considered two sentences: "Washington, D." and "C.". During evaluation, we compare the selected sentence with the answer phrase. In this case, retrieving a sentence may yield an incomplete answer to a question.

6   Conclusion

In this paper, we propose a generalist model to process questions in a variety of tasks, namely Question Topic Classification, Equivalence Question Recognition, and Question Answering. The idea is that sometimes a generalist model can be useful even when it does not beat specialist models at their own speciality.

We fine-tune SBERT as a generalist model for processing questions. We observe that one version of the generalist model QBERT-RR turns out to perform similar to the specialists in many cases except for QA retrieval on the SQuAD dataset. The specialist methods used here for comparison are SBERT models fine-tuned respectively on QT, QE (classification data) and QA (both datasets). Instead, another generalist method QBERT-OBO performs worse than the specialists on QT and QE (classification). The reasons for this performance need to be further investigated, but it may happen because the OBO curriculum results in forgetting the tasks that are learnt in the earlier training stage.

In the future, it would also be useful to experiment with more tasks that can be represented with sentence embedding.

References

1. Bowman, S., Angeli, G., Potts, C., Manning, C.D.: A large annotated corpus for learning natural language inference. In: Proceedings of the 2015 Conference on Empirical Methods in Natural Language Processing, pp. 632–642 (2015)
2. Caruana, R.: Multitask learning. Mach. Learn. **28**(1), 41–75 (1997)
3. Cer, D., Diab, M., Agirre, E., Lopez-Gazpio, I., Specia, L.: SemEval-2017 task 1: semantic textual similarity multilingual and crosslingual focused evaluation. In: Proceedings of the 11th International Workshop on Semantic Evaluation (SemEval-2017), pp. 1–14 (2017)
4. Csernai, K.: Quora question pairs (2017). https://www.quora.com/q/quoradata/First-Quora-Dataset-Release-Question-Pairs
5. Devlin, J., Chang, M.W., Lee, K., Toutanova, K.: BERT: pre-training of deep bidirectional transformers for language understanding. arXiv preprint arXiv:1810.04805 (2018)

6. Hadsell, R., Chopra, S., LeCun, Y.: Dimensionality reduction by learning an invariant mapping. In: 2006 IEEE Computer Society Conference on Computer Vision and Pattern Recognition (CVPR 2006), vol. 2, pp. 1735–1742. IEEE (2006)

7. Kingma, D.P., Ba, J.: Adam: a method for stochastic optimization. In: ICLR (2015)

8. Liu, X., He, P., Chen, W., Gao, J.: Multi-task deep neural networks for natural language understanding. In: Proceedings of the 57th Annual Meeting of the Association for Computational Linguistics, pp. 4487–4496 (2019)

9. McCann, B., Keskar, N.S., Xiong, C., Socher, R.: The natural language decathlon: multitask learning as question answering. arXiv preprint arXiv:1806.08730 (2018)

10. Radford, A., Narasimhan, K., Salimans, T., Sutskever, I.: Improving language understanding by generative pre-training (2018)

11. Raffel, C., et al.: Exploring the limits of transfer learning with a unified text-to-text transformer. J. Mach. Learn. Res. **21**(140), 1–67 (2020)

12. Rajpurkar, P., Zhang, J., Lopyrev, K., Liang, P.: SQuAD: 100,000+ questions for machine comprehension of text. In: Proceedings of the 2016 Conference on Empirical Methods in Natural Language Processing, pp. 2383–2392 (2016)

13. Redko, I., Morvant, E., Habrard, A., Sebban, M., Bennani, Y.: Advances in Domain Adaptation Theory. Elsevier, Amsterdam (2019)

14. Reed, S., et al.: A generalist agent. arXiv preprint arXiv:2205.06175 (2022)

15. Reimers, N., Gurevych, I.: Sentence-BERT: sentence embeddings using Siamese BERT-networks. In: proceedings of the 2019 Conference on Empirical Methods in Natural Language Processing and the 9th International Joint Conference on Natural Language Processing (EMNLP-IJCNLP) (2019)

16. Vaswani, A., et al.: Attention is all you need. In: Advances in Neural Information Processing Systems, pp. 5998–6008 (2017)

17. Wang, A., Singh, A., Michael, J., Hill, F., Levy, O., Bowman, S.: GLUE: a multitask benchmark and analysis platform for natural language understanding. In: Proceedings of the 2018 EMNLP Workshop BlackboxNLP: Analyzing and Interpreting Neural Networks for NLP, pp. 353–355 (2018)

18. Williams, A., Nangia, N., Bowman, S.: A broad-coverage challenge corpus for sentence understanding through inference. In: Proceedings of the 2018 Conference of the North American Chapter of the Association for Computational Linguistics: Human Language Technologies, Volume 1 (Long Papers), pp. 1112–1122 (2018)

19. Xu, Z., Howarth, A., Briggs, N., Cristianini, N., et al.: What makes us curious? Analysis of a corpus of open-domain questions. In: CS & IT Conference Proceedings, vol. 11. CS & IT Conference Proceedings (2021)

20. Yang, Y., Yih, W.T., Meek, C.: WIKIQA: a challenge dataset for open-domain question answering. In: Proceedings of the 2015 Conference on Empirical Methods in Natural Language Processing, pp. 2013–2018 (2015)

21. Yang, Z., Dai, Z., Yang, Y., Carbonell, J., Salakhutdinov, R., Le, Q.V.: XLNet: generalized autoregressive pretraining for language understanding. In: Advances in Neural Information Processing Systems, vol. 32 (2019)

22. Zhang, X., Zhao, J., LeCun, Y.: Character-level convolutional networks for text classification. In: Advances in Neural Information Processing Systems (2015)

On Compositionality in Data Embedding

Zhaozhen Xu[1]([✉]), Zhijin Guo[1], and Nello Cristianini[2]

[1] University of Bristol, Bristol, UK
{zhaozhen.xu,zhijin.guo}@bristol.ac.uk
[2] University of Bath, Bath, UK
nc993@bath.ac.uk

Abstract. Representing data items as vectors in a space is a common practice in machine learning, where it often goes under the name of "data embedding". This representation is typically learnt from known relations that exist in the original data, such as co-occurrence of words, or connections in graphs. A property of these embeddings is known as compositionality, whereby the vector representation of an item can be decomposed into different parts, which can be understood separately. This property, first observed in the case of word embeddings, could help with various challenges of modern AI: detection of unwanted bias in the representation, explainability of AI decisions based on these representations, and the possibility of performing analogical reasoning or counterfactual question answering. One important direction of research is to understand the origins, properties and limitations of compositional data embeddings, with the idea of going beyond word embeddings. In this paper, we propose two methods to test for this property, demonstrating their use in the case of sentence embedding and knowledge graph embedding.

Keywords: Embedding Compositionality · Sentence Embedding · Knowledge Graph

1 Introduction

A popular way to represent data is as vectors in an "embedding space", so the relations of interest can be represented as geometric or algebraic relations in that space. The distance between two vectors is often used to represent the presence of a relation, and their coordinates are chosen by an algorithm on the basis of the relations that we wish to incorporate. Various algorithms exist, based on different principles, to calculate "embedding vectors" for words, sentences, or entities in a knowledge graph, among other types of data.

In this approach, the values of each coordinate are not considered to be individually meaningful, so they cannot be used to interpret what information is being used by the AI system. This considerably limits the possibility of explaining the decisions of a system or to audit the system for possible biases.

© The Author(s), under exclusive license to Springer Nature Switzerland AG 2023
B. Crémilleux et al. (Eds.): IDA 2023, LNCS 13876, pp. 484–496, 2023.
https://doi.org/10.1007/978-3-031-30047-9_38

A property of certain embeddings that has the potential to help with the above concerns (as well as others) is that of "compositionality". Introduced in the domain of traditional linguistics, this property has been extended to also cover vector representations. Traditionally it refers to how the meaning of a linguistic expression results from its components. For example, the word "compositionality" can be decomposed into parts "Com+pos+ition+al+ity" that modify the meaning of the initial word stem.

In the case of vector embeddings, we substitute the "string concatenation" operation with the "vector addition" operation, so that a vector representation is compositional if it can be regarded as the sum of a small set of components (which can hopefully be interpreted and even manipulated). As an artificial example of this idea, we could imagine an embedding Φ that maps from items (tokens) to vectors in such a way that

$$\Phi(compositionality) \approx \Phi(com) + \Phi(pos) + \Phi(ition) + \Phi(ality)$$

Two important studies on word embeddings motivate our investigation.

The first [8] highlighted that word embeddings can present compositionality to such an extent that simple analogies can be performed in that representation, as in the standard examples:

$\Phi(Germany) - \Phi(Berlin) \approx \Phi(France) - \Phi(Paris)$
$\Phi(waitress) - \Phi(waiter) \approx \Phi(actress) - \Phi(actor)$

The second study [4] showed that not only is gender often present in the embedding of words, but this information is distributed across words in a biased way; for example, it was observed that certain job titles are represented in a way that aligns more with male, or female, reference words.

Both of those effects seem to originate from the way that a word embedding is learnt, which is from the distribution of words that are found in its vicinity. Finding the same effects in more general settings, such as in the embedding of Knowledge Graphs, would lead to a deeper understanding of the phenomenon.

Being able to understand and exploit compositionality as a general phenomenon would allow to manipulate these representations, to improve their usability for analogical reasoning, to remove unwanted biases, and to explain the decisions made by an algorithm that uses them. Many of the present questions in AI can benefit from a more general theory of compositionality in representations. Breaking down the representations into smaller, more manageable parts can help us understand what elements lead to the decision made by AI systems. Furthermore, some of these elements might be biased. Therefore, testing for compositionality can help design more transparent, interpretable, and accountable AI systems.

In this paper, we develop and demonstrate two different ways to test for compositionality in data embeddings, using the examples of sentence embedding and knowledge graph embedding as demonstration. In the first case, we use a pre-trained Bidirectional Encoder Representations from Transformers (BERT) [5] to embed simple sentences that are created for this purpose. Then we show that these embeddings can be decomposed into the contributions of the subject, verb and object. In the second case, we use various methods to embed the nodes of a knowledge graph (based on the IMDB movies dataset) and then examine how user behaviour descriptions can be decomposed in terms of user attributes.

Graph Embedding can be used to express multiple relations. For example, in the case of bipartite graphs, it can express the property that a certain user likes a certain movie. It is reasonable to ask if the embedding of users correlates with their age or gender, but an interesting question is how this can be rigorously measured. Follow-up questions involve the possibility of removing such bias and the trade-offs that this creates with performance in the task of "link prediction".

Sentence Embedding is currently performed in a different way through the use of Neural Networks. However, at the end, it generates embeddings of sentences, for which it is again reasonable to ask if they display the compositional property.

In this paper, we explain embedding and compositionality in Sect. 2 and how words, sentences, and graphs are embedded in Sect. 3. In Sect. 4, we introduce methods for testing the compositionality in embeddings. And we apply the methods to sentence embedding in Sect. 3.2 and knowledge graph embedding in Sect. 6.

2   Embedding and Compositionality

In machine learning, embedding is the process of mapping the elements of a set to points in a vector space. We write a set of coordinates $\mathbf{B}$ to represent the items of I as follows.

$$\mathbf{B} = \Phi(I)$$

Where Φ is the mapping function that maps the items (elements of the set) to their coordinates. This embedding function can be learned from a set of data containing those items: for words, this can be done by exploiting co-occurrence statistics between words; for elements of a graph, by exploiting the topology, therefore, the relations between different elements.

A learned representation is compositional when it can represent complex concepts or items by combining simple attributes [6]. In this paper, we mainly look into additive compositionality as follows.

$$\mathbf{b}_I = \sum_{i=1}^{N} \mathbf{x}_i$$

where I is an item that has a set of N attributes. I can be represented with embedding vector $\mathbf{b}_I$, and the attributes can be represented with $\mathbf{x}$.

3   Words, Sentences and Graphs

Three types of data that greatly benefit from embedded representation are words, sentences and graphs.

3.1   Word Embedding: GloVe

GloVe (Global Vectors for Word Representation) is an unsupervised learning algorithm for obtaining the vector representation of each word in a corpus [10]. The signal used to inform the embedding comes from the statistics of word occurrences in a reference corpus: words are represented as a vector, in such a way that the inner product of two vectors reflects the probability of these two words co-occurring near each other. This means that the representation of each word is informed by the distribution of its context words. This high-dimensional distributed vector has been proved to reveal some semantic information. Thus, we can define the distribution/semantic similarity just by calculating the distance of the high-dimensional vectors.

3.2   Sentence Embedding: BERT

To embed the sentence, we apply a pre-trained sentence embedding model called SBERT [11], which is a sentence embedding version of BERT [5]. BERT was pre-trained with Wikipedia, Bookcorpus to perform tasks such as guessing a masked word or a neighbouring sentence, so once more the signal informing the embedding is one of co-occurrence statistics, albeit obtained with a much more complex method than the simple GloVe. Differently than GloVe, BERT can also embed items that were not seen in the training set.

For our purposes, we will use SBERT as a generic embedding function, and we will just note that SBERT was trained on natural language inference corpora (NLI). NLI corpora [3,13] contain contradiction, entailment, and neural sentence pairs. SBERT is a version of BERT trained specifically for generating a sentence representation as follows,

$$\Phi_{BERT} : I \to \mathbf{B} \in \mathbb{R}^{768}$$

where I is a set of sentences and $\mathbf{B}$ is a space of 768-dimensional vectors. These representations can be compared using cosine similarity in the embedding space.

3.3   Knowledge Graph Embedding

A knowledge graph is formed of nodes and edges to represent entities and relations. In general, it is a heterogeneous directed multigraph: there can be different types of nodes and edges (heterogeneous), the edge can have a direction (directed), and multiple edges between the same nodes can exist (multigraph). As the edge is directed, these two nodes are often called "head" and "tail", and the link is called "relation". This forms a triple (head, tail, relation) also known as "a fact". For example, a triplet can express the relation: John Watched "Toy Stories", where John is a user, "Toy Stories" s a movie, and watched is a specific type of relation, with its own specific directionality.

This kind of graph can be embedded by assigning coordinates to its nodes, in such a way that related nodes (for a given type of relation) are mapped to nearby positions, in a given metric. An optimisation algorithm calculates these embeddings on the bases of triples representing either true or false "facts", such as: John, "Toy Stories", rel=watched is true, while John, "Star Wars", rel = disliked might be false. In practice, false facts are often not available, and are therefore approximated by randomly sampled facts, which are assumed to be, on average false. To summarise, a graph embedding function maps nodes to points in a vector space, that is:

$$\Phi_{KG} : V \to \mathbb{R}^d, h \in V, t \in V$$

The training algorithm simultaneously learns the embedding of each entity and the parameters for the specific metric that is used to represent each relation.

To embed the knowledge graph, we use GC-MC [1] with negative entity sampling [2]. We apply this to the IMDB movie dataset, where the relation between users and movies is given by their ratings.

4   Methods

For sentence embedding, we are interested in testing the possible approximation

$$\Phi(Sentence) \approx \Phi(Word_1) + \cdots + \Phi(Word_N)$$

In our experiments, we will generate a simple sentence corpus by creating sentences containing only subject, verb and object.

For Knowledge graph embedding, we are interested in the situation where the embedding of a user based on their behaviour can be approximated by a combination of their personal attributes.

$$\Phi(User_{Behaviour}) \approx \Phi(Age) + \cdots + \Phi(Gender) + \Phi(Occupation)$$

For this we will leverage the data from the IMDB movie recommendation website.

We use the standard statistical method of Canonical Correlation Analysis (CCA) to analyse the linear dependencies between two paired sets of vectors. We also make use of Linear Systems to refine the analysis.

4.1   Methods - 1: Canonical Correlation Analysis

CCA is a method concerned with comparing two multivariate variables, each describing different properties of the same individual item [12]. For example, a multivariate variable might contain a vector representation of a patient's clinical information, and another vector representation of the same patient's demographics. Given two such sets of variables, CCA can be used to discover if certain linear functions of one variable are correlated with certain linear functions of the other.

Suppose we are given two paired sets of vectors $\mathbf{B}_Y = \{\mathbf{b}_{y_i}\}$ and $\mathbf{B}_Z = \{\mathbf{b}_{z_i}\}$. We are interested in finding the projections $\mathbf{W}_Y$ and $\mathbf{W}_Z$ such that the Pearson correlation coefficient $Corrcoef(\mathbf{W}_Y^T \mathbf{B}_Y, \mathbf{W}_Z^T \mathbf{B}_Z)$ is maximal. And we analyse the significance of these dependencies between the vectors by comparing the Pearson correlation with randomised pairs of vectors represent different items, which is designed to deliberately violate the assumption that each pair of vectors represents the same item.

Note that we only apply the CCA method on the user embedding generated by the knowledge graph. This is because the user behaviour embeddings are generated without personal attributes. CCA can discover if the user behaviour is related to personal attributes. On the other hand, for sentence embedding, which is learned by understanding the words and content, it is not essential to validate the relation between word and sentence with CCA.

4.2   Methods - 2: Linear System

Based on the definition of additive compositionality (as seen in word embedding), we postulate the existence of (unknown) component vectors that can be used as building blocks to create the representation of a sentence or a graph node.

We will assume that the embeddings of a set of items are called $\mathbf{B}$, and the unknown components that formed them are called $\mathbf{X}$, while the information about which components are combined to form which item is stored in a composition matrix of coefficients $\mathbf{A}$, so that

$$\mathbf{AX} = \mathbf{B}$$

Assuming that there are three types of components i, j, k that determine the embedding of an item $\mathbf{b}_{i,j,k}$, we can represent them as vectors $\mathbf{x}_{i,*,*}$ or $x_{*,k,*}$ etc. that add a contribution to $\mathbf{b}_{i,j,k}$ in terms of Eq. 1.

$$\mathbf{b}_{i,j,k} = \mathbf{x}_{i,*,*} + \mathbf{x}_{*,j,*} + \mathbf{x}_{*,*,k} \tag{1}$$

This can be written in matrix form $\mathbf{AX} = \mathbf{B}$ by introducing the binary compositonal matrix $\mathbf{A}_{m,n}$, where $\mathbf{A}_{m,n}$ records whether a given building block (indicated by n) is present in a given item (indicated by m).

As the list of building blocks is formed by three lists, for three types of building blocks, the indicator n can be obtained from the indicators of these three lists. In the case where there are 10 elements or less for each type of building block, we can write: $m = 100 * i + 10 * j + k$.

$$\mathbf{A}_{m,n} = \begin{cases} 1, & \text{if } n = i \text{ or } n = 10 + j \text{ or } n = 20 + k \\ 0, & \text{otherwise} \end{cases} \tag{2}$$

Given a set of embeddings $\mathbf{B}$ for all items in a set of interest, we can find all the unknown vectors $\mathbf{x}_{i,*,*}$, $\mathbf{x}_{*,j,*}$, and $\mathbf{x}_{*,*,k}$ by solving the linear system $\mathbf{AX} = \mathbf{B}$.

4.3   Statistical Hypothesis Testing

The linear system encodes the assumption that the embeddings of the items and the components are related. We test this assumption by introducing the null hypothesis that no such relation exists and using a non-parametric test. This is done by generating randomly shuffled data, in such a way that item embeddings and those of the building blocks are randomly matched. This null hypothesis can be discarded if the test statistic computed on the original data has a value that cannot be obtained, with high probability, in the randomised data.

To test that $\mathbf{B}_{i,j,k} \approx \mathbf{x}_{i,*,*} + \mathbf{x}_{*,j,*} + \mathbf{x}_{*,*,k}$, we apply three evaluation matrices and compare the results with 100 random permutations by shuffling embedding $\mathbf{B}$. Using the randomly reshuffled pairs of (component, shuffled item) breaks down the connection between embedding and its building blocks.

Test Statistic: We use three different test statistics: (1) the loss of the linear system; (2) the cosine similarity between $\mathbf{B}$ and reconstruct embedding $\hat{\mathbf{B}}$; (3) the accuracy of retrieving $\mathbf{B}$ with $\hat{\mathbf{B}}$.

Null Hypothesis (H_0): The embeddings of an item and its components are independent. There is no benefit in the decomposing item into components.

Alternative Hypothesis (H_a): The item embeddings can be represented by adding up the components.

Significance threshold: $\alpha = 0.01$.

5   Decomposing Sentence Embeddings

When decomposing learned word embedding like word2vec [9], research [8] found that the embedding can include both semantic and syntactic relationships between words, which can be revealed using simple linear algebra. For example, $\Phi(king) - \Phi(man) + \Phi(women) \approx \Phi(queen)$ (semantic) and $\Phi(dog) - \Phi(dogs) \approx \Phi(cat) - \Phi(cats)$ (syntactic). The question here is if sentence embedding can be decomposed in a similar way. Sentences are compositional structures that are built from words. Therefore, it is natural to ask if the learned representations reflect the compositionality. We assume that there is an additive composition-ality between words and sentences so that the sentence representation can be decomposed in terms of

$$\Phi_{BERT}(Sentence) \approx \Phi(Word_1) + \cdots + \Phi(Word_N)$$

We leverage a linear system to decompose the sentence embedding into word representations to investigate the compositionality in BERT sentence embedding. To do this, we generated a sentence corpus that includes 1,000 sentences. Each sentence consists of the simplest elements required for completing a sentence: subject, verb and object.

5.1   Data Generation

We generated the sentence corpus[1] with 30 components, equally divided into subject (Sbj), verb, and object (Obj). These components were combined into

---

[1] The corpus is available at https://github.com/CarinaXZZ/On_Compositionality_in_Data_Embedding.

$10 \times 10 \times 10$ $(Sbj, Verb, Obj)$ triplets. Then the $(Sbj, Verb, Obj)$ makes up short, simple sentences with the same preposition and article. For example, for triplet (cat, sat, mat), the sentence is generated as "The cat sat on the mat.". There are 1,000 sentences in total. By creating the SVO sentence corpus, we can look into and understand each part we decompose with a linear system.

BERT tokenises the sentence into word tokens with a subword-based tokeniser. Some words such as "bookshelf" will be tokenised into two words ("book" and "shelf"). To keep all the sentences in the same number of tokens, we carefully picked the words we used to build the corpus. Note that BERT takes punctuation as an input token as well; there are 7 tokens in total for each sentence.

To construct a sentence (I), we add the subject, verb, and object phrases with indices i, j, and k, respectively. Thus, $I_{i,j,k} = Sbj_i + Verb_j + Obj_k$. We calculate sentence embedding $\mathbf{B}_{i,j,k} = \Phi_{BERT}(I_{i,j,k})$ with a fine-tuned BERT introduced in Sect. 3.2.

5.2   The Linear System and the Statistical Test

Having computed a set $\mathbf{B}$ of embeddings for all our sentences, we can find the unknown vectors $\mathbf{x}_{i,*,*}$, $\mathbf{x}_{*,j,*}$, and $\mathbf{x}_{*,*,k}$ by solving the linear system $\mathbf{AX} = \mathbf{B}$, where $\mathbf{A} \in \mathbb{R}^{1000 \times 30}$, $\mathbf{X} \in \mathbb{R}^{30 \times 768}$, and $\mathbf{B} \in \mathbb{R}^{1000 \times 768}$. This system of 1000 equations with 30 variables does not have (in general) an exact solution, so we approximate the solution by solving a linear least squares problem, using the pseudo-inverse method, as follows:

$$\mathbf{X} = (\mathbf{A}^T \cdot \mathbf{A})^+ \cdot \mathbf{A}^T \cdot \mathbf{B} \tag{3}$$

We evaluated the linear system by calculating the loss (L) as Eq. 4. The smaller the loss, the more accurately the embeddings can be represented as a composition of components $\mathbf{X}$ (which is our definition of compositionality).

$$L = \|\mathbf{AX} - \mathbf{B}\|^2 \tag{4}$$

To create the null hypothesis, the sentence embeddings are shuffled to break the link between the sentence and its embedding, and after that, we calculate the loss on the randomised data. We repeat this process 100 times.

One of the interesting challenges is if we can predict the sentence embedding $\mathbf{b}_{i,j,k}$ with the word representations solved by the linear system without seeing the actual sentences. To test this, we utilise the leave one out method to solve the linear system and reconstruct the sentence embedding by adding up the word representations we obtained with Eq. 3 so that

$$\Phi_{Composed}(I_{i,j,k}) = \Phi_{Composed}(Sbj_i) + \Phi_{Composed}(Verb_j) + \Phi_{Composed}(Obj_k)$$
$$\hat{\mathbf{b}}_{i,j,k} = \mathbf{x}_{i,*,*} + \mathbf{x}_{*,j,*} + \mathbf{x}_{*,*,k}$$
$$\tag{5}$$

By using the leave one out method, we remove the target sentence from the dataset during the learning process of the linear system. As a result, the word representations are learned without previous knowledge of the target sentence $I_{i,j,k}$. We will test the quality of these components by trying to predict the embedding of a new sentence from that of its components, and then comparing it with the BERT embedding of that sentence. This comparison is done in two ways: 1) by computing the cosine similarity between the predicted and the actual embedding, and 2) by using the reconstructed embedding to retrieve the correct one from a set of 1000 candidates. To do this, we repeat the process of leaving a sentence out, solving the linear system with the remaining 999 sentences, then using those components to predict its embedding.

5.3   Results

Figure 1 illustrates the performance of decomposing BERT sentence embedding. These results show that the BERT sentence embedding can be decomposed into three separate components: subject, verb, and object. And those components can then be used to predict the embedding of a new sentence.

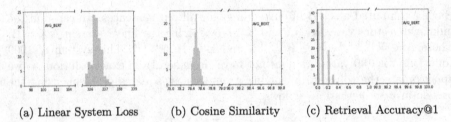

(a) Linear System Loss   (b) Cosine Similarity   (c) Retrieval Accuracy@1

Fig. 1. The test statistics for sentence embedding decomposition. AVG_BERT is the average performance of $\hat{B}$ learned from the BERT embedding. The bars are the distribution of the results from random permutations that run for 100 times.

The linear system decomposes the sentence embedding with a loss of 100.14, which is lower than the lowest loss from random permutations (335.65). Hence, the p-value for the non-parametric testing is smaller than the significance threshold ($\alpha = 0.01$), which rejects H_0. In other words, the BERT sentence embedding can be approximated as the composition of three different attributes.

Moreover, it is possible to approximately reconstruct the embedding of a sentence just from the Sbj, Verb, Obj components learned from the linear system. This reconstruction embedding $\hat{B}$ can be compared with the reconstruction that would result from using the same method but based on randomised (attributes, embeddings) pairs with cosine similarity. The cosine similarity between $\hat{B}$ and the BERT embedding is 98.44%, which is higher than any randomised trial.

The reconstructed embedding can retrieve the actual embedding with 99.5% accuracy. On the other hand, the composed embedding with randomised attribute/embedding pairs failed to retrieve the randomised embeddings, with the highest accuracy of 0.4%.

6   Decomposing Knowledge Graph Embeddings

While decomposing knowledge graph embedding, we focus on the case of movie recommendation. This knowledge graph contains the relations (ratings) between a set of users and a set of movies. The embedding function Φ_{KG} is such that users are mapped close to movies they rate highly, for an appropriate distance. This function does not make use of personal attributes of the user. We are interested in testing of these user embedding can be decomposed into components that depend on personal attributes of the user.

6.1   Dataset

Movie lens [7] is widely used for movie recommender systems, consisting of a set of movie ratings for each individual user. There are 6,040 users and approximately 3,900 movies in this dataset, and each user-movie pair is rated from five ratings (1–5). 5 represents the strongest preference and vice versa. After training on the movie lens dataset, the knowledge graph embeddings are able to predict the rating of an unseen movie from a user's history ratings.

For each user, the dataset also contains their personal attributes, such as gender, age, and occupation, which are not used to calculate the knowledge graph embedding. In this data, there are 2 genders, 7 ages, and 21 occupations. We represent them as categorical variables, and make use of indicator vectors (one-hot vectors), so that the attributes can be encoded as a Boolean vector of dimension 30, $\Phi_{Boolean} : I \to \mathbf{B} \in \mathbb{R}^{30}$. This representation will be used for the rest of the analysis, but we will repeat the tests with and without occupation.

6.2   CCA Results on Knowledge Graph Embedding

In this experiment, we apply CCA on the pairwise user representations $\Phi_{Boolean}(I)$ and $\Phi_{KG}(I)$). By calculating the Pearson correlation ρ between these two types of embeddings, we aim to understand whether the learned knowledge graph embedding contains information about the user attributes.

To perform a statistical test, we randomly shuffle the pairing between embedding and attributes of the users, $\Phi_{KG}(I)$ and $\Phi_{Boolean}(I)$. The Pearson correlation between randomised embeddings is compared with ρ in Fig. 2. It illustrates that the first Pearson correlation coefficient is larger than the largest value obtained on permuted data. Thus, we believe that the user embeddings $\Phi_{KG}(I)$ produced by the knowledge graph embedding are able to capture some of the personal attributes without having the prior information during training. These personal attributes could lead to unwanted bias because they show some private traits of users. Detecting them with CCA can help people build a more fair AI system.

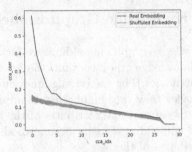

Fig. 2. Pearson value comparison for each component of CCA between knowledge graph embedding and permuted embedding.

6.3   Linear System

We then try to interpret the knowledge graph embedding with user attributes. Similar to sentence embedding, a linear system is used to calculate the representation for each user attribute. Note that not all of the combinations of attributes exist in the movie lens dataset. For each type of user, we calculate the mean of their embeddings for decomposition. We decompose the user embedding into gender and age. The results are illustrated in Fig. 3.

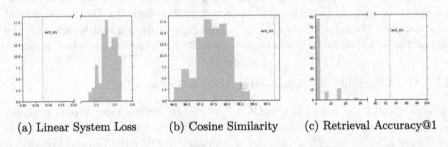

(a) Linear System Loss   (b) Cosine Similarity   (c) Retrieval Accuracy@1

Fig. 3. The test statistics for knowledge graph embedding decomposition. AVG_KG is the average performance of $\hat{\mathbf{B}}$ learned from the knowledge graph embedding. The bars are the distribution of the results from random permutations that run for 100 times. $p\_value < 0.01$.

These results show that the knowledge graph embedding can be decomposed into 2 genders and 7 age attributes, which can then be used to predict the embedding of a new user. The linear system can decompose the knowledge graph embedding with a loss of 0.37, which is lower than the random permutations. Moreover, it is possible to approximately reconstruct the embedding of a user just from the age and gender attributes obtained from the linear system. This reconstruction embedding $\hat{\mathbf{B}}$ can be compared with the reconstruction that would result from using the same method but based on randomised user-attributes/embeddings pairs using cosine similarity. The cosine similarity of the reconstruction embedding to the actual knowledge graph embedding is 99.79%, which is higher than any randomised trial (between 96.02% and 98.86%). The reconstructed embedding also retrieves the actual embedding 92.90% of the times, which outperforms the random permutations (between 0.00% and 28.57%). The p-value for the non-parametric testing is smaller than the significance threshold ($\alpha = 0.01$), thus, H_0 is discarded.

Our results are not so clear when we add more features. When we include the occupation in the representation, the linear system has decomposed the embedding into gender, age, and occupation (p-value smaller than the threshold). However, when we try to use the composed embedding $\hat{B}$ to retrieve the actual knowledge graph embedding, there is 4% accuracy in retrieving the actual user embedding. Although the result is still statistical significant compared to the random permutation, the composed embedding is not closest to the actual embedding. More work is on the way in this direction.

7 Conclusion

Both the sentence embeddings produced by BERT and the knowledge graph embeddings generated on the basis of IMDB data by GC-MC present some compositionality, that is some of the information contained in them can be explained in terms of known attributes. This creates the possibility to manipulate those representations, for the purpose of removing bias, or to explain the decisions of the algorithm using them, or to answer analogical or counterfactual questions.

A general way to decompose embedding vectors, based on solving linear systems and CCA problems, has been presented as an example here, but a number of other methods could be of use. As a matter of fact, many such methods already exist in different fields (for example methods related to the technique of "projection pursuit"). Future work should import those techniques within the field of Learning Embedding Representations.

References

1. Berg, R.V.D., Kipf, T.N., Welling, M.: Graph convolutional matrix completion. arXiv preprint arXiv:1706.02263 (2017)
2. Bose, A., Hamilton, W.: Compositional fairness constraints for graph embeddings. In: Chaudhuri, K., Salakhutdinov, R. (eds.) Proceedings of the 36th International Conference on Machine Learning. Proceedings of Machine Learning Research, vol. 97, pp. 715–724. PMLR (2019). https://proceedings.mlr.press/v97/bose19a.html
3. Bowman, S., Angeli, G., Potts, C., Manning, C.D.: A large annotated corpus for learning natural language inference. In: Proceedings of the 2015 Conference on Empirical Methods in Natural Language Processing, pp. 632–642 (2015)
4. Caliskan, A., Ajay, P.P., Charlesworth, T., Wolfe, R., Banaji, M.R.: Gender bias in word embeddings: a comprehensive analysis of frequency, syntax, and semantics. arXiv preprint arXiv:2206.03390 (2022)
5. Devlin, J., Chang, M.W., Lee, K., Toutanova, K.: BERT: pre-training of deep bidirectional transformers for language understanding. arXiv preprint arXiv:1810.04805 (2018)

6. Fodor, J.A., Lepore, E.: The Compositionality Papers. Oxford University Press, Oxford (2002)

7. Harper, F.M., Konstan, J.A.: The movielens datasets: history and context. ACM Trans. Interact. Intell. Syst. (TIIS) **5**(4), 1–19 (2015)

8. Mikolov, T., Chen, K., Corrado, G., Dean, J.: Efficient estimation of word representations in vector space. arXiv preprint arXiv:1301.3781 (2013)

9. Mikolov, T., Sutskever, I., Chen, K., Corrado, G.S., Dean, J.: Distributed representations of words and phrases and their compositionality. Adv. Neural Inf. Process. Syst. **26** (2013)

10. Pennington, J., Socher, R., Manning, C.: GloVe: global vectors for word representation. In: Proceedings of the 2014 Conference on Empirical Methods in Natural Language Processing (EMNLP), pp. 1532–1543. Association for Computational Linguistics, Doha (2014). https://doi.org/10.3115/v1/D14-1162, https://aclanthology.org/D14-1162

11. Reimers, N., Gurevych, I.: Sentence-BERT: sentence embeddings using Siamese BERT-networks. In: proceedings of the 2019 Conference on Empirical Methods in Natural Language Processing and the 9th International Joint Conference on Natural Language Processing (EMNLP-IJCNLP) (2019)

12. Shawe-Taylor, J., Cristianini, N.: Kernel Methods for Pattern Analysis. Cambridge University Press (2004). Illustrated edition. http://www.amazon.com/Kernel-Methods-Pattern-Analysis-Shawe-Taylor/dp/0521813972

13. Williams, A., Nangia, N., Bowman, S.: A broad-coverage challenge corpus for sentence understanding through inference. In: Proceedings of the 2018 Conference of the North American Chapter of the Association for Computational Linguistics: Human Language Technologies, Volume 1 (Long Papers), pp. 1112–1122 (2018)

Author Index

Printed in the United States
by Baker & Taylor Publisher Services

Printed in the United States
by Baker & Taylor Publisher Services